P9-CJT-077

Nanoscale Technology in Biological Systems

Nanoscale Technology in Biological Systems

Edited by
Ralph S. Greco
Fritz B. Prinz
R. Lane Smith

Boca Raton London New York Washington, D.C.

Library of Congress Cataloging-in-Publication Data

Catalog record is available from the Library of Congress

Direct all inquiries to CRC Press, 2000 N.W. Corporate Blvd., Boca Raton, Florida 33431.

Visit the CRC Press Web site at www.crcpress.com

International Standard Book Number 0-8493-1940-4
Printed in the United States of America 1 2 3 4 5 6 7 8 9 0
Printed on acid-free paper

Dedication

This textbook is dedicated to all of the surgical residents at the Stanford University School of Medicine and all of the graduate students in the School of Engineering at Stanford whose work has been an inspiration to the editors, in the laboratory, the clinic and in the preparation of this manuscript.

Preface

In 1959, Richard P. Feynman, Professor of Physics at the California Institute of Technology and Nobel Laureate, delivered an address at the American Physical Society, which is given the credit for inspiring the field of nanotechnology. Published in *Engineering and Science,* Feynman's address entitled "Plenty of Room at the Bottom" described a new field of science dealing with "the problem of manipulating and controlling things on a small scale."*

Feynman theorized that the development of improved electron microscopes would allow scientists to view the components of DNA, RNA, and proteins, to develop miniature computers and miniature machine systems, as well as to manipulate materials at the atomic level. "Perhaps this doesn't excite you to do it and only economics will do so. Then I want to do something; but I can't do it at the present moment, because I haven't prepared the ground. It is my intention to offer a prize of $1000 to the first guy who can take the information on the page of a book and put it on an area 1/25,000 smaller in linear scale in such a manner that it can be read by an electron microscope." Secondarily, Feynman said, "And I want to offer another prize — if I can figure out how to phrase it so that I don't get into a mess of arguments about definitions — of another $1000 to the first guy who makes an operating electric motor — a rotating electric motor, which can be controlled from the outside and, not counting the lead-in wires, is only 1/64 inch cube." In addition, he ended, "I do not expect that such prizes will have to wait very long for claimants."

He was right. His second challenge was achieved in 1960 by an engineer named William McLellan. McLellan constructed his small motor by hand using tweezers and a microscope. The nonfunctioning motor currently resides in a display at the California Institute of Technology. It took until 1985 for Thomas Newman, then a graduate student at Stanford, to achieve the first challenge by using a computer-controlled, finely focused pencil electron beam to write, in an area 5.9 micrometers square, the first page of Charles Dickens' *A Tale of Two Cities.*

In the 40 plus years since Feynman's challenges, the field of nanotechnology has advanced in many directions and at an astonishing pace. Some of the earliest advances, which made the burgeoning field feasible, were in microscopy and included not just the scanning electron microscope and the transmission electron microscope, but the scanning tunneling microscope and the atomic force microscope. With these in hand, scientists were able to begin to observe and manipulate structures at a scale measured in nanometers. The field of nanotechnology has since developed rapidly. It is considered likely by most experts that nanotechnology will influence energy more than any other industry, but that its application to biology and medicine

---

* Richard Feynman's talk at the December 29, 1959, annual meeting of the American Physical Society at the California Institute of Technology (Caltech), first published in the February 1960 issue of Caltech's *Engineering and Science*.

is inevitable. In 2000, President Bill Clinton announced the founding of the U.S. National Nanotechnology Initiative (NNI). In the last three years this national institute has grown in scope and support, with a federal budget in 2003 of $710.2 million. Governments in Europe, Japan, and other Asian nations have responded with competitive investments in programs that are national in scope. Although the era of nanotechnology is in its infancy, as it comes into full maturity there undoubtedly will be profound implications on not only many branches of science, but in all of our lives on a daily basis.

Ralph S. Greco, M.D.

Acknowledgments

We would like to express our appreciation to Stephanie Fouchy, without whose assistance the preparation of this book would not have been possible.

Editors

Ralph S. Greco is the Johnson & Johnson Distinguished Professor at the Stanford University School of Medicine. He is also Chief of the Division of General Surgery at Stanford and the Director of the General Surgery Training Program. Dr. Greco joined the faculty at Stanford in the year 2000. He graduated cum laude from Fordham University in 1964 and the Yale Medical School in 1968. His internship and residency was served at Yale New Haven Hospital, and after two years of military service, he became an Assistant Professor at the Rutgers Medical School (which later changed its name to Robert Wood Johnson Medical School) in 1975. He became Chief of the Division of General Surgery there in 1982 and Chief of Surgery at Robert Wood Johnson University Hospital in 1997.

Dr. Greco is a member of the American Surgical Association, the Society of University Surgeons, the Society for Biomaterials, the Association of Program Directors in Surgery, and the Surgical Infection Society, among many other surgical societies. He is board certified in General Surgery and is a Fellow of the American College of Surgeons. Dr. Greco has been the recipient of research grants from the National Heart, Lung and Blood Institute and served as a consultant to the NHLBI and the NSF. Dr. Greco is the recipient of six patents on various aspects of antibiotic bonding and has published more than 100 papers in the scientific literature. His research interest is focused on biomaterials, vascular grafts, the host response to implantable biomaterial surfaces, and surface modification of biomaterials. When he arrived at Stanford he began a collaboration with Friedrich Prinz and R. Lane Smith in a related, but new area, namely the nanofabrication of new biomaterial surfaces and their potential application to a new generation of biomaterials for clinical applications.

Fritz B. Prinz is the Rodney H. Adams Professor at the Standford University School of Engineering, and Department Chair, Mechanical Engineering. His current research focuses on the design and manufacturing of micro- and nanoscale devices. Examples include fuel cells and bioreactors. He is interested in materials selection, scaling theory, electro-chemical phenomena, and quantum modeling. He initiated a project on the observation of reduction-oxidation reactions in biological cells. He received his Ph.D. in Vienna in 1975.

Professor Prinz directs the Rapid Prototyping Laboratory (RPL), which is dedicated to improving product design and scientific discovery through efficient use of rapid prototyping. The RPL focuses its efforts on two different application domains. One is energy, the other biology. The RPL is exploring processing methods to build thin film solid oxide fuel cells with relatively low operating temperatures. Such fuel cells hold the promise of high efficiency and cost-effective production. The electrochemical measurement techniques available to Prinz' group, together with their ability to build sensors with nanoscale dimensions, help in observing oxidation–reduction reactions not only in fuel cells but also in biological cells. The RPL studies mass transport within and between lipid bilayers to gain insights into the physics and thermodynamics of electrochemical phenomena of thin biological membranes. RPL has a rich infrastructure and long tradition with respect to designing and manufacturing structures that are difficult, if not impossible, to make with conventional techniques. Examples include three-dimensional biodegradable tissue crafts and devices made with focused ion beam methods in a layered fashion.

R. Lane Smith is a Professor (Research) in the Department of Orthopaedic Surgery at Stanford University, Stanford, California. He has served as codirector and director of the Orthopaedic Research Laboratory at Stanford University since 1977 and currently holds a position at the Rehabilitation Research and Development Center at the VA Palo Alto Health Care System, where he is a career research scientist. He received his Ph.D. from the University of Texas at Austin in 1971. His graduate work was followed by postdoctoral study at the Friedrich Miescher Institute in Basel, Switzerland and a membrane-pathobiology fellowship at Stanford University.

Dr. Smith's research focuses on fundamental problems directed at understanding the molecular mechanisms influencing metabolism of cartilage and bone during normal homeostasis and pathogenesis.

His currently funded research examines fundamental mechanisms by which mechanical stimulation may function as a productive stimulus for tissue regeneration. His research has provided insight into how mechanical loading can function to induce increased synthesis of critical cartilage macromolecules. This work has culminated in a patent that describes a process for increasing chondrocyte matrix synthesis that has been licensed to a privately held company. The company has targeted cartilage repair as a therapeutic area for commercialization and has recently received FDA approval for phase 1 trials with their product. His experimental approach to the effects of mechanical loading on extracellular matrix has been extended to adult human mesenchymal stem cells.

Dr. Smith is on the Editorial Board of the *Journal of Biomedical Materials Research (Applied Biomaterials)* and has been a member of various national scientific review panels. He is a reviewer for numerous journals in the fields of biochemistry, biomaterials, and extracellular matrix biology. He is a member of the Orthopaedic Research Society, Society for Biomaterials, International Cartilage Repair Society, and Federation for Experimental Biology and Medicine.

Dr. Smith has published more than 104 peer-reviewed papers, 18 review articles and book chapters, and 150 meeting abstracts and presentations.

Contributors

David Altman
Graduate Student
Department of Biochemistry
Stanford University School of Medicine
Stanford, California

Seoung-Jai Bai
Graduate Student
Department of Mechanical Engineering
Rapid Prototyping Laboratory
Stanford University School of Medicine
Stanford, California

Anne-Elise Barbu
Undergraduate Student
Department of Biology
University of California at Davis
Davis, California

Stephane Barbu, M.S.
Director
High Frequency ASIC Products
Maxim Integrated Products
Sunnyvale, California

Stephan Busque, M.D. M.Sc., FRCSC
Associate Professor of Surgery
Director, Adult Kidney and Pancreas Transplantation Program
Stanford University School of Medicine
Palo Alto, California

Brent R. Constantz, Ph.D.
Consulting Associate Professor
Department of Engineering
Stanford University
Stanford, California

Christopher H. Contag, Ph.D
Assistant Professor
Departments of Pediatrics, Radiology, and Microbiology & Immunology
Molecular Imaging Program at Stanford
Stanford University School of Medicine
Stanford, California

Chris J. Elkins, Ph.D.
San Diego School of Medicine
University of California
San Diego, California

Plamena Entcheva, Ph.D.
Graduate Student
Departments of Civil and Environmental Engineering, Biological Sciences, and Geological and Environmental Sciences
Stanford University
Stanford, California

Rainer Fasching, Ph.D.
Research Associate
Department of Mechanical Engineering
Rapid Prototyping Laboratory
Stanford University
Stanford, California

Michael E. Gertner, M.D.
Lecturer in Surgery
Co-director, Surgical Innovative Program
Department of Surgery
Stanford University School of Medicine
Stanford, California

Ralph S. Greco, M.D.
Johnson & Johnson Distinguished Professor
Chief, Division of General Surgery
Stanford University School of Medicine
Stanford, California

Kyle Hammerick
Graduate Student
Department of Mechanical Engineering
Rapid Prototyping Laboratory
Stanford University
Stanford, California

Christopher R. Jacobs, Ph.D.
Associate Professor
Departments of Mechanical Engineering and Biomedical Engineering
Stanford University
Stanford, California

D. Denison Jenkins, M.D.
Resident in General Surgery
Department of Surgery
Stanford University School of Medicine
Stanford, California

Theo Kofidis, M.D.
Graduate Student
Department of Cardiothoracic Surgery
Stanford University School of Medicine
Stanford, California

Thomas M. Krummel, M.D.
Emile Holman Professor
Chair, Department of Surgery
Stanford University School of Medicine
Stanford, California

Michael D. Kuo, M.D.
Center for Translational Medical Systems: Radiology
San Diego School of Medicine
University of California
San Diego, California

Martin Morf, Ph.D.
Professor of ETH
Consulting Professor, EE Department
SSPL/Center for Integrated Systems
Stanford University
Stanford, California

Jeffrey A. Norton, M.D.
Professor of Surgery
Chief of Surgical Oncology
Division of General Surgery
Stanford University School of Medicine
Stanford, California

Fritz B. Prinz, Ph.D.
Rodney H. Adams Professor of Engineering
Chair, Department of Mechanical Engineering
Department of Materials Science and Engineering
Stanford University
Stanford, California

Robert C. Robbins, M.D.
Director, Stanford Cardiovascular Institute
Associate Professor
Department of Cardiothoracic Surgery
Stanford University School of Medicine
Stanford, California

Hootan Roozrokh, M.D.
Clinical Instructor
Department of Surgery
Stanford University School of Medicine
Stanford, California

WonHyoung Ryu
Graduate Student
Department of Mechanical Engineering
Rapid Prototyping Laboratory
Stanford University
Stanford, California

Minnie Sarwal, M.D., Ph.D., MRCP
Associate Professor of Pediatrics
Stanford University School of Medicine
Stanford, California

Renee Saville
Graduate Student
Department of Civil and Environmental Engineering
Stanford University
Stanford, California

Soni Shukla
Life Science Research Assistant
Department of Civil and Environmental Engineering
Stanford University
Stanford, California

R. Lane Smith, Ph.D.
Rehabilitation Research and Development Center
VA Palo Alto Health Care System
Professor
Department of Orthopaedic Surgery
Stanford University School of Medicine
Stanford, California

Alfred M. Spormann, Ph.D.
Associate Professor
Departments of Civil and Environmental Engineering, Biological Sciences, and Geological and Environmental Sciences
Stanford University
Stanford, California

James A. Spudich, M.D.
Douglas M. and Nola Leishman Professor of Cardiovascular Disease
Department of Biochemistry
Stanford University School of Medicine
Stanford, California

Mary X. Tang, Ph.D.
Senior Research Engineer
Stanford Nanofabrication Facility
Stanford University
Stanford, California

Eric Tao
Graduate Student
Department of Mechanical Engineering
Rapid Prototyping Laboratory
Stanford University
Stanford, California

Kai Thormann, Ph.D.
Graduate Student
Department of Civil and Environmental Engineering, Biological Sciences, and Geological and Environmental Sciences
Stanford University
Stanford, California

Peter Wagner, Ph.D
Senior Vice President and Chief Technology Office
Zyomyx, Inc.
Hayward, California

David S. Wang, M.D.
Division of Cardiovascular and Interventional Radiology
Department of Radiology
Stanford University School of Medicine
Stanford, California

Jacob M. Waugh, M.D.
Center for Translational Medical Systems: Radiology
San Diego School of Medicine
University of California
San Diego, California

Russell K. Woo, M.D.
Department of Surgery
Stanford University School of Medicine
Stanford, California

Lidan You, Ph.D.
Graduate Student
Department of Mechanical Engineering
Stanford University
Stanford, California

Table of Contents

1 Biomaterials: Historical Overview and Current Directions

Russell K. Woo, D. Denison Jenkins, and Ralph S. Greco

CONTENTS

0-8493-1940-4/05/$0.00+$1.50

1.1 INTRODUCTION

Over the last two centuries, the field of medicine has increasingly utilized biomaterials in the investigation and treatment of disease. Common examples include surgical sutures and needles, catheters, orthopedic hip replacements, vascular grafts, implantable pumps, and cardiac pacemakers. The purpose of this chapter is to provide a historical overview of biomaterials, emphasizing the evolution of three generations of materials over the last century, and detailing current trends in the development and application of biomaterials in medicine.

Most have defined biomaterials broadly. Park stated that a *biomaterial* is "a synthetic material used to replace part of a living system or to function in intimate contact with living tissue."[1] Similarly, the National Institutes of Health consensus development conference of 1982 defined a biomaterial as "any substance other than a drug, or combination of substances, synthetic or natural in origin which can be used for any period of time as a whole or as a part of the system that treats, augments, or replaces any tissue, organ or function of the body."[2] In contrast, a *biological material* is a substance produced by a living organism.[3] Muscle and bone are examples. Of note, synthetic materials that only come in contact with the skin, such as hearing aids and bandages, are generally not categorized as biomaterials.[3]

The development and application of biomaterials has been significantly influenced by advances in medicine, surgery, biotechnology, and material science. Specifically, advances in surgical technique and instrumentation have enabled the placement of implants into previously poorly accessible locations.[3] Examples include the placement of endovascular stents and the surgical insertion of mechanical heart valves. Similarly, advances in biotechnology have led to the development of scaffolds for tissue engineering.[4,5] Today, biomaterials can be found in over 8000 different types of medical devices.[6] Table 1.1 lists several current applications of biomaterials in modern medicine.

1.2 HISTORICAL BACKGROUND

The first reported clinical application of a "biomaterial" can be traced back to 1759, when Hallowell repaired an injured artery using a wooden peg and twisted thread.[7] However, it was not until the promotion of aseptic surgical techniques in the 1860s by Lister that the practical use of biomaterials became possible.[8] Before this, surgical procedures were often complicated by serious and often life-threatening infection. Foreign materials deliberately implanted into the body exacerbated this problem, often representing a nidus for infection that the body's natural immune response could not effectively penetrate.[3] As the widespread implementation of sterile techniques brought infection rates under control, the impact of the physical properties of specific medical materials on the success of implant procedures was recognized.[8]

The recognition of the therapeutic potential of biomaterials, along with advances in surgical techniques, led to increasing interest in the incorporation of synthetic materials into living tissues.[8] These early implants were largely applied to the skeletal system. In the 1900s, bone plates were used to fix long bone fractures, though many of these early plates failed as a result of poor mechanical design.[3] Also, early

TABLE 1.1
Uses of Biomaterials

Problem Area	Examples
Replace diseased or damaged parts	Soft or hard tissue prosthetic implants, cardiac valve replacements, renal dialysis machines, tissue engineering scaffolds
Assist in healing	Sutures, adhesives and sealants, bone plates, screws, and nails
Improve function	Cardiac pacemakers, intraocular lens
Correct functional abnormality	Cardiac defibrillator/pacemaker
Correct cosmetic problem	Soft tissue implants (breast, chin, calf)
Aid diagnosis of disease	Probes, catheters, and biosensors
Aid treatment of disease	Catheters, drains, implantable pumps, and controlled drug delivery systems

Source: Modified from Park.[3]

materials chosen primarily for their mechanical properties often corroded rapidly in the body and inhibited the natural healing processes.[3] The early use of Vanadium steel in orthopedic implants is an example of this.

Despite these early troubles, improved designs and more suitable materials were soon introduced. In the 1930s, the introduction of stainless steel and cobalt chromium alloys led to greater success in fracture fixation and the performance of the first joint replacement surgeries.[3] Similarly, during World War II, it was noted that retained fragments of plastic from aircraft canopies did not result in chronic adverse reactions in injured pilots.[3] Consequently, the use of plastics and polymers as biomaterials grew exponentially. Table 1.2 lists several notable events in the early history of biomaterials.

1.3 FIRST-GENERATION BIOMATERIALS (1950s–1960s)

1.3.1 General Characteristics

The experiences of the late nineteenth and early twentieth centuries led to the development of the first set of modern biomaterials in the 1950s and 1960s. These "first-generation" biomaterials were specifically designed for use inside the human body and saw application in multiple disciplines of medicine including orthopedics, cardiovascular surgery, ophthalmology, and wound healing. First-generation biomaterials were often based on commonly available materials selected from engineering practice. For example, the original dialysis tubing was made of cellulose acetate, a commodity plastic, and early vascular grafts were made of Dacron, a polymer developed in the textile industry.[6] Though these materials facilitated the treatment of disease, it became clear that they had the potential to elicit serious inflammatory reactions. Therefore, newer materials were selected for two funda-

TABLE 1.2
Notable Events in the Early History of Biomaterial Implants

Year	Investigators	Development
Late 18th–19th century		Various metal devices to fix bone fractures: wires and pins from Fe, Au, Ag, and Pt
1860–1870	J. Lister	Aseptic surgical techniques
1886	H. Hansmann	Ni-plated steel bone fracture plates
1893–1912	W.A. Lane	Steel screws and plates (Lane fracture plates)
1912	W.D. Sherman	Vanadium steel plates first developed for medical use; lesser stress concentration and corrosion (Sherman plate)
1924	A.A. Zierold	Introduced satellites (CoCrMo alloy)
1926	M.Z. Lange	Introduced 18-8sMo stainless steel, better than 18-8 stainless steel
1926	E.W. Hey-Goves	Used carpenter's screw for femoral neck fracture
1931	M.N. Sith-Petersen	First femoral neck fixation device made of stainless steel
1936	C.S. Venable, W.G. Stuck	Introduced Vitallium (19-9 stainless steel), later changed the material to CoCr alloys
1938	P. Wiles	First total hip prosthesis
1939	J.C. Burch, H.M. Carney	Introduced Tantalum (Ta)
1946	J. Judet, R. Judet	First biomechanically designed femoral head replacement prosthesis, first plastics (PMMA) used in joint replacements
1940s	M.J. Dorzee, A. Franceschetti	First use of acrylics (PMMA) for corneal replacement
1947	J. Cotton	Introduction of Ti and its alloys
1952	A.B. Vorhees, A. Jaretzta, A.B. Blackmore	First successful blood vessel replacement made of cloth for tissue ingrowth

Source: Modified from Park.[3]

mental characteristics: the ability to be tolerated by the body and the ability to reproduce the natural functions of the tissues to be augmented or replaced.[9]

The primary characteristic is often termed *biocompatibility* and refers to the acceptance of an artificial implant by the surrounding tissues and by the body as a whole.[10] A completely biocompatible material would not cause thrombogenic, toxic, or inflammatory responses when placed into living tissue. Furthermore, the material would not elicit carcinogenic, mutagenic, or teratogenic effects. While biocompatibility remains a desired characteristic for all biomaterials, it should be noted that no synthetic material is completely biologically inert. Biocompatibility is more accurately a relative term. The specific host response to biomaterials will be covered in Chapter 2.

The second characteristic, sometimes termed *biofunctionality*, refers to a biomaterial's ability to exhibit adequate physical and mechanical properties to augment or replace body tissues.[11] As expected, these properties vary greatly depending on the target tissue. For example, a material being used for bone augmentation must

TABLE 1.3
Properties of First-Generation Biomaterials

Property	Description
Biocompatibility	• Biologically "inert" • Causes little thrombogenic, toxic, or inflammatory response in host tissue • Noncarcinogenic, mutagenic, or teratogenic
Biofunctionality	• Exhibits adequate physical and mechanical properties to replace or augment the desired tissue
Practical	• Amenable to being machined or formed into different shapes • Not cost prohibitive • Readily available

exhibit a high compressive strength, while a material used for ligament replacement must exhibit a high degree of flexibility and tensile strength. Finally, for practical purposes, a biomaterial must be amenable to being machined or formed into different shapes, have relatively low cost, and be readily available.[11] Table 1.3 lists the primary properties of first-generation biomaterials.

In general, biomaterials are categorized by their origin (i.e., natural or synthetic) as well as by their chemical composition (i.e., polymers, ceramics, alloys, and composites). The following sections provide an overview of various first-generation biomaterials and their predecessors. However, it should be noted that these categories may also apply to newer generations of biomaterials and that there is significant overlap between the three generations of biomaterials highlighted in this chapter. Table 1.4 categorizes some of the most commonly used biomaterials.

1.3.2 Naturally Occurring Biomaterials

Naturally occurring biomaterials encompass biological products of nonhuman origin that are or were used in clinical applications. For example, cellulose, catgut, ivory, silk, natural rubber, glass, graphite, and several pure metals are natural biomaterials.[8] Historically, natural biomaterials were some of the first devices to be used in clinical practice. In 1860, Lister reported the use of catgut as a suture material.[8] Similarly, natural rubber was used by Horsley in the early 1900s for the development of synthetic grafts.[8] Today, natural occurring biomaterials have largely been replaced by synthetic materials deliberately designed with specific characteristics.

1.3.3 Metals and Alloys

Metals are commonly used for load-bearing implants. Specifically, orthopedic procedures utilize a variety of metals to replace or augment skeletal function. Examples range from simple plates and screws to complex joint prostheses. In addition, metals are used in cardiovascular surgery, and for dental and maxillofacial implants.[11] Overall, the biocompatibility of metallic implants is an important characteristic because these implants can corrode in the *in vivo* environment.[12] Corrosion leads to the disintegration of the implant material and the release of potentially harmful products into the surrounding tissues.[3] These issues are addressed in detail in Chapter 2.

TABLE 1.4
Examples of Biomaterials and Their Applications

Material	Principal Applications
Metals and Alloys	
316L stainless steel	Fracture fixation, stents, surgical instruments
CP–Ti, Ti–Al–V, Ti–Al–Nb, Ti–13Nb–13Zr, Ti–Mo–Zr–Fe	Bone and joint replacement, fracture fixation, dental implants, pacemaker encapsulation
Co–Cr–Mo, Cr–Ni–Cr–Mo	Bone and joint replacement, dental implants, dental restorations, heart valves
Ni–Ti	Bone plates, stents, orthodontic wires
Gold alloys	Dental restorations
Silver products	Antibacterial agents
Platinum and Pt–Ir	Electrodes
Hg–Ag–Sn and amalgam	Dental restorations
Ceramics and Glasses	
Alumina	Joint replacement, dental implants
Zirconia	Joint replacement
Calcium phosphates	Bone repair and augmentation, surface coatings on metals
Bioactive glasses	Bone replacement
Porcelain	Dental restorations
Carbons	Heart valves, percutaneous devices, dental implants
Polymers	
Polyethylene	Joint replacement
Polypropylene	Sutures
Polyamides	Sutures
PTFE	Soft-tissue augmentation, vascular prostheses
Polyesters	Vascular prostheses, drug delivery systems
Polyurethanes	Blood-contacting devices
PVC	Tubing
PMMA	Dental restorations, intraocular lenses, joint replacement (bone cements)
Silicones	Soft-tissue replacement, ophthalmology
Hydrogels	Ophthalmology, drug-delivery systems
Composites	
BIS-GMA-quartz/silica filler	Dental restorations
PMMA-glass fillers	Dental restorations (dental cements)

Note: Abbreviations: CP–Ti, commercially pure titanium; PTFE, polytetra fluoroethylenes (Teflon, E.I. DuPont de Nemours & Co.); PVC, polyvinyl chlorides; PMMA, polymethyl methacrylate; BIS-GMA, bisphenol A-glycidyl.

Source: Davis JR. Overview of biomaterials and their use in medical devices, in *Handbook of Materials for Medical Devices*, Davis JR, Ed., Materials Park, OH, ASM International, 2003, pp. 1–13.

1.3.3.1 Pure Metals

The noble metals, such as gold, silver, and platinum, represent some of the earliest used biomaterials due to their tissue compatibility and corrosion resistance.[8] For example, metal ligatures of silver, gold, platinum, and lead were utilized by Levert in 1829. Similarly, Cushing used silver clips in 1911 to control bleeding during operations to remove cerebral tumors.[8] While the pure metals were the earliest metals used as biomaterials, they have been steadily replaced by alloys engineered for improved strength and biocompatibility. In fact, titanium, which was initially used in World War II for aircraft devices, has been the only new pure metal biomaterial introduced since 1940.[8]

1.3.3.2 Alloys

Metal alloys have largely replaced pure metals as biomaterials. Although many alloys are used in medical devices, the most commonly employed are stainless steels, commercial titanium alloys, and cobalt-based alloys.[11] The first metal alloy developed specifically for human use was "Vanadium steel," which was used to manufacture bone fixation plates and screws in the early 1900s.[3] However, as mentioned earlier, Vanadium steel corrodes *in vivo*. Since then, several stainless steel alloys have been developed with greater strength and improved corrosion resistance. These include 18-8 or type 302 stainless steel as well as the later 18-8sMo or type 316 stainless steel. Of note, 18-8sMo steel was unique in that it contained a small percentage of molybdenum to improve its corrosion resistance in salt water.[3] Reduction of the carbon content from 0.08 to 0.03% led to the development of type 316L steel. Together, types 316 and 316L stainless steel are known as the austenitic stainless steels and are the most widely used alloys for biomaterial applications.[3]

Similar to stainless steel alloys, cobalt-chromium alloys were developed for commercial application in the early 1900s and subsequently utilized as biomaterials. These alloys were used as an alternative to gold alloys in dentistry and have recently been utilized in the production of artificial joints.[3] Most recently, titanium and its alloys have been used as biomaterials for implant fabrication.[13] Initially discovered by Gregor in 1791, titanium remained a laboratory curiosity until 1946, when Kroll developed a process for the commercial production of titanium by reducing titanium tetrachloride with magnesium.[14] Since then, titanium and its alloys have been widely used as biomaterials because of their relative biological inertness and superior mechanical properties.[13] Specifically, titanium is a reactive metal that forms a tenacious oxide layer on its surface when exposed to air, water, or specific electrolytes.[15] This oxide layer provides a protective coating that shields the material from chemical degradation and the biological environment.[15] Conversely, the oxide layer, which is in contact with body tissues, is essentially insoluble and does not release ions that can react with other molecules.[15] Lastly, titanium exhibits a high yield strength and low elastic modulus, properties that are desirable in orthopedic implants.[14]

1.3.3.3 Shape-Memory Alloys (SMAs)

Shape-memory alloys are a group of metals that have the interesting properties of thermal shape memory, superelasticity, and force hysteresis.[16,17] Nitinol, an approximately equiatomic allow of nickel and titanium, is the most widely used member of this group. Originally discovered at the U.S. Naval Ordinance Laboratory and then reported by Beuhler and colleagues in 1963, nitinol is relatively biocompatible and more compliant than most other alloys.[18] Currently, nitinol is used in an increasing number of surgical prostheses and disposables, including a variety of endovascular stents, intracranial aneurysm clips, and vascular suture anchors.[16]

1.3.4 Ceramics

Ceramics are one of the oldest artificially produced materials, used in the form of pottery for thousands of years. Ceramics are polycrystalline compounds including silicates, metallic oxides, carbides, and various refractory hydrides, sulfides, and selenides.[19] The first clinical application of ceramics was the use of plaster of Paris as a casting material.[20] However, until recently, the general use of ceramics as implantable biomaterials was limited due to their inherent brittleness, low tensile strength, and low impact strength.[19] In recent years, newer, "high-tech" ceramics have gained increased use as biomaterials due to their relative bioinertness and high compressive strength.[21,22] These implantable ceramics have been termed *bioceramics* and are grouped into three categories based on their biologic behavior in certain environments: the relatively bioinert ceramics, the bioreactive or surface reactive ceramics, and the biodegradable or reabsorbable ceramics.

Relatively bioinert bioceramics are nonabsorbable carbon-containing ceramics, alumina, zirconia, and silicon nitrides.[19] While in a biological host, relatively bioinert ceramics maintain their mechanical and physical properties.[21] They are used in dense and porous forms, usually have good wear, and are excellent for gliding functions.[19] For example, bioinert bioceramics are used to produce femoral head replacements. In addition, relatively bioinert materials are typically used as structural-support implants such as bone plates and bone screws.[19]

The bioactive ceramics include glass, glass-ceramics, and calcium phosphate-based materials.[19] They are characterized by their ability to provoke surrounding bone and tissue responses, which makes them advantageous for anchoring an implant or reducing its stress.[19] They have been successfully used as coatings, continuous layers, and embedded particles in orthopedics and dental surgery.[23]

Lastly, biodegradable or resorbable ceramics include aluminum calcium phosphate, coralline, plaster of Paris, hydroxyapatite, and tricalcium phosphate.[19] They differ from the bioactive ceramics in two major ways. First, they are more soluble, and consequently are degraded by surrounding tissues. Second, due to their porous structure they may stimulate tissue ingrowth and therefore offer the potential to fill or bridge defects. These materials have been used in the fabrication of various orthopedic implants as well as for solid or porous coatings on hybrid implants made of other biomaterials.[24,25]

Overall, bioceramics are unique in their ability to form porous structures. This is advantageous because a large surface area to volume ratio results in a greater tissue contact surface.[26] In addition, interconnected pores permit tissue ingrowth and may facilitate blood and nutrition delivery. Though the bioactive and biodegradable ceramics are included in this discussion, they are often classified as second-generation biomaterials due to their dynamic qualities.

1.3.5 Polymers

In the 1890s, the earliest synthetic plastics were developed using cellulose, a major structural component of plants.[27] In the 1930s, nylon, the first commercial polymer, was produced and made widely available, leading to the birth of the field of polymer science.[8] However, the field of polymer science was relatively limited until 1954, when Professor G. Natta developed a new polymerization technique that transformed random structural arrangements on noncrystallizable polymers into structures of high chemical and geometrical regularity.[28] This discovery had a significant impact on the development of propylene polymers, an inexpensive petroleum derivative used for a variety of medical applications.[8]

Since then, synthetic polymetric materials have been extensively used in a variety of biomedical applications including medical disposables, prosthetic materials, dental materials, implants, dressings, and extracorporeal devices.[29] Overall, polymeric biomaterials display several key advantages. These include ease of manufacturing into products with a wide variety of shapes, ease of secondary processability, reasonable cost, wide availability, and wide variety of mechanical and physical properties.[29]

Polymers consist of small repeating units, or isomers, strung together to form long chain molecules.[30] These long chains are formed by covalent bonding along the backbone chain and can be arranged into linear, branched, and network structures, depending on the functionality of the repeating units.[29] The long chains may be held together by a variety of chemical and ionic forces. These include secondary bonding forces, such as van der Waals forces and hydrogen bonds, and primary covalent bonding forces through cross-links between chains.[29] Such cross-linking can increase the density of materials to improve their strength and hardness. However, cross-linked materials often lose their flexibility and become more brittle.[29]

The physical properties of polymers can be deliberately changed in many ways. In particular, altering the molecular weight and its distribution has a significant effect on the physical and mechanical properties of a polymer.[10] For example, with increasing molecular weight, the chains of a polymer become longer and less mobile, resulting in a more rigid material.[29] Similarly, changing the chemical composition of the backbone or side chains can change the physical properties of the polymer.[10] For example, the substitution of a backbone carbon in a polyethylene with divalent oxygen increases the rotational freedom of the chain, resulting in a more flexible material.[10] Likewise, side chain substitution, cross-linking, and branching all affect the physical properties of polymers. Increasing the size of side groups or branches, or increasing the cross-linking of the main chains all result in a poorer degree of molecular packing. This retards the polymer crystallization rate, thereby decreasing the melting temperature of a material.[10,29] Similarly, changes in temperature can have a significant effect on the properties of polymers.[10] The *glass transition temperature* refers to the point between the temperature range in which a polymer is relatively stiff (glassy region), and the temperature range in which a polymer is very compliant (rubbery region).[10,29] Depending on the temperature, a single polymeric material can therefore take on a variety of forms.

Today, a variety of polymers are used as biomaterials. These include polyvinylchloride (PVC), polyethylene (PE), polypropylene (PP), polymethylmethacrylate (PMMA), polystyrenes (PS), fluorocarbon polymers (most notably polytetraflouroethylene or PTFE), polyesters, polyamides or nylons, polyurethanes, resins, and polysiloxanes (silicones).[8,30] These synthetic polymers have found extensive utilization in a wide variety of applications including implantable devices, coatings on devices, catheters and tubing, vascular grafts, and injectable drug delivery and imaging systems.[30–32] Table 1.5 lists the names, key properties, and traditional applications of commonly used nondegradable polymers.[30]

1.3.6 Composites

Composite biomaterials are composed of two or more distinct constituent materials, incorporating the desired physical and mechanical properties of each.[33] This results in a hybrid product whose overall properties may be significantly different from the homogenous materials. For example, rubber used in various catheters is often filled with very fine particles of silica to enhance the strength and toughness of the material.[33] Other examples of composite biomaterials in clinical use include composite resins commonly used as dental fillings (e.g., polymer matrix and barium, glass, or silica inclusions) as well as porous orthopedic and soft tissue implants.[34]

1.4 SECOND-GENERATION BIOMATERIALS (1970s–2000)

1.4.1 General Characteristics

Over the last three decades, the field of biomaterials shifted away from the traditional properties of first-generation biomaterials. While biocompatibility and biofunctionality continued to remain important, the long-standing goal of achieving a bioinert tissue response began to be replaced by the notion of developing materials that were *bioactive,* or *biodegradable*.[9] These bioactive or biodegradable biomaterials have been termed "second-generation" biomaterials and have seen increasing clinical application.

While bioinert materials were designed to elicit little or no tissue response, bioactive materials have been designed to elicit specific and controlled interactions between the material and the surrounding tissue. For example, synthetic hydroxyapatite (HA) ceramics are used as porous implants, powders, and coatings on metallic prostheses to provide bioactive fixation.[9,23] HA coatings on implants lead to a controlled tissue response in which bone grows along the coating.[9] This response, termed osteoconduction, promotes the formation of a mechanically strong interface between the implant and the native bone.[9] Such materials have seen widespread

TABLE 1.5
Chemical Names, Key Properties, and Traditional Applications of Commonly Used Nondegradable Polymers

Chemical Name and Trade Name	Key Property	Traditional Applications
Poly(ethylene) (PE) (HDPE, UHMWPE)	Strength and lubricity	Orthopedic implants and catheters
Poly(propylene) (PP)	Chemical inertness and rigidity	Drug delivery, meshes and sutures
Poly(tetrafluroethylene) (PTFE) (Teflon), extended-PTFE (Gore-Tex®)	Chemical and biological inertness and lubricity	Hollow fibers for enzyme immobilization, vascular grafts, guided tissue regeneration barrier membranes for the prevention of tissue adhesions
Poly(methymethacrylate) (Palacos®)	Hard material, excellent optical transparency	Bone cement, ocular lens
Ethylene-co-vinylacetate (EVA) (Elvax®)	Elasticity, film forming properties	Implantable drug delivery devices
Poly(dimethylsiloxane) (PDMS) (Silastic®, silicone rubber)	Ease of processing, biological inertness, excellent oxygen permeability, excellent optical transparency	Implantable drug delivery devices, device coatings, gas exchange membranes, ocular lens, orbital implants
Low MW poly(dimethylsiloxane) (Silicone oil)	Gel-like characteristics	Filler in silicone breast implants
Poly(ether-urethanes) (PU) (Tecoflex®, Tecothane®, BioSpan®)	Blood compatibility and rubber-like elasticity	Vascular grafts, heart valves, blood contacting devices, coatings
Poly(ethylene terephthalate) (PET) (Dacron®)	Fiber-forming properties and slow *in vivo* degradation	Knitted Dacron vascular grafts, coatings on degradable sutures, meshes for abdominal surgery
Poly(sulphone) (PS)	Chemical inertness, creep resistant	Hollow fibers and membranes for immobilization of biomolecules in extracorporeal devices
Poly(ethyleneoxide) (PEO, PEG)	Negligible protein adsorption and hydrogel forming characteristics	Passsivation of devices toward protein adsorption and cell encapsulation
Poly(ethyleneoxide-copropyleneoxide) (PEO-PPO), (Pluronics®)	Ampiphilicity and gel forming properties	Emulsifier
Poly(vinylalcohol)	Surfactant and gel-forming properties	Emulsifier in drug encapsulation processes and matrix for sustained drug delivery

Source: Modified from Shastri.[30]

TABLE 1.6
Properties of Second-Generation Biomaterials

Property	Description
Bioactive	• Elicits a controlled interaction between the material and surrounding tissues
Biodegradable/ bioresorbable	• Exhibits controlled chemical breakdown and reabsorption • Allows functional tissue to grow in its place

clinical use in the fields of orthopedics and dentistry.[23] Overall, bioactivity is increasingly becoming a characteristic of modern biomaterials as expanding applications call for dynamic biomaterials and devices.

Similarly, biodegradable or bioresorbable materials were designed to exhibit clinically relevant breakdown and absorption. In this manner, the issue of interface integration of the implant and surrounding tissue is addressed as the foreign material is eventually replaced by regenerating tissues. A common example is absorbable sutures. Consisting of a polymer composed of polylactic (PLA) and polyglycolic (PGA), these sutures decompose into carbon dioxide and water after a designated length of time. Consequently, the foreign material used to approximate tissue during surgery is resorbed after the tissue has naturally healed together.[35] Other applications of bioresorbable materials include resorbable fracture fixation plates and screws as well as controlled-release drug delivery systems.[9] Table 1.6 lists the properties of second-generation biomaterials. The following sections detail several classes of second-generation biomaterials, including biodegradable polymers, hydrogels, and bioactive and biodegradable ceramics.

1.4.2 Biodegradable Polymers

Many types of biomaterials utilized for soft and hard tissue repair are required only for a short time to support tissue regeneration and ingrowth. These temporary biomaterials have been constructed using biodegradable polymers that degrade when placed in the body, allowing functional tissue to grow in its place. The mechanisms for degradation for these types of polymers include both hydrolytic and enzymatic degradation.[35] These degradation processes are influenced by many factors including molecular weight and distribution of the polymer, chemical composition of the polymer backbone, polymer morphology (e.g., amorphous/crystalline structure), glass transition temperature, additives, and environmental conditions (pH, temperature, etc.).[35]

Biodegradable polymers can be natural or synthetic in origin. Natural polymers include starch, chitin, collagen, and glycosaminoglycans.[35] For example, preparations of chitin and its derivative chitosan have been developed for wound dressings and sustained-release drug delivery systems.[36–40] The first totally synthetic biodegradable polymer was polyglycolic acid which was introduced in the early 1970s. This was followed by the polylactides in 1985.[41] Both of these polymers fall under the poly(α-hydroxy acid) family of polymers which is the most widely used for the production of biodegradable biomaterials. Other biodegradable polymers include polycaprolactone, poly(ortho esters), polyanhydrides, and polyphosphazenes.[35] Currently, biodegradable polymers are used for a variety of biomaterials applications

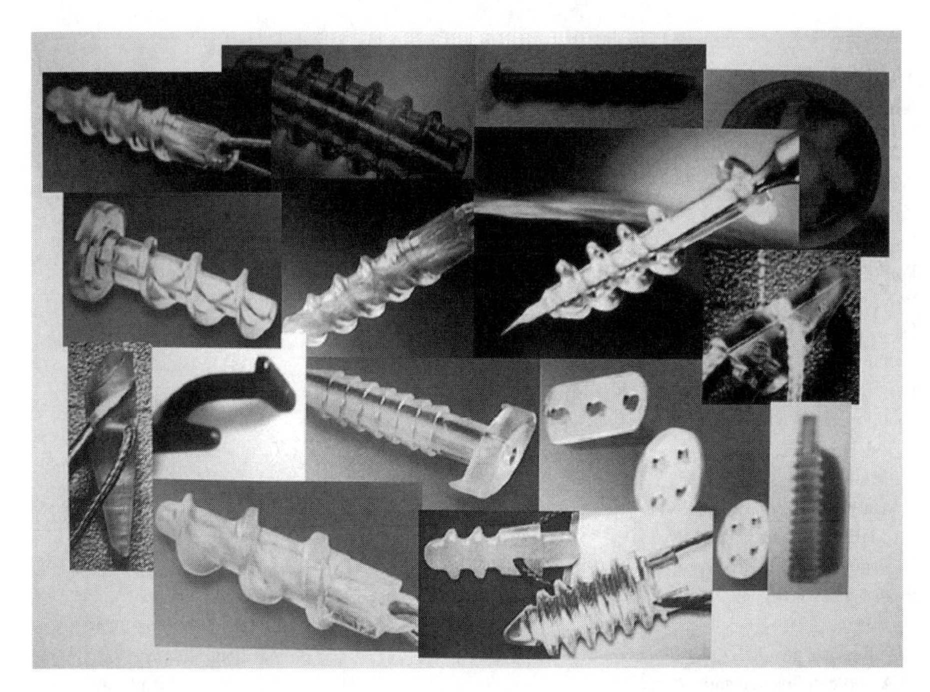

FIGURE 1.1 Various biodegradable implants. (From Sudkamp NP, Kaab MJ. Biodegradable implants in soft tissue refixation: experimental evaluation, clinical experience, and future needs, *Injury,* 2002, 33, Suppl. 2, B17–B24.)

including sutures, wound dressings, fracture plates and screws, and controlled-release delivery systems.[41] In addition, biodegradable matrices and scaffolds show a great deal of promise in the field of tissue engineering due to their ability to degrade while encouraging functional host tissue to take their place.[42,43] Figure 1.1 depicts several commercially available biodegradable implants.

1.4.3 Hydrogels

Hydrogels are cross-linked hydrophilic polymer networks that can absorb water or other biological fluids. First used for a biomedical application in the late 1950s as a soft contact lens material, hydrogels have since found widespread application as orthopedic and ophthalmic implants, controlled drug systems, and biosensors.[44] Table 1.7 lists some commonly used hydrogels and their biomedical applications. Even though hydrogels were first utilized in the 1950s, they are included here as second-generation biomaterials because of their unique properties.

Hydrogels display several modifiable characteristics that make them useful as biomaterials. These characteristics, which are primarily determined by the hydrogel's polymer network structure, include a hydrogel's swelling behavior, diffusive characteristics, and surface properties.[44] The swelling behavior of a hydrogel refers to the material's ability to absorb water or other fluids and is largely a function of its structural network properties.[44] When a biopolymer network comes in contact with an aqueous solution, the network starts to swell due to the interaction of the polymer

TABLE 1.7
Medical Applications of Hydrogel Polymers

Hydrogel Polymer	Medical Applications
Poly(vinyl alcohol) [PVA]	
Polyacrylamide [PAAm]	
Poly(N-vinyl pyrrolidone) [PNVP]	Blood-compatible hydrogels
Poly(hydroxyethyl methacrylate) [PHEMA] Poly(ethylene oxide) [PEO]	
Poly(ethylene glycol) [PEG]	
Poly(ethylene glycol) monomethyl ether [PEGME]	
Cellulose	
Poly(hydroxyethyl methacrylate) [PHEMA]	
copolymerized with:	
NVP	Contact lenses
Methacrylic acid [MAA]	
Butyl methacrylate [BMA]	
Methyl methacrylate [MMA]	
3-methoxy-2-hydroxypropylmethacrylate [MHPM]	
PHEMA/poly(ethylene terephthalate) [PTFE]	Artificial tendons
	Other medical applications
Cellulose acetate	Artificial kidney
PVA and cellulose acetate	Membranes for plasmapheresis
PNVP, PHEMA, cellulose acetate	Artificial liver
PVA and PHEMA	Artificial skin
Terpolymers of HEMA, MMA and NVP	Mammoplasty
PHEMA, P(HEMA-co-MMA)	Maxillofacial reconstruction
PVA	Vocal cord reconstruction
P(HEMA-b-siloxane)	Sexual organ reconstruction
PVA, poly(acrylic acid) [PAA], poly(glyceryl methacrylate)	Ophthalmic applications
PVA, HEMA, MMA	Articular cartilage
	Controlled drug delivery[a]
Poly(glycolic acid) [PGA], Poly(lactic acid) [PLA], PLA-PGA, PLA-PEG, Chitosan, Dextran, Dextran-PEG, polycyanoacrylates, fumaric acid-PEG, sebacic acid/1,3-bis(*p*-carboxyphenoxy) propane [P (CPP-SA)]	Biodegradable hydrogels
	Nonbiodegradable hydrogels
PHEMA, PVA, PNVP, poly(ethylene-co-vinyl acetate) [PEVAc]	Neutral
Poly(acrylamide) [PAAm], poly(acrylic acid) [PAA], PMAA, poly(diethylaminoethyl methacrylate)[PDEAEMA], poly(dimethylaminoethyl methacrylate) [PDMAEMA]	pH-Sensitive
Poly(methacrylic acid-grafted-poly(ethylene glycol)) [P(MAA-g-EG)], poly(acrylic acid-grafted-poly(ethyleneglycol)) [P(PAA-g-EG)]	Complexing hydrogels
Poly(N-isopropyl acrylamide) [PNIPAAm]	Temperature-sensitive
PNIPAAm/PAA, PNIPAAm/PMAA	pH/Temperature-sensitive

[a] These drug delivery applications have been used for the controlled release of several therapeutic agents such as contraceptives, antiarrhythmics, peptides, proteins, anticancer agents, anticoagulants, antibodies, among others. This table does not include all the copolymers of such hydrogels.

Source: Modified from Peppas.[44]

chains and water. This propensity to swell is offset by the resistive force induced by the cross-links of the polymer network.[44] When these two forces balance, an equilibrium is reached and swelling stops. In addition to the network structure of the polymer, a hydrogel's swelling behavior can be influenced by the presence of ionizable pendant groups.[44] In such cases, the forces influencing the swelling equilibrium may be altered by the localization of charges within the hydrogel.[44]

The diffusive characteristics of a hydrogel have been the basis of many of their newer applications, including drug and protein delivery.[6,45,46] While the chemical potential gradient of a solute determines its flux within a system, several important characteristics of hydrogels may serve to influence the rate and pattern of this diffusion. These include the structure and pore size of a hydrogel, the polymer composition, the water content, and the nature and size of the solute.[44] In addition, interactions between different solutes, between solutes and the gel polymers, and between solutes and solvents all play a role in determining the overall diffusion characteristics of a solute and a hydrogel.[44]

Finally, the surface properties of hydrogels may be altered to achieve a variety of unique characteristics.[44] In general, hydrogels have a complex surface structure composed of numerous dangling chains attached on one end to the polymer network. By altering these chains, the surface properties of hydrogels can be engineered to serve a number of different purposes. Hern and Hubbell incorporated adhesion-promoting oligopeptides into hydrogels, giving them the ability to mediate cell adhesion properties.[47] Similarly, by using modified lipid bilayers, hydrogel surfaces have been engineered to mimic cell membranes.[48]

By adjusting these unique characteristics, engineered hydrogels have recently been created to serve a variety of biomedical applications. Specifically, controlled drug delivery systems have been developed by altering the diffusive characteristics of hydrogels.[45,46] Such systems are able to maintain a desired blood concentration of a specific drug for a prolonged period of time and have been constructed as either matrix or reservoir configurations.[44] In a reservoir system, the agent is stored in a central core surrounded by a polymer membrane.[44] Conversely, in a matrix system, the agent is uniformly distributed throughout the material and slowly released from it.[44] In addition, these drug delivery systems can be constructed using biodegradable polymers, thereby negating the necessity of surgical removal.

More recently, hydrogel-based delivery systems have been created that release their agents in response to changes in their environment.[46,49,50] These new, bioactive systems may respond to changes in pH, temperature, ionic potentials, solvent concentrations, and magnetic or electrical fields.[51] The use of such systems is being investigated for the controlled release of a variety of agents including insulin, streptokinase, lysozyme, and salmon calcitonin.[44] In addition, a recently recognized advantage of hydrogels is that they may have the ability to protect embedded drugs, peptides, or proteins from the potentially harsh biological environment. This may enable the oral delivery of engineered proteins, which may otherwise be prematurely broken down in the digestive tract.[6]

1.4.4 Bioactive and Biodegradable Ceramics

As stated earlier, the bioactive ceramics include glass, glass-ceramics, and calcium phosphate-based materials.[19] They are characterized by their ability to provoke surrounding bone and tissue responses, which makes them advantageous for anchoring an implant or reducing its stress.[23] The biodegradable ceramics include aluminum calcium phosphate, coralline, plaster of Paris, hydroxyapatite, and tricalcium phosphate.[19] They are unique in that they are soluble and are therefore degraded and absorbed by surrounding tissues. In addition, due to their porous structure they stimulate bone ingrowth. These materials have been used in the fabrication of various orthopedic implants, as well as for solid or porous coatings on implants made of other biomaterials.[23]

1.5 THIRD-GENERATION BIOMATERIALS (2000–PRESENT)

Recently described by Hench and Polak, a new, third-generation of biomaterials is being designed.[9] For this class of biomaterials, the distinct characteristics of bioactivity and biodegradability have now converged. As opposed to second-generation biomaterials, which are generally either bioactive or biodegradable, third-generation biomaterials are being designed that display both of these characteristics.[9] Furthermore, these new biomaterials are being designed alongside advances in the fields of tissue engineering, microfabrication, and nanofabrication. In this manner, third-generation biomaterials will be created to aid in the regeneration, and not simply the replacement of injured or lost tissues. In addition, micro- and nanofabrication techniques are being utilized to create "smart" biomaterials and implants that can detect and respond to various tissue and cellular stimuli.

1.5.1 Biomaterials in Tissue Engineering

In a recent review, Griffith defined tissue engineering as "the process of creating living, physiological, three-dimensional tissues and organs utilizing specific combinations of cells, cell scaffolds, and cell signals, both chemical and mechanical."[5] As the field of tissue engineering has developed, novel biomaterials have been created to facilitate these approaches. Broadly speaking, there are currently three ways in which biomaterials have been utilized in the field of tissue engineering:[6]

1. To induce cellular migration or tissue regeneration
2. To encapsulate cells and act as an immunoisolation barrier
3. To provide a matrix to support cell growth and cell organization

The first approach involves the use of bioactive biomaterials to facilitate local tissue repair. Such materials have been developed in several forms (e.g., powders, solutions, or doped microparticles) and have been designed to release a variety of chemicals, proteins, and/or growth factors in a controlled fashion by diffusion or network breakdown.[9] These agents trigger a local cellular response leading to tissue repair and regeneration. The infusion of bone morphogenic protein into orthopedic implants[9] as well as the use of glycosaminoglycan and collagen constructs to act as an artificial skin substitute are current examples of biomaterials used in tissue engineering.[6,52]

In the second strategy, a polymer is used as an immunoisolation barrier for encapsulated cells.[6] For example, hollow fiber membranes enclosing hepatocytes have been used to construct a bioartifical liver for the treatment of liver failure.[6,53,54] Similarly, microcapsules that store and protect cellular transplants also have been developed.[6] These microcapsules protect the transplanted cells from immune cell penetration while allowing the passage of medium- and small-sized substances, such as oxygen, nutrients, and wastes.[6,35]

The third and most widespread use of biomaterials in tissue engineering involves the use of tissue scaffolds to direct the three-dimensional organization of cells *in vitro* and *in vivo*.[5] Scaffolds are porous structures created from natural or synthetic polymers.[35,55] Scaffolds can be constructed in a variety of forms including sponge-like sheets and fabrics, gels, and highly complex three-dimensional structures with detailed pores and channels.[4,56] Similar to the biomaterials of the 1950s and 1960s, the early tissue engineering scaffolds were adapted from other uses in surgery.[5] These early scaffolds, often made of surgical fabrics and mesh, provided a three-dimensional biocompatible structure onto which cells could attach.[4] However, they did not specifically interact with cells in a controlled fashion. More recently, scaffolds are being created with embedded growth factors or cellular ligands, allowing them to influence signaling pathways necessary for cell migration and proliferation.[42,57,58] In this manner, these new scaffolds can be thought of as synthesized extracellular matrix, providing tissues with the appropriate architecture and signaling pathways to influence key cell functions.[4] Currently, scaffolds have been used to engineer a variety of tissues including vascular tissue, skin, bone, and cartilage.[57,59–62] Figure 1.2 represents a porous collagen scaffold.

1.5.2 Micro/Nanotechnology and Biomaterials

In recent years, significant interest has developed surrounding the utilization of micro- and nanofabrication techniques for the construction of novel biomaterials and implants. Developed from advances in the fields of computer science, engineering, physics, and biology, these techniques have found widespread application in an array of industries including computing, consumer electronics, manufacturing, and biotechnology. While the application of these techniques toward biomaterials is relatively new, they promise to enable the creation of smaller, more active, and more dynamic implants and devices.

1.5.2.1 Microfabrication and Microtechnology

Microfabrication is the process of constructing materials and devices with dimensions in the micrometer to millimeter range.[63] Developed and refined for the processing of integrated circuits, microfabrication techniques have been utilized since the 1950s for the production of semiconductor-based microelectronics. In general,

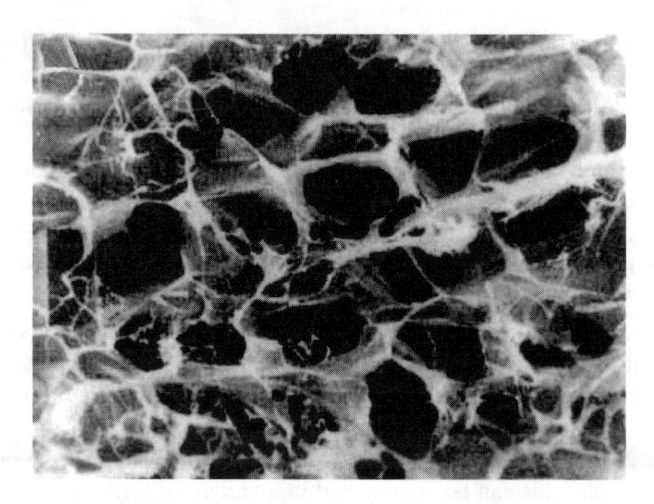

FIGURE 1.2 Collagen scaffold. (From Vats A, Tolley NS, Polak JM, Gough JE. Scaffolds and biomaterials for tissue engineering: a review of clinical applications, *Clin. Otolaryngol.*, 2003, 28(3), 165–172. With permission.)

microfabrication techniques utilize a "top down" approach for creating structures, where one takes a substrate and builds a device out of the bulk material (bulk micromachining) or on its surface (surface micromachining).[63] These techniques include unit process steps such as thin-film growth/deposition, photolithography, etching, and bonding and will be covered in detail in later chapters.[63,64]

Over the last few decades, microfabricated devices containing electrical and mechanical components, also known as microelectromechanical systems (MEMS), have made their way into a multitude of products used in daily life. For example, the air bag system in an automobile is likely to include a MEMS accelerometer to help the system detect the proper time to fire.[65] Similarly, in the healthcare arena, MEMS technology can be found in blood pressure sensors, blood chemistry analysis systems, DNA array systems, and home telemetry monitoring systems.[64,65] In the field of biomaterials and implants, MEMS technology is being applied toward the development of implantable drug delivery systems.[65] Such systems utilize micropumps that deliver regulated amounts of stored medication in a controlled fashion. The use of such systems for the delivery of insulin for the treatment of diabetes is currently being investigated.[65] Figure 1.3 and Figure 1.4 represent examples of MEMS technology in medicine.

1.5.2.2 Nanofabrication and Nanotechnology

Nanotechnology has been defined as "research and technology development at the atomic, molecular, and macromolecular levels in the length and scale of approximately 1–100 nanometer range, to provide a fundamental understanding of phenomena and materials at the nanoscale and to create and use structures, devices, and systems that have novel properties and functions because of their small and/or intermediate size."[66] On a simpler level, this definition can be broken down into two

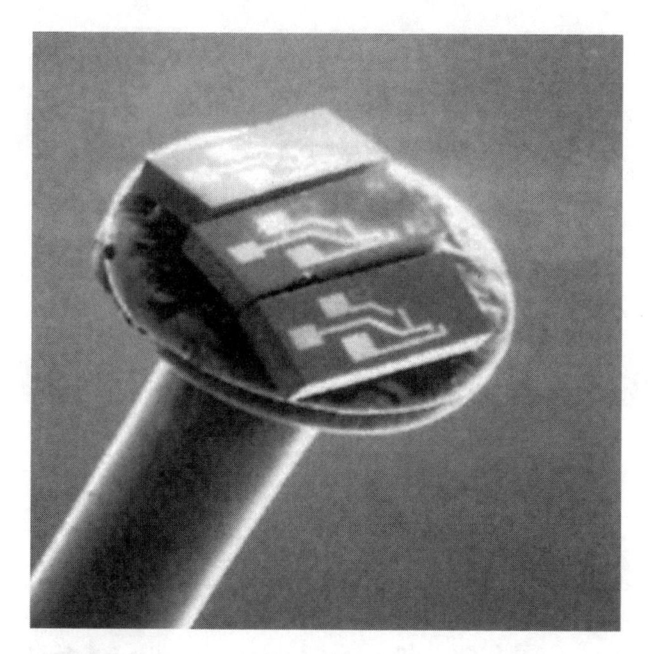

FIGURE 1.3 Catheter tip blood pressure transducer. (From Salzberg AD, Bloom MB, Mourlas NJ, Krummel TM. Microelectrical mechanical systems in surgery and medicine, *J. Am. Coll. Surg.*, 2002, 194(4), 463–476. With permission.)

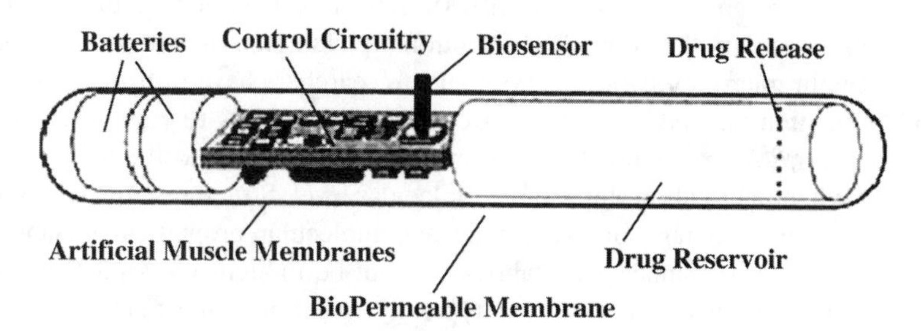

FIGURE 1.4 MEMS drug delivery system. (From Salzberg AD, Bloom MB, Mourlas NJ, Krummel TM. Microelectrical mechanical systems in surgery and medicine, *J. Am. Coll. Surg.*, 2002, 194(4), 463–476. With permission.)

specific and important characteristics that distinguish nanotechnology and nanostructures from their micro and macro counterparts. First, nanotechnology refers to the creation and manipulation of structures at the nanoscale range (1–100 nm).[67] Table 1.8 compares several nanostructures to biological structures. Second, nanotechnology is concerned with the characterization and application of the unique physical, chemical, and biological properties that nanoscale structures display because of their size.[68,69] New behavior at the nanoscale range is not necessarily predictable and may be completely different from that of the same material in its bulk, macroscopic

TABLE 1.8
Sizes of Nanostructures in Comparison to Natural Structures[67]

Structure	Size
Nanotechnology Structures	
Nanoparticles	1–100 nm
Fullerene (C60)	1 nm
Quantum dot (CdSe)	8 nm
Dendrimer	10 nm
Natural Structures	
Atom	0.1 nm
DNA (width)	2 nm
Protein	5–50 nm
Virus	75–100 nm
Bacteria	1000–10,000 nm
White blood cell	10,000 nm

form.[68] The understanding and utilization of these properties is a fundamental goal of nanoscience and nanotechnology.

Nanofabrication refers to the processes and methods employed in the creation of nanoscale materials and structures. In contrast to the microfabrication techniques developed for the semiconductor and MEMS industries, nanofabrication techniques utilize complementary "top down" and "bottom up" fabrication methods.[69] Building on advances in microfabrication, nanoscience researchers have utilized newer "top down" fabrication methods such as electron beam lithography to yield near-atomic-scale precision.[69] Alternatively, the "bottom up" creation of nanostructures is being developed where fabrication starts at the molecular level. Structures are "self-assembled" by taking advantage of the atomic and molecular properties of nanoscale materials.[67,69] In this manner, nanofabrication can be thought of as being inspired by nature, as biological structures are typically assembled and rearranged at the nanoscale range using molecular interactions such as van der Waals forces, hydrogen bonds, and electrostatic dipoles.[70]

Today, nanotechnology is an exploding field. Since the year 2000, more than 35 countries have developed programs in nanotechnology.[70] Furthermore, worldwide government funding of nanotechnology has increased approximately five-fold since 1997, exceeding 2 billion dollars in 2002.[70] In the United States alone, the National Nanotechnology Initiative, established by President Bill Clinton in 2000, has grown rapidly in both scope and support, with a 2002 federal budget award of approximately $697 million.[70] With this intense interest and development, nanotechnology and nanofabrication have already played a significant role in the development of new medical devices and materials. Today, pharmaceutical preparations are being encapsulated in a variety of nanostructures to enhance effectiveness and decrease side effects.[71] Similarly, nanoparticle-based scaffolds that facilitate bone growth have already received FDA approval for use in orthopedic surgery.[71] As we look toward the future, nanotechnology will undoubtedly play a significant role in the development and use of future generations of biomaterials.

1.6 CONCLUSION

Biomaterial science has developed significantly over the last century. Originally adopted from industrial materials never intended for biological use, biomaterials are now being developed to specifically interact with living tissue — aiding in tissue repair, regeneration, and replacement. Throughout the three generations of biomaterials covered in this chapter, a central theme of interaction between science, engineering, and medicine is evident. As new materials with new properties are engineered, new clinical applications for these materials are developed. Conversely, as newer ways to diagnose and treat disease are discovered, new materials and implants are created to facilitate these techniques. Today, the fields of tissue engineering and micro- and nanotechnology promise to revolutionize the science and practice of medicine. Their impact on the development of new biomaterials, as well as their application in multiple fields of medicine is the focus of this book.

REFERENCES

1. Park JB, Lakes RS. *Biomaterials: An Introduction*, New York, Plenum Press, 1992.
2. Williams DF. *Definitions in Biomaterials,* Vol. 4. New York, Elsevier, 1987.
3. Park JB. Biomaterials, in *The Biomedical Engineering Handbook,* Bronzino JD, Ed., Boca Raton, FL, CRC Press LLC, 1995.
4. Naughton GK. From lab bench to market: critical issues in tissue engineering, *Ann. N. Y. Acad. Sci.*, 2002, 961, 372–385.
5. Griffith LG. Emerging design principles in biomaterials and scaffolds for tissue engineering, *Ann. N. Y. Acad. Sci.,* 2002, 961, 83–95.
6. Langer R, Peppas NA. Advances in biomaterials, drug delivery, and bionanotechnology, *AICHE J.,* 2003, 49(12), 2990–3006.
7. Wesolowski SA, Dennis C. *Fundamentals of Vascular Grafting*, New York, McGraw-Hill, 1963.
8. Friedman DW, Orland PJ, Greco RS. Biomaterials: an historical perspective, in *Implantation Biology: The Host Response to Biomedical Devices,* Greco RS, Ed., Boca Raton, FL, CRC Press, 1994.
9. Hench LL, Polak JM. Third-generation biomedical materials, *Science,* 2002, 295(5557), 1014–1017.
10. Park JB, Lakes RS. Polymeric implant materials, in *Biomaterials: An Introduction*, second ed., Park JB, Lakes RS, Eds., New York, N.Y., Plenum Press, 1992.
11. Davis JR. Overview of biomaterials and their use in medical devices, in *Handbook of Materials for Medical Devices*, Davis JR, Ed., Materials Park, OH, ASM International, 2003, pp. 1–13.
12. Williams DF. Orthopedic implants: fundamental principles and the significance of biocompatability, in *Biocompatibility of Orthopedic Implants*, Vol. 1, Williams DF, Ed., Boca Raton, FL, CRC Press, 1982, 1–50.

13. Long M, Rack HJ. Titanium alloys in total joint replacement — a materials science perspective, *Biomaterials,* 1998, 19(18), 1621–1639.
14. Hammond CR. The Elements, in *Handbook of Chemistry and Physics, 84th Edition,* Lide, DR, Ed., Boca Raton, FL, CRC Press, 2003.
15. REMBAR Titanium Technical Info, http://www.rembar.com/Titanium.htm, *The REMBAR Company, Inc.* (Web site), accessed 12/30/2003.
16. Barras CD, Myers KA. Nitinol — its use in vascular surgery and other applications, *Eur. J. Vasc. Endovasc. Surg.*, 2000, 19(6), 564–569.
17. Machado LG, Savi MA. Medical applications of shape memory alloys, *Braz. J. Med. Biol. Res.,* 2003, 36(6), 683–691 (e-published June 2003).
18. Beuhler WJ, Gilfrich JV, Riley RC. Effect of low-temperature phase changes on the mechanical properties of alloys near composition Ti-Ni, *J. Appl. Phys.*, 1963, 34, 1475–1477.
19. Bajpai PK, Billote WG. Ceramic biomaterials, in *The Biomedical Engineering Handbook*, Bronzino JD, Ed., Boca Raton, FL, CRC Press, 1995.
20. Ducheyne P, Hastings GW. *Metal and Ceramic Biomaterials: Strength and Surface,* Vol. 2, Boca Raton, FL, CRC Press, 1984.
21. Marti A. Inert bioceramics (Al_2O_3, ZrO_2) for medical application, *Injury,* 2000, 31 (Suppl. 4), 33–36.
22. Catledge SA, Fries MD, Vohra YK, et al. Nanostructured ceramics for biomedical implants, *J. Nanosci. Nanotechnol.,* 2002, 2(3–4), 293–312.
23. Bauer TW, Smith ST. Bioactive materials in orthopaedic surgery: overview and regulatory considerations, *Clin. Orthop.,* 2002, 395, 11–22.
24. Gisep A. Research on ceramic bone substitutes: current status, *Injury* 2002, 33 (Suppl. 2), B88–B92.
25. Hamadouche M, Sedel L. Ceramics in orthopaedics, *J. Bone Joint Surg. Br.* 2000, 82(8), 1095–1099.
26. Davis ME. Ordered porous materials for emerging applications, *Nature,* 2002, 417(6891), 813–821.
27. Bloch B, Hastings GW. *Plastics Materials in Surgery*, Springfield, IL, Charles C. Thomas, 1972.
28. Galanti AV, Mantell CC. *Polypropylene — Fibers and Films*, New York, Plenum Press, 1965.
29. Lee HB, Kim SS, Khang G. Polymeric biomaterials, in *The Biomedical Engineering Handbook*, Bronzino JD, Ed., Boca Raton, FL, CRC Press, 1995, 581–597.
30. Shastri VP. Non-degradable biocompatible polymers in medicine: past, present and future, *Curr. Pharm. Biotechnol.,* 2003, 4(5), 331–337.
31. Beiko DT, Knudsen BE, Watterson JD, Denstedt JD. Biomaterials in urology, *Curr. Urol. Rep.* 2003, 4(1), 51–55.
32. Debry C, Schultz P, Vautier D. Biomaterials in laryngotracheal surgery: a solvable problem in the near future? *J. Laryngol. Otol.,* 2003, 117(2), 113–117.
33. Park JB, Lakes RS. Composites as biomaterials in *Biomaterials: An Introduction*, Park JB, Lakes RS, Eds., New York, Plenum Press, 1992.
34. Lakes R. Composite biomaterials, in *The Biomedical Engineering Handbook*, Bronzino JD, Ed., Baco Raton, FL, CRC Press, 1995, 598–610.
35. Piskin E. Biodegradable polymeric matrices for bioartificial implants, *Int. J. Artif. Organs,* 2002, 25(5), 434–440.
36. Alonso MJ, Sanchez A. The potential of chitosan in ocular drug delivery, *J. Pharm. Pharmacol.,* 2003, 55(11), 1451–1463.

37. Synowiecki J, Al-Khateeb NA. Production, properties, and some new applications of chitin and its derivatives, *Crit. Rev. Food Sci. Nutr.,* 2003, 43(2), 145–171.
38. Khor E, Lim LY. Implantable applications of chitin and chitosan. *Biomaterials.* 2003, 24(13), 2339–2349.
39. Hejazi R, Amiji M. Chitosan-based gastrointestinal delivery systems, *J. Control. Release,* 2003, 89(2), 151–165.
40. Kato Y, Onishi H, Machida Y. Application of chitin and chitosan derivatives in the pharmaceutical field, *Curr. Pharm. Biotechnol.,* 2003, 4(5), 303–309.
41. Sudkamp NP, Kaab MJ. Biodegradable implants in soft tissue refixation: experimental evaluation, clinical experience, and future needs, *Injury,* 2002, 33 (Suppl. 2), B17–B24.
42. Vogel V, Baneyx G. The tissue engineeting puzzle: a molecular perspective, *Annu. Rev. Biomed. Eng.,* 2003, 5, 441–463.
43. Gunatillake PA, Adhikari R. Biodegradable synthetic polymers for tissue engineering, *Eur. Cell. Mater.,* 2003, 5, 1–16, discussion 16.
44. Peppas NA, Huang Y, Torres-Lugo M, Ward JH, Zhang J. Physicochemical foundations and structural design of hydrogels in medicine and biology, *Annu. Rev. Biomed. Eng.,* 2000, 2, 9–29.
45. Peppas NA, Bures P, Leobandung W, Ichikawa H. Hydrogels in pharmaceutical formulations, *Eur, J. Pharm. Biopharm.,* 2000, 50(1), 27–46.
46. Torres-Lugo M, Garcia M, Record R, Peppas NA. pH-Sensitive hydrogels as gastrointestinal tract absorption enhancers: transport mechanisms of salmon calcitonin and other model molecules using the Caco-2 cell model, *Biotechnol. Prog.,* 2002, 18(3), 612–616.
47. Hern DL, Hubbell JA. Incorporation of adhesion peptides into nonadhesive hydrogels useful for tissue resurfacing, *J. Biomed. Mater. Res.,* 1998, 39(2), 266–276.
48. Chaikof EL. Biomaterials that imitate cell microenvironments, *ChemTech,* 1996, 26(8), 17–22.
49. Oral E, Peppas NA. Responsive and recognitive hydrogels using star polymers, *J. Biomed. Mater. Res.,* 2004, 68A(3), 439–447.
50. Peppas NA, Huang Y. Polymers and gels as molecular recognition agents, *Pharm. Res.,* 2002, 19(5), 578–587.
51. Prestwich GD, Matthew H. Hybrid, composite, and complex biomaterials, *Ann. N. Y. Acad. Sci.,* 2002, 961, 106–108.
52. Ammirati CT. Advances in wound closure materials, *Adv. Dermatol.,* 2002, 18, 313–338.
53. Lee H, Cusick RA, Utsunomiya H, Ma PX, Langer R, Vacanti JP. Effect of implantation site on hepatocytes heterotopically transplanted on biodegradable polymer scaffolds, *Tissue Eng.,* 2003, 9(6), 1227–1232.
54. Tzanakakis ES, Hess DJ, Sielaff TD, Hu WS. Extracorporeal tissue engineered liver-assist devices, *Annu. Rev. Biomed. Eng.,* 2000, 2, 607–632.
55. Vats A, Tolley NS, Polak JM, Gough JE. Scaffolds and biomaterials for tissue engineering: a review of clinical applications, *Clin. Otolaryngol.,* 2003, 28(3), 165–172.
56. Nomi M, Atala A, Coppi PD, Soker S. Principals of neovascularization for tissue engineering, *Mol. Aspects Med.,* 2002, 23(6), 463–483.
57. Mann BK, West JL. Cell adhesion peptides alter smooth muscle cell adhesion, proliferation, migration, and matrix protein synthesis on modified surfaces and in polymer scaffolds, *J. Biomed. Mater. Res.,* 2002, 60(1), 86–93.
58. Hutmacher DW. Scaffold design and fabrication technologies for engineering tissues — state of the art and future perspectives, *J. Biomater. Sci. Polym. Ed.,* 2001, 12(1), 107–124.

59. Mann BK, West JL. Tissue engineering in the cardiovascular system: progress toward a tissue engineered heart, *Anat. Rec.,* 2001, 263(4), 367–371.
60. Walgenbach KJ, Voigt M, Riabikhin AW, et al. Tissue engineering in plastic reconstructive surgery, *Anat. Rec.,* 2001, 263(4), 372–378.
61. Sipe JD. Tissue engineering and reparative medicine, *Ann. N. Y. Acad. Sci.,* 2002, 961, 1–9.
62. Hutmacher DW. Scaffolds in tissue engineering bone and cartilage, *Biomaterials,* 2000, 21(24), 2529–2543.
63. Voldman J, Gray ML, Schmidt MA. Microfabrication in biology and medicine, *Annu. Rev. Biomed. Eng.,* 1999, 1, 401–425.
64. Polla DL, Erdman AG, Robbins WP, et al. Microdevices in medicine, *Annu. Rev. Biomed. Eng.,* 2000, 2, 551–576.
65. Salzberg AD, Bloom MB, Mourlas NJ, Krummel TM. Microelectrical mechanical systems in surgery and medicine, *J. Am. Coll. Surg.,* 2002, 194(4), 463–476.
66. Interagency Working Group on Nanoscience E, and Technology. *National Nanotechnology Initiative: Leading to the Next Industrial Revolution.* Washington, D.C., Committee on Technology, National Science and Technology Council, 2000.
67. Gordon N, Sagman U. *Nanomedicine Taxonomy,* Canadian Institutes of Health and Canadian NanoBusiness Alliance, Verdun, Quebec, 2003.
68. Initiative CftRotNN. *Small Wonders, Endless Frontiers. A Review of the National Nanotechnology Initiative,* Washington, D.C., Division of Engineering and Physical Sciences, National Research Council, 2002.
69. Roukes M. Plenty of room, indeed, *Scientific American, Special Edition: The Edge of Physics,* 2001, 54–56.
70. Roco MC. Nanotechnology: convergence with modern biology and medicine, *Curr. Opin. Biotechnol.,* 2003, 14(3), 337–346.
71. Haberzettl CA. Nanomedicine: destination or journey? *Nanotechnology,* 2002, 13, R9–R13.

2 The Host Response to Implantable Devices

D. Denison Jenkins, Russell K. Woo, and Ralph S. Greco

CONTENTS

The immune system targets, destroys, and removes foreign material. It is a complex and redundant apparatus that has evolved over eons into a coordinated process to protect the host from infection. In the modern era, a detailed understanding of the immune system has coincided with the ability to surgically implant devices. Nature, however, could not have anticipated a foreign body with benevolent intent, such as a vascular graft or a hip prosthesis. Consequently, there are complex interactions between the host immune system and biomaterials. This chapter will review the host response to implantable devices in order to provide a conceptual framework for understanding immunity and inflammation within a biologic host.

2.1 OVERVIEW OF THE IMMUNE SYSTEM

The description of the biologic response to biomaterials will begin with a brief overview of immunology. Overall, the immune response attempts to protect the host from opportunistic infection. The distinction of native tissue from foreign material, or in other words, self from nonself, is a fundamental component of this process. A delicate balance is needed, as overactive immune surveillance leads to autoimmune disease, and immunodeficiency places the host at increased risk of infection. For example, multiple sclerosis and systemic lupus erythematosus (SLE) result from the inappropriate targeting of self tissue by the immune system. Conversely, acquired immunodeficiency syndrome (AIDS) and severe combined immunodeficiency (SCID) are examples of immune compromise that result from either a viral infection or a gene deficiency, respectively. Given the devastating consequences of either autoimmunity or immunodeficiency, the system is highly regulated, with extensive

0-8493-1940-4/05/$0.00+$1.50

integration and redundancy between components. In general, regulatory failure leads to loss of homeostatic integrity, systemic injury, and, if severe, the death of the host.

Two broad categorizations define the immune response: innate and acquired immunity, and cell-mediated versus humoral immunity. These categories overlap broadly, yet provide a foundation to understand the immune system from two different perspectives. Innate immunity is the nonspecific response to foreign and/or infectious material (see Table 2.1). It is an important initial line of defense, though teleologically more primitive than acquired immunity.[1,2] By definition, the innate response lacks immunologic memory, is antigen independent, and is the same each time a pathogen is encountered.[1,2] The various components of innate immunity are molecules, such as serum proteins, complement, cytokines, and certain immune cells. Upon contact with foreign material, serum proteins immediately bind to its surface,[3] a process called protein adsorption. Protein adsorption facilitates the removal of foreign material by phagocytic cells,[4] a sequence of events also known as opsonization. Common serum proteins are listed in Table 2.2. Once a phagocyte engulfs a foreign body, it is generally destroyed through intracellular exposure to a complex array of toxins,[1] preventing damage to healthy adjacent tissue.

Platelets are another component of innate immunity, as they too bind to foreign material, release inflammatory mediators, and, at times, initiate local thrombosis.[5] The complement cascade, a series of escalating reactions that results in the binding of complement to the surface of the foreign body with concurrent release of additional inflammatory mediators, is another part of the innate immune response. The complement cascade will be discussed in greater detail later in this chapter. Finally, dendritic cells, neutrophils, monocytes, macrophages, and natural killer (NK) cells make up the cellular component of innate immunity.[1] These cells recognize, engulf, and destroy foreign material. This process, namely the cellular ingestion of foreign material, is also known as phagocytosis. Other cells, for example basophils, eosinophils, and mast cells, release inflammatory mediators to induce a local or systemic

TABLE 2.1
Innate vs. Acquired Immunity

Innate	Acquired
Primitive	Advanced
Nonspecific	Specific
Nonadaptive	Adaptive
No Memory	Memory

TABLE 2.2
Common Serum Proteins

Albumin	IgG
Fibrinogen	Transferrin
C3	IgA
IgM	Haptoglobins
α-Antitrypsin	α-Macroglobulin

response to the foreign material.[1] The complex inflammatory response specifically elicited by biomaterials will be discussed in greater detail in a later section of this chapter.

Acquired immunity is the highly complex system that has evolved in vertebrates to recognize, respond, and remember a unique pathogen.[1,2] Also known as adaptive immunity, this type of immune response improves with repeated exposures to a pathogen (see Table 2.1) and generally takes place in the lymph nodes, spleen, and mucosa-associated lymphoid tissue of the host.[1] When a new pathogen is encountered, unique proteins on its cell membrane lead to a coordinated T cell–mediated response to the pathogen,[1,6] and B cells are induced to produce specific immunoglobulins to the targeted antigen(s) of the pathogen.[1,6] Finally, memory cells catalog the unique epitope(s) and produce a more rapid response to the pathogen, should it ever be encountered again.[1] Using a complex array of gene modification techniques, lymphocytes can produce an estimated 10^{15} variations of T and B cell receptors to help specifically identify a broad range of antigenic material.[1,2]

Acquired immunity has both specificity and immunologic memory; in other words, it targets and destroys specific material it recognizes — or remembers — as foreign.[1] Though separate processes, the innate and adaptive immune systems typically work in tandem to coordinate the removal of an infectious agent.[1] For example, the innate immune response may stimulate local dendritic cells to activate T cells, leading to an inflammatory response that ultimately yields adaptive immunity.[7] T cells can, in turn, modify dendritic cell responsiveness to a local foreign body.[7] Innate and adaptive immunity are coordinated and complex responses designed to protect the host from infectious disease.

The immune response may also be divided into either cell-mediated or humoral immunity. As will be apparent, this method of classification overlaps broadly with the cellular component of innate immunity and acquired immunity in its entirety. Lymphocytes are the principle effectors of both cell-mediated and humoral immunity. Specific receptors on these cells bind to antigens and, thus, initiate a complex inflammatory and immune response.[1,6] In simplified terms, T cells are responsible for cell-mediated immunity, and B cells are responsible for humoral immunity. However, similar to the coordinated integration of innate and acquired immunity, T cells and B cells act in concert with one another to promote — and improve upon — a targeted response to foreign material. T cells are generally divided into cytotoxic T cells, which destroy infected or atypical cells, and helper T cells, which coordinate the cell-mediated and humoral immune response. Conversely, activated B cells produce significant quantities of antibodies. Broadly defined, antibodies are antigen-specific proteins that circulate within bodily fluids that either sterically inhibit the attachment of a pathogen to its target cell and, therefore, prevent infection or recruit other immune cells to kill the offending agent.[6] Figure 2.1 provides a schematic representation of an antibody binding to a specific antigen present on a pathogen. As detailed below, the antibody–antigen complex is also a potent stimulant of the classic complement cascade. Humoral immunity is a highly specific response to foreign materials. Antibody–antigen interaction leads to the agglutination of bacteria, the neutralization of toxins, and the activation of nonspecific immune responses, such as phagocytosis.[6]

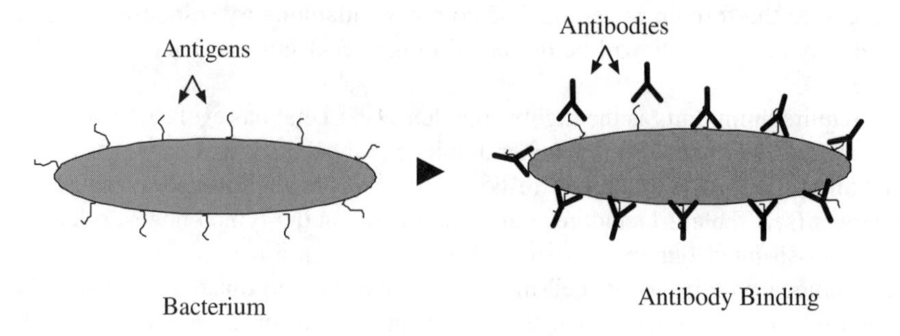

FIGURE 2.1 Antigen–antibody complex.

The complement cascade is an integrative process of the immune system. Once activated, it propagates the release of potent inflammatory mediators and chemokines, facilitates the opsonization and phagocytosis of infectious material, and ultimately integrates the innate and acquired immune responses.[8] The complement system includes more than 30 plasma proteins that circulate either as zymogens or as inactive serine proteases.[8] Many of the complement proteins are acute phase reactants and are thus present in increased quantities during a systemic inflammatory response.[8,9] There are two pathways of complement activation — the classic and the alternative — though both reactions converge into a shared common pathway (see Figure 2.2). The classic pathway is activated though contact of C1 complex with a specific region of an antibody–antigen complex. After a series of intervening enzymatic reactions, C3 convertase is activated, which converts C3 into C3a (a potent inflammatory mediator) and C3b. After another series of intervening steps, C5 convertase is formed, initiating the membrane attack complex.[8,10] The alternative pathway of complement is initiated after C3 comes in contact with foreign material. It is distinguished from the classic pathway in that it does not require a humoral reaction for initiation and, consequently, may

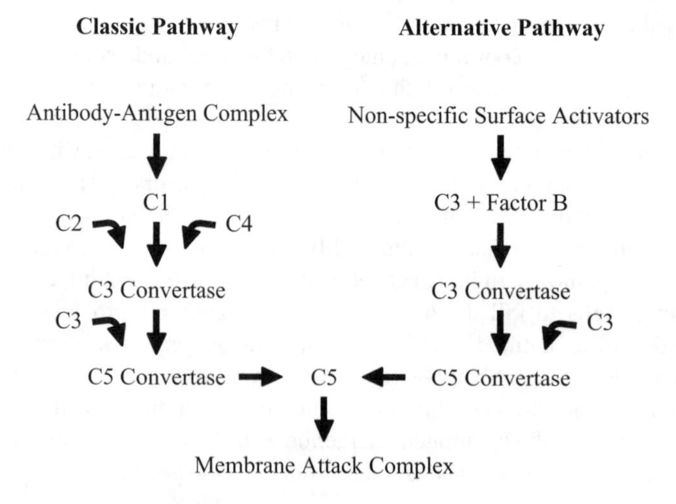

FIGURE 2.2 Simplified complement cascade.

be the more common mechanism of complement activation with implantable devices.[8,10] Other activated complement factors stimulate the release of histamine from mast cells, further propagating the inflammatory response.[1] Increased blood vessel permeability and neutrophil recruitment to the area of injury and/or inflammation may subsequently occur.[1]

The major histocompatability complex (MHC) is the primary mechanism through which the immune system distinguishes the host from foreign material. MHC class I molecules are present on the cell membrane of all nucleated cells.[1,11] In simplified terms, proteins are broken down in the cytosol of a cell and transported to the endoplasmic reticulum, where the peptide fragments bind to MHC class I molecule. The MHC-peptide complex is then transported to the plasma membrane, where it is "read" by circulating cytotoxic T cells and NK cells as a mechanism to determine self from nonself. Cells with "self" peptides are tolerated, whereas cells with foreign peptides are destroyed. For example, a cell infected with a virus will demonstrate viral peptides within its MHC complex and thus mark the infected cell for clearance by cytotoxic T cell.[1,11] In addition, cells without MHC class I complexes are also generally targeted for destruction.[1] MHC class II complexes are present on the cell membrane of antigen-presenting cells and facilitate the initiation of an adaptive immune response.[1] In brief, after a phagocyte has ingested and destroyed an invading material, various foreign particles are complexed to a MHC class II molecule on the phagocyte's cell surface. This cell then "presents" the foreign material to a helper T cell, which leads to a coordinated and targeted response to the invading pathogen.

The specific human leukocyte antigen (HLA) phenotypes of MHC are another important determinant of "self." MHC class I is subdivided into HLA-A, B, and C, and MHC class II is subdivided into HLA-DP, DR, and DQ.[1] Each individual receives two haplotypes of these genes, one from each parent, and expression is codominant. Consequently, an individual may express as many as twelve HLA proteins. These surface proteins are an important element of successful engraftment of a transplanted organ. HLA discordance leads to graft rejection in the absence of immune suppression.

The immune system is a highly complex system of coordinated molecular and cellular processes to prevent infection. In the modern era, with its significant advances in medical and surgical technology, devices and organs have been deliberately implanted for therapeutic purposes. The immune response, however, does not distinguish an infectious agent from an implanted biomaterial. Having completed a brief overview of the immune system, the host response to implanted devices will now be described.

2.2 HOST RESPONSE TO IMPLANTED DEVICES

The materials used in implantable devices are engineered to be biocompatible. However, biocompatibility is a relative term, and the immune system reacts to some degree with every surgically implanted device.[12] In fact, materials presumed in the past to be biologically inert may cause subtle, or at times even distinctive, inflammatory reactions within the recipient. Though the effects of chronic inflammation and monocyte function are certainly important components of long-term device function, the next sections will focus primarily on the acute inflammatory response, specifically the neutrophil reaction to biomaterials.

The acute host response to a device may be broken into a sequence of two overlapping events: local injury at the time of insertion and the inflammatory process.[12] Local tissue injury leads to changes in vascular permeability and a protein-rich exudate at the site of implantation.[12] Serum proteins quickly bind to the surface of the device, wrapping it in a proteinaceous film.[3,13,14] Protein adsorption — especially fibrinogen — leads to platelet adherence,[14] complement activation,[15] and in certain situations, thrombus formation. The complement cascade and platelet activation serve as potent mediators of the inflammatory response, as highlighted previously. The adsorbed proteins, attached platelets, and bound complement attract, bind, and activate neutrophils,[15] which are found on biomaterials within minutes of implantation.[16] Specifically, the 190-202 epitope of fibrinogen interacts with the Mac-1 receptor on phagocytes, leading to their recruitment to the site of implantation, their binding to the device, and the propagation of the overall inflammatory response to the biomaterials.[17] This represents the first stage of the complex cellular inflammatory response to an implanted device.

Though protein adsorption and phagocyte activation are effective in the destruction and clearance of a bacterium, they are ineffective when confronting a large, inorganic, and nonbiodegradable material. Specifically, when activated neutrophils are incapable of engulfing a foreign body, a futile process defined as "frustrated phagocytosis" occurs.[12,18,19] "Frustrated" neutrophils inappropriately release reactive oxygen species locally, rather than intracellularly.[12] The extracellular release of toxic agents designed to destroy infectious material intracellularly may cause adjacent tissue damage and chronic inflammation.[19] Furthermore, proteases and oxygen-derived free radicals released by activated inflammatory cells can contribute to the degradation of the implanted biomaterial.[12,20] There are several other untoward effects of frustrated phagocytosis. For example, neutrophils attached to biomaterials demonstrate reduced bacteriocidal activity[21] and decreased viability,[22] perhaps explaining why implanted devices are prone to infection. Research has demonstrated that the surface topography of a biomaterial influences neutrophil survival,[23] and that protein kinase inhibitors may prevent the activation of neutrophils bound to a biomaterial.[24] Figure 2.3 demonstrates an activated neutrophil adherent to a large biomaterial. Illustrative examples of common implantable devices will provide additional perspective into the host immune response.

Total hip arthroplasty uses a titanium replacement to restore a dysfunctional joint. Long-term function is common, though device design, operative technique, and mechanical factors are generally implicated in the need for revision surgery.[25] However, cell-mediated and humoral immunity have also been associated with mechanical failure after hip replacement surgery. For example, nanoscale particles dislodged during routine wear may lead to phagocyte activation, which in turn may lead to osteolysis and, if severe, aseptic loosening of the prosthesis.[25,26,27] Furthermore, antibodies that target biomaterials have been reported,[28,29] and humoral immunity has also been implicated in aseptic loosening of hip prostheses.[25] Though the

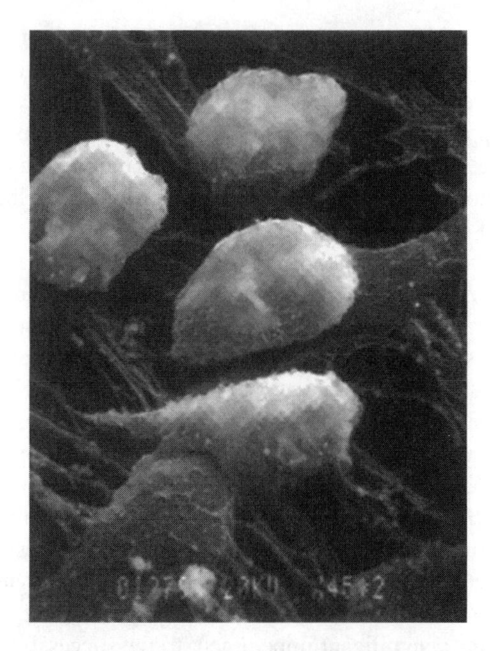

FIGURE 2.3 Electron micrograph of neutrophils attached to a large biomaterial.

materials used in this procedure are generally regarded as biocompatible, such data raise new questions regarding whether they are truly biologically inert.

Allogenic heart valves provide a second illustrative example of the immunology of implantable devices. HLA typing is not performed prior to implantation of allogeneic cryopreserved human heart valves, and recipients do not receive immunosuppresion after surgery.[30] Although the majority do well after allograft valve implantation, a humoral immune response against donor HLA may be associated with early graft failure,[30] though the functional impact of anti-HLA antibodies after valve replacement is controversial.[30,31] A separate study showed a strong antibody response to cryopreserved nonvalved allografts in a pediatric population and concluded that luminal calcification and stenosis of a graft may result from immune-mediated tissue injury.[32] Finally, the link between valvular heart disease and the immune response has a historical precedent: streptococcal pharyngitis is known to cause rheumatic heart disease as a consequence of molecular mimicry between the bacterial antigens and native heart valves.[33] Such data provide additional perspective of the host response to engineered tissues and replacement parts.

Though not classically a "device," organ transplantation provides perhaps the best illustrative overview of the host response to implanted material. This type of graft is biologic — and not mechanical — and thus commonly requires immunosuppression for long-term graft function. Furthermore, it provides a macro-level perspective of the potential host response to replacement tissues engineered from allogeneic sources. Prior to transplantation of an organ, cross matching between the donor and patient is performed to prevent hyperacute rejection of the graft, which occurs when preformed antibodies exist to the graft prior to implantation. Even with appropriate HLA matching between donor and recipient, minor antigens present on the graft may lead to rejection unless the recipient receives immunosuppression. Though the adaptive immune response is largely responsible for graft rejection, innate immunity appears to have regulatory and functional role in transplant rejection.[2,34] In this sense, successful organ transplantation requires careful and meticulous suppression of the host response. Even with adequate long-term immunosuppression, chronic rejection of a transplanted organ generally occurs.

Each of these three examples provides perspective of the immune response to implantable replacement devices. Titanium hips restore structural function, allogeneic heart valves restore mechanical function, and organ transplants restore physiologic function. Tissue engineering represents the next generation of replacement technology, as cellular constructs are integrated within a biologic (or inorganic) scaffold to repair the deficient host. Numerous technical hurdles exist. For example, cellular behavior may vary depending on the size of the engineered construct,[35] as depicted in Figure 2.4. Furthermore, the inflammatory response to a biomaterial scaffold is potentially integrated with the immune response to allogeneic or xenogeneic cells. The combined inflammatory stimulus is not necessarily additive, as a recent study revealed that coadministration of a common biomaterial enhanced the humoral response to a common antigen.[36] The immune response to engineered tissue replacements remains a significant impediment to the successful implementation of replacement biology.

As implanted devices evolve from structural to mechanical to electronic, and with an improved understanding of the intricacies of the host response, a concept of biofouling has emerged.[37] For clinical success, an implanted device must withstand long-term exposure to both the physical and the physiologic environment. For example, a titanium hip or prosthetic heart valve must maintain structural integrity

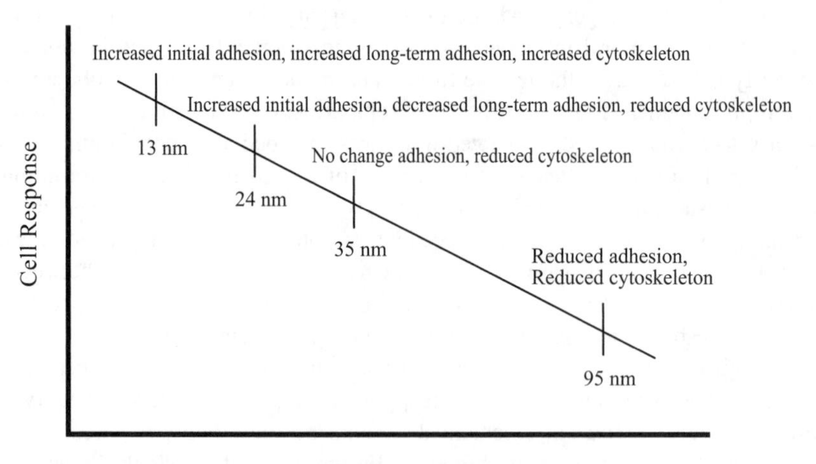

FIGURE 2.4 Differential cellular response to nanoscale scaffold. (Reprinted from *Biomaterials*. Vol. 25, Dalby et al. Rapid fibroblast adhesion to 27 nm high polymer demixed nanotopography, p. 77, ©2004, with permission from Elsevier.)

within the load dynamics of its respective mechanical environment. Furthermore, these implants must also maintain their function during the acute and chronic inflammatory response. Failure to do so is defined as biofouling,[37] and generally results from chronic oxidation, hydrolysis, mechanical loading, and mineralization of the device.[20] As implanted devices become smaller and more mechanically and electronically complex, the negative impact of the immune response could increase. Furthermore, a size threshold may exist for biomaterials that alleviates frustrated phagocytosis. Should this occur, implantable devices would move from the extracellular to the intracellular environment and thus open new complexities and uncertainties of the host response to an implantable device (see Figure 2.5).

Nanoscale particles are either the inadvertent product of biomaterial wear, or are the product of deliberate and highly technical engineering. As discussed previously, regardless of its source, nanoscale material may elicit an inflammatory response. Particle toxicology is a relatively new scientific field that studies the impact of fine and ultrafine materials on the host[38] and may clarify the immune response to small biomaterials. Ultrafine particles are defined as material less than 100 nm in diameter.[39] Nanoscale particles, especially material that is aerosolized, may have a toxic profile, even when larger particles of the same material are nontoxic.[39] The inflammatory response to ultrafine titanium dioxide[40] and 14-nm particles of carbon black[39] are examples of nontoxic materials that elicit an immune response when milled into nanoscale particles. Though the proinflammatory mechanism of ultrafine material is poorly understood, phagocyte activation after particle ingestion is believed to contribute.[27,39] Also, as highlighted earlier, as devices shrink to a size that permit effective phagocytosis, a potential new intracellular inflammatory frontier emerges. Again, see Figure 2.5.

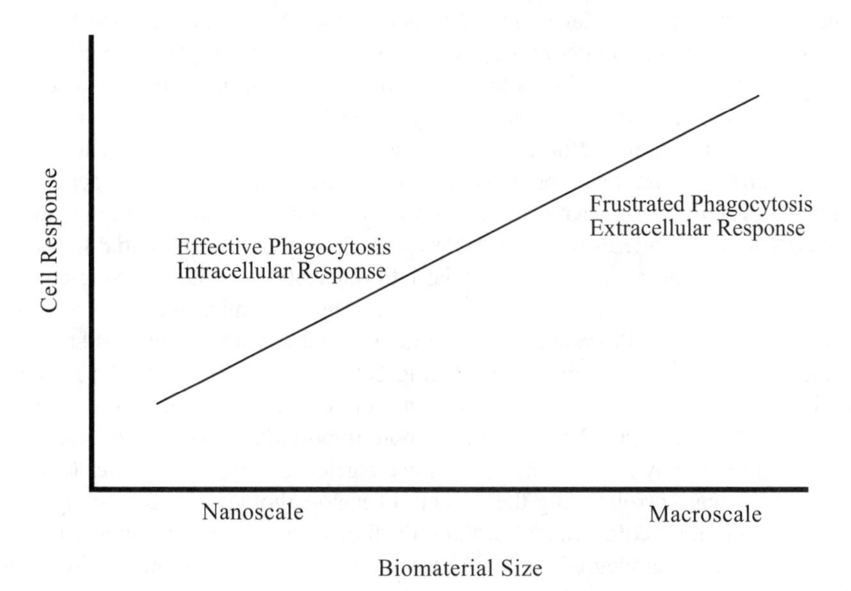

FIGURE 2.5 Differential cellular response to changes in biomaterial size.

Bioengineers attempt to minimize the untoward effects of the host inflammatory response by designing implantable devices that are relatively inert and nontoxic. The impact of material size, however, is poorly defined. As implantable devices shrink, the inflammatory profile of their components may evolve. In addition, as their function grows ever more complex, the impact of protein adsorption, cell adhesion, and long-term oxidative and hydrolytic stress may also change.

2.3 NANOTECHNOLOGY AND THE IMMUNE RESPONSE

The study of the interaction of micro- and nanoscale devices within the biologic environment is in its infancy. For example, the *in vivo* immune response to microelectromechanical (MEM) drug delivery systems, an implantable device designed to release a drug from a reservoir in a complex dosing pattern, was only recently described. As a preliminary study, the report examined the material components of a MEM system placed within a steel cage, which was subsequently implanted within the dorsum of recipient rodents.[37] Over a 21-day period, exudative fluid was tested for acute and chronic inflammatory responses; all materials generated an early neutrophil response when compared to empty cage controls but were determined to be biocompatible.[37] Long-term *in vivo* function of drug delivery was not tested. In a second study, a nanoscale scaffold made of poly-L-lactic acid resulted in increased protein adsorption and increased cell attachment *in vitro* when compared to solid-walled scaffolds of the same material.[41] Perhaps more importantly, the T and B cell receptors have been estimated to be between 60 and 170 nm in size[42] and, therefore, may not independently bind and/or recognize a nanoscale device. However, it is well known that a nonimmunogenic particle may still stimulate an inflammatory reaction when coupled with a carrier molecule.[43] As discussed below, the immune response to nanoscale particular material may be substantially more complex than previously believed, potentially involving both intra- and extracellular inflammatory reactions.

Inorganic and nonbiodegradable nanoparticles have been found in liver, kidney, and colon, leading to the definition of a new concept: nanopathology.[44,45] The toxicity of fine particulate matter is perhaps best demonstrated through the association of lung cancer with chronic exposure to airborne material less than 2.5 μm in diameter that constitute, in part, air pollution.[46] New data also suggest the potential for pathologic dissemination of inorganic material into other bodily tissues, raising concern over nonairborne environmental exposures of nano- and microscale material. For example, a patient with systemic granulomatosis and a worn dental prosthesis was found to have particulate matter measuring between 6 to 20 μm in the liver and particles measuring less than 6 μm in both kidneys.[47] Though describing a unique clinical finding, this case report, perhaps more importantly, revealed that the intestinal epithelium may not provide an effective barrier to particulate matter less than 20 μm.[47] Further corroborating the concept of nanopathology, a recent study of 18 patients with either colon cancer or Crohn's disease demonstrated nanoparticles of inorganic and nonbiodegradable materials such as silicon, aluminum, zirconium, titanium, stainless steel, and silver; none of the samples from healthy donors demonstrated any evidence of nanoscale particles.[45] Though the concept of nanopathology is controversial, such reports mandate further analysis of the inflammatory and pathologic effects of nanoscale materials. As minaturized devices are implanted into a biologic host, they must withstand the inflammatory response and maintain long-term function. If history is any guide, our understanding of the immune response will likely evolve along with this emerging technology.

2.4 CONCLUSION

The immune system is a complex process designed to protect the host from infection. Until the modern era, the foreign material encountered by host defenses consisted largely of pathogens, whereas at present, implantable devices are being frequently used for therapeutic purposes. Consequently, the immune response is increasingly being challenged by noninfectious material. Our understanding of the immune system has evolved with our ability to implant therapeutic devices, and biocompatibility is emerging as a dynamic process that changes over time. Though the data are at times controversial, knowledge of the host response to implantable devices is growing. Nanotechnology represents yet another frontier, and our understanding of immunology will undoubtedly grow as the host response to nanoscale devices is defined.

REFERENCES

1. Delves PJ, Roitt IM. The immune system: first of two parts, *N. Engl. J. Med.*, 2000, 343, 37.
2. He H, Stone JR, Perkins DL. Analysis of differential immune responses induced by innate and adaptive immunity following transplantation, *Immun.* 2003, 109, 185.
3. Andrade JD, Hlady VL. Plasma protein adsorption: the big twelve, *Ann. N. Y. Acad. Sci.*, 1987, 516, 158.
4. Aderem A, Underhill DM. Mechanisms of phagocytosis in macrophages, *Annu. Rev. Immunol.*, 1999, 17, 593.
5. Hobertt TA. Principles underlying the role of absorbed plasma proteins in blood interactions with foreign material, *Cardiovasc. Pathol.*, 1993, 2, 137S.
6. Delves PJ, Roitt IM. The immune system. Second of two parts, *N. Eng. J. Med.*, 2000, 343, 108.
7. Altmann DM, Boyton RJ. Reciprocal conditioning: T cell as regulators of dendritic cell function, *Immunology*, 2003, 109, 473.
8. Sim RB, Dodds AW. *Complement: A Practical Approach*, Oxford University Press, 1997.
9. Gabay C, Kushner I. Acute phase proteins and other systemic responses to inflammation, *N. Engl. J. Med.*, 1999, 340, 448.
10. Hakim RM. Complement activation by biomaterials, *Cardiovasc. Pathol.*, 1993, 2, 187s.
11. Hewitt EW. The MHC class I antigen presentation pathway: strategies for viral immune invasion, *Immunology*, 2003, 110, 163.
12. Anderson JM. Mechanisms of inflammation and infection in implanted devices, *Cardiovasc. Pathol.*, 1993, 2, 33S.
13. Hu W, Eaton JW, Tang L. Molecular basis of biomaterial-mediated foreign body reactions, *Blood*, 2001, 98, 1231.

14. Tsai W, Grunkemeier JM, Horbett TA. Variations in the ability of adsorbed fibrinogen to mediate platelet adhesion to polystyrene-based materials: a multivariate statistical analysis of antibody binding to the platelet binding sites of fibrinogen, *J. Biomed. Mater. Res.*, 2003, 67A, 1255.
15. Wettero J, Bengtsson T, Tengvall P. Complement activation on immunoglobulin G-coated hydrophobic surfaces enhances the release of oxygen radicals from neutrophils through an actin-dependent mechanism, *J. Biomed. Mater. Res.*, 2000, 51, 742.
16. Nygren H, Eriksson C, Lausmaa J. Adhesion of platelets and polymorphonuclear granulocyte cells at TiO_2 surfaces, *J. Lab. Clin. Med.*, 1997, 1, 35.
17. Tang L, Ugarova TP, Plow EF et al. Molecular determinants of acute inflammatory responses to biomaterials, *J. Clin. Invest.*, 1996, 97, 1329.
18. Henson PM, Johnston RB. Tissue injury in inflammation, *J. Clin. Invest.*, 1987, 79, 669.
19. Wettero J, Tengvall P, Bengtsson T. Platelets stimulated by IgG-coated surfaces bind and activate neutrophils through a selectin-dependent pathway, *Biomaterials*, 2003, 24, 1559.
20. Stokes K. Biodegradation, *Cardiovasc. Pathol.*, 1993, 2, 111S.
21. Zimmerli W, Lew PD, Waldvogel FA. Pathogenesis of foreign body infection: evidence for a local granulocyte defect, *J. Clin. Invest.*, 1984, 73, 1191.
22. Nadzam GS, De La Cruz C, Greco RS, et al. Neutrophil adhesion to vascular prosthetic surfaces triggers nonapoptotic cell death, *Ann. Surg.*, 2000, 231, 587.
23. Chang S, Popowich Y, Greco RS, et al. Neutrophil survival on biomaterials is determined by surface topography, *J. Vasc. Surg.*, 2003, 37, 1082.
24. Katz DA, Haimovich B, Greco RS, Neutrophil activation by expanded polytetrafluoroethylene is dependent on the induction of protein phosphorylation, *Surgery*, 1994, 116, 446.
25. Takayanagi S, Nagase M, Shimizu T, et al. Human leukocyte antigen and aseptic loosening in Charnley total hip arthroplasty, *Clin. Orthop.*, 2003, 413, 183.
26. Sun D, Trindade MC, Nakashima Y, et al. Human serum opsonization of orthopedic biomaterial particles: protein-binding and monocyte/macrophage activation *in vitro*, *J. Biomed. Mater. Res.*, 2003, 65A, 290.
27. Maloney WJ, Sun D, Nakashima Y, et al. Effects of serum protein opsonization on cytokine release by titanium-alloy particles, *J. Biomed. Mater. Res.*, 1998, 41, 371.
28. Schlosser M, Wilhelm L, Urban G, et al. Immunogenicity of polymeric implants: long-term antibody response against polyester (Dacron) following the implantation of vascular prostheses in LEW.1A rats, *J. Biomed. Mater. Res.*, 2002, 61, 450.
29. Human P, Zilla P. Characterization of the immune response to valve bioprostheses and its role in primary tissue failure, *Ann. Thorac. Surg.*, 2001, 71, S385.
30. Welters MJ, Oei FB, Witvliet MD, et al. A broad and strong humoral immune response to donor HLA after implantation of cryopreserved human heart valve allografts, *Human Immunol.*, 2002, 63, 1019.
31. Bechtel JF, Bartels C, Schmidtke C, et al. Does histocompatibility affect homograft valve function after the Ross Procedure? *Circulation*, 2001, 104S, I25.
32. Breinholt JP 3rd, Hawkins JA, Lambert LM, et al. A prospective analysis of the immunogenicity of cryopreserved nonvalved allografts used in pediatric heart surgery, *Circulation*, 2000, 102, Suppl. 3, III179.
33. Zabriskie JB. T-cells and T-cell clones in rheumatic fever vasculitis, *Circulation*, 1995, 92, 281.
34. McKenna RM, Takemoto SK, Terasaki PI. Anti-HLA antibodies after solid organ transplantation, *Transplantation*, 2000, 69, 319.

35. Dalby MJ, Giannaras D, Riehle MO, et al. Rapid fibroblast adhesion to 27 nm high polymer demixed nano-topography, *Biomaterials*, 2004, 25, 77.
36. Matzelle MM, Babensee JE. Humoral immune responses to model antigen co-delivered with biomaterials used in tissue engineering, *Biomaterials*, 2004, 25, 295.
37. Voskerician G, Shive MS, Shawgo RS, et al. Biocompatability and biofouling of MEMS drug delivery devices, *Biomaterials*, 2003, 24, 1959.
38. Borm PJ. Particle toxicology: from coal mining to nanotechnology, *Inhal. Toxicol.*, 2002, 14, 311.
39. Donaldson K, Brown D, Clouter A, et al. The pulmonary toxicology of ultrafine particles, *J. Aero. Med.*, 2002, 15, 213.
40. Ferin J, Oberdorster G, Penney DP. Pulmonary retention of ultrafine and fine particles in rats, *Am. J. Respir. Cell Mol. Biol.*, 1992, 6, 535.
41. Woo KM, Chen VJ, Ma PX. Nano-fibrous scaffolding architecture selectively enhances protein adsorption contributing to cell attachment, *J. Biomed. Mater. Res.*, 2003, 67A, 531.
42. Garcia KC, Teyton L, Wilson IA. Structural basis of T cell recognition, *Annu. Rev. Immunol.*, 1999, 17, 369.
43. Mitchison NA. The carrier effect in the secondary immune response to hapten-protein conjugates, *Eur. J. Immunol.*, 1971, 1, 18.
44. Gatti AM, Rivasi F. Biocompatability of micro- and nanoparticles: part I. In liver and kidney, *Biomaterials*, 2002, 23, 2381.
45. Gatti AM. Biocompatability of micro- and nano-particles in the colon. Part II, *Biomaterials*, 2004, 25, 385.
46. Pope CA 3rd, Burnett RT, Thun MJ, et al. Lung cancer, cardiopulmonary mortality, and long-term exposure to fine particulate air pollution, *JAMA*, 2002, 287, 1132.
47. Ballestri M, Baraldi A, Gatti AM, et al. Liver and kidney foreign bodies granulomatosis in a patient with malocclusion, bruxism, and worn dental prostheses, *Gastroenterology*, 2001, 121, 1234.

3 Nanobiotechnology

Peter Wagner

CONTENTS

3.1 INTRODUCTION

Nanotechnology is one of the most exciting new technologies of the twenty-first century. With its increasingly important and diverse impact on a variety of sectors such as biotechnology, energy, electronics, and consumer goods, it promises to transform our lives within this decade. As its ubiquitous prefix implies, nanotechnology entails the creation of functional materials, devices, and systems through the control of matter at the scale of 1 to 100 nanometers, as well as the exploitation of novel properties and phenomena at the same scale.[1]

The nanoworld of 1- to 100-nm dimensions is a land where fundamental properties are defined. This world resides in the gap between the scale of chemical structures and that of micromechanical and microelectronic devices. Rather than being a single technology, nanotechnology is a synonym for a multitude of approaches, devices, and systems all aimed at the understanding, control, and construction of molecular structures exhibiting new useful properties and functions.

0-8493-1940-4/05/$0.00+$1.50

Despite the current excitement and recent waves of visionary predictions, nanoscience and nanotechnology are not entirely new areas. Nanoscale materials such as carbon black (10- to 400-nm sized carbon particles) have been in use as fillers in automotive tires for more than 100 years. Carbon nanotubes, candidates for novel nanomechanical devices and nanoelectronic materials, were discovered at NEC Corporation in 1991. Scanning probe microscopes, capable of imaging and manipulating surface-bound molecules with atomic-scale precision, have been around for about 15 years and are now widely used in materials science and, to a lesser extent, in biology. Nanodispersed systems, such as lipid nanoparticles and liposomes, have been of interest for effecting the controlled delivery of active ingredients in skin care products. Nanoscale materials are now used in electronic, optoelectronic, catalytic, pharmaceutical, biomedical, personal care, energy, and materials applications. Given these broad application ranges, nanotechnology is considered to be a general-purpose technology which still has most of its experts located in academia and government-funded basic research centers.

What makes the nanobiotechnology field so interesting and revolutionary is the convergence of the angstrom-, nano- and microscale worlds (see Figure 3.1) with the exploitation of physical principles, chemical synthesis capabilities, and functional properties of biological nanostructures. Nature has made highly precise and functional nanostructures for billions of years: DNA, proteins, membranes, filaments, and cellular components. These biological nanostructures typically consist of simple

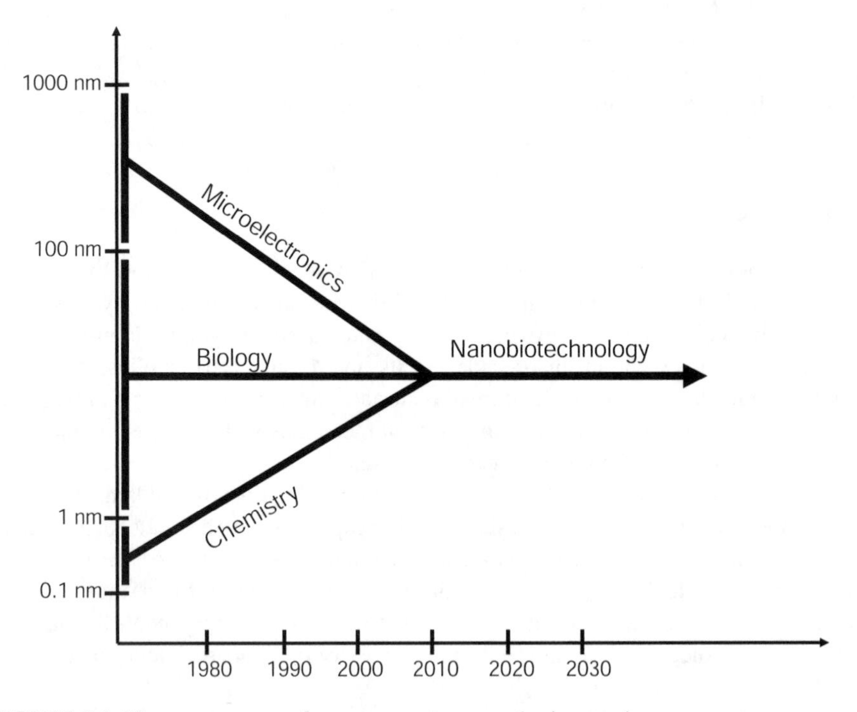

FIGURE 3.1 The convergence of angstrom-, nano-, and micro-scale.

molecular building blocks of limited chemical diversity arranged into vast numbers of complex three-dimensional architectures and dynamic interaction patterns. Nature has evolved the ultimate design principles for nanoscale assembly by supplying and transforming building blocks such as atoms and molecules into functional nanostructures and utilizing templating and self-assembly principles, thereby providing systems that can self-replicate, self-repair, self-generate, and self-destroy. In this "bottom-up" approach, molecular recognition and self-assembly of simple building blocks are key elements in the formation of nano- and microstructures of unique chemical and physical properties. Scientists and engineers in the field of nanobiotechnology (or bionanotechnology) are working to gain the ability to design synthetic materials on the nanoscale in hopes of shaping and creating the molecular architecture of biologically relevant molecules. Eventually, this will allow the integration of biological and artificial matter. In other words, bioinspired synthesis based on principles of molecular recognition, self-assembly, and templating will create biomimetic nanostructures with potentially vastly expanded chemical diversity and physical properties. The ability to design and synthesize nanoscale supramolecular structures of inorganic and organic matter is an important step toward this goal.

New approaches of a highly interdisciplinary nature are required to learn how biological nanostructures are built, function, and interact within larger biological systems, to apply tools for nanoscale analytical purposes (e.g., single molecule analysis), and to develop methods for creating devices composed of biological and inorganic matter (e.g., nanoscale lithography). Understanding the biological-inorganic interface is an important step in this direction. Today, many approaches still depend on milling down from the macroscopic level with high precision instruments. This "top down" approach utilizes physical methods to carve out desired nanostructures from materials of larger dimensions or to analyze nanoscopic properties with advanced analytical tools (e.g., scanning probe microscopy) and often occurs as a natural extension of well-known methods in microlithography and materials science.

While the realization of molecular robots and artificial organelles is far away at this time, other, less dramatic but nonetheless important, developments are in an advanced stage of implementation. This includes, for example, the fabrication of biologically relevant nanomaterials for applications in biosensing and drug delivery, the engineering of biocompatible materials, and the development of a variety of tools and methods such as ultrahigh-sensitivity bioanalytical tools, high-precision lithographic methods, and nanoscale dispensing devices. In the following overview of the field, I briefly highlight three of the most exciting areas in bionanotechnology currently under development: (1) following nature's lead; biological nanostructures as source for biomimetic and bioinspired nanomachines, (2) DNA-based nanotechnologies, and (3) nanoparticles.

Other exciting fields, equally worth mentioning, have not been covered in this brief overview (e.g., miniaturized protein biochips, nanowires, nanolithography, or scanning probe microscopy). It is beyond the scope of this chapter to provide a comprehensive review. For further reading, references 2–4 are recommended.

3.2 LEARNING FROM NATURE

Cellular nano- and microstuctures provide stimulating templates for designing artificial nanomachines. Cellular nanostructures such as the photosynthetic reaction center, the ribosome, linear and rotary molecular motors, the DNA replication complex, mitochondria, and membrane channels have been extensively studied *in vivo* and *ex vivo* by biochemists, biophysicists, and cell biologists. Comprehensive studies of the structure and function relationships of these biological nanomachines have provided substantial insight into how different mechanisms of nanomachines can be from the basic working principles of macroscopic manmade machines. Nanobiotechnology faces the challenge of emulating these unique mechanisms in an attempt to integrate them into novel artificial systems. In this section, two examples of bioinspired nanoscience are briefly discussed.

3.2.1 Linear and Rotary Molecular Motors

Molecular motors are proteins that convert chemical energy into mechanical energy (chemomechanical energy transduction) through the hydrolysis of adenosine 5′-triphosphate (ATP) or, alternatively, the translocation of ions through cellular membranes. The potential use of molecular motors as actuators and power sources for nanoelectromechanical systems (NEMS) has been widely discussed. Examples of such motors include the protein F1-ATPase,[5,6] kinesin and dynein,[7,8] myosin,[9,10] and RNA polymerase.[11]

The rotational movement and forward propulsion of molecular motors have been extensively studied and optically visualized with fluorescence-based single molecule *in vitro* motility assays over the last decade.

3.2.1.1 F1F0ATPase

The F1F0ATPase is composed of a multidomain F1 head group connected to a F0 base piece that is embedded in the mitochondrial inner membrane. The physiological function of this enzyme is the synthesis of ATP using a transmembrane pH gradient as energy source. The electrochemical gradient enables the F0 base to produce a mechanical force via a cam shaft, which drives the F1 part to undergo conformational changes that result in the conversion of ADP to ATP. The complex is reversible: ATP hydrolysis is used to pump protons into the extramitochondrial compartment with the cam shaft turning counterclockwise (instead of clockwise during ATP synthesis). The efficiency of the F1ATPase motor can be as high as 100% under viscous loading, higher than any other hydrolysis-driven molecular motor.[12]

In a remarkable experiment, Noji and coworkers visualized the rotary motion of the cam shaft by attaching a genetically modified form of F1ATPase site-specifically onto a surface (with the catalytic subunits surface bound) and following the rotation of the motor shaft with a bioconjugated, fluorescently labeled actin filament (see also Figure 3.2).[6,13,14]

This is a classic case of an experiment that has utilized the power of a multifaceted approach that combines the diverse fields of protein engineering with surface

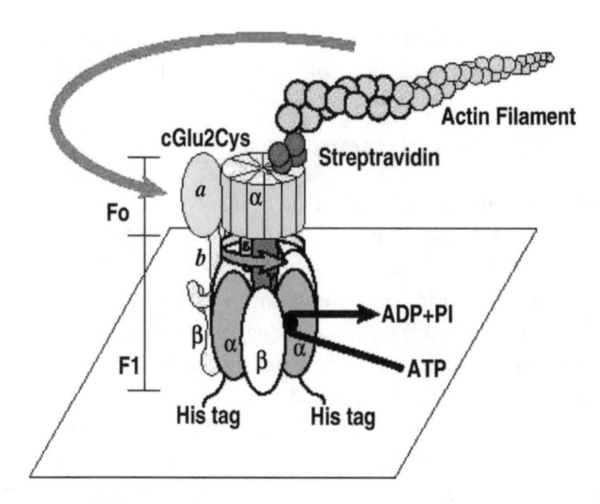

FIGURE 3.2 Illustration of c subunit rotation in F0F1 visualized by connecting a fluorescently labeled actin filament to the c subunit. (From Sambongi Y, Iko Y, Tanabe M, Omote H, Iwamoto-Kihara A, Ueda I, Yanagida T, Wada Y, Futai M. Mechanical rotation of the c subunit oligomer in ATP synthase(F0F1): direct observation, *Science*, 1999, 286, 1722–1724. With permission.)

chemistry and fluorescence detection. It seems obvious to consider modified versions of the F1ATPase as nanopropellers within two-dimensional NEMS devices.

3.2.1.2 Linear Molecular Motors

Linear molecular motors such as myosin and kinesin are responsible for muscle contraction, transport of vesicles, and cell division. These motors have been extensively studied over the last decade using a multifaceted approach. *In vitro* motility assays of surface-bound heavy meromyosin (HMM) developed 20 years ago by Spudich and coworkers, as well as single molecule laser trapping experiments, have provided substantial insight into the molecular mechanism of molecular motors.[15] It seems obvious that in a next step spatially oriented defined tracks of molecular motors could be used to facilitate the vectorial transport of other proteins or nanoscale matter along certain trajectories within artificial two-dimensional nanostructures. For further information on molecular motors the reader is referred to Chapter 8 by Altman and Spudich within this volume.

3.2.2 Abalone Shells

Abalone shells are known for their high-performance, highly resistant, composite mineral structure. They are formed of a microlaminate composed of proteins and calcium carbonate crystals with a microarchitecture that provides crack- and shatter-resistant properties. The shells are unlikely to break or shatter because, when a crack forms, it propagates along complicated paths that diffuse the crack. The underlying nanostructure has been extensively studied.[16–21]

Abalone shells have been attracting the attention of nanoscientists, who aim to replicate new materials and ceramics with similar mechanical properties.

3.3 DNA NANOTECHNOLOGY

DNA is the central biological molecule. Due to its chemical properties, DNA is also, without doubt, the work horse of *ex vivo* bio-experimentalists. Despite its classical importance in biology, DNA is increasingly explored for applications in nonbiological areas (e.g., nanoelectronics and computing). DNA can be chemically labeled, cut, ligated, copied, replicated, extended, sequenced, boiled, dried, and subjected to electrical voltages, all of which are ideal properties for serving as a construction medium for nanotechnological applications. In this section, three different approaches are summarized: (1) the formation of three-dimensional structures built from DNA building blocks, (2) the rapid, single-molecule analysis of nucleic acids using nanopores, and (3) DNA chemically linked to inorganic matter for the construction of nanoelectronic devices.

3.3.1 Structural DNA Assembly

Pioneered by Ned Seeman during the 1980s, the concept of structural nucleic acid nanotechnology based on branched DNA was introduced to form larger two-dimensional networks and three-dimensional structures of precise geometry. In this "bottom-up" approach, DNA of known sequence with defined cohesive attachment points (e.g., single-stranded, sticky ends) is hybridized via sticky-ended junctions into predictable three-dimensional nanostructures of high precision. For example, this approach has generated a truncated octahedron (molecular weight 790 kD), cube-like structures, and two-dimensional periodic arrays. One of the potential applications communicated by Seeman is the scaffolding of biomolecular crystallization using DNA as a host lattice for the organization of macromolecular guest molecules into a crystalline lattice for subsequent diffraction analysis.[22–28]

In a similar approach, DNA scaffolding has been used for the generation of two-dimensional arrays of metallic nanoparticles (1.4 nm in diameter) of precise spacing, providing a promising strategy for applications in nanoelectronics and nanomaterials synthesis.[29]

3.3.2 Nanopore DNA Sequencing

Rather than using DNA as a building block for the formation of nanostructures, new strategies continue to emerge for the rapid and sensitive detection and characterization of nucleic acids. Branton and Deamer have pioneered the use of nanopores for the analysis of single DNA and RNA molecules. In this approach, single-stranded DNA or RNA is translocated through a voltage-biased nanopore in a linear manner, thereby blocking the open-pore ionic current. A prerequisite for this method to work is that the diameter of the nanopore must accommodate the single-stranded DNA or RNA only, which presents challenges with respect to the construction of the pore. The resulting electrical signal fluctuations are used for real-time single molecule counting and sizing, providing microsecond time scale resolution of nucleic acid

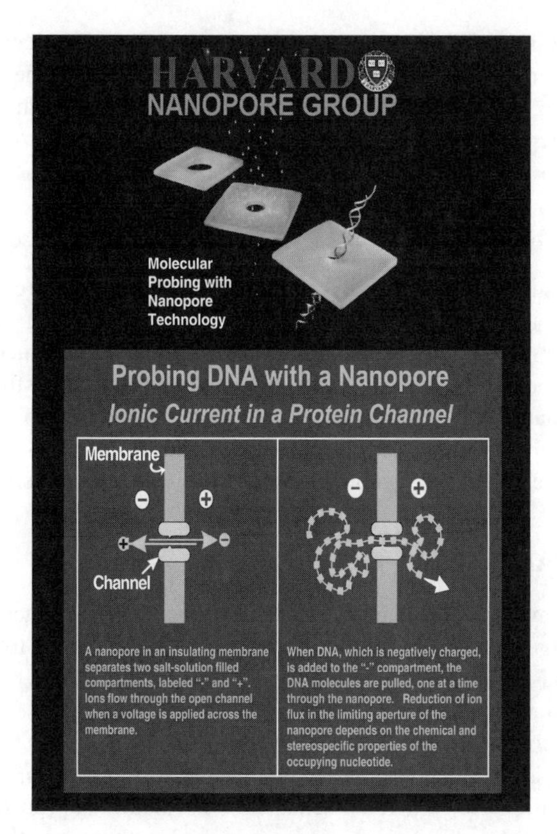

FIGURE 3.3 Principles and components of high-speed device for probing single molecules of DNA. (Courtesy D. Branton, Harvard University.)

characterization, including nucleotide composition (see Figure 3.3). A single, ultrathin hole in a solid state membrane can function as a nanopore,[30] or as extensively studied, a naturally occurring biopore such as α-hemolysine, can be utilized as a membrane channel. (α-Hemolysine is a 33-kD protein of 2.6-nm diameter secreted by *Staphylococcus aureus* and incorporated into an insulating lipid bilayer, exhibiting a channel aperture of 15 angstrom.) Newer research in this arena focuses on electron tunneling detection, rather than measurement of ionic current fluctuations. If further advanced to a multiplexed, robust device level, strategies such as this could provide dramatic improvements in ultra-high-speed DNA sequencing and single-nucleotide polymorphism analysis.[31–37]

3.3.3 DNA Coupled to Carbon Nanotubes

Carbon nanotubes, molecular-scale wires made of graphene sheets rolled into cylinders, have significant potential as building blocks for the construction of nanoelectronic and nanomechanical devices at the sublithographic scale. Functionalization of carbon nanotubes with DNA would enable the programmed assembly of the nanotubes into functional architectures taking advantage of the molecular recognition properties of the DNA. Dekker and coworkers succeeded in the covalent coupling of PNA (peptide nucleic acid, an uncharged DNA analogue) to the ends of single-walled carbon nanotubes (SWNTs) and hybridizing the adducts with complementary DNA, a first step toward DNA-directed SWNT integration into nanoelectronic devices.[38]

3.3.4 DNA-Modified Surfaces

DNA-modified surfaces play an important role in bioanalytical (e.g., gene chips) and biophysical applications (e.g., surface-based analysis using scanning probe microscopy). For micro- and nanoelectronic detection of DNA, semiconductor materials such as silicon are of high value. DNA has been successfully coupled to hydrogen-terminated silicon surfaces via covalent silicon–carbon bonds.[39–41] This strategy is very different from the less reproducible, but nevertheless well established, method of surface silane-based monolayer formation on oxidized silicon substrates.

3.4 NANOPARTICLES

Nanoparticles play a major role in nanobiotechnology. In fact, applications involving nanoparticle reagents have already been commercially developed to a high degree. Penn, Natan, and He have summarized the use of nanoparticles in biotechnology in an excellent review.[42]

Particles of controlled size with at least one dimension of less than 100 nm are considered nanoparticles. Depending on their chemical composition, their optical and electronic properties have garnered much attention in studies of properties of particles approaching molecular dimensions. At nanometer length scales, the line between colloids and molecules becomes blurred (i.e., methods for synthesis, handling, and coating of nanoparticles are derived from both colloidal and synthetic organic and inorganic chemistry). Novel synthetic strategies have created a vast number of new nanoscopic materials and fueled a significant amount of fundamental research directed at exploring the optical, magnetic, and electronic properties of the new nanoscopic materials, often in comparison with bulk samples of the same materials. Defined surface chemistry, particle morphology, and narrow size distribution are important factors regarding the utility for these applications.

3.4.1 Nanoparticles for Biological Assays

Nanoparticles have found widespread use for biological assays. They can be utilized as assay labels (tags), replacing standard fluorescent probes, or as substrates for multiplexed assays. Nanoparticles have substantial advantages over classical organic dyes because of their superior photophysical properties, which overcome many of the spectral limitations of molecular fluorophores.

Semiconductor quantum dots, for example, have tunable, narrow emission spectra with very high photostability. They have the advantage of a single exciting wavelength providing for multiple emission wavelengths.[43]

For example, a Her2 assay with quantum dots has been used for immunofluorescent imaging of breast cancer cells.[44]

Quantum dots also appear to be valuable reagents for nucleic acid bioassays including single nucleotide polymorphism (SNP) analysis.[45]

Metallic nanoparticles such as colloidal gold (1–5 nm in diameter) are popular staining agents in biological electron microscopy, appearing dark in transmission electron microscope images (due to high Au density) and bright in scanning electron microscopy images (due to high backscatter coefficient).[46]

In addition, colloidal gold has been successfully used for biological assays such as DNA hybridization. For example, gold nanoparticles show a characteristic plasmon resonance absorption which is useful for detection purposes.[47,48]

A conductivity-based method has been reported by Mirkin and coworkers, utilizing oligonucleotide functionalized gold nanoparticles that provide conductivity between microelectrodes upon the binding of the nanoparticles to the electrodes.[49] Resonance light scattering of gold nanoparticles can be also used for microarray-based DNA hybridization.[50]

In another approach, biomarkers for leukemia cell recognition were conjugated with ruthenium complexes encapsulated in silica.[51]

Nanoparticles can also be considered as substrates for assays similar to microbead assays (e.g., for immunoturbidimetric assays). Nanoparticle suspensions are more stable than microparticle suspensions and they offer substantially reduced scatter of visible light, thereby providing higher signal-to-noise rations and assay sensitivities. For example, an immunoturbidimetric assay for urine albumin was developed using 40-nm sized anti-albumin antibody coated poly(vinyl-naphthalene) particles quite some time ago.[52,53]

Superparamagnetic nanoparticles such as iron oxide particles (Fe_3O_4 or γ-Fe_2O_3) have enormous potential for separation and biomolecular assay applications. They can be magnetized in a magnetic field, but then redispersed (without agglomeration) after removal of the field. For example, microparticles coated with superparamagnetic nanoparticles have been used for cell sorting, DNA isolation, and proteomic assays.[54–56]

Natan and coworkers have pioneered the design and applicability of striped metal nanoparticles (Nanobarcodes™ particles) for multiplexed assays. Figure 3.4 shows images of Nanobarcodes™ particles and their basic manufacturing process. These particles are made of alternating layers of reflective metals (gold, silver) that function as "barcodes." The barcode provides an intrinsic encoding signal that is based on reflectivity, which allows multiplexed assays to be performed simultaneously when the barcode is surface-coated with active biomolecules such as antibodies or nucleic acids.[57,58]

3.4.2 Nanoparticles as Drug Delivery Vehicles

Top-down milling processes can be used to reformulate highly insoluble drugs as nanoparticles, thereby providing more efficient uptake. For example, Elan Pharmaceuticals has reported a method of using nanocrystalline formulations of human immunodeficiency virus (HIV) protease inhibitors.

Alternative strategies involving new types of drug delivering vehicles are in high demand.

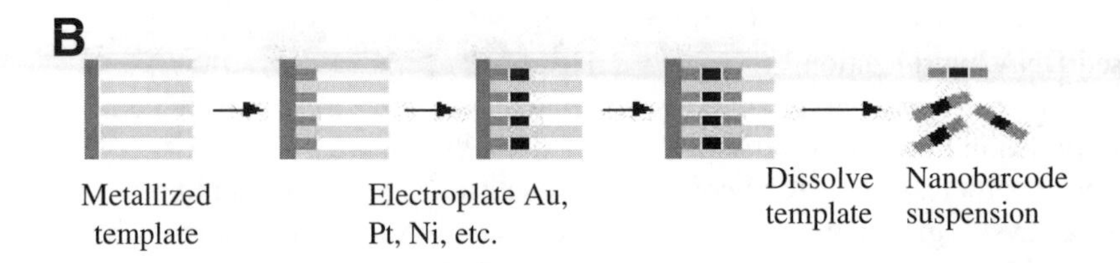

FIGURE 3.4 (A) Nanobarcodes™ particles. Nanobarcodes have specific patterns of gold (Au) and silver (Ag) stripes. (B) Production process of nanobarcodes. (Courtesy M. Natan, Nanoplex Technologies.)

Liposomes in which pharmaceutical agents have been encapsulated have also been in use. Although liposomes exhibit outstanding biocompatibility and low toxicity, problems such as drug leakage and low delivery efficiency (e.g., for gene therapy) remain major obstacles.

In a different approach, magnetically guided drug targeting makes use of microparticles composed of elemental iron particles and activated carbon with drug absorbance and release properties in order to direct the drug agents directly to the target tissue (e.g., a tumor).[59]

In addition, magnetoliposomes and bacterial magnetic particles (50–100 nm in diameter) have been evaluated for targeted drug delivery and release in antitumor chemotherapy applications.[60,61]

Another class of important nanoparticles are dendrimers. Dendrimers are spherical nanoscale, polymeric, polyvalent molecules of well-defined chemical structure consisting of multiple shells ("generations") synthesized around a small core molecule. Dendrimer nanotechnology is a rapidly expanding field with drug delivery only one of many potential applications. The first dendrimer drugs with antiviral activity (HIV, herpes simplex) are under clinical investigation. In addition, a polyanionic dendrimer with potency as anti-HIV microbicide is undergoing regulatory approval processes in the United States.[62–64]

West and Halas at Rice University, pioneered the use of nanoshells, a different type of nanoparticle, for bioassays and targeted drug delivery. Nanoshells are composed of a dielectric core (e.g., silica) and a thin metal shell (e.g., gold). They have tunable optical properties and utility for various imaging and biosensing applications, such as near infrared imaging, ELISA-based assays, and surface enhanced raman spectroscopy (SERS).[65,66] West and Halas also described the use of nanoshells for thermal ablative therapy of cancer. In this approach, antibody-conjugated nanoshells are targeted to human breast carcinoma cells, where therapeutic dosages of heat upon absorption of light in the near infrared are delivered, resulting in the localized destruction of the tumor cells.[67]

3.4.3 Nanoparticles as Contrast Agents

Nanoparticles can function as contrast agents for better image resolution and tissue-specific targeting. For example, superparamagnetic Fe_3O_4 nanoparticles provide better contrast in magnetic resonance imaging (MRI). Tissue enriched with these nanoparticles appear darker due to protons decaying from excited states more effectively via energy transfer to neighboring nuclei.[68]

3.5 PATH INTO THE FUTURE AND ECONOMIC CONSIDERATIONS

What will be the biggest breakthroughs in nanobiotechnology? Where are the challenges in transforming nanoscience into mainstream products? Why do some fear nanotechnology? Many questions exist in a field that is now widely recognized by the general public, has created excitement at the media, and has earned Wall Street's attention. Working in the field of nanobiotechnology means operating without the traditional boundaries between physics, biology, and chemistry. Research in this area is expensive and demands large amounts of funding, thereby catalyzing more and more collaboration-centered, consortium-like efforts. Most advances in the field to date have actually been in nanoscience, not in nanotechnology.

As mentioned above, nanotechnology is, or has the potential to become, a general-purpose technology (some estimates predict a market size of $1 trillion by 2015). However, landmark scientific breakthroughs will only have an impact if commercial success follows. Commercial success will only happen if compelling applications are found and major economic and technical hurdles can be overcome. What are the challenges? For example, converting innovation into implemented products with economic sustainability is capital intensive and quite challenging for an area with unfamiliar formats that are characterized by slow product adoption rates. In addition, having incremental progress only slows down the adoption process. Large-scale production can be challenging for highly integrated nanodevices, as well as bottom-up strategies that require economic ways of mass self-assembly. Moore's law has been one of the major competitive driving factors in the microelectronics industry, leading to better products over time. If there is an analogy with the computer industry — what will be Moore's law of nanobiotechnology?

Last, but not least, for certain applications in nanobiotechnology, substantial regulatory, and socioeconomic hurdles exist. While it is naïve to call all nanotechnology applications as potentially suspect, the request for an in-depth environmental analysis is reasonable (example: asbestos). The educational system and the media will be challenged to accurately address the excitement and scientific highlights in the face of an existing ambiguous fear over a potentially suspect, oversimplified technology area that is composed of many fragmented nonrelated segments. Because of this factor, discussions of certain aspects of nanobiotechnology will be challenging, much like discussions on embryonic stem cell research, genetically modified food, and somatic cell nuclear transfers.

What applications will emerge in nanobiotechnology? With an increasing demand for better healthcare and an aging population in the developed countries, healthcare will continue to be the major stimulating factor for advances in nanobiotechnology. Clever integration of biology, information science, and materials science is required to develop new generations of medical devices, diagnostic tools, and therapeutic products.

Tools to manipulate biological matter at the nanometer scale were among the first products to be developed in nanobiotechnology (e.g., atomic force microscopes, microfluidic devices), followed by reagents such as nanoparticles and biochips. Now, products using nanostructured materials for drug delivery and tissue engineering are approaching clinical test phases. Applications in bioanalysis, biomaterials, and drug delivery will predominate in the near future. For example, high-performance protein biochips for multiplexed protein profiling providing higher speeds of analysis and novel information content will accelerate research in drug development, provide better diagnostic tests, and help in the stratification of patient populations. The above-mentioned nanoparticles are extremely valuable reagents for these purposes.

The following listed applications will require substantial development breakthroughs to become reality, but appear, nevertheless, to be achievable in the not-so-distant future: (1) external tissue products (e.g., artificial skin, tissue reconstruction), (2) internal tissue products (e.g., xenotransplants from other animals or tissue grown artificially on nanostructured scaffolds), (3) *in vivo* testing devices (e.g., for the detection of cancer, infections, heart conditions), (4) smart pills for localized multiple drug bursts of high dosage, (5) nonviral nanovectors for gene therapy, (6) handheld devices for personal diagnosis, (7) polymer–drug conjugates and encapsulated drugs, and finally (8) nanosurgical particles capable of stitching tissue at the cellular level, thereby accelerating wound healing processes stemming from surgical or injury trauma. Some of the listed applications will require thorough approval processes regarding safety and bioethic standards, an immediate need given the time scale of these upcoming biomedical breakthroughs. Unfortunately, many public discussions about nanobiotechnology are instead still centered on and fueled by the exaggerated speculations of a few on the outperformance of evolution and even apocalyptic scenarios of self-replicating robots turning the earth's matter into a "grey goo" that is unable to support life. This is an unfortunate situation, since such overly dramatic speculations capture more attention than the science itself, thereby potentially causing setbacks to the short- and mid-term challenges and developments that require our utmost attention. It seems for some, nanobiotechnology is Aladdin's lamp, and for others it is Pandora's box.

REFERENCES

1. Interagency Working Group on Nanoscience, Engineering and Technology. *National Nanotechnology Initiative: Leading to the Next Industrial Revolution,* Washington, D.C., Committee on Technology, National Science and Technology Council, 2000.
2. Goodsell DS. *Bionanotechnology: Lessons from Nature*. New York, Wiley-Liss, 2004.
3. Freitas RA Jr. *Nanomedicine, Vol. I: Basic Capabilities*. Austin, TX, Landes-Bioscience, 1999.
4. Whitesides GM. The "right" size in nanobiotechnology, *Nat. Biotechnol.*, 2003, 21, 1161–1165.
5. Stock D, Leslie AGW, Walker JE. Molecular architecture of the rotary motor in ATP synthase, *Science*, 1999, 286, 1700–1705.
6. Noji H. The rotary enzyme of the cell: the rotation of F1-ATPase, *Science*, 1998, 282, 1844–1845.
7. Howard J, Hudspeth AJ, Vale RD. Movement of microtubules by single kinesin molecules, *Nature*, 1989, 342, 154–158.
8. Block SM, Goldstein LSB, Schnapp BJ. Bead movement by single kinesin molecules studied with optical tweezers, *Nature*, 1990, 348, 348–352.
9. Mehta AD, Rief M, Spudich JA, Smith DA, Simmons RM. Single-molecule biomechanics with optical methods, *Science,* 1999, 283, 1689–1695.
10. Mermall V, Post PL, Mooseker MS. Unconventional myosins in cell movement, membrane traffic, and signal transduction, *Science*, 1998, 279, 527–533.
11. Yin H, Wang MD, Svoboda K, Landick R, Block SM, Gelles J. Transcription against an applied force, *Science*, 1995, 270, 1653–1657.
12. Oster G, Wang H. Why is the mechanical efficiency of F1-ATPase so high? *J. Bioenerg. Biomembr.*, 2000, 32, 459–469.
13. Noji H, Yasuda R, Yoshida M, Kinosita K. Direct observation of the rotation of F1-ATPase, *Nature*, 1997, 386, 299–302.
14. Sambongi Y, Iko Y, Tanabe M., Omote H, Iwamoto-Kihara A, Ueda I, Yanagida,T, Wada Y, Futai M. Mechanical rotation of the c subunit oligomer in ATP synthase(F0F1): direct observation, *Science*, 1999, 286, 1722–1724.
15. Kron SJ, Spudich JA. Fluorescent actin filaments move on myosin fixed to a glass surface, *Proc. Natl. Acad. Sci. U.S.A.*, 1986, 83, 6272–6276.
16. Fritz M, Belcher AM, Radmacher M, Walters DA, PK, Stucky GD, Morse DE, Mann S. Flat pearls from biofabrication of organized composites on inorganic substrates, *Nature*, 2002, 371, 49–51.
17. *Proc. Royal Soc. London B,* 1994, 256, 17–23.
18. Zaremba CM, Belcher AM, Fritz M, Li Y, Mann S, Hansma PK, Morse DE, Speck JS, Stucky GD. Critical transitions in the biofabrication of abalone shells and flat pearls, *Chem. Mater.*, 1996, 8, 679–690.
19. Belcher AM, Wu XH, Christensen RJ, Hansma PK, Stucky GD, Morse DE. Control of crystal phase switching and orientation by soluble mollusc-shell proteins, *Nature*, 1996, *381,* 56–58.
20. Belcher AM, Hansma PK, Stucky GD, Morse DE. First steps in harnassing the potential of biomineralization as a route to new high-performance composite materials, *Acta Mater.*, 1998, 46, 733–736.
21. Fritz M, Morse DE. The formation of highly organized biogenic polymer/ceramic composite materials: the high-performance microlaminate of molluscan nacre, *Curr. Opin. Colloid Interface Sci.*, 1998, 3, 55–62.
22. Seeman NC. Nucleic acid junctions and lattices, *J. Theor. Biol.*, 1982, 99, 237–247.

23. Seeman NC. DNA in a material world, *Nature*, 2003, 421, 33–37.
24. Seeman NC. Biochemistry and structural DNA nanotechnology: an evolving symbiotic relationship, *Biochemistry*, 2003, 42, 7259–7269.
25. Seeman NC. At the crossroads of chemistry, biology and materials: structural DNA nanotechnology, *Chem. Biol.*, 2003, 10, 1151–1159.
26. Seeman NC. DNA Nanostructures for mechanics and computing: nonlinear thinking with life's central molecule, in *NanoBiotechnology*, Mirkin C, Niemeyer C, Eds., Weinheim, Wiley-VCH, 2004, pp. 308–318.
27. Rothemund PWK, Papadakis N, Winfree E. Algorithmic self-assembly of DNA Sierpinski triangles, in *DNA Computing IX*, Reif JH, Chen J. Eds., Heidelberg, Springer-Verlag, 2004.
28. Yan H, LaBean TH, Feng LP, Reif JH. Directed nucleation assembly of DNA tile complexes for barcode-patterned lattices, *Proc. Natl. Acad. Sci. U.S.A.*, 2003, 100, 8103–8108.
29. Xiao S, Liu F, Rosen AE, Hainfeld JF, Seeman NC, Musier-Forsyth K, Kiehl RA. Self-assembly of metallic nanoparticle arrays by DNA scaffolding, *J. Nanoparticle Res.*, 2002, 4, 313–317.
30. Li J, Gershow M, Stein D, Brandin E, Golovchenko JA. DNA molecules and configurations in a solid-state nanopore microscope, *Nat. Mater.*, 2003, 2, 611–615.
31. Kasianowicz JJ, Brandin E, Branton D, Deamer DW. Characterization of individual polynucleotide molecules using a membrane channel, *Proc. Natl. Acad. Sci. U.S.A.*, 1996, 93, 13770–13773.
32. Branton D, Golovchenko J. Adapting to nanoscale events. *Nature*, 1999, 398, 660–661.
33. Deamer DW, Akeson M. Nanopores and nucleic acids: prospects for ultrarapid sequencing, *Trends Biotechnol.*, 2000, 18, 147–151.
34. Wang H, Branton D. Nanopores with a spark for single-molecule detection, *Nat. Biotechnol.*, 2001, 19, 622–623.
35. Vercoutere W, Winters-Hilt S, Olsen H, Deamer D, Haussler D, Akeson M. Rapid discrimination among individual DNA hairpin molecules at single nucleotide resolution using an ion channel, *Nat. Biotechnol.*, 2001, 19, 248–252.
36. Meller A, Branton D. Single molecule measurements of DNA transport through a nanopore, *Electrophoresis*, 2002, 23, 2583–2591.
37. Sauer-Budge AF, Nyamwanda JA, Lubensky DK, Branton D. Unzipping kinetics of double-stranded DNA in a nanopore, *Phys. Rev. Lett.*, 2003, 90, 238101-1–238101-4.
38. Williams KA, Veenhuizen PTM, de la Torre BG, Eritja R, Dekker C. Carbon nanotubes with DNA recognition, *Nature*, 2002, 420, 761.
39. Wagner P, Nock S, Spudich JA, Volkmuth WD, Chu S, Cicero RL, Wade CP, Linford MR, Chidsey CED. Bioreactive self-assembled monolayers on hydrogen-passivated Si(111) as a new class of atomically flat substrates for biological scanning probe microscopy, *J. Struct. Biol.*, 1997, 119, 189–201.
40. Strother T, Cai W, Zhao X, Hamers RJ, Smith LM. Synthesis and characterization of DNA-modified silicon (111) surfaces, *J. Am. Chem. Soc.*, 2000, 122, 1205–1209.
41. Lin Z, Strother T, Cai W, Cao X, Smith LM, Hamers RJ. DNA attachment and hybridization at the silicon (100) surface, *Langmuir*, 2002, 18, 788–796.
42. Penn SG, Natan, MJ, He L. Nanoparticles in bioanalysis, *Curr. Opin. Chem. Biol.*, 2003, 7, 609–615.
43. Bruchez M, Moronne M, Gin P, Weiss S, Alivisatos AP. Semiconductor nanocrystals as fluorescent biological labels, *Science*, 1998, 281, 2013–2016.

44. Wu X, Liu H, Liu J, Haley KN, Treadway JA, Larson JP, Ge N, Peale F, Bruchez MP. Immunofluorescent labeling of cancer marker Her2 and other cellular targets with semiconductor quantum dots, *Nat. Biotechnol.*, 2003, 21, 41–46.
45. Xu, H, Sha, MY, Wong EY, Uphott J, Xu Y, Treadway JA, Truong, A, O'Brien E, Asquith S, Stubbins M, Spurr NK, Lai EH, Mahoney W. Multiplexed SNP genotyping using the Qbead™ system: a quantum dot-encoded microsphere-bead assay, *Nucleic Acids Res.*, 2003, 31, e43.
46. Hayat M. *Colloidal Gold: Principles, Methods, and Applications*, San Diego, Academic Press, 1990.
47. He L, Musick MD, Nicewarner SR, Salinas FG, Benkovic SJ, Natan MJ, Keating CD. Colloidal Au-enhanced surface plasmon resonance for ultrasensitive detection of DNA hybridization, *J. Am. Chem. Soc.*, 2000, 122, 9071–9077.
48. Elghanian R, Storhoff JJ, Mucic RC, Letsinger RL, Mirkin CA. Selective colorimetric detection of polynucleotides based on the distance-dependent optical properties of gold nanoparticles, *Science*, 1997, 277, 1078–1081.
49. Park SJ, Taton, TA, Mirkin CA. Array-based electrical detection of DNA with nanoparticle probes, *Science*, 2002, 295, 1503–1506.
50. Bao P, Frutos AG, Greef C, Lahiri J, Muller U, Peterson TC, Warden L, XieX. High-sensitivity detection of dna hybridization on microarrays using resonance light scattering, *Anal. Chem.*, 2002, 74, 1792–1797.
51. Santra S, Zhang P, Wang W, Tapec R, Tan W. Conjugation of biomolecules with luminophore-doped silica nanoparticles for photostable biomarkers, *Anal. Chem.*, 2001, 73, 4988–4993.
52. Medcalf EA, Newman DJ, Gorman, EG, Price CP. Rapid, robust method for measuring low concentrations of albumin in urine, *Clin. Chem.*, 1990, 36, 446–449.
53. Simo JM, Joven J, Clivelle X, Sans T. Automated latex agglutination immunoassay of serum ferritin with a centrifugal analyzer, *Clin. Chem.*, 1994, 40, 625–629.
54. Saiyed ZM, Telang SD, Ramchand CN. Application of magnetic techniques in the field of drug discovery and biomedicine, *Biomagn. Res. Technol.*, 2003, 1, 2.
55. Safarik I, Safarikova M. Use of magnetic techniques for the isolation of cells, *J. Chromatogr. B Biomed. Sci. Appl.*, 1999, 722, 33–53.
56. Safarik I, Safarikova M. Magnetic nanoparticles and biosciences, *Mon. Chem.*, 2002, 133, 737–759.
57. Walton ID, Norton SM, Balasingham A, He L, Oviso DF, Gupta D, Raju PA, Michael MJ, Freeman RG. Particles for multiplexed analysis in solution: detection and identification of striped metallic particles using optical microscopy, *Anal. Chem.*, 2002, 74, 2240–2247
58. Nicewarner-Pena SR, Freeman RG, Resiss BD, He L, Pena DJ, Walton IA, Cromer R, Keating CD, Natan MJ. Submicrometer metallic barcodes, *Science,* 2001, 294, 137–141.
59. Fricker J. Drugs with a magnetic attraction to tumors, *Drug Discov. Today,* 2001, 6, 387–389.
60. Babincova M, Cicmanec P, Altanerova V, Altaner C, Babinec P. AC-magnetic field controlled drug release from magnetoliposomes: design of a method for site-specific chemotherapy, *Bioelectrochemistry*, 2002, 55, 17–19.
61. Schuler D, Frankel, RB. Bacterial magnetosomes: microbiology, biomineralization and biotechnology applications, *Appl. Microbiol. Biotechnol.*, 1999, 52, 464–473.

62. Witvrouw M, Fikkert V, Pluymers W, Matthews B, Mardel K, Schols D, Raff J, Debyser Z, Clercq ED, Holan G, Pannecouque C. Polyanionic (i.e., polysulfonate) dendrimers can inhibit the replication of human immunodeficiency virus by interfering with both virus adsorption and later steps (reverse transcriptase/integrase) in the virus replicative cycle, *Mol. Pharmacol.*, 2000, 58, 1100–1108.
63. Gong Y, Matthews B, Cheung D, Tama T, Gadawski I, Leung D, Holan G, Raff J, Sacks S, Evidence of dual sites of action of dendrimers: SPL-2999 inhibits both virus entry and late stages of herpes simplex virus replication, *Antiviral Res.*, 2002, 55, 319–329.
64. Bourne N, Stanberry LR, Kern ER, Holan G, Matthews B, Bernstein DI. Dendrimers, a new class of candidate topical microbicides with activity against herpes simplex virus infection, *Antimicrobial Agents and Chemotherapy*, 2000, 44, 2471–2474.
65. Oldenburg SJ, Averitt RD, Westcott SL, Halas NJ. Nanoengineering of optical resonances, *Chem. Phys. Lett.*, 1998, 288, 243–247.
66. Jackson JB, Halas NJ. Silver Nanoshells: Variations in morphologies and optical properties, *J. Phys. Chem. B*, 2001, 105, 2743–2746.
67. Hirsch LR, Stafford RJ, Sershen SR, Halas NJ, Hazle JD, West JL. Nanoshell-assisted tumor ablation using near infrared light under magnetic resonance guidance, *Proc. Natl. Acad. Sci. U.S.A.*, 2003, 100, 113549–113554.
68. Hahn PF, Stark DD, Lewis JM, Saini S, Elizondo G, Weissleder R, Fretz CJ, Ferrucci JT. First clinical trial of a new super-paramagnetic iron oxide for use as an oral gastrointestinal contrast agent in MR imaging, *Radiology*, 1990, 175, 695–700.

4 Next Generation Sensors for Measuring Ionic Flux in Live Cells

Rainer Fasching, Eric Tao, Seoung-Jai Bai, Kyle Hammerick, R. Lane Smith, Ralph S. Greco, and Fritz B. Prinz

CONTENTS

4.1 INTRODUCTION

During the past decade, details regarding the role of intracellular ions in signaling pathways have followed technological breakthroughs in biochemistry, genetics, and developmental biology. Yet the fundamental mechanisms by which the total intracellular ionic state influences the cell response to its chemical and mechanical environment are still undetermined. This deficiency is due in part to limitations of analytical methods for assessing intracellular ion flux in living cells. Many measurement methods rely on the addition of chemical reagents such as ion binding dyes that may alter cell metabolism and provide only limited spatial resolution. Other deficiencies in sensing technologies can be attributed to the difference in size between the molecular processes taking place and the sensor. Recent advances in nanofabrication allow for the creation of arrays of sensors, each with a tip radius of

0-8493-1940-4/05/$0.00+$1.50

tens of nanometers and a spacing of a few microns between tips. The tip arrays will form the core of massively parallel electrodes directly connected to field effect transistors and operational amplifiers. These probes introduce a new methodology for the direct assessment of intracellular metabolism in living cells. Following controlled insertion into the interior environment of living cells, the probes are designed to measure membrane permeability through impedance spectroscopy and changes in the intracellular pH in the cytosol and organelles. The distribution of the nanotips within the arrays will provide a spatial resolution of no more than 2 μm. Measurements will be made using 20 to 60 active probes almost simultaneously on a single cell crosssection. Sequential probing of cells at different depth levels will establish three-dimensional maps of internal cellular environments. Microelectromechanical systems (MEMS) and complementary metal oxide semiconductor (CMOS) fabrication can be used to advance current impedance spectroscopy technology towards a concept better referred to as impedance microscopy. This probe technology will provide a framework for investigating mechanisms underlying ion-based regulatory processes that modulate phenotypic expression. The data can be correlated to the cellular reaction to externally applied physical stimuli and soluble factors in the chemical environment.

4.2 BACKGROUND

The impetus for the development of dynamic intracellular sensing probes comes from an increasing body of evidence substantiating the role of intracellular ion flux in cell metabolic control. Observing the dynamics of intracellular electrochemistry is crucial in creating a theoretical framework for understanding interactions between signaling mechanisms within a single cell and also between networks of cells. Today, the highest resolution tools investigating cell communication phenomena are cell-attached capacitance measurements using patch clamps. The revolutionary insights provided by the patch clamp sensor of Sakman, Neher, and Marty[1,2] furthered our understanding of trans-membrane signaling mechanisms. Other advanced techniques for studying intracellular pH and ion transport include fluorescent microinjection coupled with flash photolysis of caged moieties and patch amperometry.[3–5] Plasma membrane decoupling from the cytoskeleton, mechanical perturbations in the membrane, and cell ballooning are problems associated with patch clamp methods.[6] Fluorescence detection methods require the addition of exogenous compounds that may adversely affect the system that is being measured. These techniques, however, lack the capability to investigate electrochemical phenomena simultaneously at a variety of different cell locations and couple that data with force feedback measurements.

The transmission of information from the cell membrane to the nucleus involves complex chemical interactions in which pH and ion flux are key elements. Functional similarities associated with membrane impedance and intracellular pH occurs in all cell types. In epithelial tissue, tight junctions regulate paracellular conductance and ionic selectivity.[7] Recent experiments show that overexpression of human claudin-4, a membrane protein, decreases paracellular conductance by changing Na^+ and Cl^- permeability.[8] In embryonic cells, the ionic flux between the nucleus and cytosol has been hypothesized to alter chromatin condensation and influence gene expression.[9] In macrophages, an ATP binding cassette transporter, ABC1, regulates a cAMP-dependent anionic flux in the cells.[10] These results demonstrate the ability of ion flux to influence paracellular ion selectivity, cell metabolism, and potentially tissue differentiation. Physiological processes by which cells respond to environmental stimuli require autogenous metabolic reactions as well as reactivity to extrinsic factors. The host-defense cell, the neutrophil, and the articular cartilage chondrocyte represent two specific cell types whose metabolism depends on the intracellular ionic state. While one cell is phagocytic and the other is synthetic, both types of cells respond to environmental stimuli that influence intracellular ions and pH. The following sections describe fabrication methods toward the creation of nanotips and nanotip arrays. Initially, the focus will be on strategies for the realization of single probe tips, followed by a discussion of tip arrays and signal transduction.

4.3 NANOPROBES AND TIP CHARACTERIZATION

The purpose of these efforts is to develop a probe capable of electrochemically characterizing the cell membrane surface and the cell interior by piercing the cell while causing minimal damage. Previous research identified scanning electrochemical microscopy (SECM) as an important analytical tool for studying surface reactions and their kinetics down to nanoscale dimensions. The use of this method has been demonstrated in a wide range of applications, such as resolving fast heterogeneous kinetics at various material interfaces or imaging of biological molecules.[11–13] The spatial resolution of SCEM depends on shape and size of the electrochemical electrode. Ultra-micro electrodes (UME), which are tip probes having submicron electrodes on the top, are required to obtain resolution at the nanometer scale. Manufacturing approaches have been investigated, ranging from isolation of etched metal wires or scanning tunneling electron microscopy (STEM) tips for single-electrode systems to batch fabrication strategies for electrode array systems.[14–17] A combination of SECM with other scanning probe microscopy (SPM) techniques, such as atomic force microscopy (AFM), is highly desirable in order to obtain electrochemical information as well as complementary surface information simultaneously.[18] In particular, the combined use of SCEM and AFM will allow for precise positioning of probes adjacent to or inside of cells and sensing of concentrations and fluxes of electrochemically active substances. A crucial component of such a system is the specialized probe system that has to be composed of a micromechanical cantilever structure necessary for the AFM mode and an electrochemical UME tip required for a high performance SECM. Several strategies for fabrication of such transducer structures have already been reported.[19–24] While these efforts resulted in functional tip structures and combined AFM and SECM imaging, performance issues related to the tip sharpness, low aspect ratio of the tip structure, and size of the electrochemical electrode remain. The commonality for most of the fabrication technologies above described is a single-electrode production scheme, which limits miniaturization potential and fabrication of multielectrode systems. In order to overcome these limitations, enable radical miniaturization, and build high-density array probes, fabrication schemes exploiting micro- and nanofabrication technologies have been developed.

Micro- and nanofabrication technologies originate from the field of microelectronics, combining parallel processing techniques with a miniaturization potential of submicron regimes. In particular, high-aspect-ratio electrochemical tip probes embedded in silicon nitride cantilevers have been developed for simultaneous AFM and SECM analyses.[17,25] The fabrication process is based on batch processes in combination with an etch-mask technology utilizing focused ion beam (FIB) techniques to achieve both well-defined UME and sharp high-aspect-ratio tips on a single cantilever, as well as in cantilever arrays. The process has been developed on 4-in wafers and is divided into three main fabrication pathways. An overview of the fabrication process is depicted in Figure 4.1. First, high-aspect-ratio silicon (HARS) tips are shaped, combining isotropic etching with an anisotropic deep-RIE-silicon etch process. Second, silicon tips are embedded in a silicon nitride layer; electrode systems are patterned and passivated with an isolation layer. Finally, a UME on top of the tip structures is established by etching the isolation layer only at the tip. To achieve accuracy and resolution in the nanometer regime on the HARS tip, an etch mask technology has been developed utilizing a focused ion beam-based technique. Subsequently, the cantilevers with embedded electrochemical tip electrodes are shaped and released. A detailed description of the whole fabrication process can be found in previous publications.[17,25] In Figure 4.2A, an example of a finished HARS tip is shown. Tips with a shaft diameter of 1.2 μm are created using oxide caps as a mask.[17,26] HARS tips with an aspect ratio of up to 20 and a diameter in the submicron regime can be fabricated in this way. Figure 4.2B shows a detail of a sharpened tip with a tip radius smaller than 50 nm. HARS tips embedded in a 500 nm thick low-stress silicon nitride are shown in Figure 4.2C and Figure 4.2D. Application of a nonuniform coating of photo resist in combination with silicon nitride etching allows for the exposure of the silicon tips.[25,27] About 5 μm of the silicon tip at its base is anchored in the silicon nitride while the rest of the tip protrudes from the nitride layer. This processing step results in an additional reduction of the tip diameter to 600 nm and creates further sharpening of the HARS tip.[28] In Figure 4.2E and Figure 4.2F, a platinum electrode structure is patterned on the embedded tips. Using magnetron sputtering, a homogeneous metal layer is formed and sufficient side-wall coverage of the HARS tip structures is observed. This conformal metallization establishes a reliable electrical connection between an electrode on top of the tip and bonding structures on the bulk silicon. This process is not limited to single-electrode patterning and can be expanded to larger linear arrays as well as to two-dimensional arrays. A finished tip probe array is depicted in Figure 4.3A. The bending of the cantilevers is caused by small residual stresses resulting from the nitride thin film passivation layer. The platinum UMEs on top of the tips are recognizable in Figure 4.3B. The UME diameter of these probes is about 200 nm. The probes are used for the electrochemical characterization. A low-magnification optical micrograph shows a probe device with an array of four cantilever tip probes and its bonding pad system (Figure 4.4).

4.3.1 Electrochemical Characterization

Electrochemical characterization is carried out using a Solatron 1287, an electrochemical interface, in combination with a Solatron 1260 impedance/gain phase

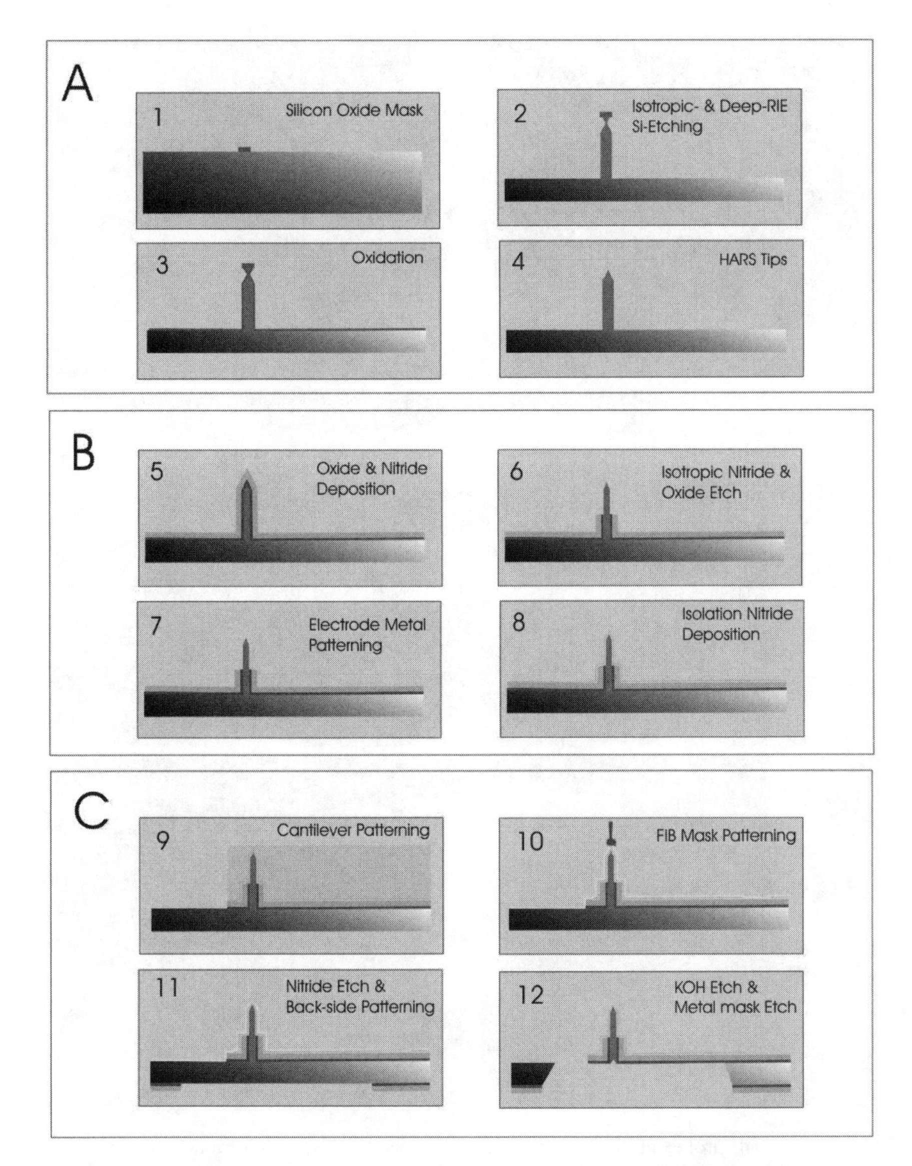

FIGURE 4.1 (A) Fabrication of HARS tips: (1) patterning of a silicon oxide etch mask; (2) isotropic silicon etching to shape the tip and deep-RIE Si-etching to shape the shaft; (3) releasing silicon oxide caps and sharpening of tips by oxidation and back etching of silicon oxide; (4) finished HARS tip. (B) Embedding of HARS tips in silicon nitride and metal patterning of the electrode system: (5) growing of silicon oxide and silicon nitride; (6) back-etching of the silicon nitride using nonuniform resist coating on high-aspect-ratio structures and release of oxide layer using wet etching; (7) patterning of the electrode system utilizing lift-off technique; (8) isolation of the electrode system and tip structure with deposition silicon nitride. (C) Fabrication of UMEs and cantilevers: (9) shaping of cantilevers; (10) FIB-patterning of etch metal mask; (11) etching of the isolation layer exposing the electrode metal layer (creation of UMEs) and patterning of back side etch mask for release; (12) releasing of cantilever structure using wet etching.

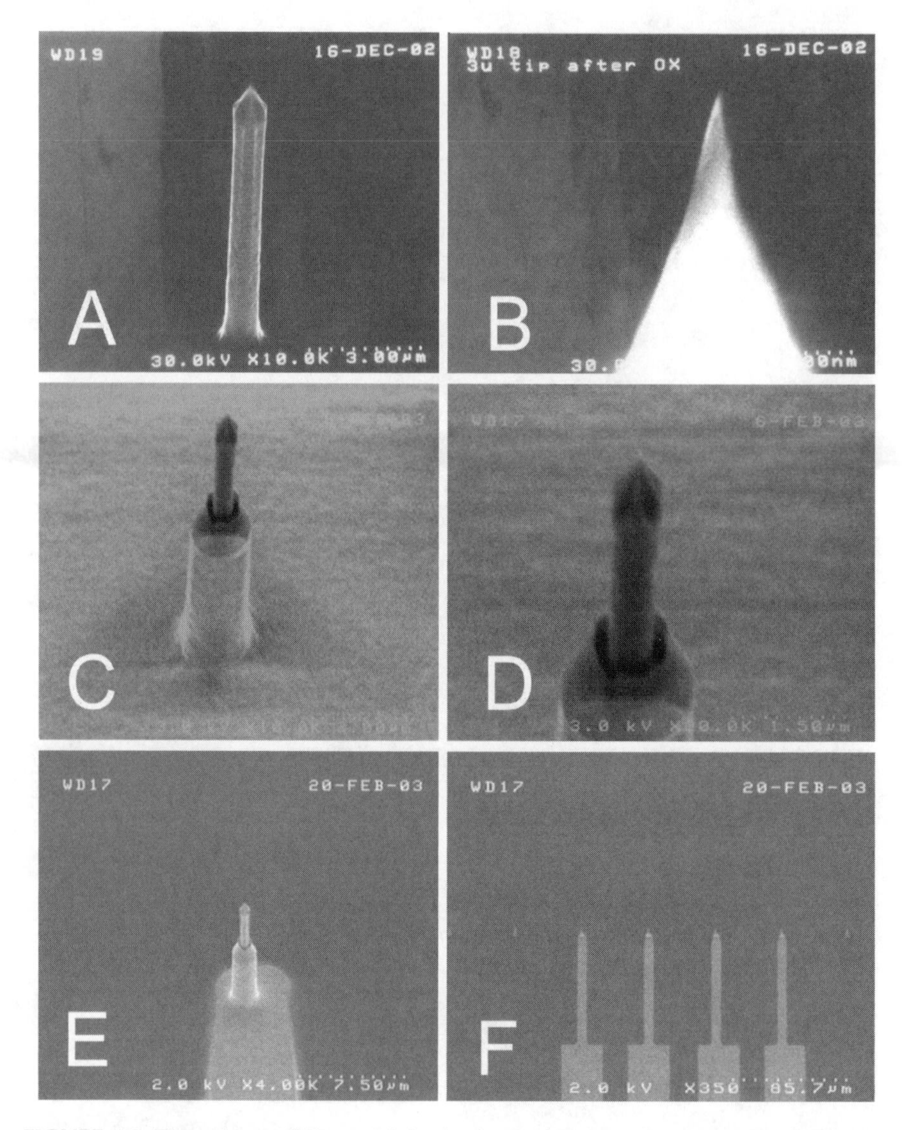

FIGURE 4.2 Tip-probes at different fabrication levels: (A) example of an etched HARS tip; (B) detail of a sharpened tip with a tip radius smaller than 50 nm; (C–D) HARS tips embedded in a 500-nm thick low stress silicon nitride; (E–F) metallized embedded HARS tips.

analyzer (Solatron Analytical). As shown in Figure 4.5, a three-electrode arrangement was employed. The cantilever device is mounted on a micromanipulator stage (PCS-6000, Burleigh Instruments) or atomic force microscope (PicoPlus, MI). The tips on the cantilever are immersed and positioned in a drop or film of electrolyte in a controlled fashion. A platinum thin film layer is used as a counter electrode, and an Ag/AgCl wire electrode functions as the reference electrode. Cyclic voltammetry of the tip probe electrode with a platinum wire reference electrode is used to characterize the electrochemical behavior of the tip probes. In these experiments,

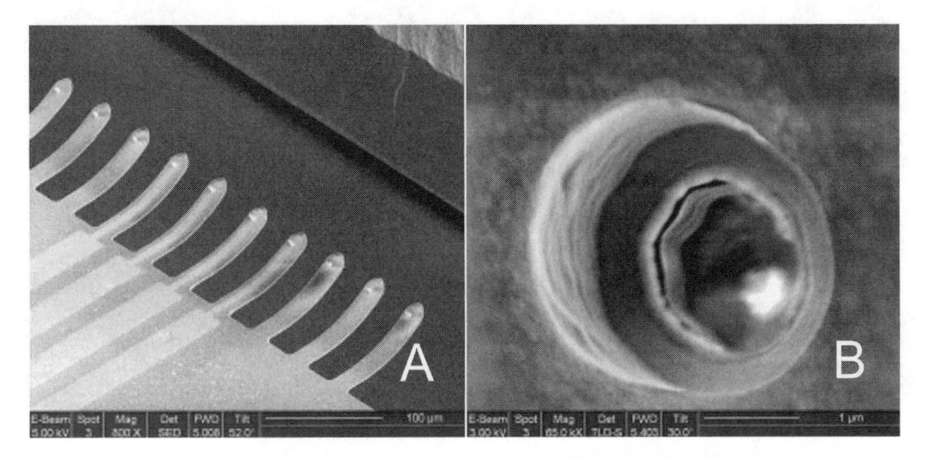

FIGURE 4.3 SEM pictures of a finished tip probe array: (A) tip probe array with four UME cantilever structures; (B) tip probe of the cantilever structures with a platinum UME (radius of 200 nm).

FIGURE 4.4 Optical low magnification micrographs of an SECM-AFM probe device.

phosphate buffer is used as an electrolyte solution. The tip probe system shows all the important electrochemical surface reactions of a platinum electrode in a phosphate buffer solution. Hexaamineruthenium (III) chloride is used to study the response of the electrode in a reversible redox system. Figure 4.6 shows voltammograms taken in an electrolyte with 10 mmol and 0 mmol concentration of $Ru(NH_3)_6^{3+}$. The tip probes in these experiments have an approximate electrode area of 0.125 μm^2. The reduction current appears in the potential regime below –200 mV related to Ag/AgCl at 0.1 mol KCl. Electrochemical impedance spectroscopy (EIS) analysis leads to a faradic impedance of the tip transducer system at a DC working potential of –300 mV of 5×10^7 Ohm for 10 mM $Ru(NH_3)_6^{3+}$ and 4×10^9 Ohm for phosphate

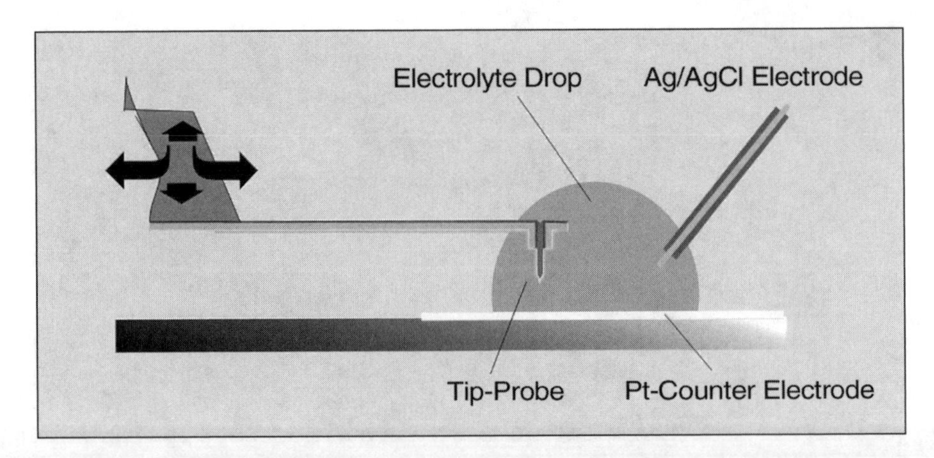

FIGURE 4.5 Schematic electrode arrangement of an electrochemical measurement setup.

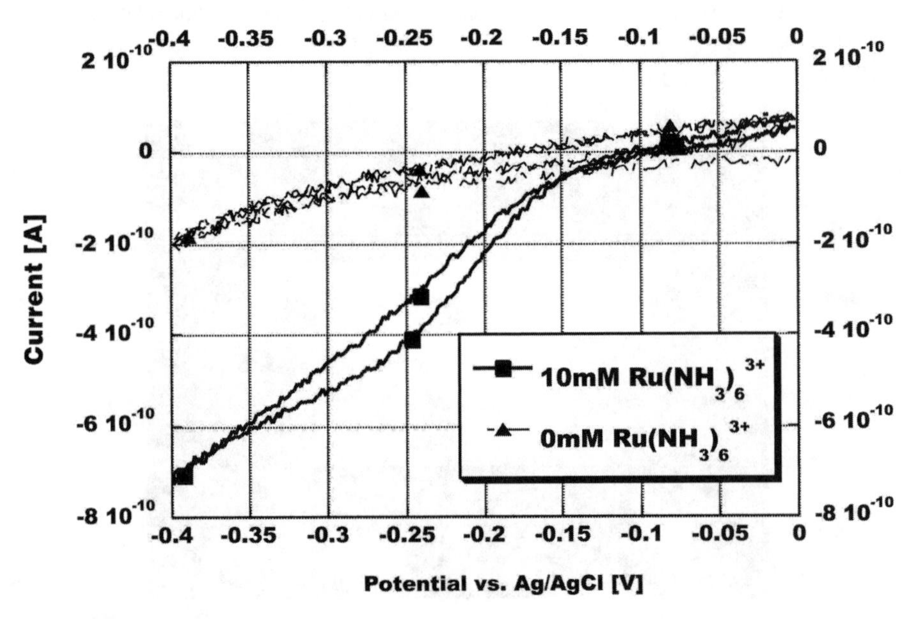

FIGURE 4.6 Cyclic voltammograms of tip probe with a platinum UEM (0.125 μm^2) in 0.1 M phosphate buffer electrolyte and in a 10 mmol $Ru(NH_3)_6Cl$ and 0.1 KCl solution. The potential is related to a 0.1 M KCl Ag/AgCl-Reference electrode. Potential sweep rate was 10 mV/s.

buffer only. Both the behavior of the reduction current and values of the faradic impedance in a $Ru(NH_3)_6^{3+}$ redox system show typical characteristics of a single ultra-micro electrode in submicron size and indicate the electrochemical functionality of the tip probe system. This electrochemical, platinum, nano-sized transducer system can be directly applied to measure ion fluxes or electrochemically active species within and between cells. Furthermore, the transducer system can be modified to detect specific ions such as sodium, potassium, or substances involved in cellular metabolic reactions or signaling pathways.[29]

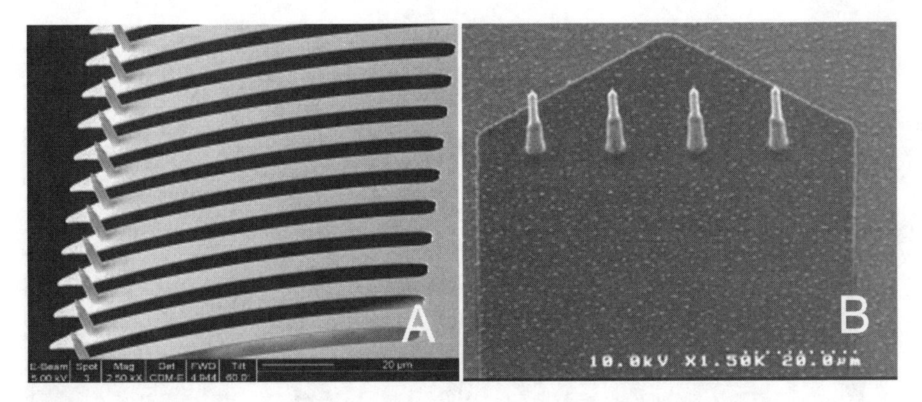

FIGURE 4.7 (A) Silicon nitride cantilever array with single high-aspect-ratio tips. The thickness of cantilevers is about 500 nm. (B) High-aspect-ratio tip array on single silicon nitride cantilever.

4.4 PROBE ARRAYS

4.4.1 Silicon Nitride Cantilever Array with Single Tips and Tip Arrays

Obtaining measurements with high spatial resolutions requires increased probe densities. Arrays of multiprobe systems on a single cantilever were designed and fabricated as shown in Figure 4.7. Silicon tips with spacings of 10 μm are embedded in silicon nitride pillars and supported by a 500-nm-thick silicon nitride cantilever. The electrically conductive tips will be connected separately to the monitoring system such that individual signals from each tip can be transmitted. The metal lines are sputtered on an isolated substrate surface and covered with another insulation layer to define the UME on the tip of the probe structure as described previously. The cantilever and its substrate are compatible with conventional AFM (atomic force microscope) systems. The probe or probe array can be mounted onto an AFM scanner to take advantage of the precise closed-loop motion control and force feedback data of the AFM system.

4.4.2 Multiple Arrays of Ultrasharp High-Aspect-Ratio Silicon Tips

Multiple arrays of ultrasharp nanotips with tip radii as small as 20 are currently being fabricated by combining two plasma etching processes on single crystalline wafers. As shown in Figure 4.8, nanotips located on top of the pillars respectively have an aspect ratio of over 10:1. Such a high density of ultrasharp tips represents a potentially powerful tool for measuring cell properties in real time. The tips can be inserted into cells such that multiple electrodes are present within a single cell or with a single tip in adjacent cells. The electrically active tips will be individually connected by conductive vias to CMOS transistors on the back side of the wafer. In this way, both individual addressability and signal processing are achieved. The

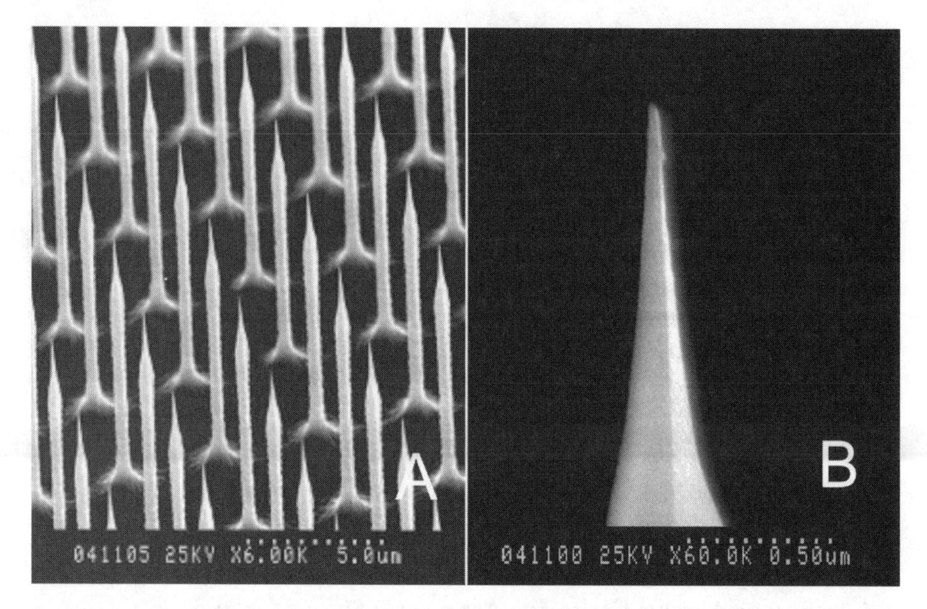

FIGURE 4.8 (A) Multi-array ultra sharp silicon tips fabricated with a combination of two plasma etching steps. (B) A closer look at the individual silicon tip with a radius less than 20 nm.

tip is defined so that most of the tip is insulated and only a small region at the top has an exposed metal electrode. This will provide a well-defined tip geometry with nanometer dimensions to further ensure localized measurements of dielectric behavior within the cell interior. The fabricated nanotip arrays on cantilever structures integrate well with the AFM stage for precise positioning relative to cell surface and subsequent penetration of the cell membrane.

4.4.3 Fabrication of Silicon Via Structures

Limitations of connection density of current technologies are imposed by space restrictions resulting from designing and fabricating both connection structures (bonding pads, flip chip pumps) and functional devices (MEMS or CMOS) on a single wafer side. In order to overcome these limitations and open the use of both sides of a wafer independently, electrical through-wafer interconnections are necessary. The through-wafer interconnection concept is based on multilayer deposition techniques. Openings in a double-sided polished wafer are created by applying a high-density inductively coupled plasma (ICP) etch technique.[30] Hole structures with a diameter of 20 μm are formed through a 350-μm thick wafer. A multilayer system of up to eight layers consisting of alternating conducting layers (N-type doped poly-silicon) and isolating layers (silicon-oxide) are grown until the vias are filled. The silicon-oxide and poly-silicon layers are grown using a low-pressure chemical vapor deposition process to achieve a high deposition uniformity. Subsequently, all layers on the wafer surface are removed in a chemical mechanical polishing process. In this way, a multiconnection, through-wafer structure can be fabricated. Figure 4.9 shows an example of such a

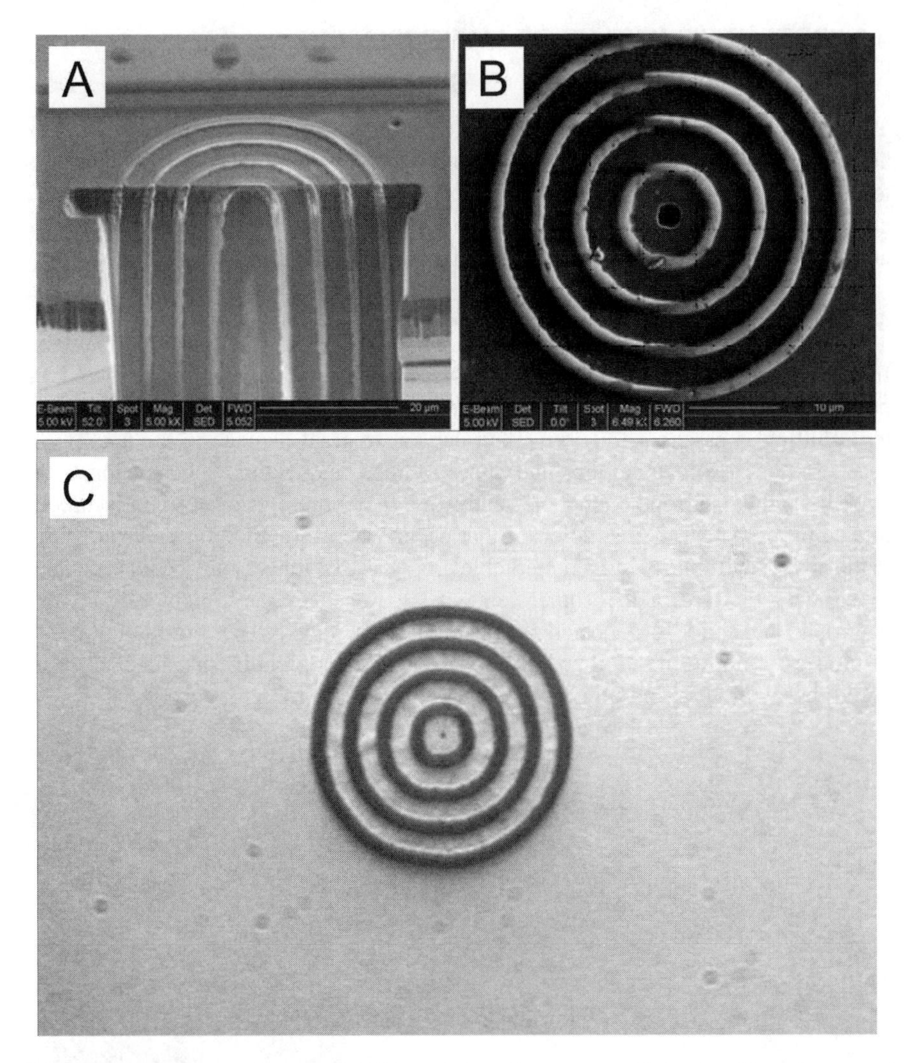

FIGURE 4.9 A 40-µm-diameter ETWI consisting of four 2-µm-thick poly-silicon layers and four 1-µm-thick thermal silicon-oxide layers: (A) SEM image of the cross section; (B) SEM image of the top view (poly-silicon and silicon appears as dark, silicon-oxide appears as bright); and (C) optical microscope image of the top view (poly-silicon and silicon appears as bright, silicon-oxide appears as dark).

multilayered interconnection structure. The applied low-pressure chemical vapor deposition techniques guarantee a sufficient homogenous coating outside and inside of the entire structure to a minimum layer thickness of one micrometer. The connection quality has been examined combining impedance spectroscopy and focused ion beam technology. Depending on the geometry and the doping profile of the poly-silicon layers, a connection resistance of less than 80 ohms can be achieved with sufficient DC isolation. This technique is compatible with high-temperature processes and is suitable for MEMS as well as CMOS applications.

4.5 POTENTIAL IMPACT OF NANOPROBE MEASUREMENTS

The probes can contribute to a fundamental understanding of cell signaling mechanisms at unprecedented levels of spatial and temporal resolution. Two examples of terminally differentiated cell systems that can be probed with the nanotips are chondrocytes and neutrophils. Figure 4.10 shows a specific example of a tip probing an individual chondrocyte with an ultrasharp planer silicon nitride tip probe.

4.5.1 NEUTROPHILS

In the complex repertoire of host defenses, neutrophils play a central role with both positive and negative consequences for the organism. The neutrophil's role in bacterial killing is essential to survival but can cause great damage to the host in the case of trauma, organ failure, and sepsis. Greco and Haimovich developed the hypothesis that neutrophil adherence/activation to nonbiological substrata is directly related to the defect in host defenses that occur when foreign bodies are implanted in the host. Neutrophils have a remarkable propensity to adhere and become activated by surfaces that are inert to other cell types.[31] Recent research demonstrates that neutrophil adhesion to uncoated or plasma coated polystyrene, glass, Teflon®, or Dacron®, but not to immobilized fibrinogen or immunoglobulin, triggers cell death within hours *in vitro* and *in vivo*.[32]

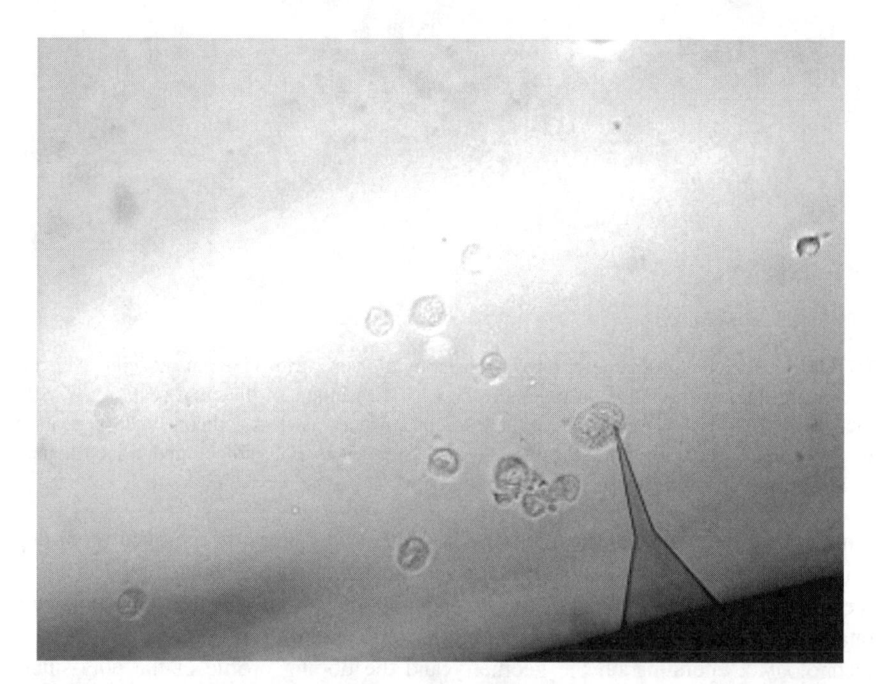

FIGURE 4.10 Ultrasharp planer silicon nitride tip fabricated with a combination of microfabrication technique and FIB (focused ion beam) techniques penetrating individual chondrocytes.

A number of investigators have studied the biochemical basis of neutrophil activation and adhesion.[33–35] Polymorphonuclear leukocytes (PMNs) residing on immobilized fibronectin have been shown to respond to TNF-α with an intense Cl^- efflux that leads to a decrease in the unusually high resting Cl^- content of phagocytes. The finding that Cl^- efflux depends on $\beta 2$ integrin engagement suggests that PMN agonists are also required to trigger Cl^- efflux. Anti-CD18 monoclonal antibodies trigger a marked release of Cl^- ions, which is accompanied by neutrophil spreading and respiratory burst. Cl^- efflux appears to be independent of either calcium ion signaling or plasma membrane potential but is sensitive to changes in pH. Thus, it is now known that:

- Cross-linking of $\beta 2$ integrins triggers Cl^- efflux.
- $2Cl^-$ efflux regulates neutrophil spreading and respiratory burst.
- Cl^- efflux is independent of calcium ion alterations and changes in plasma membrane potential.
- The activation of Cl^- efflux may be regulated by a rise in pH and is independent of protein tyrosine phosphorylation and cAMP.

In the past, measurement of pH was a static, discrete measurement. Improvements described recently include pH-responsive probes in size fractionated liposomes and transfection of pH-sensitive variants of green fluorescent protein (GFP).[36,37] Grinstein and others utilizing macroscale techniques have shown that cell spreading of neutrophils on substrates results in rapid cytosolic alkalinization mediated by Na^+/H^+ exchange and Cl^- efflux associated with pH changes.[35] Nanotip arrays can be used to study cell signaling during this obligate neutrophil pathway and determine the relationship between CD18 receptor engagement and subsequent changes in pH and ion flux. This investigation can elucidate pathways that lend themselves to manipulation in order to obviate or enhance various neutrophil functions.

4.5.2 Chondrocytes

Chondrocytes respond to a variety of loading conditions both *in vitro* and *in vivo*.[38–40] Experimental studies in animal models and human clinical trials confirm an essential role for mechanical loading in the preservation of cartilage.[41–46] Finite element studies of cartilage loading predict that deviatoric and dilatational stresses influence matrix thickness.[47–50] Smith et al. demonstrate that shear stress and intermittent hydrostatic pressure applied *in vitro* alter chondrocyte gene expression.[51–55] Despite the lack of data directly linking the effects of these applied loads to ion flux, chondrocyte cellular signaling resulting from exposure to mechanical stress appears to involve electrical gradients and changes in intracellular ion densities.[56–58] Investigations of mechanical loading effects on cartilage implicate Na^+/H^+ flux and shifts in pH as factors in chondrocyte metabolism.[59–61] In calf articular cartilage, external acidity, pHo, alters intracellular pH and reduces matrix synthesis.[62] Articular chondrocytes exposed to increased load also show decreased extracellular pH affecting an increase in extracellular cation concentration and osmolality.[63,64] In addition to intracellular changes in pH, stress on the cells or the localized outer surface of the cells may be associated with changes in the cytoskeleton and intracellular electrical potentials.[65] The precise measurement of intracellular potential differences due to applied loads has not been attained. Mechanical load induced changes in intracellular ions and pH can be examined in greater depth than current methodologies by using a nanoprobe array to correlate ion flux with chondrocyte extracellular matrix synthesis.

4.6 SUMMARY AND FUTURE DIRECTIONS

An electrochemical cantilever transducer system with platinum electrodes in submicron regimes was developed. Electrochemical investigations showed full functionality of the probe system. Due to the high-aspect-ratio topography of the tip structure and low spring constant of silicon nitride cantilevers, these probes are particularly well suited for combined high-resolution SECM and AFM analysis in living cells. Furthermore, this technology allows for a production of both linear probe arrays and two-dimensional probe arrays.

Probe arrays with characteristic feature dimensions well below the cell diameter are expected to provide new insights into the dynamics of cell metabolism. Such insights will include rate, magnitude, and localization of oxidation–reduction reactions (ORRs) in the cytosol and possibly inside organelles. The probes are anticipated to reveal ORRs adjacent to the inside surface of the cell membrane, including charge transfer resistance and membrane impedance. Challenges are anticipated during insertion of the probe into the cell interior without inflicting damage to the cell structure or chemical composition of the cytosol, and further research on probe surface coating to minimize such damage is envisioned. In particular, atomic layer deposition (ALD) processing is envisioned to play a role for surface coating in future probe generations. Ultrathin material layers, furthermore, allow for the fabrication of ion selective membranes with high species sensitivity and short response times. Ultimately, these probes can be used to transduce signals from intracellular ion responses in many cell systems including chondrocytes and polymorphonuclear leukocytes.

REFERENCES

1. Sakmann, B., and Neher E., *Single Channel Recording*, 2nd ed., Plenum Press, New York, 1995.
2. Neher, E., and Marty, A., Discrete changes of cell membrane capacitance observed under conditions of enhanced secretion in bovine adrenal chromaffin cells, *Proc. Natl. Acad. Sci. U.S.A.*, 79, 6712–6716, 1982.
3. Seksek, O., Biwersi, J., and Verkman, A.S., Direct measurement of trans-Golgi pH in living cells and regualtion by second messengers, *J. Biol. Chem.*, 270, 4967–4970, 1996.
4. Chow, R.H., Kingauf, J., and Neher, E., Time course of Ca^{2+} concentration triggers exocytosis in neuroendocrine cells, *Proc. Natl. Acad. Sci. U.S.A.*, 91, 12765–12769, 1994.

5. Albillos, A., Dernick, G., Horstmann, H., Almers, W., Alvarez de Toledo, G., and Lindau, M., The exocytotic event in chromaffin cells revealed by patch amperometry, *Nature*, 389, 509–512, 1997.
6. Hamill, O.P., and McBride, D.W., Jr., Induced membrane hypo/hyper-mechanosensitivity, *Annu. Rev. Physiol.*, 59, 621–631, 1997.
7. Ropke, M., Unmack, M.A., Willumsen, N.J., and Frederiksen, O., Comparative aspects of actions of a short-chain phospholipid on epithelial Na+ channels and tight junction conductance, *Comp. Biochem. Physiol. A Physiol.,* 118, 211–214, 1997.
8. Van Itallie, C., Rahner, C., and Anderson, J.M., Regulated expression of claudin-4 decreases paracellular conductance through a selective decrease in sodium permeability, *J. Clin. Invest.,* 107, 1319–1327, 2001.
9. Olovnikov, A.M., Towards the quantitative traits regulation, fountain theory implications in comparative and developmental biology, *Int. J. Dev. Biol.*, 41, 923–931, 1997.
10. Becq, F., Hamon, Y., Bajetto, A., Gola, M., Verrier, B, and Chimini, G., ABC1, an ATP binding cassette transporter required for phagocytosis of apoptotic cells, generates a regulated anion flux after expression in *Xenopus laevis* oocytes, *J. Biol. Chem.*, 272, 2695–2699, 1997.
11. Bard, A.J., Fan, F.-R.F., Pierce, D.T., Unwin, P.R., Wipf, D.O., and Zhou, F., Chemical imaging of surfaces with the scanning electrochemical microscope, *Science*, 254, 68–74, 1991.
12. Mirkin, M.V., High resolution studies of heterogeneous processes with the scanning electrochemical microscope, *Mikrochim. Acta*, 130, 127–153, 1999.
13. Bard, A.J., Fan, F.-R., and Mirkin, M.V., Scanning electrochemical microscope, in *Electroanalytical Chemsitry*, 18, A.J. Bard, Ed., Marcel Dekker, New York, 1994, 243–273.
14. Fan, F.-R.F., and Demaille, C., The preparation of tips for scanning electrochemical microscopy, in *Scanning Electrochemical Microscopy*, A.J. Bard and Michael V. Mirfin, Eds., Marcel Dekker, New York, 2001, 75–107.
15. Shao, Y., and Mirkin, M.V., Nanometer-sized electrochemical sensors, *J. Anal. Chem.*, 69, 1727–1634, 1997.
16. Sun, P., Zhang, Z., Guo, J., and Shao, Y., Fabrication of nanometer-sized electrodes and tips for scanning electrochemical microscopy, *J. Anal. Chem.,* 73, 5346–5351, 2001.
17. Fasching, R., Tao, Y., Hammerick, K., and Prinz, F., A pencil probe system for electrochemical analysis and modification in nanometer dimension, *Proc. SPIE, 5116, Smart Sensors, Actuators, and MEMS — Microtechnologies for the New Millenium*, Maspalonas, Gran Canaria, Spain, 128–135, 2003.
18. Bonnell, D., Ed., *Scanning Probe Microscopy and Spectroscopy, Theory, Techniques and Applications*, Wiley-VHX, New York, 2000.
19. Ludwig, M., Kranz, C., Schumann, W., and Gaub, H.E., Topography feed mechanism for the scanning electrochemical microscope based on hydrodynamic forces between tip and sample, *Rev. Sci. Intrum.,* 66, 2857–2860, 1995.
20. Macpherson, J.V., and Unwin, P.R., Combined scanning eletrochemical-atomic force microscope, *J. Anal. Chem.,* 72, 276–285, 2000.
21. James, P.J., Garfias-Mesias, L.F., Moyer, P.J., and Smyrl, W.H., Scanning electrochemical microscopy with simultaneous independent topography, *J. Electrochem. Soc.*, 145, L64, 1998.
22. Buechler, M., Kelley, S.C., and Smyrl, W.H., Scanning electrochemical microscopy with shear force feedback investigation of the lateral resolution of different experimental configurations, *Electrochem. Solid State Lett.,* 3, 35–38, 2000.

23. Kranz, C., Friedbacher, G., Mizaikoff, B., Lugstein, A., Smoliner, J., and Bertagnolli, E., Integrating an ultramicroelectrode in an AFM cantilever: combined technology for enhanced information, *J. Anal. Chem.,* 73, 2491–2500, 2001.
24. Shi, L., Kwon, O., Miner, A.C., and Majumdar, A., Design and batch fabrication of probes for sub-100nm scanning thermal microscopy, *J. Microelectromech. Syst.*, 10 (3), 370–378, 2004.
25. Fasching, R., Tao, Y., and Prinz, F., Fabrication of an electrochemical tip-probe system embedded in SiNx-cantilevers for simultaneous SECM and AFM analysis, *Proc. SPIE, 5341, Micromachining and Microfabrication Process Technology IX*, West Photonica, CA, 53–64, 2004.
26. Boisen, A., Hansen, O., and Boutwstra, S. AFM probes with directly fabricated tips, *J. Micromech. Microeng*., 6, 58–62, 1996.
27. Grow, R.J., Minne, S.C., Manalis, S.R., and Quate, C.F., Silicon nitride cantilevers with oxidation-sharpened silicon tips for atomic force microscopy, *J. Microelectromech. Syst.*, 11 (4), 317–321, 2002.
28. Marcus, R.B., Ravi, T.S., Gmitter, T., Chin, K., Lui, D., Orvis, W.J., Ciarlo, D.R., Hunt, C.E., and Trujillo, J., Formation of silicon tips with $<$ 1nm radius, *J. Appl. Phys. Lett*., 56, 236–238, 1990.
29. Wang, J., in *Analytical Electrochemistry*, 2nd ed., 171–197, Wiley-VCH, NewYork, 2000.
30. Klaasen, E.H., Petersen, K., Noworolski, J.M., Logan, J., Maluf, N.I., Brown, J., Storment, C., McCulley, W., and Kovacs, G.T.A., Silicon fusion bonding and deep reactive ion etching: a new technology for microstructures, *Sensors and Actuators A-Physical,* 52, 132–139, 1996.
31. De La Cruz, C., Haimovich, B., and Greco, R.S., Immobilized IgG and fibrinogen differentially affect the cytoskeletal organization and bactericidal function of adherent neutrophils, *J. Surg. Res*., 80, 28–34, 1998.
32. Nadzam, G.S., De La Cruz, C., Greco, R.S., and B. Haimovich. Neutrophil adhesion to vascular prosthetic surfaces triggers nonapoptotic cell death, *Ann. Surg*., 231, 587–599, 2000.
33. Menegazzi, R., Busetto, S., Decleva, E., Cramer, R., Dri, P., and Patriarca, P., Triggering of chloride ion efflux from human neutrophils as a novel function of leukocyte β2 integrins: relationship with spreading and activation of the respiratory burst, *J. Immunol*., 162, 423–434, 1999.
34. Demaurex, N., Downey, G.P., Waddell, T.K., and Grinstein, S., Intracellular pH regulation during spreading of human neutrophils, *J. Cell Biol*., 133, 1391–1402, 1996.
35. Aharonovitz, O., Zaun, H. C., Balla, T., York, J.D., Orlowski, J., and Grinstein, S., Intracellular pH regulation by Na^+/H^+ exchange requires phosphatidylinositol 4,5-bisphosphate, *J. Cell Biol*., 150, 213–224, 2000.
36. Seksek, O., Biwersi, J., and Verkman, A.S., Evidence against trans-Golgi acidification in cystic fibrosis, *J. Biol. Chem*., 271, 15542–15548, 1996.
37. Llopis, J., McCaffery, J.M., Miyawaki, A., Farquhar, M.G., and Tsien, R.Y., Measurement of cytosolic, mitochondrial, and Golgi pH in single living cells with green fluorescent proteins, *Proc. Natl. Acad. Sci. U.S.A*., 95, 6803–6808, 1998.
38. Wong, M., Wuethrich, P., Buschmann, M.D., Eggli, P., and Hunziker, E., Chondrocyte biosynthesis correlates with local tissue strain in statically compressed adult articular cartilage, *J. Orthop. Res.,* 152, 189–196, 1997.
39. Hall, A.C., Urban, J.P., and Gehl, K.A., The effects of hydrostatic pressure on matrix synthesis in articular cartilage, *J. Orthop. Res.*, 91, 1–10, 1991.

40. Akeson, W.H., Amiel, D., Abel, M.F., Garfin, S.R., and Woo, S.L., Effects of immobilization on joints, *Clin. Orthop.*, 219, 28–37, 1987.
41. Radin, E.L., Parker, H.G., Pugh, J.W., Steinberg, R.S., Paul, I.L., and Rose, R.M., Response of joints to impact loading. 3. Relationship between trabecular microfractures and cartilage degeneration, *J. Biomech.*, 61, 51–57, 1973.
42. McDevitt, C.A. and Muir, H., Biochemical changes in the cartilage of the knee in experimental and natural osteoarthritis in the dog, *J. Bone Joint Surg. Br.*, 581, 94–101, 1976.
43. Langenskiold, A., Michelsson, J.E., and Videman, T., Osteoarthritis of the knee in the rabbit produced by immobilization. Attempts to achieve a reproducible model for studies on pathogenesis and therapy, *Acta Orthop. Scand.*, 501, 1–14, 1979.
44. Videman, T., Eronen, I., and Friman, C., Glycosaminoglycan metabolism in experimental osteoarthritis caused by immobilization. The effects of different periods of immobilization and follow-up, *Acta Orthop. Scand.*, 521, 11–21, 1981.
45. Moskowitz, R.W., Goldberg, V.M., and Malemud, C.J., Metabolic responses of cartilage in experimentally induced osteoarthritis, *Ann. Rheum. Dis.*, 406, 584–592, 1981.
46. Pelletier, J.P., Martel-Pelletier, J., Ghandur-Mnaymneh, L., Howell, D.S., and Woessner, J.F., Role of synovial membrane inflammation in cartilage matrix breakdown in the Pond-Nuki dog model of osteoarthritis, *Arthritis Rheum.*, 285, 554–561, 1985.
47. Carter, D.R., Orr, T.E., Fyhrie, D.P., and Schurman, D.J., Influences of mechanical stress on prenatal and postnatal skeletal development, *Clin. Orthop.*, 219, 237–250, 1987.
48. Carter, D.R., Rapperport, D.J., Fyhrie, D.P., and Schurman, D.J., Relation of coxarthrosis to stresses and morphogenesis. A finite element analysis, *Acta Orthop. Scand.*, 586, 611–619, 1987.
49. Carter, D.R., and Wong, M., The role of mechanical loading histories in the development of diarthrodial joints, *J. Orthop. Res.*, 66, 804–816, 1988.
50. Wong, M., and Carter, D.R., Theoretical stress analysis of organ culture osteogenesis, *Bone,* 112, 127–131, 1990.
51. Smith, R.L., Trindade, M.C., Ikenoue, T., Mohtai, M., Das, P., Carter D.R., Goodman, S.B., and Schurman, D.J., Effects of shear stress on articular chondrocyte metabolism, *Biorheology*, 371, 95–107, 2000.
52. Mohtai, M., Gupta, M.K., Donlon, B., Ellison, B., Cooke, J., Gibbons, G., Schurman, D.J., and Smith, R.L., Expression of interleukin-6 in osteoarthritic chondrocytes and effects of fluid-induced shear on this expression in normal human chondrocytes *in vitro*, *J. Orthop. Res.*, 141, 67–73, 1996.
53. Smith, R.L., Donlon, B.S., Gupta, M.K., Mohtai, M., Das, P., Carter, D.R., Cooke, J., Gibbons, G., Hutchinson, N., and Schurman, D.J., Effects of fluid-induced shear on articular chondrocyte morphology and metabolism *in vitro*, *J. Orthop. Res.*, 136, 824–831, 1995.
54. Smith, R.L., Rusk, S.F., Ellison, B.E., Wessells, P., Tsuchiya, K., Carter, D.R., Caler, W.E., Sandell, L.J., and Schurman, D.J., *In vitro* stimulation of articular chondrocyte mRNA and extracellular matrix synthesis by hydrostatic pressure, *J. Orthop. Res.*, 141, 53–60, 1996.
55. Das, P., Schurman, D.J., and Smith, R. L., Nitric oxide and G proteins mediate the response of bovine articular chondrocytes to fluid-induced shear, *J. Orthop. Res.*, 151, 87–93, 1997.
56. Urban, J.P., Present perspectives on cartilage and chondrocyte mechanobiology. *Biorheology*, 37, 185–190, 2000.

57. Wilkins, R.J., Browning, J.A., and Urban, J.P., Chondrocyte regulation by mechanical load, *Biorheology*, 37, 67–74, 2000.
58. Wang, D., Canaff, L., Davidson, D., Corluka, A., Liu, H., Hendy, G.N., and Henderson, J.E., Alterations in the sensing and transport of phosphate and calcium by differentiating chondrocytes, *J. Biol. Chem.*, 276, 33995–34005, 2001.
59. Browning, J.A., Walker, R.E., Hall, A.C., and Wilkins, R.J., Modulation of Na^+ x H^+ exchange by hydrostatic pressure in isolated bovine articular chondrocytes, *Acta Physiol. Scand.*, 166, 39–45, 1999.
60. Wilkins R.J., and Hall, A.C., Control of matrix synthesis in isolated bovine chondrocytes by extracellular and intracellular pH, *J. Cell Physiol.*, 164, 474–481, 1995.
61. Dascalu, A., Korenstein, R., Oron, Y., and Nevo, Z., A hyperosmotic stimulus regulates intracellular pH, calcium, and S-100 protein levels in avian chondrocytes, *Biochem. Biophys. Res. Commun.*, 227, 368–373, 1996.
62. Gray, M.L., Pizzanelli, A.M., Grodzinsky, A.J., and Lee, R.C., Mechanical and physiochemical determinants of the chondrocyte biosynthetic response, *J. Orthop. Res.*, 6, 777–792, 1988.
63. Garcia, A.M., Black, A.C., and Gray, M.L., Effects of physicochemical factors on the growth of mandibular condyles *in vitro*, *Calcif. Tissue Int.*, 54, 499–504, 1994.
64. Urban, J.P., The chondrocyte: a cell under pressure, *Br. J. Rheumatol.*, 33, 901–908, 1994.
65. Lee, H.S., Millward-Sadler, S.J., Wright, M.O., Nuki, G., and Salter, D.M., Integrin and mechanosensitive ion channel-dependent tyrosine phosphorylation of focal adhesion proteins and beta-catenin in human articular chondrocytes after mechanical stimulation, *J. Bone Miner. Res.*, 15, 1501–1509.

5 Synthesis of Cell Structures

Kyle Hammerick, WonHyoung Ryu, Rainer Fasching, Seoung-Jai Bai, R. Lane Smith, Ralph S. Greco, and Fritz B. Prinz

CONTENTS

5.1 INTRODUCTION

The shortage of organs for transplantation has provided the impetus for increasing investigation into the development of artificial organs. Organs themselves consist of three-dimesional structures made up of cells with different degrees of specialization, an extracellular matrix, and a variety of tubes and ducts to transport their products and waste into the blood stream. One can imagine the need for an artificial pancreas

0-8493-1940-4/05/$0.00+$1.50

to treat patients with type I diabetes mellitus, an artificial liver to treat acute fulminant hepatitis and as a bridge to transplantation for other liver diseases, an artificial kidney to replace dialysis, artificial hearts and lungs, artificial skin, artificial endocrine glands, artificial reproductive organs, and artificial vision. The inherent complexity of each of these devices is related to both the incorporation of cells and our ability to create three-dimensional structures for them. Autologous cells, which would not require immunosuppression, create a need for complex processes to harvest the host's own cells, when available, and require additional interventions. The use of homologous cells or xenogeneic cells creates a level of complexity, immunologically, for which there clearly is not a simple solution. Nevertheless, the quest for artificial organs will continue unabated, and the possibility that microfabrication and nanotechnology can contribute to this process is substantial.

5.2 MULTIDIMENSIONAL ORGANIZATION OF TISSUES AND ORGANS

The functional properties of tissues and organs represent the summation of activities of the various types of differentiated cells within a three-dimensional extracellular matrix. Early studies on the culture of cells removed from the normal *in vivo* environment introduced the concepts of the necessity for specific nutrients, vitamins, and gaseous exchange.[1] Knowledge regarding various nutrient demands was followed by the discovery of soluble growth factors accessible to cells within tissues that were critical for inducing and maintaining specialized cell functions.[2] However, in some cases, the delivery of the growth factor alone was insufficient to support cell function, and it became apparent that macromolecules within the extracellular matrix were required to establish the differentiated state.[3] One prominent example of the necessity for cell-to-cell contact has emerged from immunology, where it has been shown that immune cell differentiation of T cells does not proceed in the absence of occupancy of cell surface receptors for some period of time.[4] Other recent studies show that stem cell differentiation is dependent on interactions between cell surface receptors and extracellular matrix proteins that create the proper microenvironment for cells. For instance, the expansion of hematopoietic stem cells in an irradiated animal requires homing of the cells to the appropriate stromal cell microenvironment.

The extracellular macromolecules that support the cellular specialization are often derived from the cells themselves. In most tissues, collagen serves as a major element in the extracellular matrix by providing tensile strength to the tissue. Although type I collagen is predominant in skin, bone, and tendon, the types of collagen vary from among the more than twenty genes coding for variants that contribute to organization of the extracellular matrix. Other molecules combine with the collagens to form specialized extracellular structures that contribute to tissue organization, including laminin, vitronectin, cell-surface proteoglycans, and fibrillins.[5] Awareness of the role of extracellular matrix in cellular differentiation has generated a subspecialty within biomaterial science that focuses on the development of three-dimensional scaffolds for cellular deposition. The current state of fabrication remains relatively crude when compared to the molecular heterogeneity that is established during embryogenesis. In the embryo, three discrete germ layers form as the developing cells undergo a series of transformations from epithelial tissue with cell–cell links to mesenchyme tissue. As development continues, the cells that are free to migrate often shift back from loosely held contacts to form epithelial sheets that exhibit folding, invagination, and evagination. The results of these transitions create complicated structures of mixed types of cells and structures. Examples of this include the pancreas, liver, and connective tissues such as cartilage and ligaments.

The challenge facing biologists, materials scientists, and engineers in creating regenerative tissues and organs lies in establishing cellular microenvironments with pathways for cellular communication that enhance cell function. Knowledge gained from genetic experiments in which extracellular matrix molecules, integral membrane proteins, and intracellular signaling molecules have been either knocked out or over expressed confirms that competency for cell differentiation and phenotypic expression requires each of these three levels of cellular connectivity. A close examination of functional attributes of pancreas, liver, and cartilage are demonstrative of ways in which fabrication of materials that establish atomic, molecular, and cell-sized surface features that mimic natural ligands or position tissue structures will be necessary to stimulate cellular function. A number of studies show that loss of either extracellular matrix proteins or cell surface receptors such as integrins leads to malformation or lack of function of a number of different organs. Therefore, design of tissue-engineered substrates will require either surfaces that naturally adhere extracellular matrix molecules or reproduce the high affinity binding sites for these cell-associated receptors to reproduce the natural tissue organization observed in the pancreas, liver, and cartilage, for example.

5.2.1 Pancreas

The first of these tissues to be examined, the pancreas, is an example where two major functions are provided for the organism. Each function is a product of selective organization of differentiated cell types. The alkaline secretions that are released following a meal aid in the breakdown of proteins, carbohydrates, fats, and nucleic acids. In addition, the gland produces two hormones, insulin and glucagon, that regulate carbohydrate metabolism. The pancreas is composed of a loose connective tissue stroma that is interposed among the epithelial acinar cells. The acinar cells have polarity so that secretions are released into a series of intralobular ducts that merge together prior to emptying into the lumen of the duodenum. The primary endocrine function of the pancreas is provided by islets of Langerhans, which are isolated groups of cells present within the body of the pancreas that release insulin, glucagon, and somatostatin. The cells of the islets are clustered together, and the groups are scattered among the acinar cells. The cells of the islets can be subdivided into four types that are distinguished by size, electron density, and granule content. The α cells are associated with release of glucagon, the β cells release insulin, the δ cells release somatostatin, and the PP cells release the pancreatic polypeptide. The close proximity of each of the different cell types leads to interdependence in the regulation of α, β, and δ cell products. The coordinated regulation of the levels of glucagon, insulin, and somatostatin then contributes to physiological homeostasis with respect to glucose metabolism in the whole organism. Loss of this functional regulation is the basis of type I diabetes, where autoimmune processes have resulted in islet cell destruction.

5.2.2 Liver

In contrast to the pancreas, the liver is the largest compound gland in the body. It is the primary site for detoxification of circulating substances and serves as a reservoir for carbohydrates, a source of glucose, plasma proteins, and lipoproteins. The liver also exhibits secretory activity through the release of bilirubin, IgA, and bile salts. The predominant cells of the liver, the hepatocytes, are secretory epithelial cells. Hepatocytes are organized into rows that converge toward a central vein creating sinusoids where the cell surface is exposed to blood plasma. The proximity of the cells to blood plasma facilitates the uptake of compounds and release of secretory products. Recent studies show that hepatocyte function can be improved in a coculture system in which the hepatocytes are placed in proximity to fibroblasts. Interactions between the hepatocytes and fibroblasts increase albumin production and urea synthesis when compared to the cultures of hepatocytes maintained alone. These observations are consistent with a tissue organization where plates of hepatocytes are localized in a cage of supporting reticuloendothelial cells. The reticuloendothelial cell meshwork includes endothelial cells that form the walls of the sinusoids, macrophages that are anchored in the sinusoidal space, and stellate cells or lipocytes that store fat. The reticuloendothelial cells communicate with the hepatocytes and with each other, in part by release of cytokines such as hepatocyte growth factor. A major effort in attempts to stimulate liver regeneration is tied to the development of substrates that will maintain the hepatocytes within a plate-like structure and reduce the propensity for the cells to form spheroids that cause hypoxia and prevent nutrient flow to the interior cells. Technology for development of three-dimensional scaffolds will contribute greatly to liver regeneration either for tissue-engineering applications or for *in vivo* implantation to aid in recruitment of putative liver stem cells.

5.2.3 Cartilage

The development of cartilage, bone, and tendon for repair of damaged tissues provides a demonstration of the importance of extracellular matrix in tissue specialization. The use of autologous chondrocyte implantation for cartilage repair has demonstrated the need for adequate support scaffolds.[6] In this procedure, cartilage cells are isolated from a surgically collected specimen, expanded in culture, and placed back in the joint as an aggregate of cells.[7] The cells are held in place by a graft of periosteal tissue. In some cases, the outcomes show restoration of matrix, but often the implanted tissue degenerates to a fibrocartilage phenotype. One hypothesis for the change in matrix synthesis is that the isolated chondrocytes do not have an organized extracellular matrix that can support the loads arising within the joint.

As a result, the inappropriate loads may induce a shift in chondrocyte metabolism and alter the pattern of collagen expression. An alternative hypothesis is that the original cells may not survive and are replaced by cells that penetrate the subchondral bone to generate the fibrocartilage phenotype. To circumvent this problem, approaches are under development to use resorbable matrices that can be preloaded with mesenchymal stem cells to generate repair cartilage. The use of mesenchymal stem cells has been proposed for bone replacement and tendon repair under conditions where the scaffolds include ceramics and bioglasses to mimic mineralized matrix, or where collagen is used with other proteins to provide for tensile properties. The ultimate mechanical performance of the graft is strongly dependent on the structure and degradation of the resorbable matrices. Both of these properties can be tailored with micro- and nanoengineering efforts.

5.3 CELL PATTERNING

These examples of cellular organization in native tissues invite the question of how to impart organized structures to engineered constructs. It is apparent that recreating the complex biological organization demanded by natural tissues can benefit from technologies suited to highly redundant structures, vastly different length scales, and very small feature sizes. These biological systems require microscale and nanoscale engineering efforts with resolutions that are on the order of cellular and subcellular length scales. The positioning of cells or parts of cells with respect to cell structures or topologically complex substrates involves very high degrees of precision in initial patterning of cells and control of cell growth. At the minimum length scales, this can correlate to nanometers in the case of precise positioning of neurite growth cones and biomolecules responsible for cellular outgrowth. At the maximum required length scales, the precisions can be as large as hundreds or thousands of micrometers for placing capillary ducts relative to islets of Langerhans in pancreatic tissue engineering. Although traditional machining can be quite good at attaining feature sizes of several micrometers, reliable indexing at submicron length scales requires more advanced technologies traditionally seen in the semiconductor fabrication field of VLSI processing. These technologies more than allow for the creation of very small feature sizes but have also allowed for the creation of features spanning length scales five to seven orders of magnitude within the same device. The application of these advanced nano- and microfabrication methodologies with new shape transfer technologies and materials to biological structures have now made the precise positioning of cells very possible. This spatial control of tissue composition at the cellular and subcellular levels can be leveraged to provide the necessary complexity and physical interrelationships between cells for advanced tissue engineering endeavors. Moreover, many of the technologies currently available can also be used to pattern entities smaller than cells, such as proteins and single molecules, thereby creating precise arrangements of relevant biomolecules with respect to cells and multicellular structures. This can be used to regulate the chemical and biological environments surrounding cells and tissues at very precise scales.

5.3.1 Cell Patterning Method

Initial attempts to engineer spatial complexity into biological systems originated from the need to investigate the natural organization of cells and tissues during embryogenesis and tissue regeneration. More specifically, this research led to investigations of an individual cell's ability to respond to its microenvironment. Carter first pioneered a method of metallic evaporation to control axonal outgrowth in neurites. This method consisted of creating a substrate not conducive to cell adhesion in the form of cellulose acetate that was melt-affixed to glass surfaces. The acetate surfaces were subsequently evaporated with palladium and shadow masked in a number of ways to create two-dimensional patterns of cell adhesive material. The palladium surface created areas of stronger cell adhesion than the surrounding cellulose acetate. This allowed for the observation of cellular motility up a gradient of cell adhesive material, first termed haptotaxis by Carter.[8] These pioneering investigations were some of the first cell patterns and laid the basis for directing cell outgrowth through altering substratum adhesiveness. A number of other investigators have followed suit, trying to impart cellular organization through cell adhesion or directed cell outgrowth.[9–16] In addition to altering substratum adhesiveness, other researchers have looked toward mechanical cues to drive cell migration and orientation. Rovensky et al. used grooves as an initial attempt to guide fibroblast growth and proliferation, discovering that cell orientation could be influenced by purely mechanical factors.[17] These preliminary investigations set the stage for ensuing microengineering of individual cellular environments mechanically and chemically.

In more recent efforts, advanced cell patterning systems have evolved from lithographic methods for photo-ablating proteins preadsorbed on surfaces[18] and immobilizing proteins on patterned thiol terminated siloxane films.[19] Kleinfeld et al. introduced the powerful technique of functionalizing surfaces with silane chemistry–based lithography techniques. Silicon or silicon dioxide surfaces are coated with photoresist. The photoresist is exposed in regions, and the open areas are bound with alkyl-trichlorosilane chains. The resist-covered areas are then revealed and coated with amino-trihydroxysilane. This creates a heterogenous surface with two distinct regions of differing cell adhesiveness. The trichlorosilane areas constitute low cell adhesive regions, and the amine derivatized silcon surfaces are highly cell adhesive.[20] A similar modification technique relies on amine derivatization without active modification of the surrounding areas with trichlorosilanes. In this alternative, the aminated areas are exposed to cells under serum free conditions to allow adhesion to only the aminated areas. Subsequently a second type of cell is applied to the surface under serum conditions. Serum proteins passively adsorb onto the bare surfaces, and the second cell type adheres to these regions surrounding the first cell type.[21,22] Whitesides further developed the surface-selective modification technique by removing the direct lithographic patterning steps and replacing them with silicone elastomeric mold technologies for microcontact printing, μcp.[23–27] Microcontact printing creates a more economically viable way to pattern cells by avoiding many of the costly lithography steps. It is also a soft lithographic technique that is compatible with natural biomolecules and a number of surfaces other than silicon. Photolithographic methods are not well suited to nonplanar substrates, and they are often chemically specific. The soft lithography microcontact printing method does require at least one lithographic step in the creation of a master mold from which the subsequent stamps are derived. In a typical μcp process, traditional microfabrication methodologies, using photoresist-based lithography steps and dry or wet etching, first shape silicon wafers. This silicon wafer constitutes the master mold from which stamps can be cast using silicone elastomers. Variations on this process include replica molding or using the cast negative to reconstitute a replica of the original master in a urethane or other polymer. In either case, the stamp is wet with a solution containing biomolecules and is then brought into conformal contact with the substrate to be printed. The chemicals or molecules are patterned on the surface in a mirror image of the stamp pattern. Subsequently, cells can be applied to the printed surface, adhering to these regions through selective processes or adhering indirectly to the surfaces after further modification.

5.3.2 Nanopatterning Technologies

Recent enhancements in patterning technologies have led to even more precise technologies for patterning features as small as tens of nanometers with nanoscribing, focused ion beam, and atomic force microscopy-based lithographies.[28–34] Nanoscribing makes use of a relatively high-aspect-ratio AFM tips to locally and discretely derivatize a surface to which proteins can be conjugated. The capillary forces of the aqueous solution keep the chemical reaction fairly tightly associated with the probe tip while it traverses the surface in complex patterns. All of these techniques have the advantage of very precise spatial control of the organization of cells and molecules, although individual techniques may have advantages in certain respects as far as resolution or speed. These advanced techniques lend the biologist new tools to probe cell reactions and the engineer new methods to create complex organic systems.

5.4 BIOARTIFICIAL DEVICES

Many of these patterning technologies can be employed in the pursuit of creating more complex systems. While two-dimensional culture techniques within petri dishes have elucidated many of the basic functions of cell and molecular biology, more realistic culture environments will serve to advance the field of bioartificial devices. Monolayer culture with defined media and single-cell-type culture systems have been the staple of cell biological investigations for the past 30 years. Unfortunately, two-dimensional culture does not often reflect *in vivo* cell growth conditions. One of the most significant differences between the *in vivo* and *in vitro* systems is the actual physical shape of the cells in culture. Monolayer culture on tissue culture plastic tends to produce cells with flattened morphologies, whereas the *in vivo* environments often maintain cells suspended in three-dimensional configurations of extracellular matrix or other cells. In fact, Folkman and Moscona were the first to demonstrate a clear connection between cell morphology and phenotypic expression correlating the proliferation of cells to their shape.[35] These fundamental observations have laid the groundwork for current more sophisticated investigations into cell response in coculture and multiple cell culture environments with complex geometries and spatial relationships. Guguen-Guillouzo et al. were able to maintain long-term hepatocyte cultures with enhanced albumin production through a coculture environment with fibroblasts.[36,37] This cellular reaction was further investigated more recently by exploiting cell patterning methods to arrange specific geometries of cocultured hepatocytes and nonperenchymal cells.[38] Cells *in vivo* also alter their metabolism based not just on their cellular morphology but on their cell-to-cell contact and cell-to-extracellular-matrix contact.[39–41] These examples of organized two-dimensional culture show vast improvements over traditional coculture environments on tissue culture plastic and in sandwich collagen preparations. The natural extension of this concept to the third dimension has been met with both engineering and biological impediments. Engineering manufacturing methods for processing biological components in an appropriate manner are limited in temperature regimes and chemical components. The biological implications of creating complex three-dimensional cultures have involved the lack of nutrient and oxygen support to cell aggregates. As the size of the cell structures grow, nutrients and oxygen can no longer support the internal cells through passive diffusion, and an active support system must be engineered within at the time of the cell patterning. Some investigators are already pursuing the combination of organization within perfused bioreactor devices.[42,43] Ultimately, systems that can extend the longevity of cells in artificial environments or enhance functionality of the cellular components will create devices that function more like the natural tissues they are engineered to approximate.

5.5 MICROFABRICATION FOR ADVANCED CELL CULTURE SYSTEMS

Three-dimensional culture, for the most part, consists of embedding cells haphazardly in cross-linked collagens, hydrogels, or commercial extracellular matrix analogs such as matrigel. This more closely approximates the suspension of cells in extracellular matrix and often maintains cell shape, but lacks the intercellular organizations inherent in native tissues. Layered fabrication methodologies and other rapid prototyping techniques such as three-dimensional printing promise significant contributions to the field. If *in vivo* cellular contacts and spatial resolutions can be engineered to more closely resemble the natural microenvironment, it is logical to assume that this improved cell culture model will more accurately metabolically reflect the real environment. Extending these concepts of improved culture models to bioartificial devices, cell and biomolecule patterning technologies can give rise to more efficient and enhanced functionality in artificial organ devices. Most of these advanced methodologies for cell organization rely in some way on microfabrication techniques.

5.5.1 SILICON MICROMACHINING

Microfabrication technologies, originating in the microelectronics industry, have not only made significant progress in achieving feature sizes in the nanometer scale but also have extended process materials to others outside of silicon and silicon-compatible materials. Novel pattern transfer techniques, such as low cost LIGA, micromolding processes, and ultraprecision manufacturing, allow for shaping ceramics and polymers in submicron dimensions with high aspect ratios.[44] These improvements in micromanufacturing allow for the shaping of three-dimensional structures such as cavities, channels, and pillars with high densities and feature sizes much smaller than single cells. These powerful techniques are now used both for basic research in biology and for the development of new biological devices.[45,46]

The fabrication of the silicon structures is based primarily on subtractive manufacturing techniques, whereby material is selectively removed from the substrate until the desired structure is achieved. In general, these silicon etch techniques can be divided into wet and dry etch processes. The most popular wet-etch process is based on a mixture of potassium hydroxide, KOH, and water as an etch solution, and it exhibits etch rates that are dependent on the crystallographic direction of silicon.[47,48] An etch rate of 1 μm per minute at a temperature of 85°C in the <100> crystallographic direction is typical, whereas almost no etching in <111> direction occurs. Hence, (111) crystallographic planes in silicon serve as etch stops so that V-shaped grooves and cavities can be shaped (Figure 5.1).

Deep reactive ion etch (DRIE) techniques, the most recently developed anisotropic dry etch methods, enable etch rates of up to 10 μm/min,[49] and the etch directionality is independent of the crystallographic direction of silicon. These processes allow for the shaping of structures with high aspect ratios of up to 20 without limitations on the shape. One widely applied process in microelectromechanical systems (MEMS) production is the Bosch advanced silicon etch process.[50] This process is based on alternating process gases between a passivation and an etch cycle. This sequential application of an isotropic etch cycle based on SF_6 chemistry and a passivation cycle results in highly anisotropic etch behaviors. Unfortunately, a scalloping effect on the etched walls results from the alternating etch cycles and can be problematic if very smooth surfaces are required (Figure 5.2).

Because of the lack of anisotropic etch methods for ceramics and polymers, most shaping methods for these materials are based on shape transfer techniques from molds. The mold tools are usually fabricated by applying silicon processing, plating techniques, or ultraprecise manufacturing. The pattern transfer processes can be divided among liquid resin molding, injection molding, and hot embossing.

5.5.2 Biodegradable Material Processing

As illustrated above, silicon is a well-defined medium for microfabrication. And while silicon and its native oxide are biocompatible, they are not very natural substances

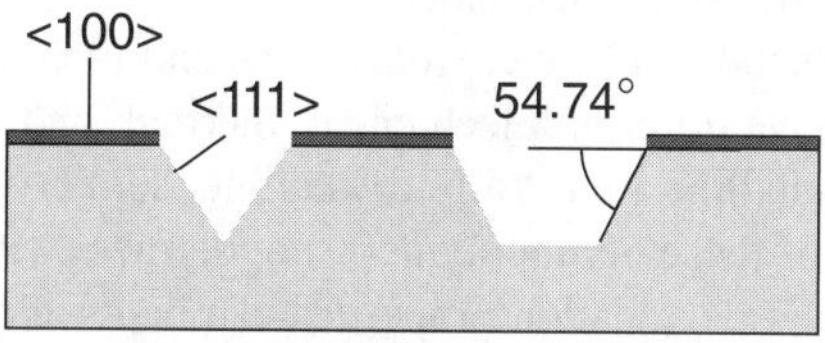

FIGURE 5.1 Anisotropic wet etching: typical etched features in a <100> wafer.

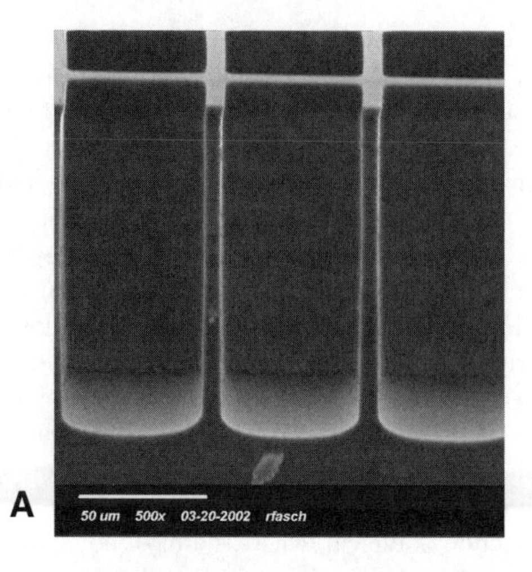

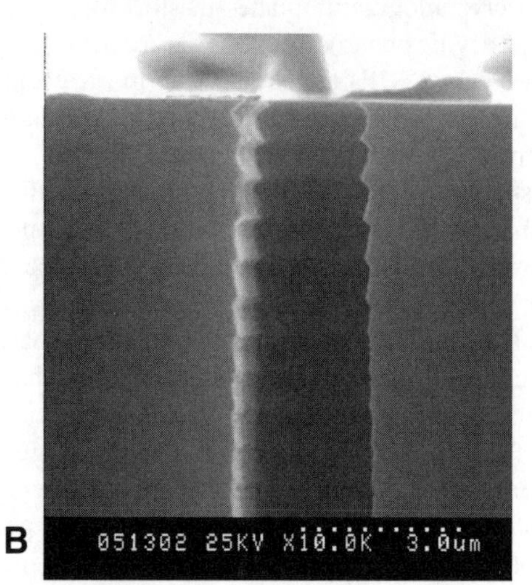

FIGURE 5.2 Typical etch features achieved with a Bosch advanced deep silicon etch process, (A) High aspect Si-wall structure, (B) Rough surfaces due to scalloping effect.

conducive to cell differentiation, proliferation, and function. In the pursuit of more natural cell and tissue function, many biodegradable materials have been investigated within the biomedical research field for the past four decades. The most popular polymers are glycolide/lactide-based biodegradable linear aliphatic polyesters, such as polyglycolide, polylactide, poly(ε-caprolactone), and their copolymers. These biodegradable polymers have different mechanical, thermal, and degradation properties, depending on their crystalline state, fully crystalline, semicrystalline, or amorphous. With the combination of these homopolymers, copolymers with precisely engineered mechanical, thermal, and degradation properties can be developed. This flexibility in material properties and biocompatibility of the degradable polymers has led to the development of various medical devices, such as bioabsorbable sutures, dental void fillers, dental guided-tissue-regeneration membranes, orthopedic fixation devices, vascular grafts and stents, and tissue-engineering scaffolds, as well as drug delivery systems.

Along with the advantages and the advances of synthetic biodegradable polymers, it has become a significant challenge in tissue engineering to process the biodegradable polymers into useful forms for either seeding cells *in vitro* or *in vivo* recruiting of the desirable cell types from the host organism.[51] The importance of geometry on engineered tissue has been addressed in the respect that geometries with sharp angles have been shown to provoke more inflammatory response and enzyme activity around implanted polymeric devices.[52] However, more importantly for cell survival and function, the size of internal pore structure, relative distances between pores, and distribution of pores are major controlling metrics for cell communication, nutrients supply, and cell migration within the structures.

Biodegradable polymer-based scaffolds have specific advantages over nondegradable yet biocompatible materials such as silicon or Teflon®. The degradation times of the device can be programmed for each application by altering the composition of the copolymers with different degradation mechanisms. Mechanical properties such as stiffness or strength can be tuned by altering compositions as well. Additional surgical interventions are eliminated if the device can degrade *in vivo* rather than require removal. More importantly, many biodegradable polymers are already widely used as drug delivery mechanisms and have FDA approval. Appropriate drugs can be locally embedded within the polymers in precise amounts and can be designed to control their release spatially and temporally via microstructured scaffolds.

Researchers have made attempts to control spatial organization of cells in tissue scaffolds using biodegradable materials. With aid of silicon microstructuring technology, photolithographically microstructured patterns can be transferred into polymers by solvent casting and injection molding.[53,54] Although these methods result in micron-sized features on a two-dimensional surface and often serve to orient cells in the patterns, they still lack true three-dimensional structures. To construct three-dimensional structures, several techniques have been used such as fiber bonding,[55] particulate leaching,[56] gas foaming,[57] and phase separation.[58] These methods produce highly interconnected porous structures. However, the random geometry of pores, random spatial organization of pores, and strong dependence of processing on solvents or thermal cycles limit the application of fabricated polymer constructs based on these approaches.

Mimicking the biological structural environment will require more precise design of compartments for proper cell morphology, migration pathways for cell recruitment, and routes for oxygen and nutrient supply. Hot embossing of biodegradable polymers with microfabricated embossing tools delivers precisely constructed microstructures for cell compartments, interconnected pathways for cell migration, and pipelines for nutrient and oxygen diffusion or convection. Conventional hot embossing is conducted at relatively high temperatures. Alternatively, since most biodegradable polymers have relatively low glass transition temperatures, it is possible to process biodegradable polymers without denaturing either the biodegradable polymer itself, or thermally sensitive embedded substances such as drugs or growth factors. Additionally, while precise microfabrication relies on very expensive photolithographic steps for each construct, hot embossing requires only a single silicon-based fabrication of the original embossing tool. In spite of the attractive advantages of precise spatial organization of microstructures, low processing temperatures, minimal use of solvents, and low-cost processing, layer-based three-dimensional fabrication introduces the technological hurdle of joining each layer to the next. Though thermal fusion or ultrasonic bonding have been tested, they often lead to destruction of the microstructures in each layer. Use of bio-adhesive or other secondary polymer as an adhesive can be adopted; however, this adds complexity to the chemical composition of the tissue-engineered construct. Bio-adhesives such as cyano-arcrylate esters and fibrin sealants have been used to a limited degree; however, they have been reported to have some side effects in clinical trials.[59] General adhesive material is not desirable either, since it may confer toxicity to the implant. Current investigations into solvent melting on microstructured surfaces shows that the deformation of microstructures can be mitigated while still adequately bonding the stacked layers. (Figure 5.3).

Many techniques have been investigated in recent years to construct three-dimensional structures for biological applications such as tissue engineering or drug delivery. As delineated above, each of the methods still has limitations that hinder immediate adoption of their use in the medical field due to too large scale structuring, lack of three-dimensional structures, solvent residue from processing, thermal degradation, and high production costs.

5.6 SYNTHESIZING CELL-PATTERNED DEGRADABLE POLYMER STRUCTURES

In an example of a process aimed at accurately recreating three-dimensional environments, complex structures have been microfabricated from a number of biodegradable polymers to impart complexity to the extracellular matrix. Cell patterning methods have been incorporated as well to create cellular organization within these scaffolds. Cell patterning of hepatocytes is not a new investigation,[60,61] but surprisingly, the use of patterned cocultures in bioartificial devices has been fairly underused. The work here is an attempt to leverage the newly discovered advantages of organized cocultures in a bioartificial device. Microfabrication methods are particularly relevant to these biological structures because they are able to resolve cellular and subcellular feature sizes and create massively redundant architectures. The fabrication process involves the creation of master molds from silicon fabrication technologies. Then polymeric cell support layers are created by compression embossing. Cell patterning is applied to the individual layers through both fluidic and shadow masking methods. Finally, these cell layers can be combined into complex three-dimensional structures with internal voids and cell support structures as well as patterned cellular components. A number of manifestations of these technologies have been fabricated, including thin-film silicon membranes with patterned cells, thin-film polymeric membranes with patterned cell structures, interconnected silicon structures with patterned cells, and ultimately a patterned thin-film layered bioreactor. These structures have vast importance in the creation of bioartificial devices and perhaps tissue engineered implantable devices.

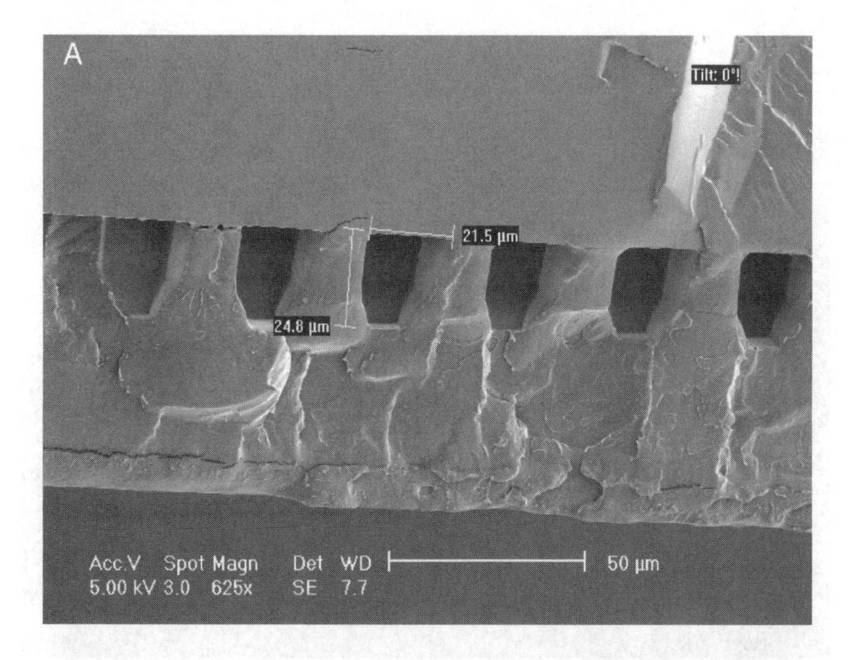

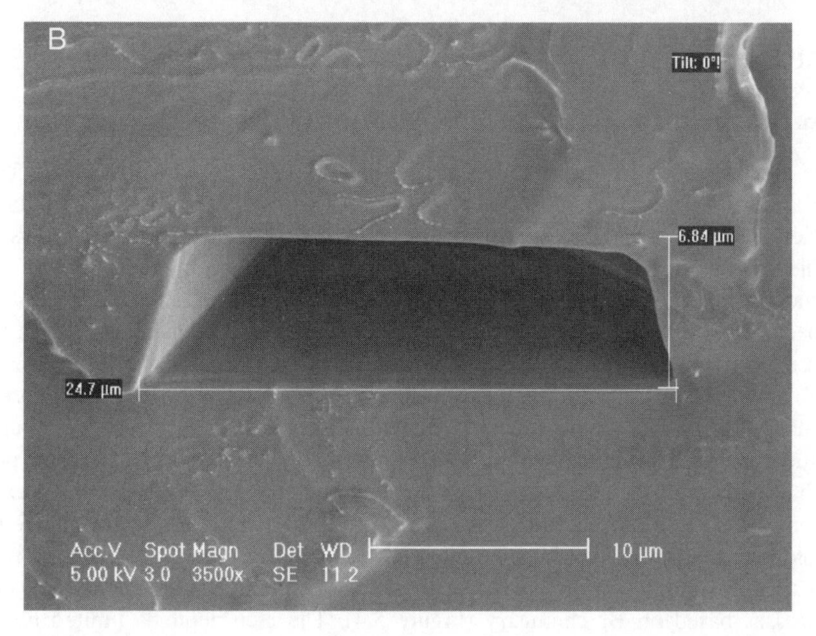

FIGURE 5.3 Cross section SEM images of bonded 50/50 poly(D,L-lactide-co-glycolic acid) layers. (A) Solvent spray bonded nonembossed (top) and embossed (bottom) layers. Solid bonding between two layers without damage in microstructures is shown. Scale bar is 50 μm. (B) Thermal fusion bonded structure of both embossed layers. Aspect ratio (height over width of structure) after bonding is shown to be less than half compared to initial aspect ratio of 2 before bonding. Scale bar is 10 μm. (C) Ultrasonically bonded structure. Microstructures in bonded area were completely destroyed. Scale bar is 500 μm.

FIGURE 5.3 (continued)

5.6.1 Structured Polymeric Thin Films with Cells

The silicon fabrication processes in this work are based on combinations of both wet- and dry-etch techniques. Micro mold tools can be fabricated by applying a variety of techniques. However, the challenge in the fabrication of these tools is the achievement of high-aspect-ratio structures and low surface roughnesses to aid in demolding. While the Bosch DRIE process allows for etching structures with high aspect ratios, it usually produces poor surface qualities. The scalloping effect discussed earlier causes rough sidewall surfaces. As a result, this etch process is problematic for use in the fabrication of micro molds. Therefore, in this work, a combination of RIE etch processes were used to achieve a surface topography without undercut structures while maintaining smooth surfaces. Because of the poor etch selectivity of the standard RIE processes for photoresists, a thick silicon oxide layer is deposited on the silicon as a mask material. The features are transferred photolithographically into the silicon oxide. The structures are shaped using a combination of isotropic dry-etch processes based on SF_6 chemistry and anisotropic etch processes based on Br chemistry (Figure 5.4). The etch depth is limited by the consumption of the silicon oxide mask during the etch process. As an example, a silicon oxide layer of 2 μm results in a 50 μm structure depth. Either the hard silicon mold itself can be used for embossing, or a soft polymer mold, silicone rubber, or epoxies, can be cast from the silicon master and used as an embossing tool (Figure 5.5). This microforming tool can now be applied to the polymeric biomaterials to create complex shapes by low-temperature forming.

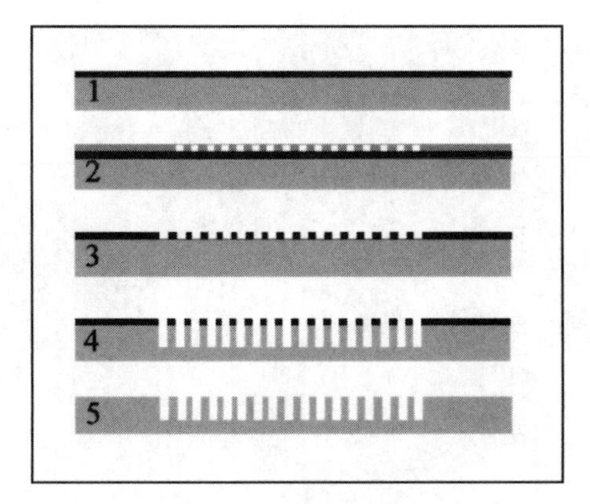

FIGURE 5.4 Fabrication of silicon molding tools. Fabrication scheme: (1) silicon wafer with thick silicon oxide layer, (2–3) pattern transfer of tool pattern in silicon oxide, (4) etching deep in silicon wafer with combined dry-etch processes, (5) cleaning of oxide residues with HF.

The fabrication process for the biodegradable scaffolds is a combination of hot embossing and compression molding. Individual layers are shaped via pattern transfer from the master mold to the polymeric biodegradable materials. The silicon tooling is surface treated with low-surface-energy materials, such as silicone rubber or Teflon, to ensure mold release. After the individual embossed layers are formed, stacking of the patterned layers can be employed to create truly three-dimensional structures with complex inner volumes. The initial forming requires stock materials in sheet or powder form. To obtain powder, biodegradable polymers are ground to micron-sized particles at cryogenic temperatures. These micron-sized particles are compressed in a hydraulic press at temperatures above their glass transition temperature, and the patterns on the forming tool are transferred to the thin biodegradable polymer films. The transferred patterns have various geometries and three-dimensional shapes. Well-type cavities with through-layer structures and differing feature densities as well as varying angles of cavity walls have been made with these processes (Figure 5.6).

The cellular patterning was undertaken in a manner that avoids direct lithography using purely mechanical techniques similar to previous work with cellular masking.[62,63] The delicate nature of these structures eliminates the possibility of any printing-based cell patterning technique, while the topographical features and nature of the materials make photo resist–based lithographic techniques untenable as well. For that reason, cells are screen-printed onto the biomaterial film surfaces through the use of through-structured PDMS sheets or thin silicon films, depending on the resolution required. Thin silicone elastomer films are formed from spin casting PDMS prepolymer onto SU-8 structured silicon. The polymerized silicone thin films are removed from the silicon wafers and washed in toluene to remove any unreacted polymer groups. The PDMS films are floated onto the surface of collagen-adsorbed degradable polymer membranes under asceptic conditions. After drying, these composite structures are placed in machined polycarbonate pump housing with hepatocyte growth medium, DMEM/F12 (Gibco BRL) containing 5 ng/ml transferrin, 20

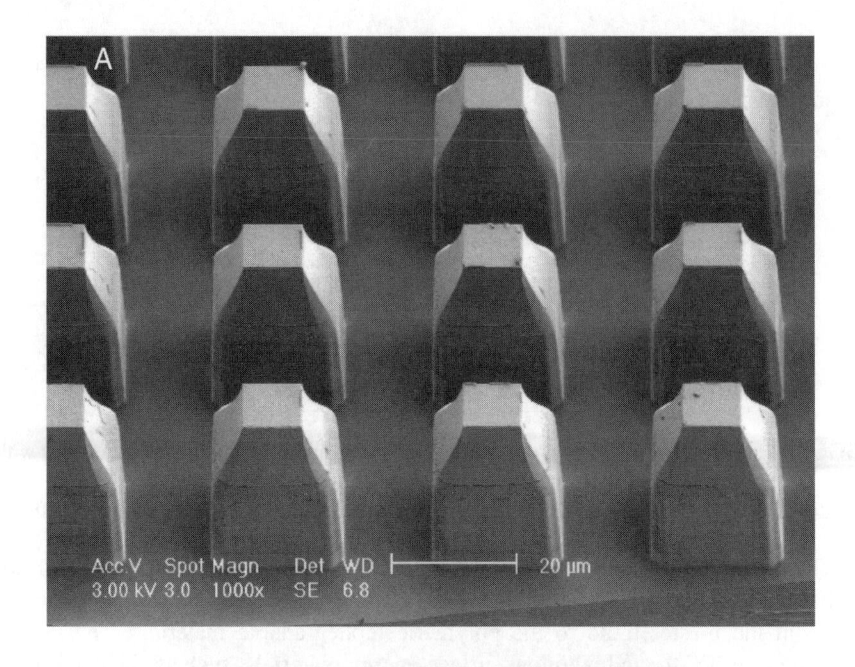

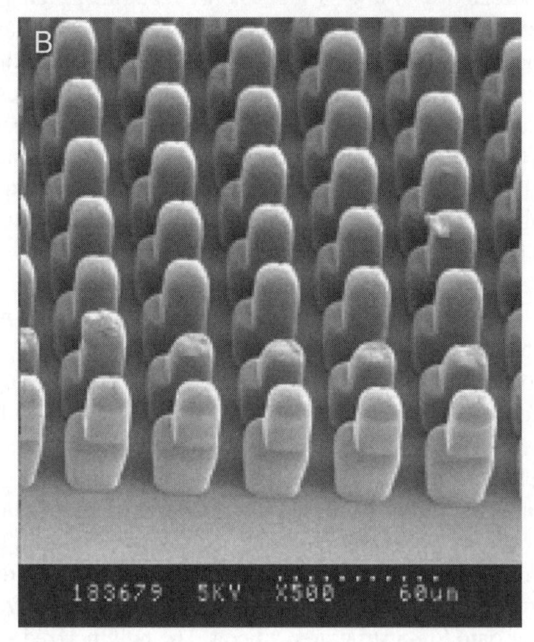

FIGURE 5.5 SEM image of microfabricated (A) hard silicon embossing mold. Scale bar is 20 μm. (B) Soft silicone rubber embossing mold. Feature sizes are 15 μm × 15 μm by 30 μm high.

FIGURE 5.6 SEM images of embossed 50/50 poly(D,L-lactide-co-glycolic acid) thin film. (A) Top view of well type cavities with tapered angle. (B) Vertically interconnected openings through a single layer. Scale bar is 50 μm.

ng/ml epidermal growth factor, 60 μg/ml L-proline, 7.5 μg/ml hydrocortisone, 0.5 U/ml insulin, 25 μg/ml ascorbic acid, 40 ng/ml dexamethasone, and 50 μg/ml gentamycin and primary rat hepatocytes at a concentration of 1×10^6 cells/ml. The films are perfused at a rate of 5 sccm, and after 10 minutes the films are removed and placed in culture at 37°C and 5% CO_2. After a 24-hour incubation period, the PDMS masking layers are aseptically removed and the substrates are washed to remove nonadherent hepatocytes. The biodegradable substrate features are fabricated such that there is an apical and basal polarity, so that the apical side of the sheet has a larger opening than the basal side. The size of the structured wells is such that a single cell was captured in each 20 μm well. In the future such structures could be used to recreate the architecture of bile canaliculi-sinusoid polarities (Figure 5.7).

5.6.2 Structured Silicon Thin Films with Cells

This process starts with structuring one side of a double-side polished wafer with the desired cellular pattern, in this case an array of 20-μm features. The pattern is etched using DRIE, and the etch depth determines the thickness of the membrane. Next, a silicon nitride layer resistant to KOH etch is deposited using low-pressure chemical vapor processes on both sides. A second lithography step opens a window on the backside of the wafer that is registered to the feature set on the front side. The size of this etch window ultimately determines the size of the silicon membrane.

FIGURE 5.7 Fluorescent image of primary rat hepatocytes residing in 20-μm cell structures in a polymer membrane.

Then the wafer is wet etched using a KOH process until the etched pattern of the front side is reached. The silicon structure of the front side, meanwhile, is protected by the silicon nitride layer. Removal of the silicon nitride with hot phosphoric acid is the last step of the fabrication process (Figure 5.8). The silicon substrates are then surface modified in a manner similar to Gölander and Eriksson to make them amenable to cell adhesion.[64] Briefly, the silicon substrates are cleaned of residual organic materials such as photoresists in a piranha solution. The silicon films are then exposed to an oxygen plasma to hydroxylate the silicon dioxide surface. This silicon dioxide is subsequently derivatized by placing it in a basic carbonate buffer with polyethyleneimine. The polyethyleneimine chemisorbs to the hydroxylated silicon surface creating an amine rich surface layer to better support cell adhesion. We as well as others have observed that these aminated surfaces show significantly better cellular adhesion than serum-incubated bare silicon surfaces (Figure 5.9). For hepatocyte patterning, these aminated surfaces were then incubated in a 1-mg/ml Type I collagen solution. The collagen adsorbed on the silica surfaces is allowed to air dry and kept desiccated until use. PDMS masking layers are floated onto the silicon membranes in a manner similar to the polymer membrane fabrication. Cells are pumped through these structures at 5 sccm. Then the structures are removed from the flow device, and the cells are allowed to adhere. The PDMS mask layers are aseptically removed, and the substrates are washed to remove nonadherent cells. Figure 5.10 shows the cellular fidelity to the elastomeric masks at this stage of fabrication. The hepatocyte-patterned silicon masks are then plated with NIH 3T3 fibroblasts (American Type Culture Collection, Manassas, VA), and these cells adhere in the spaces between the hepatocyte islands.

5.6.3 Interconnected Silicon Structures with Patterned Cells

For some applications, such as those that involve layered fabrication, scaffolds must provide recessed cavities to protect cells during layer bonding. It is also important to provide interconnections between those cells extending in three dimensions that allow for cell motility or the supply of nutrients. Such a complex morphology in silicon has been created by applying a combined wet- and dry-etch process. In the first step, a pattern is transferred into a silicon nitride layer using RIE. Then by applying DRIE this pattern is etched deep into the double polished silicon wafer. By repeating this process on the opposing side, the structure can be created such that it connects both sides of the wafer. Choosing certain ratios of distance between features, wafer thickness, and etching depth, allows for control of the feature connectivity by using a wet KOH etch step. The release of the silicon nitride layer is the last step in the fabrication process of the three-dimensional interconnected container structure (Figure 5.11). In this case, a two-stage mask process composed of both silicon and elastomeric masking was used to pattern two levels of cellular resolution into the structure simultaneously. A thin silicon membrane was used for high-resolution patterning of cells into the interconnected features. Many of these cell-loading processes and layered object manufacturing processes require that the features within the each layer be aligned with accuracy on the order of 10% of the feature size. Maintaining alignment during the cell-loading procedure during the

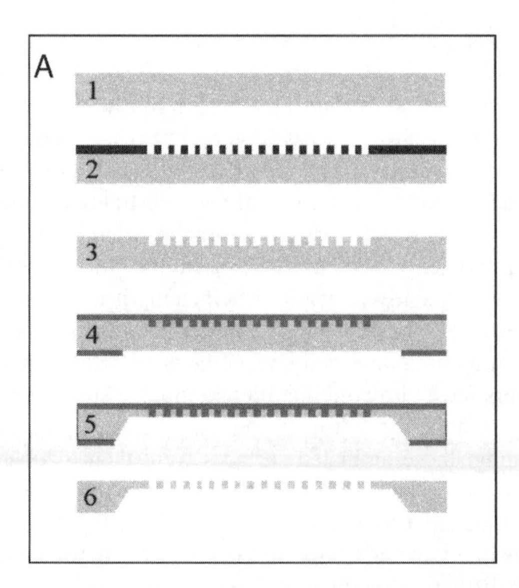

FIGURE 5.8 Fabrication of membrane masks. (A) Fabrication scheme: (1–2) photo lithography of mask pattern on one side of a double-side polished wafer, (3) DRIE of pattern in silicon wafer (etch depth determines the membrane thickness), (4) deposition of silicon nitride and opening of an etch window on the nonpatterned wafer side, (5–6) KOH etching and removal of silicon nitride. (B) Finished membrane mask.

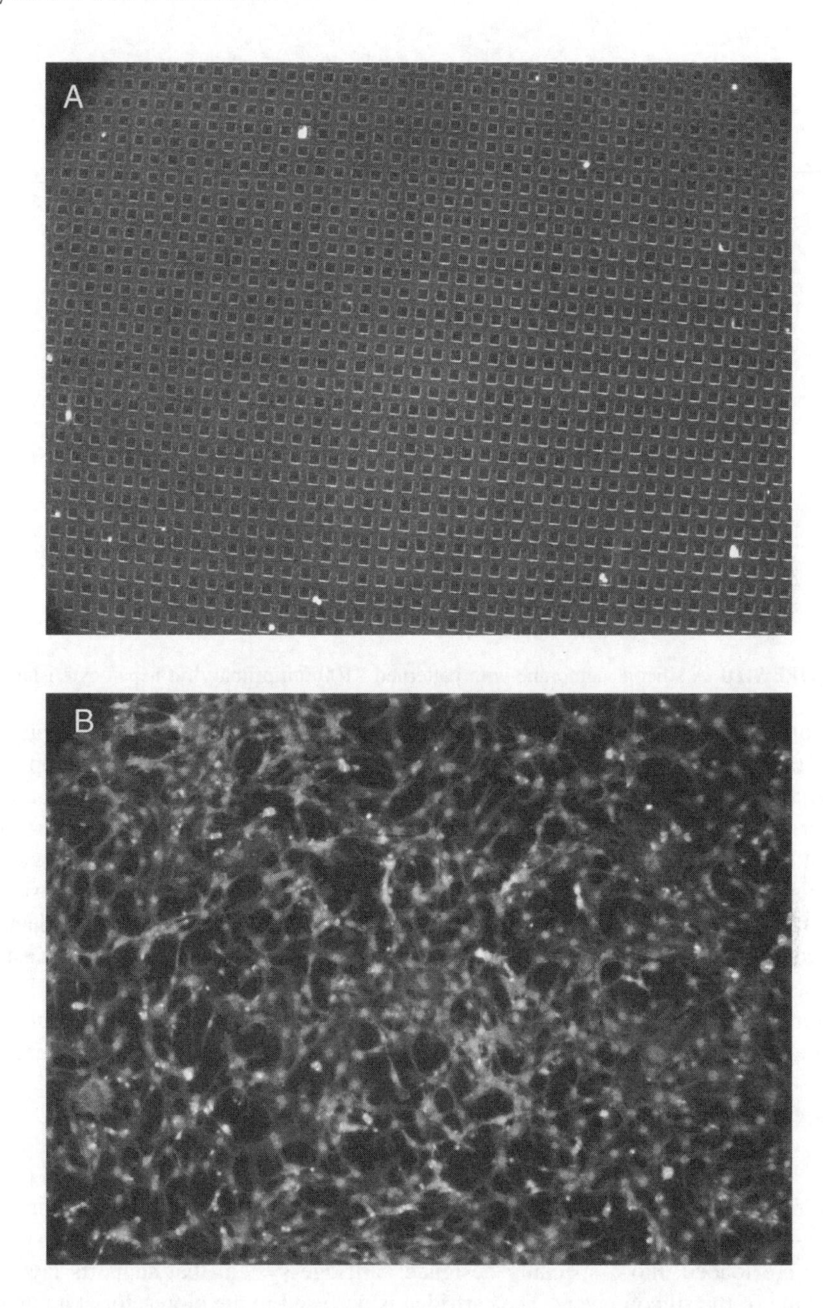

FIGURE 5.9 Silicon membrane that has been (A) incubated with 10% FBS serum only, or (B) first aminated then incubated with serum. The cells are stained with FITC-Phalloidin and DAPI. The rat dermal fibroblasts clearly adhere more readily to the aminated silicon surface.

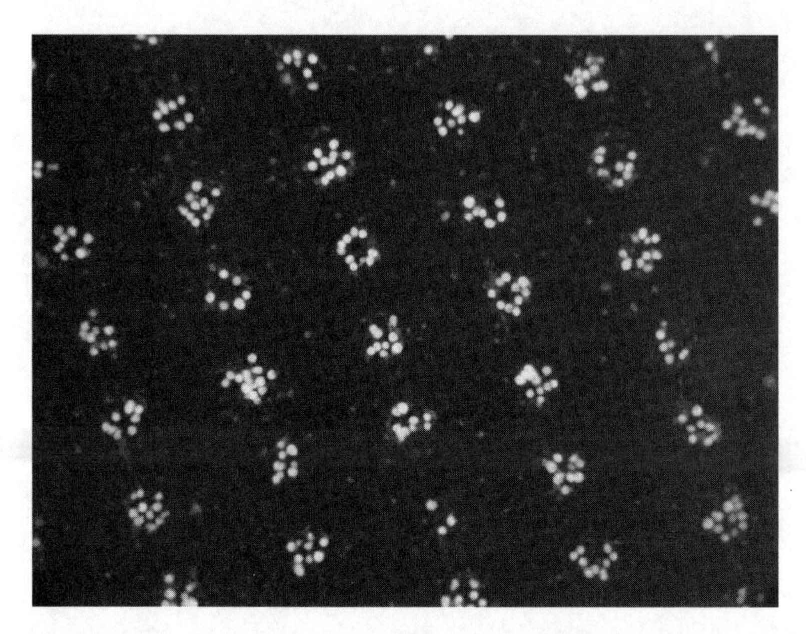

FIGURE 5.10 A silicon membrane with patterned ~100-μm primary rat hepatocyte islands.

silicon-based shadow mask technique is a necessity. By using mechanical alignment structures, such as alignment holes or etched V-grooves in combination with optical fibers, an accuracy of 10 μm is feasible.[65] Very precise alignment accuracy on the order of submicron and micrometer dimensions can be achieved using optical alignment techniques. In this case, an optical alignment approach in combination with a specially designed clamping mechanism to hold the mask and the container substrate in place is applied. In this way, the required alignment accuracy greater than 5 microns was guaranteed during the cell patterning process and the additional transfer steps. A silicone elastomeric mask was used to pattern the lower-resolution large features of hepatocytes (Figure 5.12). Stacking of these patterned wafers results in a three-dimensional microfluidic network with interconnected cell compartments.

5.6.4 Patterned Multiple Layers: A Cartridge-Based Parallel Plate Bioartificial Liver

The final application for these cell-patterned thin films is to combine them into a three-dimensional perfused device. In this experiment thin silicon membrane structures are loaded into a specially designed cartridge system that supports layered stacking of the silicon layers. The cartridge is perfused in the bioreactor at a rate of 2 sccm with a peristaltic pump. Parallel plate bioreactors have the advantage of a high surface area exposed to the flowing media for nutrient and cell metabolite exchange. However, they often suffer from very high fluid-induced shear forces that can alter or suppress cellular function. This parallel plate bioreactor was patterned so that the cells resided on the inner surface of the silicon membranes, out of the direct influence of flow. Once patterned, individual layers of silicon or thin-film

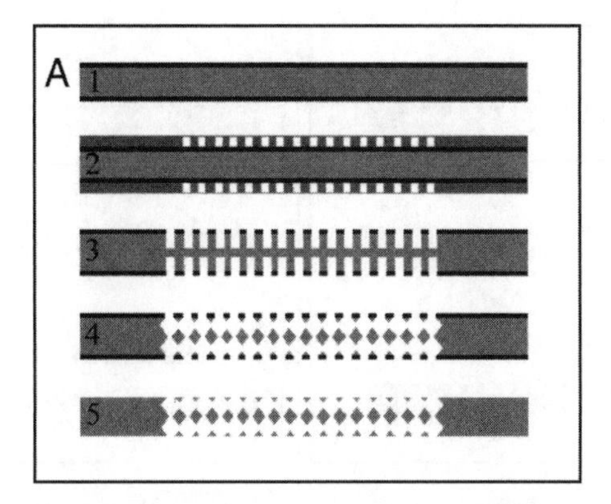

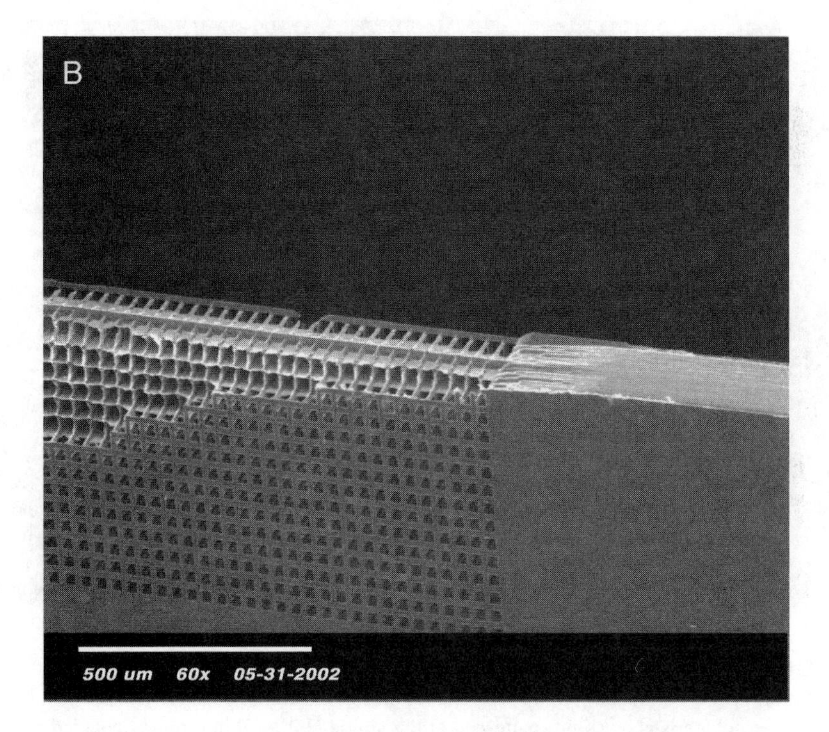

FIGURE 5.11 Fabrication of stackable silicon cell container structures. (A) Fabrication scheme: (1) Double-side polished silicon wafer with silicon nitride layers; (2) pattern transfer of container structure in silicon nitride; (3) deep etching of the container cavities in the wafer; (4) KOH wet-etch step creates three-dimensional interconnections; (5) release of silicon nitride with hot phosphoric acid. (B) Finished container structure.

polymer can be assembled to create complex structures through a rapid prototyping technique known as layered object manufacturing (LOM). The microfabricated structures in these substrates act as microconduits to allow for nutrient flow and cell metabolic product equilibration. The individual layers of the bioartficial device are

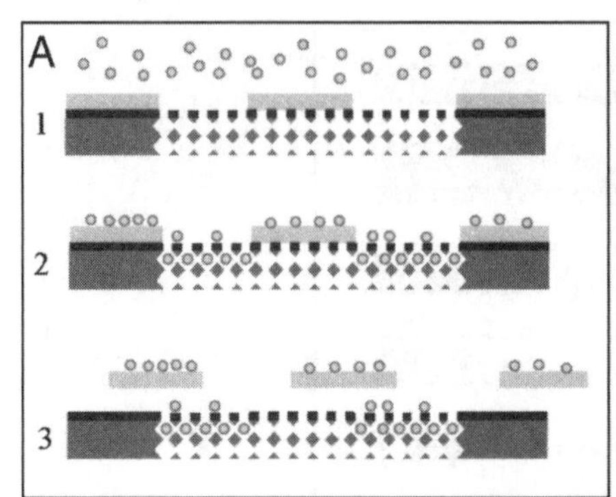

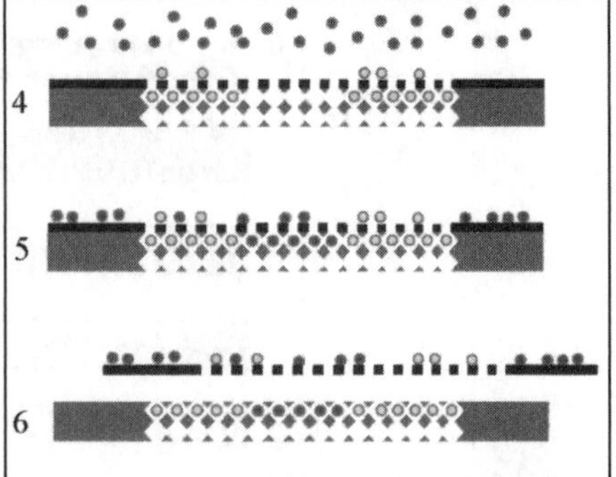

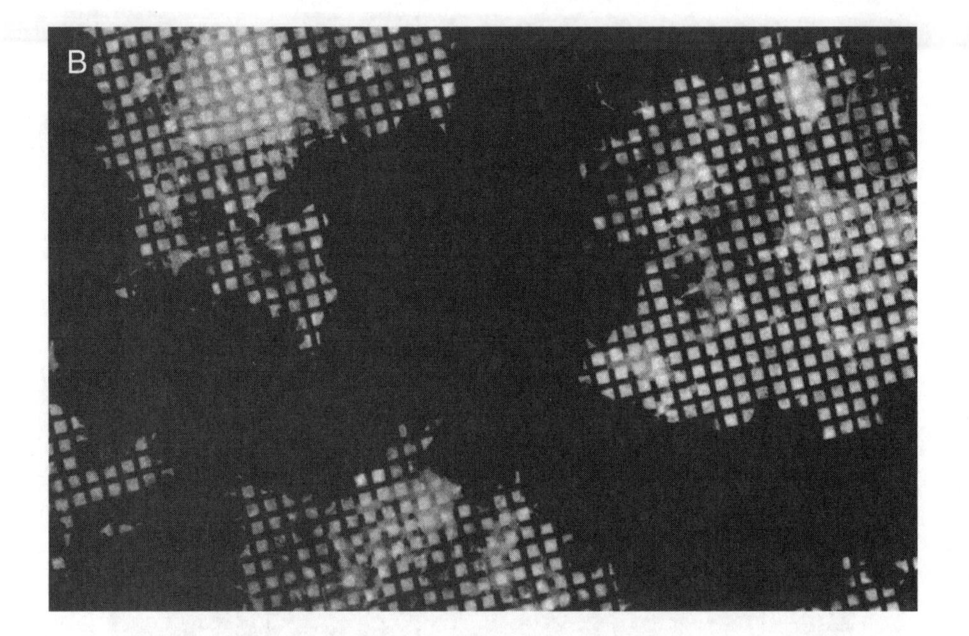

FIGURE 5.12 (A) Two-part process for patterning cells with a hard silicon mask for high-resolution, single-cell features and a low-resolution soft elastomeric mask. (B) Low-resolution mouse hepatocyte islands (light) patterned at the single-cell level into interconnected silicon microcavities surrounded by mouse dermal fibroblasts.

20 to 50 μm thick, on the order of cellular dimensions, to allow for significant cell–cell interactions. The flow system has a medium reservoir to allow for gas exchange with the incubator environment and for medium collection for analysis. The medium was collected at regular intervals and assessed for albumin content. In this manifestation of a layered cell bioreactor, albumin production by the contained hepatocytes was maintained for the duration of the experiment of one week. This bioreactor shows the successful culmination of a number of techniques for patterning cells on topologically complex substrates and subsequently bringing those substrates together to form a functional three-dimensional device.

5.7 CONCLUSIONS

It is conceivable that the ordered three-dimensonal biocompatible and biodegradable cell scaffolds fabricated here would have a variety of applications in tissue engineering, biomedical research, and regenerative medicine. This procedure is one of the first times that topographically complex substrates have been patterned with cells. The application of photoresist in traditional lithography to previously structured surfaces results in discontinuities and poor resolutions, making lithographic protein and cell patterning on structured surfaces difficult. Also, the solvents and optical techniques relevant to traditional semiconductor-based lithography are not well characterized for thin biodegradable polymer films. The masking technologies of using hard silicon masks and soft elastomeric masks for various length scales to pattern high-resolution features with hard masks and lower-resolution features with soft masking devices allow for substrate chemistry-independent patterning technologies. The polymeric thin films lack mechanical strength and do not readily withstand printing technologies without destruction of the layers or the features within the materials. These patterned thin membrane layers with cells can be stacked together to create three-dimensional devices such as a liver bioreactor that in this case maintained albumin production.

Ordered cellular structures will play significant roles in developing biomedical engineering pursuits. Currently researchers can extend the longevity and increase the productivity of cells by engineering their microenvironments. Researchers will be able to achieve even greater control over cell function by engineering complex cell-to-cell organizations as well. To some extent, research is limited now to primary cell sources; in the future, advances in stem cell technologies will certainly improve our access to totipotent and multipotent cellular sources. These stem cells will hopefully be able to fill in the gaps of cell function that are often lost in engineered devices through dedifferentiation and other factors affecting the attenuation of primary cell function. However, there will always be some need for engineered scaffolds to deliver these cells in appropriate manners. This is where significant contributions in cell patterning, material processing, and material science can be made to enhance cell delivery and provide support for cell function through organization.

ACKNOWLEDGMENTS

This work was supported by the Stanford Office of Technology and Licensing's Gap Fund, as well as Johnson & Johnson.

REFERENCES

1. Eagle, H., Nutrition needs of mammalian cells in tissue culture, *Science,* 122, 501, 1955.
2. Barnes, D., and Sato, G., Methods for growth of cultured cells in serum-free medium, *Anal. Biochem.*, 102, 255, 1980.
3. Schuppan, D., et al., The extracellular matrix in cellular proliferation and differentiation, *Ann. N.Y. Acad. Sci.*, 733, 87, 1994.

4. Lee, K.-H. et al., T cell receptor signaling precedes immunological synapse formation, *Science*, 295, 1539, 2002.
5. Bosman, F.T., and Stamenkovic, I., Functional structure and composition of the extracellular matrix, *J. Pathol.,* 200, 423, 2003.
6. Grande, D.A., et al., Evaluation of matrix scaffolds for tissue engineering of articular cartilage grafts, *J. Biomed. Mater. Res.,* 34, 211, 1997.
7. Brittberg, M., et al., Treatment of deep cartilage defects in the knee with autologous chondrocyte transplantation, *N. Engl. J. Med.,* 331, 889-895, 1994.
8. Carter, S.B., Principles of cell motility: the direction of cell movement and cancer invasion, *Nature,* 208, 1183, 1965.
9. Ivanova, O.Y., and Margolis, L.B., The use of phospholipids film for shaping cell cultures, *Nature,* 242, 200, 1973.
10. Letourneau, P.C., Cell-to-substratum adhesion and guidance of axonal elongation, *Dev. Biol.,* 44, 92, 1975.
11. Cooper, A., Munden, H.R., and Brown, G.L., The growth of mouse neuroblastoma cells in controlled orientations on thin films of silicon monoxide, *Exp. Cell. Res.,* 103, 435, 1976.
12. Furshpan, E.J., MacLeish, P.R., O'Lague, P.H., and Potter, D.D., Chemical transmission between rat sympathetic neurons and cardiac myocytes developing in microcultures: evidence for cholinergic, adrenergic, and dual-function neurons, *Proc. Natl. Acad. Sci.*, 73, 4225, 1976.
13. Albrecht-Beuhler, G., The angular distribution of directional changes of guided 3T3 cells, *J. Cell Biol.,* 80, 53, 1979.
14. Hammarback, J.A., and Letorneau, P.C., Neurite extension across regions of low cell-stratum adhesivity: implications for the guidepost hypothesis of axonal pathfinding, *Dev. Biol.*, 117, 655, 1985.
15. Ireland, G.W., Dopping-Hepenstal, P., Jordan, P., and O'Neil, C., Effects of patterned surfaces of adhesive islands on the shape, cytoskeleton, adhesion, and behavior of Swiss mouse 3T3 fibroblasts, *J. Cell Sci. Suppl.*, 8, 19, 1987.
16. Letorneau, P.C., Regulation of neuronal morphogenesis by cell-substratum adhesion, in *Society for Neuroscience Symposium*, Cowan, W.M., and Ferrendelli, J.A. Eds., p. 67, Society for Neuroscience, Bethesda, MD, 1976.
17. Rovensky, Y.A., Slavnaja, I.L., and Vasiliev, J.M., Behavior of fibroblast-like cells on grooved surfaces, *Exp. Cell Res.*, 65, 193, 1971.
18. Hammarback, J.A., Palm, S.L., Furch, L.T., and Letourneau, P.C., Guidance of neurite outgrowth by pathways of substratum adsorbed laminin, *J. Neurosci. Res.,* 13, 213, 1985.
19. Bhatia, S.K., Hickman, J.J., and Ligler, F.S., New approach to producing patterned biomolecular assemblies, *J. Am. Chem. Soc.,* 114, 4432, 1992.
20. Kleinfeld, D., Kahler, K.H., and Hockberger, P.E., Controlled outgrowth of dissociated neurons on patterned substrates, *J. Neurosci.,* 8, 4098, 1988.
21. Britland, S., Perez-Arnaud, E., Clark, P., McGinn, B., Connolly, P., and Moores, G., Micropatterning proteins and synthetic peptides on solid supports: a novel application for microelectronics fabrication technology, *Biotechnol. Prog.,* 8, 155, 1992.
22. Lom, B., Healy, K.E., and Hockberger, P.E., A versatile technique for patterning biomolecules onto glass coverslips, *J. Neurosci. Meth.*, 50, 385, 1993.
23. Kumar, A., and Whitesides, G.M., Features of gold having micrometer to centimeter dimensions can be formed through a combination of stamping with an elastomeric stamp and an alkanethiol "ink" followed by chemical etching, *Appl. Phys. Lett.,* 63, 2003, 1993.

24. Xia, Y., Kim, E., Zhao, X.-M., Rogers, J.A., Prentiss, M., and Whitesides, G.M., Complex optical surfaces by replica molding against elastomeric masters, *Science,* 273, 347, 1996.
25. Zhao, X-M, Xia, Y., and Whitesides. G.M., Fabrication of three-dimensional microstructures; microtransfer molding, *Adv. Mater.,* 8, 837, 1996.
26. Kim, E., Xia, Y., and Whitesides, G.M., Polymer microstructures formed by molding in capillaries, *Nature,* 376, 581, 1995
27. Xia, Y., and Whitesides, G.M., Soft lithography, *Ann. Rev. Mater. Sci.*, 28, 153, 1998.
28. Lee, K., Park, S., Mirkin, C.A., Smith, J.C., and Mrksich, M., Protein nanoarrays generated by Dip-Pen nanolithography, *Science,* 295, 1702, 2002.
29. MacBeath, G., and Schreiber, S.L., Printing proteins as microarrays for high-throughput function determination, *Science,* 289, 1760, 2000.
30. Bergman, A.A., Buijs, J., Herbig, J., Mathes, D.T., Demarest, J.J., Wilson, C.D., Reimann, C.T., Baragiola, R.A., Hull, R., and Oscarsson, S.O., Nanometer-scale arrangement of human serum albumin by adsorption on defect arrays created with a finely focused ion beam, *Langmuir,* 14, 6785, 1998.
31. Wadu-Mesthrige, K., Xu, S., Amro, N.A., and Liu, G.Y., Fabrication and imaging of nanometer-sized protein patterns, *Langmuir,* 15, 8580, 1999.
32. Kenseth, J.R., Harnisch, J.A., Jones, V.W., and Porter, M.D., Investigation of approaches for the fabrication of protein patterns by scanning probe lithography, *Langmuir,* 17, 4105, 2001.
33. Wadu-Mesthrig, K., Amro, N.A., Garno, J.C., Xu, S., and Liu, G.Y., Fabrication of nanometer-sized protein patterns using atomic force microscopy and selective immobilization, *Biophys. J.,* 80, 1891, 2001.
34. Wilson, D.L., Martin, R., Hong, S., Cronin-Golomb, M., Mirkin, C.A., and Kaplan, D.L., Surface organization and nanopatterning of collagen by dip-pen nanolithography, *PNAS,* 98, 13660, 2001.
35. Folkman, J., and Moscona, A., Role of cell shape in growth control, *Nature* 273, 345, 1978.
36. Clement, B., Guguen-Guillouzo, C., Campon, J.P., Glaise, D., Bourel, M., and Guillouzo, A., Long-term co-cultures of adult human hepatocytes with rat liver epithelial cells: modulation of albumin secretion and accumulation of extracellular material, *Hepatology,* 4, 373, 1984.
37. Guguen-Guillouzo, C., Clement, C., Baffet, G., Beaumont, C., Morel-Chany, E., Glaise, D., and Guillouzo, A., Maintenance and reversibility of active albumin secretion by adult rat hepatocytes co-cultured with another liver epithelial cell type, *Exp. Cell Res.,* 46, 47, 1983.
38. Bhatia, S.N., Yarmush, M.L., and Toner, M., Controlling cell interactions by micropatterning in co-cultures: hepatocytes and 3T3 fibroblasts, *J. Biomed. Mater. Res.*, 34, 189, 1997.
39. Bhatia, S.N., Balis, U.J., Yarmush, M.L., and Toner, M., Probing hertotypic cell interactions: hepatocyte function in microfabricated co-cultures, *J. Biomater. Sci. Polymer Ed.,* 9, 1137, 1998.
40. Bhatia, S.N., Balis, U.J., Yarmush, M.L., and Toner, M., Effect of cell–cell interactions in preservation of cellular phenotype: cocultivation of hepatocytes and nonparenchymal cells, *FASEB J.*, 13, 1883, 1999.
41. Berthiaume, F., Moghe, P.V., Toner, M., and Yarmush, M.L., Effect of extracellular matrix topology on cell structure, function, and physiological responsiveness: hepatocytes cultured in sandwich configuration, *FASEB J.,* 10, 1471, 1996.

42. Powers, M.J., Domansky, K., Kaazempur-Mofrad, M.R., Kalezi, A., Capitano, A., Upadhyaya, A., Kurzawski, P., Wack, K.E., Stolz, D.B., Kamm, R., and Griffith, L.G., A microfabricated array bioreactor for perfused 3-D liver culture, *Biotechnol. Bioeng.,* 78, 257, 2002.
43. Liu, V., and Bhatia, S., Three-dimensional photopatterning of hydrogels containing cells, *Biomedical Microdevices*, 4, 257, 2002.
44. Madou, M., *Fundamentals of Microfabrication: The Science of Miniaturization,* Boca Raton, CRC Press, 2002.
45. Koehler, M,. Ed., *Microsystemtechnology, A Powerful Tool for Biomolecular Studies*, Birkhaeuser, Berlin, 1999.
46. Manz, A., What can chips technology offer for next century's chemistry and life sciences?, *Chima* 50, 140, 1996.
47. Seidel, H., Nasschemische Tiefaetztechnick, in *Mikromechanik*, A. Heuberger, Ed., Heidelberg, Springer Verlag, 1989.
48. Elwenspoek, M., and Jansen, H.V., *Silicon Micromachining*, Cambridge, Cambridge University Press, 1998.
49. de Boer, M. J., Gardeniers, J.G.E., Jansen, H.V., Smulders, E., Gilde, M.-J., Roelofs, G., Sasserath, J.N., and Elwenspoek, M., Guidelines for etching silicon MEMS structures using fluorine high-density plasmas at cryogenic temperatures, *J. Micro Elec. Mech. Syst.*, 11, 385, 2002.
50. Ayon, A.A., Braff, R., Lin, C.C., Sawin, H.H., and Schmidt, M.A., Characterization of a time multiplexed inductively coupled plasma etcher, *J. Electochem. Soc.*, 146, 339, 1999.
51. Mooney, D.J., and Langer. R.S., Engineering biomaterials for tissue engineering: the 10–100 micron size scale, in *The Biomedical Engineering Handbook*, 2nd Ed., Bronzino, J.D., Ed., Boca Raton, CRC Press LLC, 2000.
52. Matlaga B.F., Yasenchak, L.P., and Salthouse, T.N., Tissue response to implanted polymers: the significance of sample shape, *J. Biomed. Mater. Res.*, 10, 391, 1976.
53. Kapur, R., Spargo, B.J., Chen, M., Calvert, J.M., and Rudolph, A.S., Fabrication and selective surface modification of three dimensionally textured biomedical polymers from etched silicon substrates, *J. Biomed. Mater. Res.*, 33, 205, 1996.
54. Miller, C., Shanks, H., Witt, A., Rutkowski, G., and Mallapragda, S., Oriented schwann cell growth on micropatterened biodegradable polymer substrates, *Biomaterials,* 22, 1263, 2001.
55. Mooney, D.J., Mazzoni, C.L., Breuer, G., McNamara, K., Hern, D., Vacanti, J.P., and Langer, R., Stabilized polyglycolic acid fibre-based tubes for tissue engineering, *Biomaterials,* 17, 115, 1996.
56. Mikos, A.G., Sarakinos, G., Leite, S.M., Vacanti, J.P., and Langer, R., Laminated three dimensional biodegradable foams for use in tissue engineering, *Biomaterials,* 14, 323, 1993.
57. Mooney, D.J., Baldwin, D.F., Suh, N.P., Vacanti, J.P., and Langer, R., Novel approach to fabricate porous sponges of poly(D,L-lactide-co-glycolic acid) without the use of organic solvents, *Biomaterials,* 17, 1417, 1996.
58. Nam, Y.S., and Park, T.G., Biodegradable polymeric microcellular foams by modified thermally induced phase separation method, *Biomaterials,* 20, 1783, 1999.
59. Smith, D.C., Adhesives and sealants, *Biomaterials Science: An Introduction to Materials in Medicine*, Ratner, B.D., Hoffman, A.S., Schoen, F.J., and Lemons, J.E., Eds., Academic Press, San Diego, 1997, 319.

60. Miyamoto, S., Ohashi, A., Kimura, J., Tobe, S., and Akaike, T., A novel approach for toxicity sensing using hepatocytes on a collage-patterned plate, *Sensors and Actuators B,* 13, 196, 1993.
61. Bhatia, S.N., Toner, M., Tompkins, R.G., and Yarmush, M.L., Selective adhesion of hepatocytes on patterned surfaces, *Ann. N.Y. Acad. Sci.,* 745, 187, 1994.
62. Jackman, R.J., Duffy, D.C., Cherniavskaya, O., and Whitesides, G.M., Using elastomeric membranes as dry resists for dry lift-off, *Langmuir,* 15, 2973, 1999.
63. Folch, A., Jo, B.-H., Hurtado, O., Beebe, D.J., and Toner, M., Microfabricated elastomeric stencils for micropatterning cell cultures, *J. Biomed. Mater. Sci.,* 52, 346, 2000
64. Gölander, C.-G., and Eriksson, J.C., ESCA studies of the adsorption of polyethyleneimine and glutaraldehyde-reacted polyethyleneimine on polyethylene and mica surfaces, *J. Coll. Int. Sci.*, 119, 38, 1987.
65. Shoaf, S.E., and Feinerman, A.D., Aligned Au-Si eutectic bonding of silicon structures, *J. Vac. Sic. Technol.*, A12, 19, 1994.

6 Cellular Mechanotransduction

Lidan You and Christopher R. Jacobs

CONTENTS

0-8493-1940-4/05/$0.00+$1.50

6.1 INTRODUCTION

For many cells, the ability to respond to mechanical perturbation in their environment is critical to their function. Therefore, an understanding of cellular mechanosensitivity is critical to the success of biomedical devices designed on a nanoscale.

Mechanotransduction is the process by which cells convert mechanical stimuli into a chemical response. It can occur both in cells specialized for sensing mechanical cues such as mechanoreceptors, and in cells whose primary function is not mechanosensory. Here we review cell mechanotransduction in multiple tissues, highlighting the increasing understanding of this process in bone cells.

The mechanical environment of the cell includes a wide variety of physical signals that potentially regulate cell metabolism. Examples include tissue deformation–induced direct strain, tissue deformation–induced fluid flow, streaming potentials (ion movement along the cell membrane), the piezoelectric field effect, and pressure changes. Hair cells and touch receptors experience direct deformation from the extracellular matrix (ECM). Endothelial cells are exposed to shear stress and can have adaptive response to stress in the range of 1 to 2 Pa.[1] The mechanical environment around bone cells includes direct strain (0.04–0.3%),[2] pressure, fluid flow–induced shear stress (0.8–3 Pa),[3] and the concomitant streaming potentials. Chondrocytes are predominantly exposed to 3–10 MPa of pressure[4] and 10–15% of compression.[5–7] Fibroblasts are mostly exposed to stretch 1–5%,[8] while muscle cells can be stretched up to 100%.[9] Furthermore, different cell types can have different responses to the same mechanical stimuli. For example, in response to cyclic pressure (10–40 kPa, 1 Hz), osteoblasts increased their proliferation transcription and translation of alkaline phosphatase (ALP), whereas endothelial cells have no response.[10] And to the intermittent hydrostatic pressure and pulsatile fluid flow (PFF), osteocytes are much more responsive, compared to osteoblasts, in terms of release of prostaglandin E_2 (PGE_2).[11]

The mechanism for detection of mechanical cues is highly complex. Cells detect mechanical cues mainly by two mechanisms: mechanosensitive channels/receptors on the membrane, where mechanical stimuli are directly translated into chemical signals such as Ca^{2+} influx and kinase activation, and cytoskeleton-integrin connection, where mechanical forces are transduced into cells, which then induce further mechanical/chemical changes (e.g., cytoskeleton reorganization, focal adhesion complex formation, and activation of a kinase cascade) inside the cells. Another possible pathway is that mechanical forces may be detected in an inside-out manner. For example, the mechanical disturbance could be transferred to the cytoskeleton directly, which will then deform the membrane so that the mechanosensitive ion channel/receptors on the membrane are activated. Furthermore, the process may involve cross talk between elements in multiple signaling pathways, so that distinguishing between different pathways is very challenging.

Once an external mechanical signal is detected, a cell can respond in a rapid (milliseconds), intermediate (minutes), or longer (hours–days) time frame. The response might include reorganization of the cytoskeleton, activation of second messengers such as Ca^{2+} and cyclic adenosine monophosphate (cAMP), activation/inhibition of certain genes, and release of autocrine/paracrine factors (e.g., PGE_2) into extracellular space. Different mechanical stimuli may cause different cellular responses. Multiple pathways may be involved in the cellular response to a single mechanical cue. Despite extensive efforts, the cellular mechanism of mechanotransduction is largely unknown. In this chapter, we explore the response of cells to various mechanical stimuli in a diverse range of tissues, with the main focus on bone tissue. We then review the regulation mechanism and the roles of several well-studied intracellular messengers (e.g., Ca^{2+}, cAMP, NO, and PGE_2) and the regulation of gene transcription in response to mechanical cues.

In this chapter, we first explore mechanical stimuli in cells. Following that, we will focus on the detection mechanism. Finally we will present information about cellular response to mechanical load. Although other cells and tissues will be described, our primary focus will be bone cell mechanosensitivity.

6.2 MECHANICAL STIMULI ON CELLS

Both *in vivo* and *in vitro* approaches have been exploited to study the cell's response to mechanical loading. However, it is difficult to isolate a specific mechanical factor to which cells respond in *in vivo* study. For example, when bone is mechanically loaded, multiple cell-level physical stimuli can be induced on bone cells, including direct strain, shear stress, pressure changes, streaming potentials, and the piezoelectrical effect. Hence, *in vivo* it is hard to study the mechanism of response to a specific physical stimulus. *In vitro* studies, however, provide an ideal approach in which a specific physical stimulus can be applied to cells, and mechanism of response can be studied. Therefore, to gain understanding of the effect of a specific mechanical stimulation, we employ *in vitro* studies to explore the mechanotransduction mechanism.

6.2.1 PRESSURE

Pressure is one of the main mechanical stimuli that chondrocytes are subjected to. It has been recognized that hydrostatic pressure can induce changes in chondrocyte metabolism, and the response is magnitude and frequency dependent.[12–16] For example, 20 h of cyclic pressure at 5 MPa and 0.25 to 0.5 Hz increased sulfate incorporation, whereas lower frequencies or shorter durations inhibited incorporation.[13]

Intermittent hydrostatic pressure has been shown to have significant effects on aggrecan and type II collagen expression and glycosaminoglycan synthesis.[14,15] It seems that intermittent hydrostatic pressure was more beneficial than constant pressure.[15]

Pressure has also been investigated in bone. There are three porosities of bone associated with bone fluid: the vascular porosity, the lacunar-canalicular porosity (the space around the cell bodies and cell body protrusions in bone matrix), and the collagen-apatite porosity. Only the pressure in the vascular porosity has been measured and is in the range of 10 to 100 mm Hg due to the blood circulation. Mechanical loading has been shown to enhance the intramedulary pressure greatly.[17] It is estimated that pressure in vascular porosity for daily activities of humans falls in the range of 10 to 40 kPa.[18] The peak pore pressure in lacunar-canalicular system is predicted to be at least one order of magnitude larger than in vascular porosity due to the different resistance arising from the different geometry dimensions of the two porosities.[19]

In vitro studies have shown that bone cells can respond to hydraulic pressure. Ozawa et al. showed that constant pressure at 300 kPa caused osteoblasts to produce and secret PGE_2;[20] intermittent hydrostatic pressure (13 kPa, 0.3 Hz) promoted the osteoblastic phenotype in osteoblasts including upregulated ALP activity, collagen expression, and actin expression, but decreased collagen expression in osteoprogenitor cells.[21] Cyclic pressure (10–40 kPa at 1 Hz) on osteoblasts enhances osteoblast functions associated with new bone formation by stimulating both synthesis and deposition of collagen.[22] Pressure of 13 kPa at 0.3 Hz has been shown to stimulate prostaglandin production in osteocytes and osteoblasts.[11] Cyclic pressure affected various genes in osteoblasts differently. For example, exposure of osteoblasts to cyclic pressure (10–40 kPa, at 1.0 Hz) resulted in enhanced transcription and translation of ALP after 5 days but did not affect osteopontin (OP) mRNA expression during the same time period.[10] Taken together, hydrostatic pressure can induce multiple responses in bone cells. However, for most studies the pressures employed are relatively low compared to the estimated physiological level of pressure in the lacunar-canalicular system. Therefore, it is not clear whether pressure plays an important role in mechanotransduction in bone cells.

Relatively few studies have been conducted on the effect of pressures on stem cells. Exposure of bone marrow cells, which include stem cells and progenitors of multiple musculoskeletal cells such as osteoblasts, osteoclasts, and chondrocytes to cyclic pressure (10–40 kPa at 1 Hz) resulted in decreased formation of osteoclastic cells from progenitors and decreased bone resorptive activity by osteoclasts.[18] Repetitive pressure of 3–10 MPa on stem cells has been shown to enhance chondrogenic differentiation.[23]

Pressure has also been shown to regulate the metabolism of nonskeletal tissues. For example, endothelial cells were not responsive to cyclic pressure, 10–40 kPa, 1 Hz, whereas fibroblast proliferation increased under similar test conditions.[10] However, Shin et al. showed that endothelial cells can sense and respond to physiologic levels of cyclic pressure (20–140 mmHg, 1 Hz) by altering cell proliferation and apoptosis.[24] It seems that, at the same level of pressure, different cell types will exhibit different responses. Moreover, for the same type of cell, different levels of pressure can induce different responses.

6.2.2 Strain

6.2.2.1 Hair Cell

The stereocilia of inner hair cells have been extensively studied because they respond to acoustic waves and sense the directionality of linear and angular momentum needed for balance.[25] The stereocilia also have a highly specialized structure in which they are arranged in rows of ascending height, much like steps, in which their tips and upper lateral margins are cross-linked by thin filaments, which link adjacent stereocilia across rows in the direction of the steps[26] (see Figure 6.1). Hudspeth[25] also showed that deflections of the stereocilia across the rows in the direction of increasing height led to ion channel opening and a depolarization of the cells,

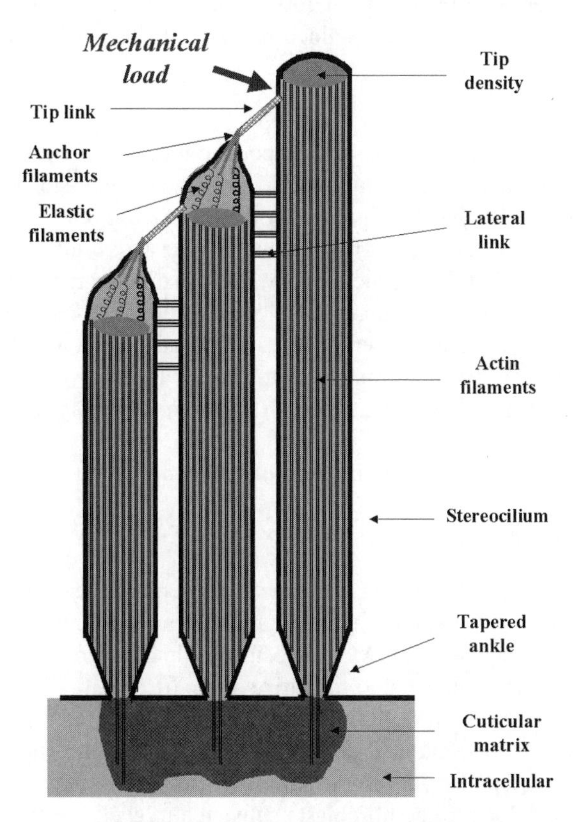

FIGURE 6.1 Schematic cross section of stereocilia on one hair cell. Stereocilia are arranged in rows of ascending height, resulting in a staircase-like pattern. They are connected to each other by tip and lateral links. When a mechanical load is applied stereocilia pivot as a coupled unit about tapered ankles at their base (From Weinbaum et al., *Biorheology*, 38, 119–142, 2001. With permission.)

whereas a bending in the opposite direction led to a hyperpolarization. In contrast, bending along a row causes no electrical response. In the case of the hair cell this depolarization leads to neurotransmitter release at synapses where the afferent nerves carry electrical signals to the brain.

The stereocilia on hair cells are 5–10 μm in length and significantly taper at their base from a diameter that is typically 0.125 μm to roughly one fourth this diameter. This allows the stereocilia to torsionally pivot at their base when they are subjected to transverse flow. The initial mechanical models for their bending treated the stereocilia as rigid levers with torsional springs at their base and elastic gating springs at their tips.[27,28] The gating spring, which is associated with the tip link filament, is critical to the mechanoelectrical transduction. Hudspeth has shown that they transmit a force that leads to the opening and closing of membrane ion channels that are in close proximity to the attachment site at the stereocilia tip.[29] Kachar et al.[30] suggested that the tip links are rigid filaments and that the elastic component arises from the tent-like membrane deformation of the microvillus tip at the point of tip attachment above a densely stained region called the tip density.[30]

6.2.2.2 Chondrocyte

It has been shown that chondrocytes can respond to mechanical strain in a magnitude-dependent manner. At low magnitude, both tension and compression have been shown to have anti-inflammatory effect on chondrocytes; 3–8% of cyclic tension (equibiaxial strain) acts as a potent antagonist of IL-1β and TNF-α and results in the inhibition of nitric oxide (NO) and PGE_2 production as well as matrix degradation.[31–34] Similarly, physiological levels of cyclic compression (15%) inhibit IL-1β-induced PGE_2 and NO production.[35,36] However, at higher magnitudes, tensile or compressive strain acts differently. Instead of being anti-inflammatory, both types of strain will induce the synthesis of proinflammatory mediators, such as NO and PGE_2, and may thus aggravate the effects of IL-1β and TNF-α.[37,38] Taken together, it seems that the response of chondrocytes to mechanical strain is magnitude dependent.[38]

6.2.2.3 Fibroblast

Extensive studies have shown that fibroblasts can sense force-induced deformation in the ECM and convert the mechanical signal into multiple intracellular responses and finally induce changes in ECM composition. Almekinders et al. (1993) showed that human fibroblasts subjected to repetitive strain (0.25%, 0.17 Hz and 1 Hz) will release PGE_2.[39] Banes et al. (1995) showed that in avian tendon fibroblasts, cyclic strain (5%, 1 Hz, 8 h) stimulated phosphorylation of tyrosine residues in multiple proteins, including pp60src, a protein kinase that phosphorylates receptor protein tyrosine kinases.[40] Rat cardiac fibroblasts subjected to cyclic strain (10%, 1 Hz, 24 h) release several-fold increased levels of active TGF-β into the medium, which induces procollagen α1(I) gene transcription.[41,42] The mRNA level of CTGF, a transcription factor known to stimulate expression of ECM proteins, was elevated more than fourfold in human lung fibroblasts when subjected to tensile stress.[43] When chick fibroblasts (15%, 0.3 Hz) were stretched, within 3–6 h mRNA for another matrix protein (tenascin-C) was induced two- to fourfold compared to cells at rest; elevated secretion of the protein could be observed after 6 h.[44] Collectively, these findings demonstrate that fibroblasts have the ability to respond to physiological levels of tensile strain with multiple responses and ultimately modify their extracellular environment by secreting ECM proteins.

6.2.2.4 Bone Cells

It has been well documented that bone cells can respond to mechanical strain. Substrate deformation has been shown to influence DNA synthesis, second messenger production (cyclic AMP), release of paracrine factors (PGE_2), redistribution of integrins, the activity of enzymes important to mineralization (alkaline phosphatase), and both collagenous and noncollagenous matrix protein synthesis.[45–48] However, the flexible membranes utilized as substrates in these studies exposed the cells to deformations of between 5-fold and 125-fold greater than that observed to occur in bone during routine physical activity (0.04–0.3%).[2] To apply smaller deformations, some investigators have utilized systems based on bending of rigid substrates such as glass slides.[49–54] Unfortunately this approach exposes the cells to significant fluid shear stress due to the lateral motion of the slide through the bathing media. By deforming flexible-bottomed plates (Flexercell Strain Unit, Flexercell Corp., McKeesport, PA), several studies have shown that bone cells can respond to rather small deformations, 0.2–0.4%, which fall in the range of physiological level of strain.[55–57] However, it is not clear what level of fluid shear stress was induced while the strain was created in these studies. Computational studies have shown that the reactive fluid shear stress could be induced due to coupled motion of the overlying culture medium.[58,59] In a recent study by You et al. (2000),[60] a novel stretch device was used, by which a 1% strain created only 0.0001 Pa shear stress, so that the effect of stretch alone on osteoblastic cells could be determined. Using this system, intracellular calcium concentration was not affected at strains of 5% strain or less (Figure 6.2). Furthermore, OP mRNA levels were not affected up to 0.5% strain, suggesting that osteoblastic cells in culture do not respond to physiological levels of strain.[60] Similar

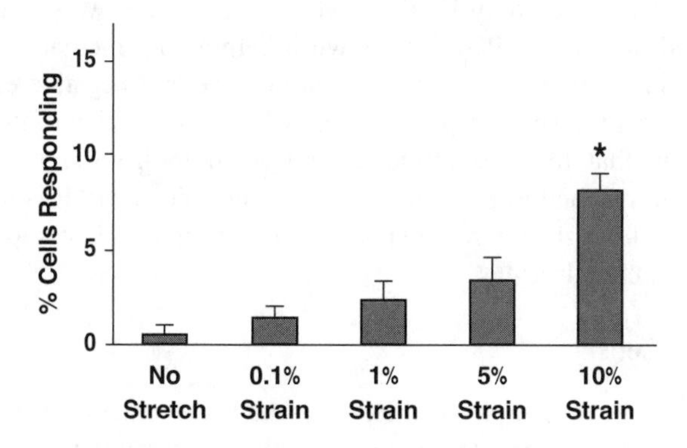

FIGURE 6.2 Fraction of bone cells responding with an increase in $[Ca^{2+}]_i$ at different substrate strains. (From You et al., *J. Biomed. Eng.*, 122, 387–393, 2000. With permission from ASME.)

conclusion was drawn by two other studies,[52,61] where comparison was made on effect of physical level shear stress and the physical level strain induced by either stretching of substrate[61] or four-point bending of plastic plates where the thickness of the plate was used to change the strain magnitude with the fluid shear stress being kept the constant.[52] Taken together, it seems that physiological levels of strain do not play an important role in bone cell mechanotransduction.

6.2.3 Fluid Flow–Induced Shear Stress

6.2.3.1 Endothelial Cell

Endothelial cells are constantly exposed to fluid flow–induced shear stress. It is not surprising that these cells can respond and adapt to the fluid shear stress they are exposed to. Extensive studies have been done in this field. The very early cellular response to shear stress involves elevation of the intracellular Ca^{2+} concentration ($[Ca^{2+}]_i$)[62–64] and ion fluxes across the cell membrane.[63,65,66] Elevation of $[Ca^{2+}]_i$ is believed to play a key role in signal transduction events elicited by fluid shear stress. For instance, it is found that increased $[Ca^{2+}]_i$ stimulates the Ca^{2+}-dependent synthesis of vasodilators such as NO[67] and prostacyclin.[68,69] Other responses to fluid shear stress include NO release,[67,70] increased expression of endothelial NO synthase,[71,72] and actin cytoskeleton rearrangement.[73] Taken together, it has been amply demonstrated that fluid shear stress is a potent physical signal in regulating endothelial function.

6.2.3.2 Chondrocyte

Fluid flow has been used in a number of studies as a physiologically relevant stimulus for chondrocytes.[74–78] Fluid-induced shear stress applied to both human and bovine articular chondrocytes for 24–72 h induced morphological and metabolic changes including elongation, reorientation, and changes in proteoglycan synthesis and size.[74,75] In Smith et al. (1995)[74] human articular chondrocytes were placed in culture, and fluid shear was applied using a cone viscometer. The synthesis of glycosaminoglycans (GAGs) increased by 100%; however, this increase was concomitant with a 10- to 20-fold increase in PGE_2, a known inflammatory mediator. This suggests that fluid flow–induced shear may have both positive and negative effects. Bovine articular chondrocytes respond rapidly to steady flow and oscillatory fluid flow (OFF) by a mechanism that involves intracellular Ca^{2+} mobilization.[76–78] It seems that chondrocytes can respond to physiological relevant fluid shear. It is not clear, however, whether the fluid shear stress plays a more important role compared to hydrostatic pressure in chondrocytes.

6.2.3.3 Bone Cells

The realization that external physical loading of a bone organ leads to cellular-level fluid flow results from a consideration of the microanatomy that was made over 25 year ago.[79] Osteocytes, the most numerous cell type in bone, are found in the mineralized matrix in defined spaces known as lacunae. These spaces are intercon-

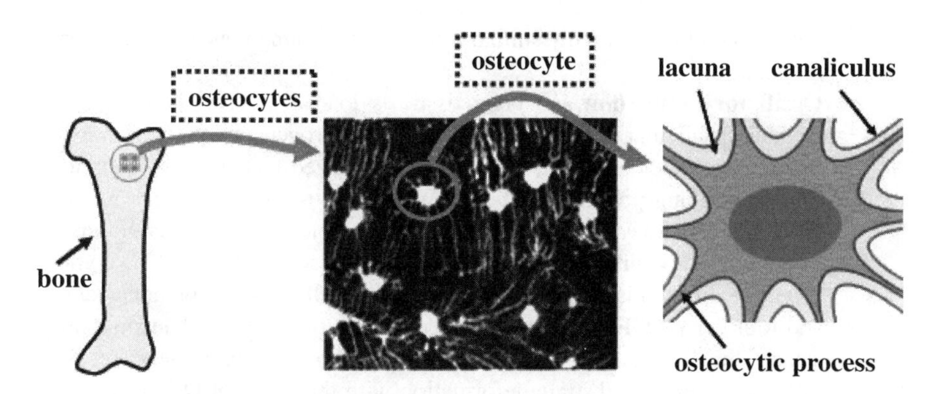

FIGURE 6.3 Sketch of bone cells embedded in bone matrix. (Reprinted from *J. Biomech.*, 34, You et al., pp. 1375–1386, ©2001, with permission from Elsevier.)

nected via long tunnels, known as canaliculae, inhabited by extensions of the osteocytes (Figure 6.3). It has been shown that the lacunar-canalicular network contains a mobile extracellular fluid which communicates with the vascular space.[80,81] In the original work by Piekarske and Munro (1977),[79] it was hypothesized that when bone tissue is loaded, fluid was forced out of the lacunar-canalicular space and into the vascular space, and as in a periodic activity such as ambulation, when the tissue is unloaded the fluid would return from the vascular space.

Due to the technical difficulty, directly quantifying loading-induced lacunar-canalicular fluid flow rates has not been feasible. Detailed theoretical models of this phenomenon have been constructed,[3,82] and they have been verified in two important ways. Qualitatively, injected tracers have been found in the lacunar-canalicular space, and the rate of transfer is enhanced by mechanical loading at the whole-organ level.[80,81] Second, flow-induced streaming potentials predicted by the models have been observed to be accurate.[82–84] Weinbaum et al. (1994) predicted a 0.8–3 Pa cell membrane shear stress in the lacunar-canalicular system due to typical mechanical loading.[3] Bone cells exposed to fluid flow in this range of shear stress in flow chambers of various designs have been shown to respond with a wide range of biologic outcomes.

- **Steady flow.** Since the early 1990s, many studies have shown that bone cells can respond to steady flow. It has been reported that applying steady flow to bone cells can induce intracellular Ca^{2+} mobilization,[85,86] cAMP production,[87] rapidly release of NO,[61,88] PGE_2,[61,89,90] increased expression of transforming growth factor beta 1 (TGFβ-1),[91] expression of cyclooxygenase-2 (COX-2) and c-Fos,[92] enhanced cell–cell communications,[93] and activation of ERK and $\alpha_v\beta_3$ integrin recruitment.[94]
- **Pulsatile fluid flow.** Although dynamic in nature, PFF is still unidirectional. Studies on the effect of PFF on bone cells showed that it down-regulates the ALP mRNA expression,[95] increases PGE_2, PGI_2,[96–98] and NO production,[99] and upregulates intercellular cell–cell communication.[100] It

seems that PFF has largely similar affects in regulating bone cell behavior as steady flow.

- **Oscillatory fluid flow.** As bone tissue is loaded *in vivo*, fluid in the lacunar-canalicular network experiences a heterogeneous pressurization in response to the deformation of the mineralized matrix. This leads to fluid flow along pressure gradients. When loading is removed, pressure gradients and flows are reversed. These fluid motions are dynamic and oscillatory (flow direction reverses) in nature. Jacobs et al. (1998) compared the differential effect of steady versus oscillating flow on bone cells and found that OFF stimulates bone cells via a different mechanism than steady flow.[101] A later study by You et al. (2000) compared the effect of different levels of substrate deformation with the effect of OFF on bone cells, finding that, at physiological relevant levels, OFF is more stimulatory to bone cells than substrate deformation.[60]

Bone cells have been found to be able to respond to OFF with intracellular Ca^{2+} mobilization.[60,101–108] The response was found to be age, frequency, and shear stress dependent.[102] Intercellular communication seems have no effect on Ca^{2+} response.[103] Donahue et al. (2003) reported that bone cells exhibit a refractory period in intracellular calcium signaling. It was found that the cells were dramatically less responsive immediately after the first flow exposure, but that responsiveness was gradually regained (~900 s) until they were as responsive as during the initial flow exposure.[106] Building on these observation, Batra et al. (2004) found that periodic insertion of a rest period appears to enhance intracellular calcium signaling.[109] A dramatically higher percentage of cells responded with a higher magnitude transient increase when a 10-s or 15-s rest period was inserted (Figure 6.4). Taken together, intracellular Ca^{2+} increase has been shown to be one of the early signals induced by OFF, and the responsiveness depends on multiple parameters of flow regime that bone cells are exposed to.

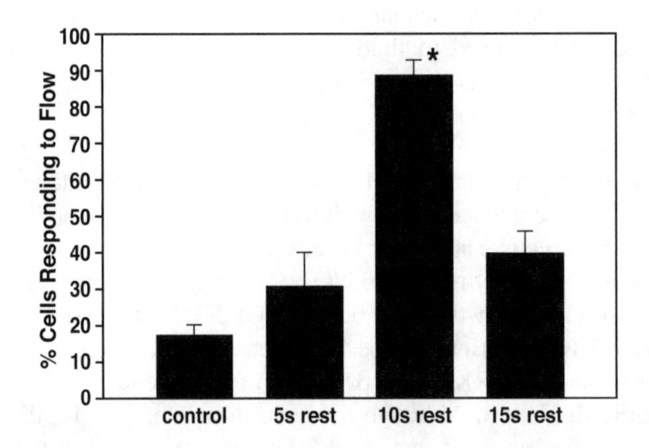

FIGURE 6.4 Insertion of 10-s and 15-s rest periods into the flow profile led to significantly higher percent of cells responding compared to control ($p < 0.05$). Shear stress = 1 Pa.

Bone cells have also been shown to be able to response to OFF with changes in other intracellular and extracellular molecules. Ogata (1997) reported that dish shaking induced OFF enhanced Erg1 (an early response gene to mechanical stimuli) mRNA expression.[110] Osteopontin (the most abundant noncollagenous matrix protein in bone) mRNA expression is activated by OFF[60,104] and is dependent on Ca^{2+} signaling and p38 and ERKs pathways.[104] TNF-α-induced NF-κB has been shown to be inhibited by OFF via an IκB kinase pathway.[111] Several studies have shown that OFF induces PGE_2 release, and this release is dependent on gap junction communication.[103,107] OFF has also been shown to upregulate gap junction expression by an ERK1/2 MAPK dependent mechanism.[103,107,112]

6.2.3.4 Bone Marrow Stromal Cells

Loading-induced OFF may also regulate the formation of new bone cells. Our lab employed a commercially available (BioWhittaker, Walkersville, MD) cell culture model of human bone marrow stromal cells, which is capable of forming the tissues of mesenchymal origin.[113] These cells were obtained from the bone marrow from healthy volunteers and subjected to isolation by density centrifugation and selective culturing conditions.[114–117] We found that exposure of these cells to loading-induced OFF at 2 Pa, 1 Hz is able to trigger calcium signaling (Figure 6.5). When the cells were exposed to OFF for 2 h at 1 Pa, 1 Hz, the proliferation rate was increased by almost 60% (Figure 6.6). Also, OP and osteocalcin expression levels were increased by roughly 50%, whereas the Cbfa1 and Col I expression was not shown to be affected (Figure 6.7). Taken together, these studies suggest that marrow stromal cells are mechanosensitive to OFF, and OFF can increases osteogenic differentiation as well as cell proliferation.[113]

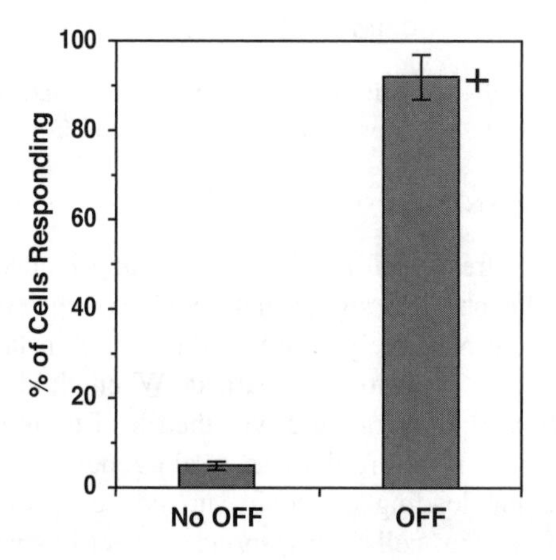

FIGURE 6.5 Percent of cells exhibiting an $[Ca^{2+}]_i$ increase in response to OFF. + indicates a significant difference from control.

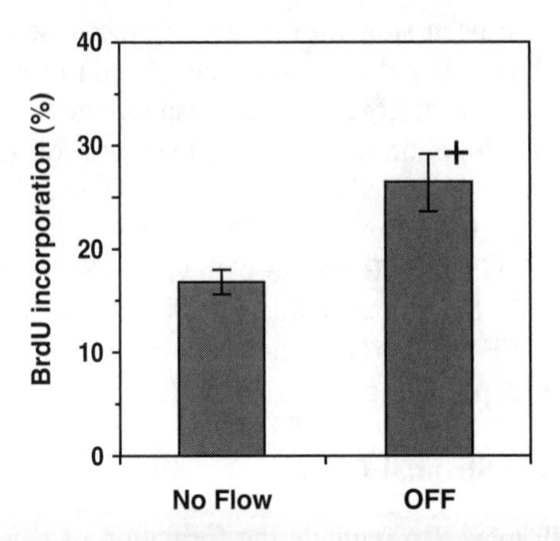

FIGURE 6.6 OFF increases BrdU incorporation for bone marrow stromal cells. BrdU incorporated 24 hours after OFF.

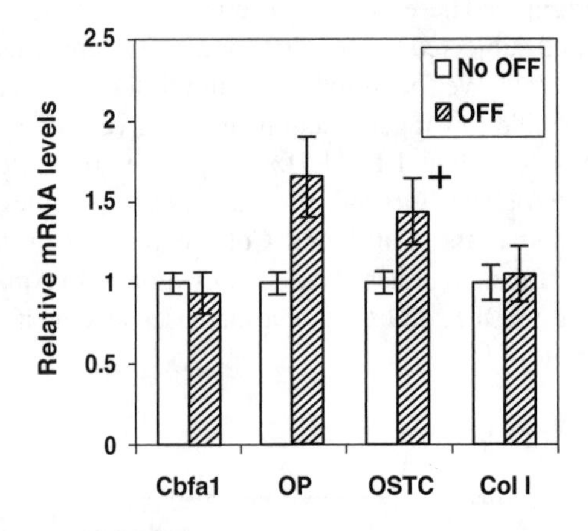

FIGURE 6.7 Effects of OFF on relative mRNA levels for bone marrow stromal cells: mRNA isolation 24 h after OFF. OP: osteopontin. OSTC: osteocalcin. Col I: collagen I.

6.2.4 Streaming Potentials

In addition to the shear stress resulting from the loading-induced fluid flow in bone, there is another cellular physical signal that could be induced due to flow. As the bone matrix proteins are negatively charged, the very thin layer of fluid near the surface of these proteins is positively charged. When fluid flow is induced, the movement of this charged layer of fluid will then lead to an electric field in bone tissue. This phenomenon is referred to as strain-generated streaming potentials. Studies have shown that loading on bone can induce streaming potentials.[118–121] Studies have also shown that cells can respond to electric stimuli.[122–141] Therefore, it has been proposed that loading-induced streaming potentials could be one of the physical signals that cells use to detect the fluid flow. However, loading-induced streaming potentials are relatively small. Several studies have shown that when these relatively small physiologic electric fields are applied to bone cells in culture, little or no change in cellular behavior is found.[11,87,142,143]

6.2.5 Chemotransport

In addition to its direct effects on cells, loading-induced fluid flow may be a mechanism for increasing nutrient exchange in bone.[79,144,145] Experimentally, physiologic levels of mechanical loading applied to bones *in situ* have been shown to increase the levels of fluorescent tracers accumulated in bone tissue,[80] indicating that chemotransport is mechanically regulated *in vivo*. Thus, if fluid flow–induced chemotransport can be shown to influence cell metabolism, it may have a significant role in mechanotransduction distinct from flow-induced shear stress. However, experiments aimed at determining the role of chemotransport are contradictory. Jacobs et al. (1998) found that a greater response was observed at all frequencies in cells exposed to a net fluid transport (steady and pulsing) relative to cells exposed to no net fluid transport (oscillating), despite the fact that the peak shear stress was the same for all flow rates. Furthermore, in the case of dynamic flow, a trend of decreasing numbers of responding cells with increasing frequency was observed. Thus they conclude that chemotransport of a serum factor may play an important role in the response of bone cells to fluid flow.[101] Allen et al. (2000) have demonstrated that shear flow in serum-supplemented medium produces enhanced intracellular calcium responses in individual cells compared to serum-free flow stimulation.[146] You et al. (2002) reported that OFF in the absence of serum failed to increase $[Ca^{2+}]_i$ in MC3T3-E1 cells, but by adding ATP or UTP to serum-free media, the ability of fluid flow to increase $[Ca^{2+}]_i$ was restored, suggesting that ATP or UTP in the extracellular medium may mediate the effect of fluid flow on $[Ca^{2+}]_i$.[105] A recent study by Donahue et al. (2003) investigated the role of chemotransport in the OFF-induced response in bone cells.[108] They showed that both the percentage of cells responding with a $Ca^{2+}{}_i$ oscillation and the production of PGE_2 increased with increasing flow rate. In addition, depriving the cells of nutrients during fluid flow resulted in an inhibition of both Ca^{2+} mobilization and PGE_2 production. However, in contrast to the above studies, by employing fluid media of various viscosities modified with the addition of neutral dextran, several groups found that bone cell responsiveness, in terms of cAMP, NO, and PGE_2 production, depended on shear stress level, but not on fluid flow rate when cells were exposed to steady flow or pulsating fluid flow.[87,143,147]

Taken together, it seems that chemotransport plays different roles in different cell signaling pathways. If one looks at intracellular Ca^{2+} mobilization, chemotransport seems an essential mediator. However, for cell response in terms of cAMP, NO, and PGE_2, the role of chemotransport is not clear. It is worth noting that, due to the different experimental parameters (cell line, medium, flow regimes, shear stress magnitude, flow rate, shear stress frequency, flow exposure time, method of analyzing the response), one has to be cautious in interpreting and comparing the results from different studies. For example, both using PGE_2 as the measure of the cell response, two groups reached very different conclusions. Donahue et al. (2003) found that, given the same shear stress, increased flow rate results in increased PGE_2

production.[108] In contrast, Bakker et al. (2001) reported no difference in PGE_2 production with increasing flow rate.[147] However, the two groups employed very different flow regimes and methods of measuring PGE_2. Donahue et al. (2003) exposed the MC3T3-E1 cells to a 1 Hz, 2 Pa OFF at 43, 28, and 18 ml/min flow rate for an hour, and assayed the PGE_2 production at 1 h post flow, whereas Bakker et al. (2001) employed 1.2 Pa, 3/9 Hz PFF at 0.63 and 0.20 cm^3/s flow rate to primary cells isolated from mouse long bone, and the PGE_2 release was measured at 5, 10, and 15 min post flow. Therefore, it is hard to compare the results from these two studies, although the same parameter, PGE_2 release, was used in both studies.

6.3 CELL DETECTION OF MECHANICAL STIMULI

Cells can detect a mechanical disturbance by diverse means, including mechanosensitive channels (e.g., stretch-activated ion channels), integrin linkages from the ECM to the cytoskeleton, and through cadherins in cell–cell connections.

6.3.1 Mechanosensitive Channel/Mechano-Gated Channel

It has been shown that ion channels activity can be altered when the cell membrane is subjected to mechanical stimuli. These channels are called mechanosentive or mechano-gated channels. Such channels have been observed in both sensory and nonsensory cells in almost all living species. The influx of ions from the extracellular environment to the intracellular compartment can serve as an important intracellular signal. Thus, the mechanical cue can be translated into a chemical signal through this mechanism. It is well accepted that mechanosensitive/mechano-gated channels play an important role in mechanotransduction.

6.3.1.1 Endothelial Cells

Stretch-activated cation channels have been suggested to be responsible for Ca^{2+} influx in response to the mechanical stimuli in some cells. Such stretch-activated channels have been identified in vascular endothelial cells and therefore might act as microsensors for mechanical forces.[148–150] Brakemeier et al. (2002) proposed that increased levels of fluid shear stress can also lead to the activation of stretch-activated channels.[151] For example, it was reported in Groschner et al. (2002) that a specific nonselective cation channel is able to transduce shear stress into local intracellular Ca^{2+} signals.[152]

6.3.1.2 Bone Cells

One of the possible mechanisms for bone cells to detect mechanical stimuli is that the mechanical stress directly activates the mechanically sensitive calcium channel and causes membrane depolarization. This depolarization activates the voltage-sensitive calcium channel to increase intracellular calcium, resulting in the activation of calcium-dependent kinases. Using gadolinium chloride, which blocks the stretch-activated membrane channel, the calcium response to steady flow was inhibited.[86,153] However, when bone cells are exposed to OFF, blocking of the stretch-activated channels did not have an effect,[104] suggesting that steady flow stimulates the stretch-activated channels whereas OFF does not. It was also found that the L-type VOCC membrane channel was involved in the calcium response to OFF,[104] whereas in steady flow experiments, these membrane channels were not found to be important.[86,153] Taken together, it seems that the biochemical mechanism of $Ca^{2+}{}_i$ mobilization involves great differences in nonreversing steady flow, PFF, and OFF.

6.3.2 Mechanosensitive Receptors

There are many transmembrane proteins that serve as receptors for the chemical signals from the extracellular environment. The conformational changes of these receptors after ligand-binding play a critical role in the signal transduction. For example, for a G protein–linked receptor, binding of extracellular signal molecules will change the relative orientation of several transmembrane α helices, so that the position of the cytoplasmic loops relative to each other is shifted and the G protein is activated. It has been speculated that mechanical force can directly or indirectly change the conformation of the receptors. Therefore, a possible mechanism for cells to detect mechanical stimuli is that the mechanical force may cause the conformational change in mechanosensitive receptors or their associated proteins, which will then induce subsequent cell responses including changes in kinase activity, release of paracrine/autocrine factors, and so on.

For example, it is well documented that ATP and UTP induce a broad spectrum of cell responses including $Ca^{2+}{}_i$ mobilization. Many animal cells release ATP and/or UTP into the extracellular environment, and often this release is mechanically mediated.[154–162] It has been demonstrated that this release requires activation of a signal-transduction pathway involving G proteins.[163,164] It has been shown that G proteins can be directly activated by fluid shear stress through G protein–linked receptors.[165] Therefore, it is possible that mechanical force can be transduced by G protein–linked receptors, directly activate the G protein, which leads to the ATP and or UTP release, and initiate further cell response through the autocrine signaling mechanism.

6.3.3 Extracellular Matrix, Integrin, Cytoskeleton Complex

ECM is physically connected to the intracellular compartment through transmembrane proteins. The primary proteins responsible for this physical link are the integrins, which are essential for transducing mechanical forces from the outside to the inside of the cell. The role of integrins in cell signaling is well established. Immediate consequences of ECM-derived forces transmitted to integrins could be a Rho-dependent assembly and growth of focal adhesion complex at these sites,[166] an increase in cytoskeletal tractional force,[167] and a triggering of MAP kinase[168,169] and/or NF-κB[170] pathways within the cells. It has also been speculated that extracellular forces might be directly transmitted to nucleus through the ECM-integrin-cytoskeleton complex and directly initiate gene transcription.[171,172] Since many signaling molecules reside on the cytoskeleton,[168,173,174] it is also possible that force-induced reorganization of the cytoskeleton may directly mediate the protein activity by physically changing the relative position of related proteins.[171,175]

6.3.3.1 Pericellular Matrix as a Transducer of Shear Stress to Intracellular Cytoskeleton

Many cells, including endothelial cells and bone cells, have densely packed pericellular matrix on or near their surfaces.[176–181] The components of the pericellular matrix are mostly proteoglycans, with the glycosaminoglycans attached to the core proteins that are attached to the cell surface (Figure 6.8). Weinbaum et al. (2003) proposed that this structure can serve as a site of mechanotransduction by coupling the mechanical load-induced dynamic response of the surface layer to the stresses and deformations induced in the underlying cytoskeleton.[182] In endothelial cells exposed to shear stress of 1 Pa, it is predicted that the bush-like core protein structure in Figure 6.8 experiences a drag force of 1.9×10^{-2} pN, a force that would result in a 6-nm lateral displacement of the actin boundary. Furthermore, the bending moment on the entire bush, 6.2 pN/nm, can be induced by the 1 Pa fluid flow. The long lever arm provided by the core proteins leads to a mechanical advantage that substantially amplifies the drag forces on the core protein tips when they are transmitted to the cytoskeleton, and this could be the initial activating step in intracellular signaling.[182] Since bone cells also experience fluid flow–induced shear stress, it is possible that a similar mechanism could be exploited by bone cells. This hypothesis is indirectly supported by a recent study by Reilly et al. (2003), which reported that removing glycocalyx on the surface of bone cells reduced the fluid flow–induced PGE_2 release.[176]

6.3.3.2 Strain Amplification Model in Bone Cell Mechanotransduction

Similar to above model, it was proposed by You et al. (2001) that the pericellular matrix can serve as a mechanotransducer for osteocytes.[183] However, in this model, the mechanocoupling is realized in a different manner. Figure 6.9 shows the structure of the PM surrounding the osteocyte process and the connection between the peri-

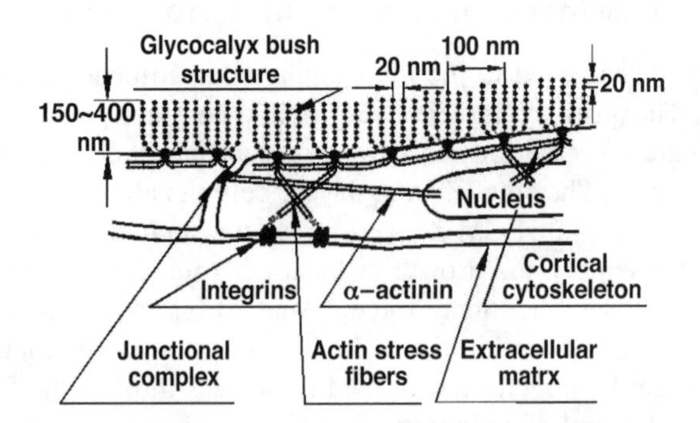

FIGURE 6.8 Sketch of pericellular matrix showing core protein arrangement and their relationship to actin cortical cytoskeleton. (From Weinbaum et al., *Proc. Natl. Acad. Sci. U.S.A.*, 100, 7988–7995, ©2003, National Academy of Sciences, U.S.A., with permission.)

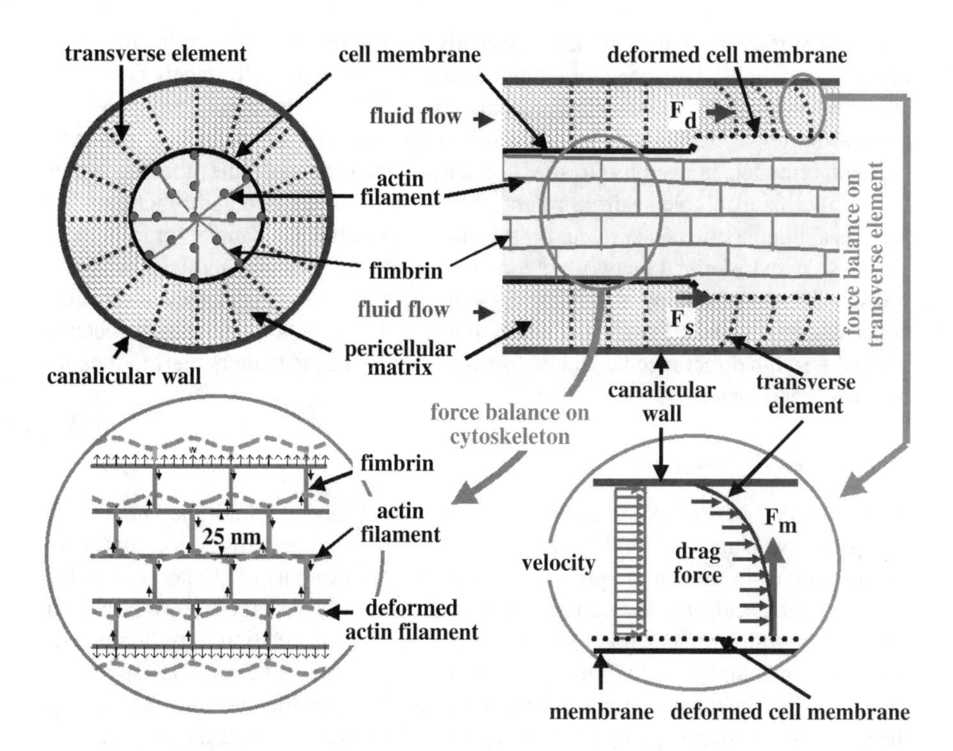

FIGURE 6.9 Schematic model showing the structure of the pericellular matrix, the actin cytoskeleton inside the process and the connection between the pericellular matrix and the cytoskeleton. Upper left, transverse cross section of canaliculus. Upper right, longitudinal cross section before and after the transverse elements are deformed by the flow. Lower left, schematic of the cell process cytoskeletal structure in longitudinal axial section. Lower right, force balance on a transverse element. (Reprinted from *J. Biomech.*, 34, You et al., pp. 1375–1386, ©2001, with permission from Elsevier.)

cellular matrix and the intracellular cytoskeleton. As shown in Figure 6.9, the pericellular matrix surrounding the osteocyte matrix is attached not only to the cell surface, but also to the canalicular wall. Therefore, the drag force induced by the fluid flow on the pericellular matrix will induce a stretch in the cell process membrane in the hoop direction. In this model, it is predicted that mechanical loading-induced fluid flow in lacuno-canalicular system under routine physical activity can produce very large deformations on osteocyte processes that are at least one order of magnitude larger than bone tissue deformations. These deformations may then be detected by the bone cells either through stretch-activated channel or direct cytoskeleton deformation-induced response.[183]

6.3.3.3 Tensegrity Mechanism

Ingber et al.[171,184–187] proposed a model in which cells are predicted to be hard-wired to respond immediately to mechanical stresses. In this model, forces are transmitted over cell surface receptors, which physically couple the cytoskeleton to the ECM or to other cells.[171] This concept emerged from studies with cell models built from sticks and string using tensegrity architecture.[184,185,188] Therefore, this model is termed the tensegrity model. It also offers a mechanism to explain how the cytoskeleton remodels in response to stress, and how signaling molecules that are immobilized on this insoluble scaffold might change their distribution and function when force is applied to the cell surface. By this mechanism, the mechanical force-induced changes in cytoskeletal network geometry may alter cell metabolism or signal transduction by changing the relative position of different regulatory molecules, hence altering their ability to chemically interact. Transduction also may be accomplished through direct mechanical distortion of molecules that alters thermodynamic or kinetic parameters.[171]

6.3.3.4 Mechanosomes

A recent study by Pavalko et al.[189] proposed a novel model, which they named the "mechanosomes model." In this model, extracellular loading will induce conformational changes in membrane proteins. Some of these proteins are linked to a solid-state signaling scaffold. The conformational change will cause the release of protein complexes, which are capable of carrying mechanical information into the nucleus. Such protein complexes are termed mechanosomes by these authors. They suggested that, after the mechanosomes enter the nucleus, they can then produce changes in the geometry of a target gene. The key aspect of this model involves the load-induced formation of mechanosomes, multiprotein complexes comprised of focal adhesion-associated or adherens junction-associated proteins. The mechanosomes contain a nucleo-cytoplasmic shuttling DNA-binding protein that typically moves between the adhesion complexes and the nucleus and one or more adaptor proteins acquired from the focal adhesions or adherens junctions. Mechanosomes then translocate to the nucleus and there initiate changes in gene transcription.

In addition to ECM-integrin-cytoskeleton complex, mechanotransduction may also occur through cell–cell contact.[171,184–187,190,191] By this mechanism, the mechanical disturbance can be transduced through the adhesion junctions (the anchoring junctions connecting adjacent cells formed by cadherins, which are transmembrane adhesion proteins), the intracellular anchor proteins, and the intracellular actin filaments network. For example, β-catenin is an adhesion-related protein. Most of a cell's β-catenin is located close to cell–cell adherens junctions, where it is associated with cadherins. The β-catenin in these junctions helps link the cadherins to the actin cytoskeleton. The mechanosome model predicts that, once the mechanical disturbance is focused on the adhesion junctions, the β-catenin is released, associates with the LEF-1/TCF transcription factor, translocates, and enters the nucleus so that gene transcription can be initiated.[189]

6.4 CELLULAR RESPONSE TO MECHANICAL STIMULI

Once the mechanical signal is detected, multiple intracellular events are initiated, including $[Ca^{2+}]_i$ elevation, cAMP increase, cytoskeleton reorganization, protein phosphorylation, protein–protein associations (e.g., integrin-cytoskeleton-FAK), activation of enzymes, production of extracellular signal molecules (e.g., NO, PGE_2), and activation of gene transcription. Although extensive studies have been done on cellular response to mechanical stimuli, due to the complex nature of this process, much remains unclear. For example, as described in Section 6.2, different cells may have different responses to the same mechanical stimuli. And different patterns of the signal, such as frequency and duration, may also induce different responses. In addition, one mechanical stimulus may trigger multiple pathways and/or cross talk between the pathways. In this section, we focus on several key components in the cellular response. We first look at second messengers, which are activated immediately after the signal is detected. We then describe cytoskeletal reorganization and activation of intracellular signaling proteins and other downstream events.

6.4.1 Activation of Second Messengers

6.4.1.1 Ca^{2+}

Intracellular Ca^{2+} is an important second messenger. It is derived from two sources: extracellular Ca^{2+} entering the cell through membrane channels, and Ca^{2+} released from internal Ca^{2+} stores, endoplasmic or sarcoplasmic reticulum. With respect to cell membrane channels for Ca^{2+}, there are three types: voltage-operated Ca^{2+} channels (VOCC), stretch-activated Ca^{2+} channels, and receptor-operated Ca^{2+} channels. With regard to stretch-activated channels, in addition to mediating direct Ca^{2+} influx, they may also indirectly activate voltage-gated Ca^{2+} channels by causing depolarization. The Ca^{2+} entry through stretch-activated channels and/or VOCC may then activate phospholipase C activity and IP_3-sensitive Ca^{2+} release, thereby further amplifying the increase in $[Ca^{2+}]_i$. Changes in $[Ca^{2+}]_i$ regulate a wide variety of cellular processes, including cell growth and differentiation, cell motility and contraction, intercellular coupling, synaptic transmission, fertilization, apoptosis, and necrosis. It is therefore significant that mechanical stimulation triggers elevation in $[Ca^{2+}]_i$ in many cells.

Extensive studies in bone cells have shown that the $[Ca^{2+}]_i$ can be dramatically increased in response to fluid flow–induced shear stress.[60,85,86,101,104–106,108,142,146] Figure 6.10 shows an example of such response, in which the intracellular Ca^{2+} level in cells before (left) and after (right) flow exposure are compared, demonstrating that flow exposure dramatically increased the intracellular Ca^{2+} concentration (see color insert following page 204). Several studies have also shown that this response may be modulated by chemotransport.[60,101,105,108,146] Different flow patterns may induce a Ca^{2+} response through different pathways.[86,104,153] For example, the stretch-activated channel is involved in steady-flow but not OFF-induced Ca^{2+} response. Conversely, the L-VOCC is known to play an role in OFF but not in the steady flow–induced Ca^{2+} response.[86,104,153] Intracellular Ca^{2+} mobilization plays an important role in mediating a variety of subsequent responses such as OP,[60,104] COX-2,[153] c-fos,[153] and Egr-1[110] expression; PGE_2,[89,98] NO,[143] and TGFβ-1[91] release; and cytoskeleton reorganization.[153]

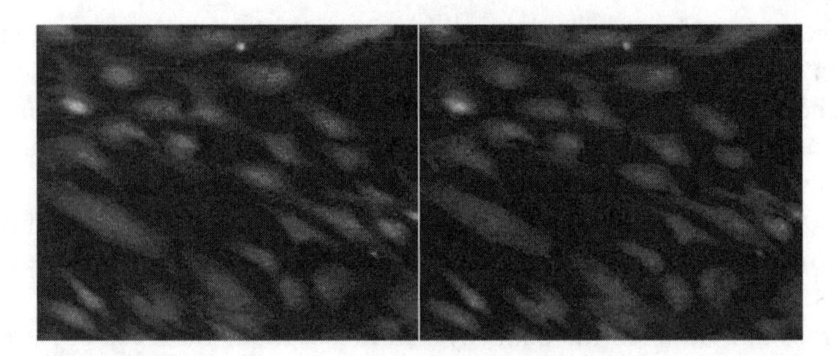

FIGURE 6.10 Image of the intracellular Ca^{2+} level in cells before (left) and after (right) flow exposure. The color reflects the $[Ca^{2+}]_i$, with the red indicating a high level and the green for a low level. (See color insert following page 204.)

6.4.1.2 cAMP

cAMP is an important second messenger in many cells. It is synthesized from ATP by a plasma membrane–bound enzyme, adenylyl cyclase. In most animal cells, cAMP exerts its effects mainly by activating cAMP-dependent protein kinase, which then either induces a metabolic response or stimulates gene transcription. Many extracellular signal molecules work by increasing cAMP, and they do so by increasing the activity of adenylyl cyclase. Adenylyl cyclase has at least eight isoforms in mammals, most of which are regulated by both G proteins and Ca^{2+}. All receptors that act via cAMP are coupled to a stimulatory G protein (G_s), which activates adenylyl cyclase and thereby increases cAMP concentration.

cAMP has been implicated in bone remodeling and turnover.[192] Increased levels of cAMP have been found to increase the proliferation of osteoblasts *in vitro*.[193,194] However, cAMP has also been linked to increased bone resorption.[46] Therefore, it is not yet clear what the role of cAMP in bone adaptation and mechanotransduction might be.

Mechanical stress has been shown to stimulate cAMP in bone cells *in vitro*. Binderman et al. (1984) observed a two- to three-fold increase in cAMP in osteoblasts due to stretch, and this response was mediated by PGE_2 synthesis.[195] Using newborn rat calvarial osteoblasts, Reich and associates found that steady fluid shear stress upregulates cAMP levels up to 12-fold, and this response is completely abolished by the inhibition of cyclooxygenase, indicating the cAMP response is mediated by prostaglandins.[87] Thus, it appears that mechanical stimuli can upregulate cAMP in bone cells, and this process is mediated by prostaglandins.

6.4.2 Cytoskeleton Reorganization

The cytoskeleton is physically connected to the ECM through transmembrane proteins, such as integrins, and to neighboring cells through adhesion molecules. It is also connected, directly or indirectly, to intracellular structures (e.g., phospholipases, protein kinases). Thus, the cytoskeleton provides an important route of transduction of mechanical stimuli. As a component of cytoskeleton, actin filaments have been found to play an important role in the fluid flow–induced intracellular response in bone cells.[92,153] It has been observed that bone cells can respond to steady flow with cytoskeletal reorganization (stress fiber formation).[92,153] Disruption of actin filaments completely abolished the steady flow–induced COX-2 and c-fos gene expression in osteoblastic cells [92,153] and PFF-induced PGE_2 and PGI_2 release in osteocytes,[96] suggesting that expression of these genes is dependent on the formation of polymerized actin stress fibers. Similarly, exposing cells to oscillatory flow also led to increases in COX-2[196] and PGE_2.[103,107,108,196] However, in a recent study by our group, we found that application of OFF to osteoblastic cells did not induce the development of F-Actin stress fibers as was observed under chronic unidirectional steady flow.[197] This suggests that oscillatory flow–induced upregulation of COX-2 and PGE_2 in osteoblasts is not dependent on F-Actin cytoskeletal reorganization, and that multiple pathways may regulate the flow-induced prostaglandin release observed in these cells.

6.4.3 Intracellular Signaling Proteins

The activities of many intracellular proteins can be affected by mechanical stimuli. These in turn regulate the activity of other proteins in a signaling cascade that can ultimately lead to a change in gene expression. Most of the proteins are activated by one of two mechanisms: protein phosphorylation, catalyzed by a protein kinase, and protein dephosphorylation, catalyzed by a protein phosphatase. Some proteins themselves are kinases, and after being activated they will then activate other proteins by phosphorylation, so that a cascade of signal transduction is initiated until the target protein or gene transcription factor is reached.

Protein phosphorylation is critical in many responses to mechanical stimuli. For example, protein phosphorylation is required in mechanical stimuli–induced NO formation,[198] altered gene expression, and rearrangement of the cytoskeleton.[94,98,104,110,112,199] Cell mechanotransduction involves a cascade of regulatory phosphorylations. Many early responses, such as elevation of $[Ca^{2+}]_i$, trigger phosphorylation. Several types of protein kinases, such as PKA (cAMP-dependent protein kinase), PKC (Ca^{2+}-dependent protein kinase), and MAPK (mitogen-activated protein kinase), have been extensively studied in mechanotransduction of cells *in vitro*. Next, we review studies on MAPK as an example.

- **MAPK.** Activity of MAPKs is important for regulating cell differentiation and apoptosis by transmitting extracellular signals to the nucleus.[200,201] MAPK family members, including extracellular signal-regulated kinase (ERK), c-Jun N-terminal kinase (JNK), and p38 MAP kinase, have been shown to be important signaling components linking mechanical stimuli to cellular responses, including cell growth, differentiation, and metabolic regulation, in endothelial cells, smooth muscle cells, myocytes, and bone cells.[94,104,112,202–206] Steady flow has been shown to be able to induce intracellular response in bone cells through the ERK pathway.[94] Similarly, ERK and p38 activation, but not JNK, were found to be involved in inducing OP gene expression when bone cells are subjected to OFF. Furthermore, a recent study by Alford and associates showed that OFF modulated gap-junction communication through ERK pathways.[112] These studies indicate that MAPKs play an important role in regulating mechanotransduction in bone.

- **Phosphatase.** In contrast to kinase, phosphatase catalyzes protein dephosphorylation. Mechanical stimuli have also been shown to be able to affect some phosphatase activity. One of the well-studied protein phosphatases in bone is ALP. ALP activity is required for bone matrix mineralization (the incorporation of the mineral to the newly formed bone matrix).[207] It is a biphasic marker of osteogenic differentiation in the sense that it increases activity immediately after proliferation stops, but then decreases immediately prior to mineralization.[115,116,208–211] *In vivo* studies have shown that mechanical loading regulation of bone remodeling involves changing ALP activity.[212–214] Hillsley and Frangos[95] found that PFF downregulated ALP expression in osteoblast cells. They suggest that the decrease of ALP activity could be because that flow affects the cell cycle of osteoblasts, or it may indicate accelerated differentiation of osteoblasts toward a more mature cell in response to fluid flow.[95] A recent study by our group has found that OFF decreased ALP by roughly 30% and demonstrated that OFF can regulate ALP activity and this may result in higher levels of osteogenesis.[113]

In addition to kinases and phosphatases, the activities of other enzymes have also been found to be affected by mechanical stimuli. For example, COX-2 is a key enzyme in the production of prostaglandins. It is associated with the synthesis and release of PGE_2 and PGI_2 in bone cells in response to mechanical stimuli *in vitro*.[11,87,92,153] In rats subjected to tibial four-point binding, periosteal bone formation is inhibited by NS-398, a specific COX-2 blocker.[215] These observations indicate that COX-2 responds quickly in mechanically loaded bone and osteogenic cells, and its expression may be essential to the mechanically induced response of bone.

6.4.4 Gene Transcription

Three different mechanisms underlie gene regulation by mechanical signals. In a primary response, a cellular mechanotransduction pathway could activate an already available transcription factor that binds to a "mechano-responsive" regulatory element in the gene promoter. In a secondary response, a mechanical signal would first induce the transcription and synthesis of a nuclear factor, which transactivates a specific gene. As a third possibility, mechanical stress might induce the synthesis and/or secretion of a growth factor or other messenger molecules that indirectly regulates gene expression via an auto- or paracrine feedback loop.[44]

- **Egr-1.** Egr-1 is a transcription factor, which can be induced by growth factors or serums. Studies have shown that it also plays a critical role in mechanotransduction in the sense that its mRNA can be increased severalfold within minutes (via a ras/Erk-1/2 pathway) after applying shear or tensile stress to cultured cells.[110,216] After being activated, it can then activate other genes such as fibronectin[217] and TGF-β1.[218] Egr-1 had been shown to play a role in skeletal differentiation or proliferation. A study by Ogata (1997) has shown that exposure of MC3T3-E1 osteoblastic cells to dynamic flow for even only 1 min can induce Egr-1, and this process involves tyrosine kinase, intracellular Ca^{2+} mobilization, and serum, whereas cAMP and prostaglandins are not involved.[110]

- **c-fos.** c-fos is an early response gene involved in the cellular effect of a wide variety of stimuli including many mitogens and mechanical signals. It is expressed by nearly all cell types, including osteoblasts. In bone, it is a transcription factor required for osteoclast differentiation.[219] Transgenic studies on mice have shown that overexpression of c-fos causes osteosarcomas and chondrosarcomas,[220,221] whereas knock-out of c-fos leads to osteopetrosis,[222] indicating the critical role of c-fos in bone physiology. *In vivo* studies show that c-fos is upregulated in response to mechanical loading.[223,224] *In vitro* studies have shown that fluid flow can induce c-fos expression in bone cells within minutes[11,92,153] and this response depends on IP_3-mediated intracellular Ca^{2+} release.[153] These observations demonstrate that expression of c-fos plays an important role in the mechanically induced response of bone.

6.4.5 Release of Autocrine/Paracrine Factors

6.4.5.1 NO

NO is a highly diffusible extracellular signal molecule that can pass readily across the target-cell membrane and directly bond to guanylyl cyclase, which then produces the intracellular mediator cGMP. In mammals, NO is synthesized from L-arginine in the presence of NO synthase. Studies have shown that NO can regulate bone remodeling by inhibiting osteoclast resorptive activity and possibly by stimulating osteoblast proliferation.[225–228]

It has been shown that fluid flow–induced shear stress stimulates the production of NO in endothelial cells and bone cells.[61,67,88,99,143,147] When subjected to steady flow, bone cells can rapidly (in seconds) and continuously (up to 70 h) release NO into the extracellular environment,[61,88] and this response is found to be G protein and Ca^{2+} independent.[143] Conversely, when cells were subjected to transient flow, release of NO has been found to be G protein and Ca^{2+} dependent,[143] indicating that different flow patterns induce NO release through different pathways. NO can also be produced when cells are exposed to PFF.[99,147] It has been found that NO release is shear stress dependent,[147] and the response can be modulated by parathyroid hormone (PTH),[99] suggesting that mechanical loading and PTH interact at the level of mechanotransduction.

6.4.5.2 PGE_2

PGE_2 is a small extracellular signal molecule formed by the action of COX on arachidonic acid (a membrane phospholipid). In bone, it is produced by osteoblasts and osteocytes and is a mediator of bone remodeling. The release of PGE_2 can promote recruitment of osteoblast precursor cells, increased osteoblast proliferation, ALP activity, and collagen synthesis.[229–231]

In vitro studies have shown that bone cells release PGE_2 in response to steady flow,[61,87,89,90,232] PFF,[11,96–99,147] and OFF.[103,107,108] To understand the cellular mechanism of PGE_2 production, researchers have used selective blockers of prostaglandin production in order to inhibit specific aspects of this pathway. For instance, steady flow–induced PGE_2 release can be strongly inhibited by PKC inhibitor,[90] G protein inhibitors GDPβS (by 83%) and pertussis toxin (by 72%),[89] chelation of extracellular calcium by EGTA (by 87%), and intracellular calcium by quin-2/AM (by 67%).[89] Furthermore, the steady flow–induced upregulation of COX-2 mRNA expression is found to be dependent on cytoskeletal reorganization.[92] These findings suggest that steady flow may induce PGE_2 release through one or a combination of two pathways: one involving G protein-linked mechanosensors in the cell membrane, Ca^{2+}, and PKC; and the other involving the actin cytoskeletal reorganization.

Similarly, PFF-induced PGE_2 production is markedly decreased by applying specific inhibitors of Ca^{2+}-activated phospholipase C, PKC, and phospholipase A2, by blocking Ca^{2+} channels and intracellular Ca^{2+} release,[98] and by disruption of the actin cytoskeleton with cytochalasin B.[96] These findings suggest that PFF-induced PGE_2 release occurred through increased $[Ca^{2+}]_i$ from ion channels and IP_3-induced Ca^{2+} release from intracellular stores. Ca^{2+} and PKC could then stimulate phospholipase A2 activity, arachidonic acid production, and finally PGE_2 release. In this process, the actin cytoskeleton is also involved; however, the mechanism underlying it is still not clear.

For OFF, Saunders and associates[103,107] found that PGE_2 release is mediated by functional gap junction communication. However, in contrast to steady flow and PFF studies,[89,98,153] they found OFF-induced PGE_2 is independent of cytosolic calcium,[107] suggesting that bone cells may respond to OFF with PGE_2 release via a different pathway (e.g., cAMP).

6.4.5.3 Extracellular Nucleotides

ATP (adenosine triphosphate) and UTP (uridine triphosphate) are locally released, short-lived, yet potent extracellular signaling molecules. They can induce a broad spectrum of cell responses including $Ca^{2+}{}_i$ mobilization. These ligands exert their effect through a large family of receptors — the P2 receptors (formerly termed purinoceptors), which are subdivided into P2Y (G protein–linked receptors) and P2X (cationic-gated channels) subtypes. P2X receptors are activated exclusively by adenine nucleotides, whereas P2Y receptors are responsive to ATP and/or UTP. Many animal cells release ATP and/or UTP into the extracellular environment, and often this release can be greatly enhanced by mechanical stress.[154–162] It has been demonstrated that this release requires activation of a signal transduction pathway involving G protein.[163,164]

Nucleotides are important local signaling molecules in bone.[105,233–235] Both osteoblasts and osteoclasts are targets for ATP/UTP.[236–242] These molecules can induce multiple signal transduction events including IP_3-mediated intracellular calcium release[105] and upregulation of c-fos.[233] As a result, these molecules can modulate osteoblast proliferation,[243] osteoblast-induced bone formation,[235] osteoclastogenesis, and bone resorption.[240,244] *In vitro* studies have shown that release of ATP by osteoblasts is enhanced significantly by fluid shear stress[245] and hypotonic stress.[161] Taken together, these observations indicate that mechanical load on bone can regulate bone remodeling by modulating ATP/UTP release by bone cells. Thus, ATP and UTP are likely to be important autocrine/paracrine mediators in bone mechanosensitivity.

6.4.5.4 Growth Factors

Growth factors are a group of extracellular polypeptide signal molecules that can stimulate a cell to grow or proliferate. Growth factor release has been observed as one of the early proliferation responses of many cells to mechanical loading.[51,55,75,91,246,247] Bone contains a number of growth factors including TGF-β1 and TGF-β2, platelet-derived growth factors (PDGFs), basic and acidic fibroblast growth factors (bFGF, aFGF), insulin-like growth factors I and II (IGF-I, IGF-II), and bone morphogenetic proteins (BMPs). These growth factors stimulate the proliferation and differentiation of the osteoblasts and promote bone formation.

Osteoblasts are capable of producing TGF-β1, TGF-β2, IGF-I, and IGF-II.[248,249] Studies have shown that mechanical loading on bone cells can promote the release of these growth factors.[51,55,91] Applying steady flow–induced fluid shear stress (1.7–2.0 Pa) to human osteosarcoma cells increased TGF-β1 expression, and the response was modulated by cation channel blockers.[91] These observations suggest that a physiological level of fluid shear stress induces the production of TGF-β1, which may promote bone formation. When rat long bone–derived osteoblast-like cells were subjected to dynamic strain (0.34%, 1 Hz), IGF-II production was increased, whereas IGF-I level was unaffected.[51] In contrast, newborn rat calvarial cells, which are in the late osteoblast/early osteocyte stage of differentiation, responded to mechanical stretch (0–3%) with an increase in IGF-I production.[55] The discrepancy may be the result of the differing systems used to induce mechanical strain, the differences in calvarial versus long bone responses to mechanical strain,[250,251] or the variable age-related responsiveness of cell cultures to mechanical stimulation.[48,102]

6.4.6 Production of Matrix Proteins

One of the outcomes of a cell's response to mechanical stimuli is that it can change its local environment via release of certain matrix proteins into the extracellular space. These proteins will further affect cell metabolism by either changing the mechanical properties of the ECM or directly interact with the cell. Here, we review studies of two important bone matrix proteins produced in response to mechanical loading.

- **Osteopontin.** OP is a mineral-binding, noncollagenous bone matrix protein. In addition to being a structural component of bone matrix, is has also been shown to be involved in regulating bone cell attachment, osteoclast function, and mineralization, suggesting an important role in regulation of bone remodeling.[212,252–255] *In vivo* studies indicate that it is an important factor in triggering bone remodeling caused by mechanical stress.[256] There is strong evidence suggesting that OP is an important factor in loading-induced changes in bone cell metabolism.[47,257,258] *In vitro* studies have shown that OFF applied to bone cells can trigger OP expression,[60,104] and this response is regulated via intracellular calcium mobilization and activation of ERK and p38 MAP kinase.[104] Studies on stem cells have shown that OFF upregulates OP expression by roughly 50%,[113] suggesting that OFF may regulate the formation of new osteoblasts from stem cells.

- **Osteocalcin.** Osteocalcin is a noncollagenous protein and has only been found in the bone ECM.[259] It is thought to be upregulated at the onset of mineralization.[209] A recent study by our group has found that OFF can upregulate osteocalcin by around 50% in stem cells, suggesting the loading-induced dynamic flow in bone may play an important role in regulating stem cell differentiation toward an osteogenic pathway.[113]

6.4.7 Gap Junction

Gap junctions are membrane-spanning channels facilitating intercellular communication by allowing the passage of small molecules (e.g., calcium, inositol phosphates, cyclic nucleotides) from cell to cell. They play an important role in intercellular communication. Furthermore, they may also provide a means by which cells can be electrically coupled. Morphological evidence shows that gap junctions exist between bone cells *in vivo*.[260] Using the murine osteocytic cell line MLO-Y4 and osteoblastic cells MC3T3-E1, Yellowley and associates showed that deformation-induced calcium signals in MLO-Y4 were passed to neighboring MC3T3-E1 cells via gap junctions, suggesting that bone cells are functionally coupled to one another via gap junctions.[261] Steady flow and PFF has been shown to stimulate bone cells in terms of the opening of gap junctions,[93,100] and this response has been found to be mediated by PGE_2.[93] Similarly, OFF also increased gap junction communication, which was regulated by an ERK1/2 MAP kinase dependent mechanism.[112] Furthermore, in response to OFF, gap junction communications has been found to modulate PGE_2 release, but does not contribute to the Ca^{2+} response.[103,107]

6.5 SUMMARY

It has been long been recognized that mechanical cues play a critical role in cell metabolism. However, how physical signals are translated into chemical signals and finally affect cell behavior is unclear. Extensive studies have been done to help to understand this process. In this chapter, we reviewed the current knowledge on cellular mechanotransduction: the mechanical environment that cells are exposed to, the mechanisms of cell detection of physical stimuli, and the cell's responses to these stimuli. It is clear that cells are capable of responding to a wide variety of physical stimuli, such as direct deformation, fluid shear stress, pressure, and streaming potentials. However, it is less clear which of these stimuli are induced at the cellular level *in vivo* in response to mechanical loading of the tissue. Different cells may also respond differently to the same type of mechanical force. Furthermore, the same type of mechanical force on the same type of cell can induce a different cellular response as a function of magnitude, frequency, and duration. Alternative models of detection of mechanical signals have been proposed, including membrane mechanosensitive ion channels, mechanosensitive receptors, the tensegrity mechanism, and the mechanosome. Cell responses, including elevation of $[Ca^{2+}]_i$, cAMP level, release of NO, PGE_2, and activation of gene transcription, have been identified in many cell types in response to distinct mechanical stimuli. Different pathways have also been identified in mechanotransduction. It has been found that multiple pathways could be involved in the response to a given mechanical stimuli and cross talk between the pathways might also be involved in some cases. With all this knowledge, however, understanding of the mechanotransduction mechanism is remarkably incomplete, and often controversial. Further systematic studies are required to gain a deeper understanding of this critical aspect of tissue metabolism. In this endeavor advancements in nanoscale fabrication will be crucial for a better understanding of the cell-level micromechanical environment and our ability to manipulate it in a controlled experimental setting.

REFERENCES

1. Kamiya, A., R. Bukhari, and T. Togawa, Adaptive regulation of wall shear stress optimizing vascular tree function, *Bull. Math. Biol.*, 1984, 46(1), 127–137.
2. Fritton, S.P., K.J. McLeod, and C.T. Rubin, Quantifying the strain history of bone: spatial uniformity and self-similarity of low-magnitude strains, *J. Biomech.,* 2000, 33(3), 317–325.
3. Weinbaum, S., S.C. Cowin, and Y. Zeng, A model for the excitation of osteocytes by mechanical loading-induced bone fluid shear stresses, *J. Biomech.*, 1994, 27(3), 339–360.
4. Mow, V.C., A. Ratcliffe, and A.R. Poole, Cartilage and diarthrodial joints as paradigms for hierarchical materials and structures, *Biomaterials*, 1992, 13(2), 67–97.
5. Guilak, F., A. Ratcliffe, and V.C. Mow, Chondrocyte deformation and local tissue strain in articular cartilage: a confocal microscopy study, *J. Orthop. Res.*, 1995, 13(3), 410–421.
6. Guilak, F., et al., Mechanically induced calcium waves in articular chondrocytes are inhibited by gadolinium and amiloride, *J. Orthop. Res.*, 1999, 17(3): p. 421–429.
7. Guilak, F., Compression-induced changes in the shape and volume of the chondrocyte nucleus, *J. Biomech.*, 1995, 28(12), 1529–1541.
8. Ker, R.F., X.T. Wang, and A.V. Pike, Fatigue quality of mammalian tendons, *J. Exp. Biol.*, 2000, 203(8), 1317–1327.
9. Tidball, J.G., Energy stored and dissipated in skeletal muscle basement membranes during sinusoidal oscillations, *Biophys. J.*, 1986, 50(6), 1127–1138.
10. Nagatomi, J., et al., Frequency- and duration-dependent effects of cyclic pressure on select bone cell functions, *Tissue Eng.*, 2001, 7(6), 717–728.
11. Klein-Nulend, J., et al., Sensitivity of osteocytes to biomechanical stress *in vitro*, *FASEB J.*, 1995, 9(5), 441–445.

12. Hansen, U., et al., Combination of reduced oxygen tension and intermittent hydrostatic pressure: a useful tool in articular cartilage tissue engineering, *J. Biomech.*, 2001, 34(7), 941–949.
13. Parkkinen, J.J., et al., Effects of cyclic hydrostatic pressure on proteoglycan synthesis in cultured chondrocytes and articular cartilage explants, *Arch. Biochem. Biophys.*, 1993, 300(1), 458–465.
14. Smith, R.L., et al., Time-dependent effects of intermittent hydrostatic pressure on articular chondrocyte type II collagen and aggrecan mRNA expression, *J. Rehabil. Res. Dev.*, 2000, 37(2), 153–161.
15. Smith, R.L., et al., *In vitro* stimulation of articular chondrocyte mRNA and extracellular matrix synthesis by hydrostatic pressure, *J. Orthop. Res.*, 1996, 14(1), 53–60.
16. Shieh, A.C. and K.A. Athanasiou, Principles of cell mechanics for cartilage tissue engineering, *Ann. Biomed. Eng.*, 2003, 31(1), 1–11.
17. Qin, Y.X., et al., Fluid pressure gradients, arising from oscillations in intramedullary pressure, is correlated with the formation of bone and inhibition of intracortical porosity. *J. Biomech.*, 2003, 36(10), 1427–1437.
18. Nagatomi, J., et al., Effects of cyclic pressure on bone marrow cell cultures, *J. Biomech. Eng.*, 2002, 124(3), 308–314.
19. Wang, L., et al., On bone adaptation due to venous stasis, *J. Biomech.*, 2003, 36(10), 1439–1451.
20. Ozawa, H., et al., Effect of a continuously applied compressive pressure on mouse osteoblast-like cells (MC3T3-E1) *in vitro*, *J. Cell Physiol.*, 1990, 142(1), 177–185.
21. Roelofsen, J., J. Klein-Nulend, and E.H. Burger, Mechanical stimulation by intermittent hydrostatic compression promotes bone-specific gene expression *in vitro*, *J. Biomech.*, 1995, 28(12), 1493–1503.
22. Nagatomi, J., et al., Cyclic pressure affects osteoblast functions pertinent to osteogenesis, *Ann. Biomed. Eng.*, 2003, 31(8), 917–923.
23. Angele, P., et al., Cyclic hydrostatic pressure enhances the chondrogenic phenotype of human mesenchymal progenitor cells differentiated *in vitro*, *J. Orthop. Res.*, 2003, 21(3), 451–457.
24. Shin, H.Y., M.E. Gerritsen, and R. Bizios, Regulation of endothelial cell proliferation and apoptosis by cyclic pressure, *Ann. Biomed. Eng.*, 2002, 30(3), 297–304.
25. Hudspeth, A.J., Models for mechanoelectrical transduction by hair cells, *Prog. Clin. Biol. Res.*, 1985, 176, 193–205.
26. Tilney, L.G., M.S. Tilney, and D.A. Cotanche, Actin filaments, stereocilia, and hair cells of the bird cochlea. V. How the staircase pattern of stereociliary lengths is generated, *J. Cell Biol.*, 1988, 106(2), 355–365.
27. Howard, J. and A.J. Hudspeth, Mechanical relaxation of the hair bundle mediates adaptation in mechanoelectrical transduction by the bullfrog's saccular hair cell, *Proc. Natl. Acad. Sci. U.S.A.*, 1987, 84(9), 3064–3068.
28. Pickles, J.O., A model for the mechanics of the stereociliar bundle on acousticolateral hair cells, *Hear. Res.*, 1993, 68(2), 159–172.
29. Hudspeth, A.J., Hair-bundle mechanics and a model for mechanoelectrical transduction by hair cells, *Soc. Gen. Physiol. Ser.*, 1992, 47, 357–370.
30. Kachar, B., et al., Three-dimensional analysis of the 16 nm urothelial plaque particle: luminal surface exposure, preferential head-to-head interaction, and hinge formation, *J. Mol. Biol.*, 1999, 285(2), 595–608.
31. Gassner, R., et al., Cyclic tensile stress exerts antiinflammatory actions on chondrocytes by inhibiting inducible nitric oxide synthase, *J. Immunol.*, 1999, 163(4), 2187–2192.

32. Xu, Z., et al., Cyclic tensile strain acts as an antagonist of IL-1 beta actions in chondrocytes, *J. Immunol.*, 2000, 165(1), 453–460.
33. Long, P., R. Gassner, and S. Agarwal, Tumor necrosis factor alpha-dependent proinflammatory gene induction is inhibited by cyclic tensile strain in articular chondrocytes *in vitro*, *Arthritis. Rheum.*, 2001, 44(10), 2311–2319.
34. Agarwal, S., et al., Cyclic tensile strain suppresses catabolic effects of interleukin-1beta in fibrochondrocytes from the temporomandibular joint, *Arthritis. Rheum.*, 2001, 44(3), 608–617.
35. Lee, D.A., et al., The influence of mechanical loading on isolated chondrocytes seeded in agarose constructs, *Biorheology*, 2000, 37(1–2), 149–161.
36. Lee, D.A., et al., Dynamic mechanical compression influences nitric oxide production by articular chondrocytes seeded in agarose, *Biochem. Biophys. Res. Commun.*, 1998, 251(2), 580–585.
37. Fermor, B., et al., The effects of static and intermittent compression on nitric oxide production in articular cartilage explants, *J. Orthop. Res.*, 2001, 19(4), 729–737.
38. Deschner, J., et al., Signal transduction by mechanical strain in chondrocytes, *Curr. Opin. Clin. Nutr. Metab. Care*, 2003, 6(3), 289–293.
39. Almekinders, L.C., A.J. Banes, and C.A. Ballenger, Effects of repetitive motion on human fibroblasts, *Med. Sci. Sports Exerc.*, 1993, 25(5), 603–607.
40. Banes, A.J., et al., PDGF-BB, IGF-I and mechanical load stimulate DNA synthesis in avian tendon fibroblasts *in vitro*, *J. Biomech.*, 1995, 28(12), 1505–1513.
41. Gutierrez, J.A. and H.A. Perr, Mechanical stretch modulates TGF-beta1 and alpha1(I) collagen expression in fetal human intestinal smooth muscle cells, *Am. J. Physiol.*, 1999, 277(5 Pt 1), G1074– G1080.
42. Lindahl, G.E., et al., Activation of fibroblast procollagen alpha 1(I) transcription by mechanical strain is transforming growth factor-beta-dependent and involves increased binding of CCAAT-binding factor (CBF/NF-Y) at the proximal promoter, *J. Biol. Chem.*, 2002, 277(8), 6153–6161.
43. Schild, C. and B. Trueb, Mechanical stress is required for high-level expression of connective tissue growth factor, *Exp. Cell Res.*, 2002, 274(1), 83–91.
44. Chiquet, M., et al., How do fibroblasts translate mechanical signals into changes in extracellular matrix production? *Matrix Biol.*, 2003, 22(1), 73–80.
45. Buckley, M.J., et al., Osteoblasts increase their rate of division and align in response to cyclic, mechanical tension *in vitro*, *Bone Miner.*, 1988, 4(3), 225–236.
46. Rodan, G.A., et al., Cyclic AMP and cyclic GMP: mediators of the mechanical effects on bone remodeling, *Science*, 1975, 189(4201), 467–469.
47. Toma, C.D., et al., Signal transduction of mechanical stimuli is dependent on microfilament integrity: identification of osteopontin as a mechanically induced gene in osteoblasts, *J. Bone Miner. Res.*, 1997, 12(10), 1626–1636.
48. Wozniak, M., et al., Mechanically strained cells of the osteoblast lineage organize their extracellular matrix through unique sites of alphavbeta3-integrin expression, *J. Bone Miner. Res.*, 2000, 15(9), 1731–1745.
49. Peake, M.A., et al., Selected contribution: regulatory pathways involved in mechanical induction of c-fos gene expression in bone cells, *J. Appl. Physiol.*, 2000, 89(6), 2498–2507.
50. Stanford, C.M., J.W. Stevens, and R.A. Brand, Cellular deformation reversibly depresses RT-PCR detectable levels of bone-related mRNA, *J. Biomech.*, 1995, 28(12), 1419–1427.
51. Zaman, G., et al., Early responses to dynamic strain change and prostaglandins in bone-derived cells in culture, *J. Bone Miner. Res.*, 1997, 12(5), 769–777.

52. Owan, I., et al., Mechanotransduction in bone: osteoblasts are more responsive to fluid forces than mechanical strain, *Am. J. Physiol.*, 1997, 273(3 Pt 1), C810– C815.
53. Cheng, M., et al., Mechanical strain stimulates ROS cell proliferation through IGF-II and estrogen through IGF-I, *J. Bone Miner. Res.*, 1999, 14(10), 1742–1750.
54. Pitsillides, A.A., et al., Mechanical strain-induced NO production by bone cells: a possible role in adaptive bone (re)modeling? *FASEB J.*, 1995, 9(15), 1614–1622.
55. Mikuni-Takagaki, Y., et al., Distinct responses of different populations of bone cells to mechanical stress, *Endocrinology*, 1996, 137(5), 2028–2035.
56. Kawata, A. and Y. Mikuni-Takagaki, Mechanotransduction in stretched osteocytes — temporal expression of immediate early and other genes, *Biochem. Biophys. Res. Commun.*, 1998, 246(2), 404–408.
57. Miyauchi, A., et al., Parathyroid hormone-activated volume-sensitive calcium influx pathways in mechanically loaded osteocytes, *J. Biol. Chem.*, 2000, 275(5), 3335–3342.
58. Brown, T.D., et al., Loading paradigms — intentional and unintentional — for cell culture mechanostimulus, *Am. J. Med. Sci.*, 1998, 316(3), 162–168.
59. Brown, T.D., et al., Development and experimental validation of a fluid/structure-interaction finite element model of a vacuum-driven cell culture mechanostimulus system, *Comput. Methods Biomech. Biomed. Engin.*, 2000, 3(1), 65–78.
60. You, J., et al., Substrate deformation levels associated with routine physical activity are less stimulatory to bone cells relative to loading-induced oscillatory fluid flow, *J. Biomech. Eng.*, 2000, 122(4), 387–393.
61. Smalt, R., et al., Induction of NO and prostaglandin E2 in osteoblasts by wall-shear stress but not mechanical strain, *Am. J. Physiol.*, 1997, 273(4 Pt 1), E751– E758.
62. Schwarz, G., et al., Shear stress-induced calcium transients in endothelial cells from human umbilical cord veins, *J. Physiol.*, 1992, 458, 527–538.
63. Mo, M., S.G. Eskin, and W.P. Schilling, Flow-induced changes in Ca^{2+} signaling of vascular endothelial cells: effect of shear stress and ATP. *Am. J. Physiol.*, 1991, 260(5 Pt 2), H1698–H1707.
64. Shen, J., et al., Fluid shear stress modulates cytosolic free calcium in vascular endothelial cells, *Am. J. Physiol.*, 1992, 262(2 Pt 1), C384–C390.
65. Olesen, S.P., D.E. Clapham, and P.F. Davies, Haemodynamic shear stress activates a K+ current in vascular endothelial cells, *Nature*, 1988, 331(6152), 168–170.
66. Hoyer, J., R. Kohler, and A. Distler, Mechanosensitive Ca^{2+} oscillations and STOC activation in endothelial cells. *FASEB J.*, 1998, 12(3), 359–366.
67. Kuchan, M.J. and J.A. Frangos, Role of calcium and calmodulin in flow-induced nitric oxide production in endothelial cells, *Am. J. Physiol.*, 1994, 266(3 Pt 1), C628–C636.
68. Rubanyi, G.M., J.C. Romero, and P.M. Vanhoutte, Flow-induced release of endothelium-derived relaxing factor, *Am. J. Physiol.*, 1986, 250(6 Pt 2), H1145– H1149.
69. Falcone, J.C., L. Kuo, and G.A. Meininger, Endothelial cell calcium increases during flow-induced dilation in isolated arterioles, *Am. J. Physiol.*, 1993, 264(2 Pt 2), H653–H659.
70. Kanai, A.J., et al., Shear stress induces ATP-independent transient nitric oxide release from vascular endothelial cells, measured directly with a porphyrinic microsensor, *Circ. Res.*, 1995, 77(2), 284–293.
71. Ranjan, V., Z. Xiao, and S.L. Diamond, Constitutive NOS expression in cultured endothelial cells is elevated by fluid shear stress, *Am. J. Physiol.*, 1995, 269(2 Pt 2), H550–H555.

72. Ziegler, T., et al., Nitric oxide synthase expression in endothelial cells exposed to mechanical forces, *Hypertension*, 1998, 32(2), 351–355.
73. Davies, P.F. and S.C. Tripathi, Mechanical stress mechanisms and the cell. An endothelial paradigm, *Circ. Res.*, 1993, 72(2), 239–245.
74. Smith, R.L., et al., Effects of fluid-induced shear on articular chondrocyte morphology and metabolism *in vitro*, *J. Orthop. Res.*, 1995, 13(6), 824–831.
75. Mohtai, M., et al., Expression of interleukin-6 in osteoarthritic chondrocytes and effects of fluid-induced shear on this expression in normal human chondrocytes *in vitro*, *J. Orthop. Res.*, 1996, 14(1), 67–73.
76. Yellowley, C.E., et al., Effects of fluid flow on intracellular calcium in bovine articular chondrocytes, *Am. J. Physiol.*, 1997, 273(1 Pt 1), C30–C36.
77. Yellowley, C.E., C.R. Jacobs, and H.J. Donahue, Mechanisms contributing to fluid-flow-induced Ca^{2+} mobilization in articular chondrocytes, *J. Cell. Physiol.*, 1999, 180(3), 402–408.
78. Edlich, M., et al., Oscillating fluid flow regulates cytosolic calcium concentration in bovine articular chondrocytes, *J. Biomech.*, 2001, 34(1), 59–65.
79. Piekarski, K. and M. Munro, Transport mechanism operating between blood supply and osteocytes in long bones, *Nature*, 1977, 269(5623), 80–82.
80. Knothe Tate, M.L., U. Knothe, and P. Niederer, Experimental elucidation of mechanical load-induced fluid flow and its potential role in bone metabolism and functional adaptation, *Am. J. Med. Sci.*, 1998, 316(3), 189–195.
81. Wang, L., et al., Delineating bone's interstitial fluid pathway *in vivo*. *Bone*, 2004, 34(3), 499–509.
82. Cowin, S.C., S. Weinbaum, and Y. Zeng, A case for bone canaliculi as the anatomical site of strain generated potentials, *J. Biomech.*, 1995, 28(11), 1281–1297.
83. Dillaman, R.M., R.D. Roer, and D.M. Gay, Fluid movement in bone: theoretical and empirical, *J. Biomech.*, 1991, 24 Suppl 1, 163–177.
84. Keanini, R.G., R.D. Roer, and R.M. Dillaman, A theoretical model of circulatory interstitial fluid flow and species transport within porous cortical bone, *J. Biomech.*, 1995, 28(8), 901–914.
85. Hung, C.T., et al., Real-time calcium response of cultured bone cells to fluid flow, *Clin. Orthop.*, 1995, 313, 256–269.
86. Hung, C.T., et al., Intracellular Ca^{2+} stores and extracellular Ca^{2+} are required in the real-time Ca^{2+} response of bone cells experiencing fluid flow, *J. Biomech.*, 1996, 29(11), 1411–1417.
87. Reich, K.M., C.V. Gay, and J.A. Frangos, Fluid shear stress as a mediator of osteoblast cyclic adenosine monophosphate production, *J. Cell Physiol.*, 1990, 143(1), 100–104.
88. Johnson, D.L., T.N. McAllister, and J.A. Frangos, Fluid flow stimulates rapid and continuous release of nitric oxide in osteoblasts, *Am. J. Physiol.*, 1996, 271(1 Pt 1), E205–E208.
89. Reich, K.M., et al., Activation of G proteins mediates flow-induced prostaglandin E2 production in osteoblasts. *Endocrinology*, 1997, 138(3), 1014–1018.
90. Reich, K.M. and J.A. Frangos, Protein kinase C mediates flow-induced prostaglandin E2 production in osteoblasts, *Calcif. Tissue Int.*, 1993, 52(1), 62–66.
91. Sakai, K., M. Mohtai, and Y. Iwamoto, Fluid shear stress increases transforming growth factor beta 1 expression in human osteoblast-like cells: modulation by cation channel blockades, *Calcif. Tissue Int.*, 1998, 63(6), 515–520.
92. Pavalko, F.M., et al., Fluid shear-induced mechanical signaling in MC3T3-E1 osteoblasts requires cytoskeleton-integrin interactions, *Am. J. Physiol.*, 1998, 275(6 Pt 1), C1591–C1601.

93. Cheng, B., et al., PGE(2) is essential for gap junction-mediated intercellular communication between osteocyte-like MLO-Y4 cells in response to mechanical strain, *Endocrinology,* 2001, 142(8), 3464–3473.
94. Weyts, F.A., et al., ERK activation and alpha v beta 3 integrin signaling through Shc recruitment in response to mechanical stimulation in human osteoblasts, *J. Cell Biochem.*, 2002, 87(1), 85–92.
95. Hillsley, M.V. and J.A. Frangos, Alkaline phosphatase in osteoblasts is down-regulated by pulsatile fluid flow, *Calcif. Tissue Int.*, 1997, 60(1), 48–53.
96. Ajubi, N.E., et al., Pulsating fluid flow increases prostaglandin production by cultured chicken osteocytes — a cytoskeleton-dependent process, *Biochem. Biophys. Res. Commun.*, 1996, 225(1), 62–68.
97. Klein-Nulend, J., et al., Pulsating fluid flow stimulates prostaglandin release and inducible prostaglandin G/H synthase mRNA expression in primary mouse bone cells, *J. Bone Miner. Res.*, 1997, 12(1), 45–51.
98. Ajubi, N.E., et al., Signal transduction pathways involved in fluid flow-induced PGE2 production by cultured osteocytes, *Am. J. Physiol.*, 1999, 276(1 Pt 1), E171–E178.
99. Bakker, A.D., et al., Interactive effects of PTH and mechanical stress on nitric oxide and PGE2 production by primary mouse osteoblastic cells, *Am. J. Physiol. Endocrinol. Metab.*, 2003, 285(3), E608–E613.
100. Cheng, B., et al., Expression of functional gap junctions and regulation by fluid flow in osteocyte-like MLO-Y4 cells, *J. Bone. Miner. Res.*, 2001, 16(2), 249–259.
101. Jacobs, C.R., et al., Differential effect of steady versus oscillating flow on bone cells, *J. Biomech.*, 1998, 31(11), 969–976.
102. Donahue, S.W., C.R. Jacobs, and H.J. Donahue, Flow-induced calcium oscillations in rat osteoblasts are age, loading frequency, and shear stress dependent, *Am. J. Physiol. Cell. Physiol.*, 2001, 281(5), C1635–C1641.
103. Saunders, M.M., et al., Gap junctions and fluid flow response in MC3T3-E1 cells, *Am. J. Physiol. Cell. Physiol.*, 2001, 281(6), C1917–C1925.
104. You, J., et al., Osteopontin gene regulation by oscillatory fluid flow via intracellular calcium mobilization and activation of mitogen-activated protein kinase in MC3T3-E1 osteoblasts, *J. Biol. Chem.*, 2001, 276(16), 13365–13371.
105. You, J., et al., P2Y purinoceptors are responsible for oscillatory fluid flow-induced intracellular calcium mobilization in osteoblastic cells, *J. Biol. Chem.*, 2002, 277(50), 48724–48729.
106. Donahue, S.W., H.J. Donahue, and C.R. Jacobs, Osteoblastic cells have refractory periods for fluid-flow-induced intracellular calcium oscillations for short bouts of flow and display multiple low-magnitude oscillations during long-term flow, *J. Biomech.*, 2003, 36(1), 35–43.
107. Saunders, M.M., et al., Fluid flow-induced prostaglandin E(2) response of osteoblastic ROS 17/2.8 cells is gap junction-mediated and independent of cytosolic calcium, *Bone*, 2003, 32(4), 350–356.
108. Donahue, T.L., et al., Mechanosensitivity of bone cells to oscillating fluid flow induced shear stress may be modulated by chemotransport, *J. Biomech.*, 2003, 36(9), 1363–1371.
109. Batra, N.N., et al., Effects of short term recovery periods on fluid induced calcium oscillations in osteoblastic cells, *J. Biomech.*, 2004, in press.
110. Ogata, T., Fluid flow induces enhancement of the Egr-1 mRNA level in osteoblast-like cells: involvement of tyrosine kinase and serum, *J. Cell. Physiol.*, 1997, 170(1), 27–34.

111. Kurokouchi, K., C.R. Jacobs, and H.J. Donahue, Oscillating fluid flow inhibits TNF-alpha-induced NF-kappa B activation via an Ikappa B kinase pathway in osteoblast-like UMR106 cells, *J. Biol. Chem.*, 2001, 276(16), 13499–13504.
112. Alford, A.I., C.R. Jacobs, and H.J. Donahue, Oscillating fluid flow regulates gap junction communication in osteocytic MLO-Y4 cells by an ERK1/2 MAP kinase-dependent mechanism small star, filled, *Bone*, 2003, 33(1), 64–70.
113. Li, Y.J., et al., Effects of loading induced oscillatory fluid flow on human marrow stromal cell proliferation and differentiation, *J. Orthop. Res.*, in press.
114. Haynesworth, S.E., et al., Characterization of cells with osteogenic potential from human marrow, *Bone*, 1992, 13(1), 81–88.
115. Jaiswal, N., et al., Osteogenic differentiation of purified, culture-expanded human mesenchymal stem cells *in vitro*. *J. Cell. Biochem.*, 1997, 64(2), 295–312.
116. Pittenger, M.F., et al., Multilineage potential of adult human mesenchymal stem cells, *Science*, 1999, 284(5411), 143–147.
117. Bruder, S.P., N. Jaiswal, and S.E. Haynesworth, Growth kinetics, self-renewal, and the osteogenic potential of purified human mesenchymal stem cells during extensive subcultivation and following cryopreservation, *J. Cell. Biochem.*, 1997, 64(2), 278–294.
118. Salzstein, R.A. and S.R. Pollack, Electromechanical potentials in cortical bone — II. Experimental analysis, *J. Biomech.*, 1987, 20(3), 271–280.
119. Scott, G.C. and E. Korostoff, Oscillatory and step response electromechanical phenomena in human and bovine bone, *J. Biomech.*, 1990, 23(2), 127–43.
120. Otter, M.W., V.R. Palmieri, and G.V. Cochran, Transcortical streaming potentials are generated by circulatory pressure gradients in living canine tibia. *J. Orthop. Res.*, 1990, 8(1), 119–26.
121. Otter, M.W., et al., A comparative analysis of streaming potentials *in vivo* and *in vitro*, *J. Orthop. Res.*, 1992, 10(5), 710–719.
122. Jaffe, L.F., Electrophoresis along cell membranes, *Nature*, 1977, 265(5595), 600–602.
123. Poo, M. and K.R. Robinson, Electrophoresis of concanavalin A receptors along embryonic muscle cell membrane, *Nature*, 1977, 265(5595), 602–605.
124. Poo, M.M., W.J. Poo, and J.W. Lam, Lateral electrophoresis and diffusion of Concanavalin A receptors in the membrane of embryonic muscle cell, *J. Cell. Biol.*, 1978, 76(2), 483–501.
125. Poo, M., In situ electrophoresis of membrane components, *Annu. Rev. Biophys. Bioeng.*, 1981, 10, 245–276.
126. Ryan, T.A., et al., Molecular crowding on the cell surface, *Science*, 1988, 239(4835), 61–64.
127. McLaughlin, S. and M.M. Poo, The role of electro-osmosis in the electric-field-induced movement of charged macromolecules on the surfaces of cells, *Biophys. J.*, 1981, 34(1), 85–93.
128. Lee, R.C., et al., Cell shape-dependent rectification of surface receptor transport in a sinusoidal electric field, *Biophys. J.*, 1993, 64(1), 44–57.
129. Giugni, T.D., D.L. Braslau, and H.T. Haigler, Electric field-induced redistribution and postfield relaxation of epidermal growth factor receptors on A431 cells, *J. Cell Biol.*, 1987, 104(5), 1291–1297.
130. Tank, D.W., et al., Electric field-induced redistribution and postfield relaxation of low density lipoprotein receptors on cultured human fibroblasts, *J. Cell Biol.*, 1985, 101(1), 148–157.
131. Cho, M.R., et al., Induced redistribution of cell surface receptors by alternating current electric fields, *FASEB J.*, 1994, 8(10), 771–776.

132. Onuma, E.K. and S.W. Hui, A calcium requirement for electric field-induced cell shape changes and preferential orientation, *Cell. Calcium*, 1985, 6(3), 281–292.
133. Onuma, E.K. and S.W. Hui, Electric field-directed cell shape changes, displacement, and cytoskeletal reorganization are calcium dependent, *J. Cell Biol.*, 1988, 106(6), 2067–2075.
134. Cho, M.R., et al., Reorganization of microfilament structure induced by ac electric fields. *FASEB J.*, 1996, 10(13), 1552–1558.
135. Luther, P.W., H.B. Peng, and J.J. Lin, Changes in cell shape and actin distribution induced by constant electric fields, *Nature*, 1983, 303(5912), 61–64.
136. Laub, F. and R. Korenstein, Actin polymerization induced by pulsed electric stimulation of bone cells *in vitro*, *Biochim. Biophys. Acta*, 1984, 803(4), 308–313.
137. Robinson, K.R., The responses of cells to electrical fields: a review, *J. Cell Biol.*, 1985, 101(6), 2023–2027.
138. Blackman, C.F., et al., Effects of ELF (1–120 Hz) and modulated (50 Hz) RF fields on the efflux of calcium ions from brain tissue *in vitro*, *Bioelectromagnetics*, 1985, 6(1), 1–11.
139. Walleczek, J. and R.P. Liburdy, Nonthermal 60 Hz sinusoidal magnetic-field exposure enhances $45Ca^{2+}$ uptake in rat thymocytes: dependence on mitogen activation, *FEBS Lett.*, 1990, 271(1–2), 157–160.
140. Walleczek, J., Electromagnetic field effects on cells of the immune system: the role of calcium signaling. *FASEB J.*, 1992, 6(13), 3177–3185.
141. Carson, J.J., et al., Time-varying magnetic fields increase cytosolic free Ca^{2+} in HL-60 cells, *Am. J. Physiol.*, 1990, 259(4 Pt 1), C687–C692.
142. Hung, C.T., et al., What is the role of the convective current density in the real-time calcium response of cultured bone cells to fluid flow? *J. Biomech.*, 1996, 29(11), 1403–1409.
143. McAllister, T.N. and J.A. Frangos, Steady and transient fluid shear stress stimulate NO release in osteoblasts through distinct biochemical pathways, *J. Bone Miner. Res.*, 1999, 14(6), 930–936.
144. Johnson, M.W., Behavior of fluid in stressed bone and cellular stimulation, *Calcif. Tissue Int.*, 1984, 36 Suppl 1, S72–S76.
145. Kufahl, R.H. and S. Saha, A theoretical model for stress-generated fluid flow in the canaliculi-lacunae network in bone tissue, *J. Biomech.*, 1990, 23(2), 171–180.
146. Allen, F.D., et al., Serum modulates the intracellular calcium response of primary cultured bone cells to shear flow, *J. Biomech.*, 2000, 33(12), 1585–1591.
147. Bakker, A.D., et al., The production of nitric oxide and prostaglandin E(2) by primary bone cells is shear stress dependent, *J. Biomech.*, 2001, 34(5), 671–677.
148. Hoyer, J., R. Kohler, and A. Distler, Mechanosensitive cation channels in aortic endothelium of normotensive and hypertensive rats, *Hypertension*, 1997, 30(1 Pt 1), 112–119.
149. Lansman, J.B., T.J. Hallam, and T.J. Rink, Single stretch-activated ion channels in vascular endothelial cells as mechanotransducers? *Nature*, 1987, 325(6107), 811–813.
150. Naruse, K. and M. Sokabe, Involvement of stretch-activated ion channels in Ca^{2+} mobilization to mechanical stretch in endothelial cells, *Am. J. Physiol.*, 1993, 264(4 Pt 1), C1037–C1044.
151. Brakemeier, S., et al., Up-regulation of endothelial stretch-activated cation channels by fluid shear stress, *Cardiovasc. Res.*, 2002, 53(1), 209–218.
152. Groschner, K., Two ways to feel the pressure: an endothelial $Ca(^{2+})$ entry channel with dual mechanosensitivity, *Cardiovasc. Res.*, 2002, 53(1), 9–11.

153. Chen, N.X., et al., Ca($^{2+}$) regulates fluid shear-induced cytoskeletal reorganization and gene expression in osteoblasts, *Am. J. Physiol. Cell Physiol.*, 2000, 278(5), C989–C997.
154. Nakamura, F. and S.M. Strittmatter, P2Y1 purinergic receptors in sensory neurons: contribution to touch-induced impulse generation, *Proc. Natl. Acad. Sci. U.S.A.*, 1996, 93(19), 10465–10470.
155. Schlosser, S.F., A.D. Burgstahler, and M.H. Nathanson, Isolated rat hepatocytes can signal to other hepatocytes and bile duct cells by release of nucleotides, *Proc. Natl. Acad. Sci. U.S.A.*, 1996, 93(18), 9948–9953.
156. Wang, Y., et al., Autocrine signaling through ATP release represents a novel mechanism for cell volume regulation, *Proc. Natl. Acad. Sci. U.S.A.*, 1996, 93(21), 12020–12025.
157. Grygorczyk, R. and J.W. Hanrahan, CFTR-independent ATP release from epithelial cells triggered by mechanical stimuli, *Am.* J. *Physiol.*, 1997, 272(3 Pt 1), C1058–C1066.
158. Lazarowski, E.R., et al., Direct demonstration of mechanically induced release of cellular UTP and its implication for uridine nucleotide receptor activation, *J. Biol. Chem.*, 1997, 272(39), 24348–24354.
159. Burnstock, G., Release of vasoactive substances from endothelial cells by shear stress and purinergic mechanosensory transduction, *J. Anat.*, 1999, 194 (Pt 3), 335–342.
160. Ostrom, R.S., C. Gregorian, and P.A. Insel, Cellular release of and response to ATP as key determinants of the set-point of signal transduction pathways, *J. Biol. Chem.*, 2000, 275(16), 11735–11739.
161. Romanello, M., et al., Mechanically induced ATP release from human osteoblastic cells, *Biochem. Biophys. Res. Commun.*, 2001, 289(5), 1275–1281.
162. Loomis, W.H., et al., Hypertonic stress increases T cell interleukin-2 expression through a mechanism that involves ATP release, P2 receptor, and p38 MAPK activation, *J. Biol. Chem.*, 2003, 278(7), 4590–4596.
163. Sprague, R.S., et al., Participation of cAMP in a signal-transduction pathway relating erythrocyte deformation to ATP release, *Am. J. Physiol. Cell Physiol.*, 2001, 281(4), C1158–C1164.
164. Sprague, R.S., et al., The role of G protein beta subunits in the release of ATP from human erythrocytes, *J. Physiol. Pharmacol.*, 2002, 53(4 Pt 1), 667–674.
165. Gudi, S., J.P. Nolan, and J.A. Frangos, Modulation of GTPase activity of G proteins by fluid shear stress and phospholipid composition, *Proc. Natl. Acad. Sci. U.S.A.*, 1998, 95(5), 2515–2519.
166. Geiger, B. and A. Bershadsky, Assembly and mechanosensory function of focal contacts, *Curr. Opin. Cell Biol.*, 2001, 13(5), 584–592.
167. Choquet, D., D.P. Felsenfeld, and M.P. Sheetz, Extracellular matrix rigidity causes strengthening of integrin-cytoskeleton linkages, *Cell*, 1997, 88(1), 39–48.
168. Schmidt, C., et al., Mechanical stressing of integrin receptors induces enhanced tyrosine phosphorylation of cytoskeletally anchored proteins, *J. Biol. Chem.*, 1998, 273(9), 5081–5085.
169. MacKenna, D., S.R. Summerour, and F.J. Villarreal, Role of mechanical factors in modulating cardiac fibroblast function and extracellular matrix synthesis, *Cardiovasc. Res.*, 2000, 46(2), 257–263.
170. Xu, J., et al., A three-dimensional collagen lattice activates NF-kappaB in human fibroblasts: role in integrin alpha2 gene expression and tissue remodeling, *J. Cell Biol.*, 1998, 140(3), 709–719.

171. Ingber, D.E., Tensegrity: the architectural basis of cellular mechanotransduction, *Annu. Rev. Physiol.*, 1997, 59, 575–599.
172. Maniotis, A.J., C.S. Chen, and D.E. Ingber, Demonstration of mechanical connections between integrins, cytoskeletal filaments, and nucleoplasm that stabilize nuclear structure, *Proc. Natl. Acad. Sci. U.S.A.*, 1997, 94(3), 849–854.
173. Janmey, P.A., The cytoskeleton and cell signaling: component localization and mechanical coupling, *Physiol. Rev.*, 1998, 78(3), 763–781.
174. Mochly-Rosen, D., Localization of protein kinases by anchoring proteins: a theme in signal transduction, *Science*, 1995, 268(5208), 247–251.
175. Shafrir, Y. and G. Forgacs, Mechanotransduction through the cytoskeleton, *Am. J. Physiol. Cell Physiol.*, 2002, 282(3), C479–C486.
176. Reilly, G.C., et al., Fluid flow induced PGE_2 release by bone cells is reduced by glycocalyx degradation whereas calcium signals are not, *Biorheology*, 2003, 40(6), 591–603.
177. Luft, J.H., Fine structures of capillary and endocapillary layer as revealed by ruthenium red, *Fed. Proc.*, 1966, 25(6), 1773–1783.
178. Wassermann, F. and J. Yaeger, Fine structure of osteocyte capsule and of wall of lacunae in bone, *Zeitschrift fur Zellforschung*, 1965, 67(5), 636–652.
179. Sauren, Y.M., et al., An electron microscopic study on the presence of proteoglycans in the mineralized matrix of rat and human compact lamellar bone, *Anat. Rec.*, 1992, 232(1), 36–44.
180. Shapiro, F., et al., Transmission electron microscopic demonstration of vimentin in rat osteoblast and osteocyte cell bodies and processes using the immunogold technique, *Anat. Rec.*, 1995, 241(1), 39–48.
181. Aarden, E.M., et al., Immunocytochemical demonstration of extracellular matrix proteins in isolated osteocytes, *Histochem. Cell. Biol.*, 1996, 106(5), 495–501.
182. Weinbaum, S., et al., Mechanotransduction and flow across the endothelial glycocalyx, *Proc. Natl. Acad. Sci. U.S.A.*, 2003, 100(13), 7988–7895.
183. You, L., et al., A model for strain amplification in the actin cytoskeleton of osteocytes due to fluid drag on pericellular matrix, *J. Biomech.*, 2001, 34(11), 1375–1386.
184. Ingber, D.E., Cellular tensegrity: defining new rules of biological design that govern the cytoskeleton, *J. Cell Sci.*, 1993, 104 (Pt 3), 613–627.
185. Ingber, D.E., et al., Cellular tensegrity: exploring how mechanical changes in the cytoskeleton regulate cell growth, migration, and tissue pattern during morphogenesis, *Int. Rev. Cytol.*, 1994, 150, 173–224.
186. Ingber, D.E., Tensegrity II. How structural networks influence cellular information processing networks, *J. Cell Sci.*, 2003, 116(Pt 8), 1397–1408.
187. Ingber, D.E., Tensegrity I. Cell structure and hierarchical systems biology, *J. Cell Sci.*, 2003, 116(Pt 7), 1157–1173.
188. Wang, N., J.P. Butler, and D.E. Ingber, Mechanotransduction across the cell surface and through the cytoskeleton, *Science*, 1993, 260(5111), 1124–1127.
189. Pavalko, F.M., et al., A model for mechanotransduction in bone cells: the load-bearing mechanosomes. *J. Cell Biochem.*, 2003, 88(1), 104–112.
190. Ko, K.S., P.D. Arora, and C.A. McCulloch, Cadherins mediate intercellular mechanical signaling in fibroblasts by activation of stretch-sensitive calcium-permeable channels, *J. Biol. Chem.*, 2001, 276(38), 35967–35977.
191. Potard, U.S., J.P. Butler, and N. Wang, Cytoskeletal mechanics in confluent epithelial cells probed through integrins and E-cadherins, *Am. J. Physiol.*, 1997, 272(5 Pt 1), C1654–C1663.

192. Peck, W.A. and S. Klahr, Cyclic nucleotides in bone and mineral metabolism, *Adv. Cyclic Nucleotide Res.*, 1979, 11, 89–130.
193. van der Plas, A., J.H. Feyen, and P.J. Nijweide, Direct effect of parathyroid hormone on the proliferation of osteoblast-like cells; a possible involvement of cyclic AMP, *Biochem. Biophys. Res. Commun.*, 1985, 129(3), 918–925.
194. Lerner, U.H., B.B. Fredholm, and M. Ransjo, Use of forskolin to study the relationship between cyclic AMP formation and bone resorption *in vitro*, *Biochem. J.*, 1986, 240(2), 529–539.
195. Binderman, I., Z. Shimshoni, and D. Somjen, Biochemical pathways involved in the translation of physical stimulus into biological message, *Calcif. Tissue Int.*, 1984, 36 Suppl 1, S82–S85.
196. You, J., et al., Oscillatory flow stimulates prosaglandin E2 release via protein kinase A in MC3T3-E1 osteoblasts involving cyclooxegenase-2, in *Transactions 47th Orthopedic Research Society*, 2001, 47, 0326.
197. Jair, K.Y., et al., Dynamics of F-actin cytoskeleton reorganization under oscillatory fluid flow in MC3T3-E1 osteoblastic cells, 2004, in preparation.
198. Dimmeler, S., et al., Activation of nitric oxide synthase in endothelial cells by Akt-dependent phosphorylation, *Nature*, 1999, 399(6736), 601–605.
199. Langille, B.L., Morphologic responses of endothelium to shear stress: reorganization of the adherens junction, *Microcirculation*, 2001, 8(3), 195–206.
200. Seger, R. and E.G., Krebs, The MAPK signaling cascade, *FASEB J.*, 1995, 9(9), 726–735.
201. Kyriakis, J.M., et al., The stress-activated protein kinase subfamily of c-Jun kinases, *Nature*, 1994, 369(6476), 156–160.
202. Rizzo, V., et al., Rapid mechanotransduction in situ at the luminal cell surface of vascular endothelium and its caveolae, *J. Biol. Chem.*, 1998, 273(41), 26323–26329.
203. Jo, H., et al., Differential effect of shear stress on extracellular signal-regulated kinase and N-terminal Jun kinase in endothelial cells. Gi2- and Gbeta/gamma-dependent signaling pathways, *J. Biol. Chem.*, 1997, 272(2), 1395–1401.
204. Yan, C., et al., Fluid shear stress stimulates big mitogen-activated protein kinase 1 (BMK1) activity in endothelial cells. Dependence on tyrosine kinases and intracellular calcium, *J. Biol. Chem.*, 1999, 274(1), 143–150.
205. Li, C., et al., Cyclic strain stress-induced mitogen-activated protein kinase (MAPK) phosphatase 1 expression in vascular smooth muscle cells is regulated by Ras/Rac-MAPK pathways, *J. Biol. Chem.*, 1999, 274(36), 25273–25280.
206. Liang, F. and D.G. Gardner, Mechanical strain activates BNP gene transcription through a p38/NF-kappaB-dependent mechanism, *J. Clin. Invest.*, 1999, 104(11), 1603–1612.
207. Atmani, H., et al., Phenotypic effects of continuous or discontinuous treatment with dexamethasone and/or calcitriol on osteoblasts differentiated from rat bone marrow stromal cells, *J. Cell Biochem.*, 2002, 85(3), 640–650.
208. Aubin, J.E., Regulation of osteoblast formation and function, *Rev. Endocr. Metab. Disord.*, 2001, 2(1), 81–94.
209. Mizuno, M. and Y. Kuboki, Osteoblast-related gene expression of bone marrow cells during the osteoblastic differentiation induced by type I collagen. *J. Biochem. (Tokyo)*, 2001, 129(1), 133–138.
210. Pavlin, D., et al., Mechanical loading stimulates differentiation of periodontal osteoblasts in a mouse osteoinduction model: effect on type I collagen and alkaline phosphatase genes, *Calcif. Tissue Int.*, 2000, 67(2), 163–172.

211. Prockop, D.J., Marrow stromal cells as stem cells for nonhematopoietic tissues, *Science*, 1997, 276(5309), 71–74.
212. Zhang, R., et al., Rat tail suspension reduces messenger RNA level for growth factors and osteopontin and decreases the osteoblastic differentiation of bone marrow stromal cells, *J. Bone Miner. Res.*, 1995, 10(3), 415–423.
213. Keila, S., et al., Bone marrow from mechanically unloaded rat bones expresses reduced osteogenic capacity *in vitro*, *J. Bone Miner. Res.*, 1994, 9(3), 321–327.
214. Machwate, M., et al., Skeletal unloading in rat decreases proliferation of rat bone and marrow-derived osteoblastic cells, *Am. J. Physiol.*, 1993, 264(5 Pt 1), E790–E799.
215. Forwood, M.R., Inducible cyclo-oxygenase (COX-2) mediates the induction of bone formation by mechanical loading *in vivo*, *J. Bone Miner. Res.*, 1996, 11(11), 1688–1693.
216. Schwachtgen, J.L., et al., Fluid shear stress activation of egr-1 transcription in cultured human endothelial and epithelial cells is mediated via the extracellular signal-related kinase 1/2 mitogen-activated protein kinase pathway, *J. Clin. Invest.*, 1998, 101(11), 2540–2549.
217. Liu, C., et al., The transcription factor EGR-1 directly transactivates the fibronectin gene and enhances attachment of human glioblastoma cell line U251, *J. Biol. Chem.*, 2000, 275(27), 20315–20323.
218. Dey, B.R., et al., Repression of the transforming growth factor-beta 1 gene by the Wilms' tumor suppressor WT1 gene product, *Mol. Endocrinol.*, 1994, 8(5), 595–602.
219. Majeska, R.J., Cell biology of bone, in *Bone Mechanics Handbook*, S.C. Cowin, Ed., 2001, CRC Press LLC, Boca Raton, FL, 2-1–2-24.
220. Grigoriadis, A.E., et al., Osteoblasts are target cells for transformation in c-fos transgenic mice, *J. Cell. Biol.*, 1993, 122(3), 685–701.
221. Grigoriadis, A.E., et al., c-Fos: a key regulator of osteoclast-macrophage lineage determination and bone remodeling, *Science*, 1994, 266(5184), 443–448.
222. Wang, Z.Q., et al., Bone and haematopoietic defects in mice lacking c-fos, *Nature*, 1992, 360(6406), 741–745.
223. Turner, C.H., Y. Tu, and J.E. Onyia, Mechanical loading of bone *in vivo* caused bone formaiton through early induction of c-fos, but not c-jun or c-myc, *Ann. Biomed. Eng.*, 1996, 24, S-74.
224. Chow, J.W. and T.J. Chambers, Indomethacin has distinct early and late actions on bone formation induced by mechanical stimulation, *Am. J. Physiol.*, 1994, 267(2 Pt 1), E287–E292.
225. Kasten, T.P., et al., Potentiation of osteoclast bone-resorption activity by inhibition of nitric oxide synthase, *Proc. Natl. Acad. Sci. U.S.A.*, 1994, 91(9), 3569–3573.
226. MacIntyre, I., et al., Osteoclastic inhibition: an action of nitric oxide not mediated by cyclic GMP, *Proc. Natl. Acad. Sci. U.S.A.*, 1991, 88(7), 2936–2940.
227. Riancho, J.A., et al., Expression and functional role of nitric oxide synthase in osteoblast-like cells, *J. Bone Miner. Res.*, 1995, 10(3), 439–446.
228. Riancho, J.A., et al., Mechanisms controlling nitric oxide synthesis in osteoblasts, *Mol. Cell. Endocrinol.*, 1995, 107(1), 87–92.
229. Vogel, J.M. and M.W. Whittle, Bone mineral changes: the second manned Skylab mission, *Aviat. Space Environ. Med.*, 1976, 47(4), 396–400.
230. Cui, L., et al., Cancellous bone of aged rats maintains its capacity to respond vigorously to the anabolic effects of prostaglandin E2 by modeling-dependent bone gain, *J. Bone Miner. Metab.*, 2001, 19(1), 29–37.
231. Yao, W., et al., Anabolic effect of prostaglandin E2 on cortical bone of aged male rats comes mainly from modeling-dependent bone gain, *Bone*, 1999, 25(6), 697–702.

232. Reich, K.M. and J.A. Frangos, Effect of flow on prostaglandin E2 and inositol trisphosphate levels in osteoblasts, *Am. J. Physiol.*, 1991, 261(3 Pt 1), C428–C432.
233. Bowler, W.B., et al., Signaling in human osteoblasts by extracellular nucleotides. Their weak induction of the c-fos proto-oncogene via Ca^{2+} mobilization is strongly potentiated by a parathyroid hormone/cAMP-dependent protein kinase pathway independently of mitogen-activated protein kinase, *J. Biol. Chem.*, 1999, 274(20), 14315–14324.
234. Yu, H. and J. Ferrier, Osteoblast-like cells have a variable mixed population of purino/nucleotide receptors, *FEBS Lett.*, 1993, 328(1–2), 209–214.
235. Jones, S.J., et al., Purinergic transmitters inhibit bone formation by cultured osteoblasts, *Bone*, 1997, 21(5), 393–399.
236. Kumagai, H., B. Sacktor, and C.R. Filburn, Purinergic regulation of cytosolic calcium and phosphoinositide metabolism in rat osteoblast-like osteosarcoma cells, *J. Bone Miner. Res.*, 1991, 6(7), 697–708.
237. Reimer, W.J. and S.J. Dixon, Extracellular nucleotides elevate $[Ca^{2+}]i$ in rat osteoblastic cells by interaction with two receptor subtypes, *Am. J. Physiol.*, 1992, 263(5 Pt 1), C1040–C1048.
238. Bowler, W.B., et al., Identification and cloning of human P2U purinoceptor present in osteoclastoma, bone, and osteoblasts, *J. Bone Miner. Res.*, 1995, 10(7), 1137–1145.
239. Gallinaro, B.J., W.J. Reimer, and S.J. Dixon, Activation of protein kinase C inhibits ATP-induced $[Ca^{2+}]_i$ elevation in rat osteoblastic cells: selective effects on P2Y and P2U signaling pathways, *J. Cell Physiol.*, 1995, 162(3), 305–314.
240. Arnett, T.R. and B.F. King, ATP as an osteoclast regulator? *J. Physiol.*, 1997, 503 (Pt 2), 236.
241. Naemsch, L.N., et al., P2X(4) purinoceptors mediate an ATP-activated, non-selective cation current in rabbit osteoclasts, *J. Cell Sci.*, 1999, 112 (Pt 23), 4425–4435.
242. Wiebe, S.H., S.M. Sims, and S.J. Dixon, Calcium signalling via multiple P2 purinoceptor subtypes in rat osteoclasts, *Cell Physiol. Biochem.*, 1999, 9(6), 323–337.
243. Shimegi, S., ATP and adenosine act as a mitogen for osteoblast-like cells (MC3T3-E1), *Calcif. Tissue Int.*, 1996, 58(2), 109–113.
244. Morrison, M.S., et al., ATP is a potent stimulator of the activation and formation of rodent osteoclasts, *J. Physiol.*, 1998, 511 (Pt 2), 495–500.
245. Bowler, W.B., et al., Release of ATP by osteoblasts: modulation by fluid shear forces, *Bone*, 1998, 22(3, Supplement), 3S.
246. Malek, A.M., et al., Fluid shear stress differentially modulates expression of genes encoding basic fibroblast growth factor and platelet-derived growth factor B chain in vascular endothelium, *J. Clin. Invest.*, 1993, 92(4), 2013–2021.
247. Ohno, M., et al., Fluid shear stress induces endothelial transforming growth factor beta-1 transcription and production. Modulation by potassium channel blockade, *J. Clin. Invest.*, 1995, 95(3), 1363–1369.
248. Baylink, D.J., R.D. Finkelman, and S. Mohan, Growth factors to stimulate bone formation, *J. Bone Miner. Res.*, 1993, 8 Suppl. 2, S565–S572.
249. Ehrlich, P.J. and L.E. Lanyon, Mechanical strain and bone cell function: a review, *Osteoporos. Int.*, 2002, 13(9), 688–700.
250. Zaman, G., et al., Mechanical strain stimulates nitric oxide production by rapid activation of endothelial nitric oxide synthase in osteocytes, *J. Bone Miner. Res.*, 1999, 14(7), 1123–1131.
251. Rawlinson, S.C., et al., Calvarial and limb bone cells in organ and monolayer culture do not show the same early responses to dynamic mechanical strain, *J. Bone Miner. Res.*, 1995, 10(8), 1225–1232.

252. Reinholt, F.P., et al., Osteopontin — a possible anchor of osteoclasts to bone, *Proc. Natl. Acad. Sci. U.S.A.*, 1990, 87(12), 4473–4475.
253. Giachelli, C.M. and S. Steitz, Osteopontin: a versatile regulator of inflammation and biomineralization, *Matrix Biol.*, 2000, 19(7), 615–622.
254. Frank, O., et al., Real-time quantitative RT-PCR analysis of human bone marrow stromal cells during osteogenic differentiation *in vitro*, *J. Cell Biochem.*, 2002, 85(4), 737–746.
255. Butler, W.T., A.L. Ridall, and M.D. McKee, Osteopontin, in *Principles of Bone Biology*, J.P. Bilezikian, Ed., 1996, Academic Press, San Diego. pp. 167–181.
256. Terai, K., et al., Role of osteopontin in bone remodeling caused by mechanical stress, *J. Bone Miner. Res.*, 1999, 14(6), 839–849.
257. Harter, L.V., K.A. Hruska, and R.L. Duncan, Human osteoblast-like cells respond to mechanical strain with increased bone matrix protein production independent of hormonal regulation, Endocrinology, 1995, 136(2), 528–535.
258. Kubota, T., et al., Influence of an intermittent compressive force on matrix protein expression by ROS 17/2.8 cells, with selective stimulation of osteopontin, *Arch. Oral Biol.*, 1993, 38(1), 23–30.
259. Ducy, P. and G. Karsenty, Skeletal gla proteins: gene structure, regulation of expression and function, in *Principles of Bone Biology*, J.P. Bilezikian, Ed., 1996, Academic Press, San Diego. pp. 183–195.
260. Doty, S.B., Morphological evidence of gap junctions between bone cells, *Calcif. Tissue Int.*, 1981, 33(5), 509–512.
261. Yellowley, C.E., et al., Functional gap junctions between osteocytic and osteoblastic cells, *J. Bone Miner. Res.*, 2000, 15(2), 209–217.

7 Nanoarchitectures, Nanocomputing, Nanotechnologies and the DNA Structure

S. Barbu, M. Morf, and A. E. Barbu

CONTENTS

7.1 INTRODUCTION AND BACKGROUND

As we approach the nanoscale, advances in molecular biology, genomics, electrical engineering, computer science, SUSY* physics, and mathematics are experiencing

* SUSY — SUper-SYmmetry. A field of modern physics. Such symmetry principles are also part of mathematics, and they are starting to be used in engineering for modeling hybrid systems, complex flows, and advanced signal processing.

0-8493-1940-4/05/$0.00+$1.50

a convergence of processes in all these fields that contain both synergy and chaos. A quarter century ago we reached the microscale with microtechnology, when Mead and Conway[1] envisioned a roadmap for very-large-scale integrated (VLSI) systems design as a multidisciplinary endeavor. Now we need a roadmap for modeling, design, and fabrication of nanosystems. This book describes many of the necessary steps in such a roadmap, and in this chapter we discuss some of the nanoarchitecture and related aspects that will play an increasing role in nanosystems.

Today's applications and fabrication processes involve mostly the microscale. However, as we did a quarter century back, we have relied on the previous scale to provide the embedding technology. For instance, in electronic systems, printed-circuit boards were used to embed or hold the next generation VLSI chips. Today we embed a whole DNA lab-analysis function on a chip. This is the first step toward the next level, where complex molecules such as DNA become the new nanoscale embedding technology. Several current proposals contain such ideas, such as using DNA as a quantum-computing substrate.

In this chapter we argue that DNA can be used not only as a *passive* mechanical or memory nanosubstrate, but also as an *active* mechanical, electrical, photonic, and information-flow network, based on classical or quantum distributed parallel nano-computing. Similarly, microtubules have already been shown to provide electro-mechanical molecule-transport services on webs of microtubules. The underlying electron-transport dipole–dipole or multipole wave-propagation interactions also implement electronic finite-state machines suitable for active switching and routing networks. The microtubules coated with layers of polar water molecules have a nanoarchitecture ideal for photonic (e.g., EH-mode) wave-guides, similar to carbon nanotubes. Microtubules also make up nerve fibers that are now acknowledged to implement point-to-point, broadcast, and packet-switched connections from the brain to control extremities such as the fingers. This process was described to the second author a quarter century ago by a medical researcher as inconsistent behavior, and packet switching was rejected. Similar descriptions of the molecular transport by microtubules are today still labeled as "random." Such simplifying assumptions were made to model Internet traffic, but some now prefer chaotic models as will be discussed in the dynamical system models for biological processes.

Natural sciences, such as biology, are driven by large amounts of data that should be processed even in simple experiments in molecular biology and genomics. This data abundance demands new tools that have to be produced in cooperation with other fields of science. Today, teams of people versed in several sciences drive large programs developed for biomedical data-processing applications, such as DNA analysis. However, programmers who are versed in systems concepts that are not yet part of classical biology often decide important lower-level data-interpretation and functionality issues.

Most of the computer-science concepts, such as software, hardware, firmware, emulation, and memory architecture, exist in the biomedical area. For example, the brain works like a computer, but not the way one thinks. The implementations of the loop concepts are very different. Engineers tend to carefully design loops. They either encapsulate them (as in a memory-cell) or avoid them (to avoid CMOS latch-up, a destructive short circuit). In contrast, biological systems seem to be based on many interacting loops, making an accurate analysis very difficult. Neighboring cells exhibit cooperative phenomena, best described by adaptive control theory as nested (nonlinear) loops. From an architectural standpoint, one should ask what the design principles are that govern the encapsulation in a cell. The cooperative phenomena act like adaptive communication protocols. An interesting example is the communication between local cells, foreign cells, and policing T cells. In this example, lateral inhibitory mutual responses create loops of inverters (flip-flops) that synchronize the behavior of cells. A T cell interacting with a neighboring foreign cell makes a friend-or-foe decision and creates a "don't recognize you, shut down your computing process and recycle yourself" signal that initiates an apoptosis response. The AIDS virus interferes with this process by disabling the T cell's policing function.

Our aim here is to show how nanotechnologies could influence and help the research process in the biology field at many levels: concepts, architectures, systems, information flow, processes, analysis, synthesis, and tools. Early interactions between biology and the other fields have already occurred in the twentieth century. The interdisciplinary field of neuroscience is one of the early examples of developing models, such as the Hopfield model,[2] of neuron activity at the macrobiological level. This activity helped generate new applications in the communication-networking field, such as packet transmission, firing neurons* (burst transmission of messages), long-duration transmission models (streaming audio/video), and concurrent learning or system identification of a network. Today, no one could imagine a simple piece of communication equipment, like the now-standard Asymmetric Digital Subscriber Line (ADSL) modems, that would not incorporate a calibrating or training mechanism that learns the phase or delay characteristics of a communication channel. Such learning or adaptation opens a complicated, layered suite of protocols that establishes the information-transmission processes. Training mechanisms are essential, even vital, today and hence unavoidable, to ensure that such transmissions are fast, robust, error-free, and globally efficient. Adaptive mechanisms have been studied intensively in electronics research starting in the early 1970s and continuing to the 1980s. Physics, and especially geophysics, inspired the use of scattering methods for source and channel modeling. This proved to be a powerful set of tools that combined the advantages of *local linearity* with *nonlinearity* at the system level; for example, tools that used bilinear or continued-fraction modeling approaches, which have many modeling applications in communications, biological systems, and other systems.[3] Important early work on nonlinear least-squares methods, and later work on Ultra-Wide-Band (UWB) communication by Swerling at Rand was largely ignored or classified. The linear case was later rediscovered and made famous by Kalman (the "Kalman Filter") and others. The new technical or computing tools that the above

* In the early days of medical research on neural activities at Stanford, the "inconsistent behavior of branching neurons" was described to us. Our explanation of packet switching became accepted only much later.

theoretical fields enabled have enhanced discoveries in medicine and biology via imaging, modeling, and control. These tools generally addressed the biology field at the macro level. Today, the main issue is to extend these insights, tools, and models down to the micro- and nanotechnology levels.

We are on the verge of a new industrial revolution due to the new domain of atomic interactions between molecular biology and genomics and the complex field of disciplines mentioned above. In this chapter we ask, what are the key advances in fundamental electronics that were enabled by advances in the other fields, and how can these advances be used to induce new ways of thinking and modeling at the nanoscale in molecular biology and genomics?

7.2 DNA AND RELATED STRUCTURES

The contributions to the understanding of the DNA and related structures that were made by the field of mathematics have been outlined well,[4,5] and further modeling related to the chemical structure has been proposed.[6–9] For definitions of DNA- and mathematics-related terms (a veritable Rosetta stone) see Carbone and Gromov.[4] Extensions to computer architecture and engineering are still forthcoming and beyond our present scope of outlining a roadmap. We endeavor here to point out major cross-links, but more complete connections have to be deferred to future publications.[T45]

The DNA structure itself has been the main target of scientific activity. Two main lines of investigation that are of interest here have been conducted: (1) the Genome study and its functions, and (2) computing and studies of new ways of computing using the DNA strings.[10,11] A third line of investigation is possible, using communication and computing concepts to describe the DNA structure (DNA architecture). This third approach creates a simultaneous bridge between the two former lines of investigation. We start this third line on the DNA architecture from a dynamic-system viewpoint, in which the concept of *saddle points* is fundamental in the evolution of the dynamic system as sources of *bifurcation phenomena.* Such phenomena include the concepts of *system evolution* (in the biological sense) and *information-increase and entropy* in communication-systems theory, applied along a DNA strand or even to proteins and cells. This view can be extended to the computing aspects within DNA, in which saddle points are seen as information-exchange gates* (including Boolean gates) that are building complex information networks at different scales.[12,13,T37,T40,T41] Before discussing this subject in more

* Information eXchange gates, or X-type gates, are elementary routing, switching, Boolean, and unitary quantum-gate computing elements. They are related to saddle points (mountain passes) that redirect source-flows from neighboring mountain peaks to sinks of these flows into neighboring valleys. Thus, they are related to topological landscapes that appear in many fields of science, including physics, mathematics, biology, and engineering. The local concept of watersheds and the dual global concept of continental divides are the simplest to illustrate the fundamental role of underlying advanced topological concepts and their robustness and computational aspects. More below; see also references.[3,34]

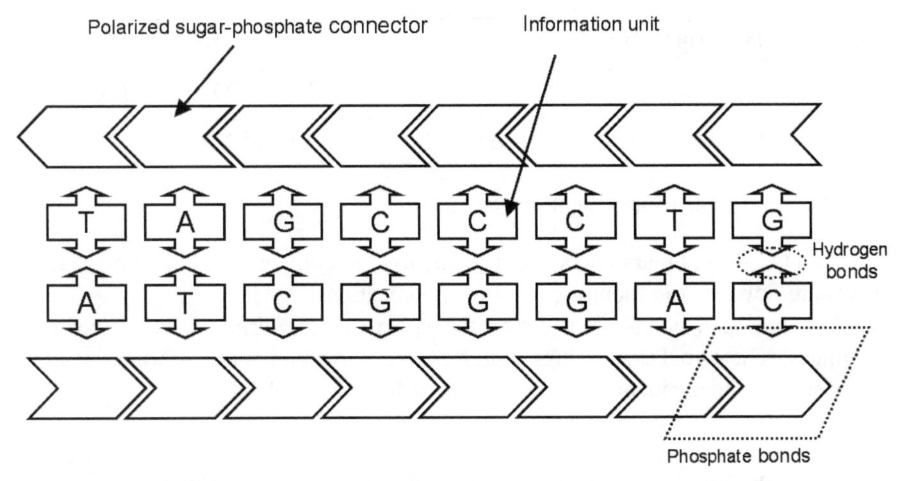

FIGURE 7.1 DNA representation with connectors and information units.

detail, we first establish the basic conventions used here and a few simple steps of abstract modeling as presentation aids.

7.2.1 DNA Conventions

DNA is viewed as a double, very long, helicoidal string (seen as a *word* in computing) of four nucleotides A (adenine), T (thymine), C (cytosine), G (guanine). They are formed by a common, naturally polarized structure, the sugar-phosphate group, which we call the *connector*, and a different base for every nucleotide, which we call the *information unit.* The connectors are linked together, creating the backbone of each helix in the DNA structure. The information units are linked together only in very specific ways (covalent bonds), and they are responsible for both the global information-system aspects in the DNA and its geometrical shape. The information units could be coupled together in specific pairs: A–T, C–G. The linking of the information units is alternating along the two individual strings (Figure 7.1).

The number of covalent bonds gives the strength of the coupling. The A–T coupling, which is governed by two covalent bonds, is somewhat weaker then the C–G coupling, which is governed by three covalent bonds. This model corresponds to the well-known Watson–Crick chemical representation of the complementary bases. The links between the bases are very weak hydrogen bonds. The phosphate bonds for the connectors are an order-of-magnitude higher in strength, which we call *intermediate strength.* The double-carbon covalent bond is the strongest bond (Figure 7.2). The *p*-electron states are responsible for the intermediate and strongest bonds, compared to the *s*-electrons that are part of the weak bonds. These different strengths will be related to multiple information-levels or logic-levels, beyond the

Hydrogen bond	1-5 Kcal/Mol
Phosphate bond	50 Kcal/Mol
Carbon-Carbon double bond	200 Kcal/Mol

FIGURE 7.2 Sample bond strengths (energy).

common two-level binary encoding typically used in computers; three or more logic levels are now used in memories and communications.

The hydrogen bonds are creating the effect of *hybridization,* as is known in genomics. The two DNA strings could couple complementary information units between two independent strings or create a local loop of the same string, called *self hybridization.* This phenomenon we prefer to call *complementary networking* or *local-loop network.* This terminology enables the use of networking concepts. Consequently, Figure 7.1 could represent both types of organizations, local-loop or complementary networking, depending on the connector being from the same or a different strand. These two different logical views do not change the fundamental physical structure. However, they introduce a new logical-level notion of information processing and exchange at the unit level of information (the A, C, G, T base level). In previous work, the sets of base pairs were considered only to contain the elementary bits of information. The string of bases was then considered to form the information *word,* using a four-letter alphabet. This is a static view. In contrast, a network concept implies a dynamic exchange of information, one that uses the alphabet to articulate complex words.

7.2.2 DNA Network Structure

The physical-scale dimensions involved in biology are diverse. They range from a few angstroms for a hydrogen bond, to 1 nm for a nucleotide, to 1 μm for a bacterium, to 2 m for DNA strings in a human cell, to 10^{14} m for the total amount of DNA in the human body. From a system-architecture viewpoint, the span of these physical-scale dimensions is very high,* so global phenomena can be very different than local phenomena. A network-view approach will encompass size scales at all levels; thus *renormalization processes* will have to occur that bring the system onto a tractable and manageable path.

DNA is differently organized for different organisms. Bacteria cells (prokaryotic cells) have a unique circular DNA or supercoil. Human DNA (or generally eukaryotic cells) have complex folding and spatial organizations known as chromosomes. Intuitively, the occurrence of different spatial organizations is not too surprising, considering the fact that the one-dimensional DNA, viewed as a program (a linear sequence of instructions) will ultimately create a three-dimensional structure. The

* For example, the ratio between the largest and the smallest scale is on the order of 10^{24}, comparable to the Avogadro constant for gases, or the size ratio of a transistor to a global communication network, hinting at the complexity of DNA models.

complex mappings required in this *dimensional expansion* process were among the first issues that attracted our interest. Hence, folding and coiling are only the first steps in this *computational origami* process, a computing term due to A. Huang.[14] It is probably not a coincidence that folding and coiling of repeated and possibly reflected strings of dipoles leads to more stable strings and coils of multipoles, a three-dimensional manifold of multipoles resulting, for example, in lower and more-stable energy states.

The human genome has neither the highest number of chromosomes nor the highest number of information bits. For example, a chicken (*Gallus domesticus*) has 39 chromosomes versus the 23 human chromosomes, with only 1.2 Gb of information versus the 3 Gb the human DNA has. A simpler vegetal form, the onion (*Allium cepa*) has 15 Gb of information with only 8 chromosomes. Certainly, we have to consider ideas such as organizational efficiency, efficient coding, and robustness in order to explain the apparent contradictions mentioned above. Nonetheless, this difference in efficiency still needs to be explained. Well-structured computer programs tend to be composed of smaller modules, and regular structures and architectures tend to be favored, such as tables or gate-arrays reminiscent of Figure 7.3, which are more amenable to implementing error-detection and correction.

DNA has a chemical relative, RNA. In RNA, the information units are A, C, G, and U, where U (uracil) replaces the T (thymine). RNA structures are not as long as DNA structures. RNA occurs in a multitude of segments in the cells, and they all have different roles. The *messenger* RNA (mRNA) helps the interaction of the DNA with proteins. The *transfer* RNA (tRNA) and the *ribosomal* RNA (rRNA) are active in the protein synthesis implementing the information carried by the mRNA. The mRNA represents 80% of the total RNA for a mammalian, followed by tRNA at 15% and rRNA at 5%. The RNA has two important properties: (1) to be a simpler keeper of information (like decoded instructions in the instruction stream of a computer's central processing unit), and (2) its spatial organization, which is related to its autocatalytic properties. RNA's behavior is that of an enzyme (protein). The three-dimensional organization is still a challenging mathematical problem, in particular the charge distribution on a three-dimensional surface.*

Information in DNA can be copied onto the RNA string with the help of other proteins such as the RNA polymerase. The resulting RNA string can be further spliced into *introns* (large subsegments) or litigated to form *exons*. This creates topologically and logically complete operations, which enable computing-like processes.

From a network viewpoint, RNA is a complex read–store computing machine. The same RNA string can splice differently near a DNA site, depending on the global conditions of the cell or local conditions near the splicing site. This acts as a site-specific, conditional program read from memory with global and/or local

---

* This suggests that a programmable structure could be made that "emulates" a desired charge distribution. In aerospace design there exists a similar concept of smart surfaces (e.g., to induce a fully laminar flow boundary surface). A combination of fluid and ionic flow is important in bone growth (see Chapter 6 on Cellular Mechanotransduction).

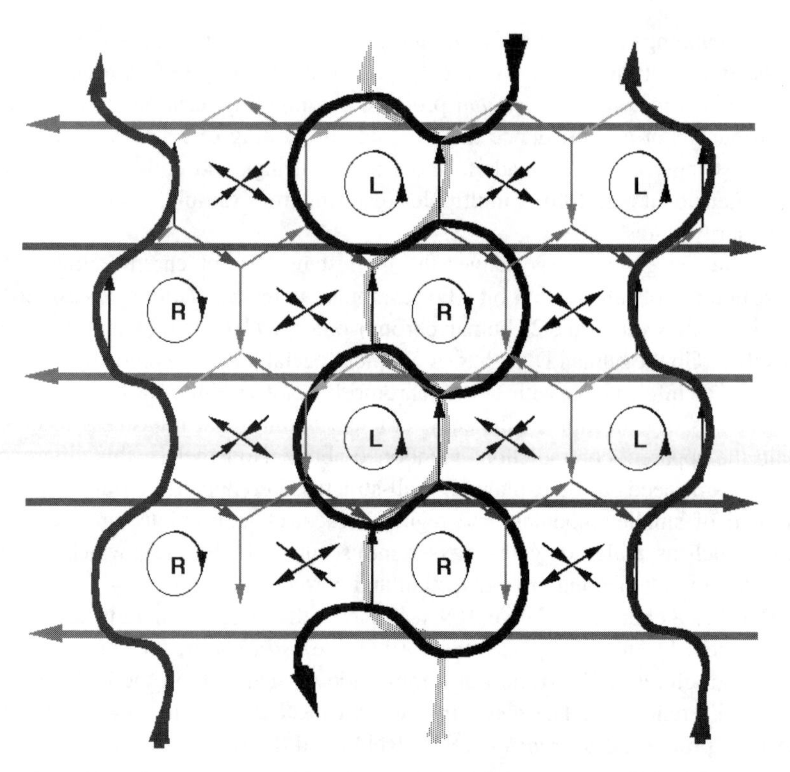

FIGURE 7.3 Slow North–South flow, a form of topological symmetry-breaking.

control. Such a DNA site is called a *gene*. There are also regulatory genes that control the mechanism of transcription that could be situated far away from the read site (the normal gene). The intergenic space between the genes is usually considered junk DNA. This is not the case from a network viewpoint, in which the spatial and temporal aspects make the *coupling* of the processing units as important as the processing units themselves.

DNA folding in chromosomes is essential to create a spatial correlation between distant information units (similar to a flat pattern for cutting the cloth for sewing clothes for a three-dimensional human). A local transcription is now possible from far-away units along the string, hence distributed information units that are correlated (connected) create the robust coding of DNA. More generally, this relates to stochastic systems and the complexity of the information coding and entropy in the sense of Kolmogorov.[4] We consider this reference a first major step in advanced mathematical modeling of the DNA structure. However, we can envision a number of alternative approaches that will have to be sorted out in the future. For example, the use of abstract free groups is very general, akin to the implementation-technology-independent use of Boolean functions in computing. But when physical (circuit or nanotechnology) constraints are used, more specific *architectures* of groups appear, reminiscent of switch-level circuits and physical-scattering theory. For example, the double and triple hydrogen bonds in the bases suggest the use 3-input/3-output D3 group-based, logically complete, and invertible X-type gate models, which are related to free energy and adiabatic processes. In addition, energy, topology, and other invariants are fundamentally linked to new homotopy theorems about information flows, [15,T45] with the advantage that complex flow models, even for quantum physics, are easy to visualize in flat-land by applied scientists, engineers, and the public at large.

From an engineering perspective, one would expect DNA coding to be related to the time, location, and function of the biological system. For example, how does the biological global positioning system (GPS) work? Tantalizing pieces of all these aspects have been slowly emerging, especially now that we are able to go more easily down to the nanoscale. Developmental biology has shown how an embryo first develops as a linear sausage-like structure,* in which the partitions are determined by the ratio of the gradients of "food sources" at both ends. At the molecular scale, this amounts to measuring relative arrival rates of different molecules, or time-difference-of-arrival (TDOA) measurements. Delosme and Morf[T17] have shown how to derive robust location information from noisy TDOA measurements. From such results it is clear that if the "food sources" lie on the periphery (i.e., on the convex hull of all locations to be determined), such location estimates are robust. Early developmental foldings contribute to robust location. However, when limbs and other extremities are required, this process can fail, because only a fraction of all desirable locations of partitions or organs can be determined robustly. We note that the "food sources" that help develop the wings of insects are not distributed only along the periphery of the wings but over a spatial area, the necessary and sufficient conditions of robust location for architectural features.

7.3 DYNAMIC AND COMPUTER SYSTEMS VIEW

A dynamic system exhibits different forms of behavior in the phase space. An important subspace of the phase space in an invariant manifold (i.e., a surface with the property that all orbits starting on it throughout the total evolution of the dynamic system remain on that surface, regardless of the direction of time). It can also be seen as a collection of orbits that are dense on that surface. A trajectory can leave a surface crossing a boundary, so that the boundary location plays a role similar to a membrane in a biological cell. The manifolds can be stable or unstable, depending on the conversion of orbits on that surface in direct-time or in reverse-time. Moreover, it is possible that these manifolds have a more complex architecture, in the sense that they are *fibered* or *foliated* by lower-dimension submanifolds.

* This sausage or submarine-like structure actually can be recognized as a *Riemannian sphere* or manifold if the communication within this structure works like a non-Archimedian network, where equi-levels have equi-delays; see Figure 7.4. A Riemannian sphere is deformable by stretching and folding analogous to a growing embryo.

7.3.1 Differentiable Invariant Manifolds

The study of manifolds are important in fields like fluids, mechanics, theoretical chemistry, molecular reaction dynamics, chaotic scattering, and mathematical biology. Nature constructs objects such as nanotubes[16,17] and DNA tori,[18] which are juxtapositions of tori representing invariant manifolds. The torus is the most typical type of manifold in KAM theory for stochastic flows.[19] Other basic invariant manifolds include equilibrium points and periodic orbits. Although periodic orbits seem very different from equilibrium points, they are similar in the homological sense. With respect to homology, the quasi-periodic, or almost-periodic orbits, the invariant tori, represent a tensorial product of two circles, hence a generalized homological operation at higher level for the given system. In this composition case, the product system is assumed to continue respecting all the conditions required by the KAM theory with respect to differentiability required for the subsystem components, the individual circles.

Two types of actions could occur with respect to the invariant manifolds: *small perturbations*, and *bifurcations*. Different mathematical methods have been used to understand the dynamics near a stable, unstable, or center manifold, including *linearized sub-bundle* methods based on foliation by Fenichel,[20–22] *Lyapounov analytic functional* methods, and the *Lie transform*. The last two methods are also frequently used in coding, adaptive filtering, and networking. The analytic behavior of discrete and continuous system representations can be bridged by *hybrid systems*, pioneered by Irwin.[23,24] He was using a generalized form of implicit function in the space of sequences. We have high expectations that this last method will produce important results in the complexity analysis of DNA words.

In Fenichel's theory, a general, autonomous, ordinary differential equation can be associated with a flow generated by the solutions passing through a point at time $t = 0$. The flow might not be defined for all points in time or all points in the space. It is called an *overflowing invariant manifold* if the flow associated with this manifold will cross strictly outward the boundary. The rates of convergence of the trajectory near the manifold will be associated with *generalized Lyapounov-type numbers*. They have several interesting properties: they are constant along the orbits, and they are independent of the choice of system matrix on the manifolds. Only in simple cases can these numbers be computed analytically as functions. The main result that Fenichel obtained was: *To every overflowing invariant manifold and the associated flow, a corresponding perturbed overflowing invariant manifold is obtained through a perturbation of the initial flow.* This kind of flow notion has a global behavior, but the condition is actually a local one. The difference between the initial flow and perturbed flow should respect strong conditions of closeness. This contrasts with the dual situation in chaotic systems, in which small perturbations have large global effects, suggesting a decomposition approach.

Fenichel's theorem covers the stability aspects of overflowing manifolds under *small perturbations*, but it does not cover the bifurcation aspects (bifurcations enable switching- and computing-like processes). However, techniques developed in differential topology can deal with such bifurcations. Two important notions are *trans-*

versality and *cobordism* related to domains which share the same border — essentially the mathematical concept of a window, a cell boundary, or a tornado wall. The notion of *vector bundle structure that can be fibered* is used in both theories. Dynamic behavior in these vector bundles, especially in orientated vector bundles (*chiral* in the physical sense), will result in closed vortices under certain conditions. We can attach a tubular neighborhood to vector bundles using the *grassmanian bundle* approach (wedge products and differential forms). Examples of collars and tubular neighborhoods of net submanifolds are given in these tubular manifolds. They could be used to calculate intersection numbers and Euler characteristics of complicated manifolds, so we can associate arithmetic numbers and arithmetic properties to such structures. A good introduction could be found in Wiggins[25] and Hirsh.[26] The tori, or "rings," are related to irrational numbers in the KAM theory. This links algebraic L-forms to arithmetic functions, geometry, and dynamic systems. The appropriate methods and algorithms are still to be developed for efficient computations of such objects.

The DNA structure could offer a two-way bridge to solve such problems. The resulting models could explain long-term behavior of biological systems. The long-time behavior of the trajectories near a stable manifold could explain important aspects of biological evolution. If the resolution of DNA differentiation were completely stable, then evolution would not be possible. The unstable manifolds could explain the transitions necessary to create the bifurcation phenomena, or computing-like processes, that create large variations of species. The Cambrian explosion appears to be such an unstable phenomenon, a computing-like process that seems more complex than a simple trial-and-error model. It seems more like a chaotic model that, within the bounds of its attractor, enumerates or explores all possibilities through perfect mixing. Even more interesting is the subsequent *Cambrian implosion*, or the dying off after the explosion, when the real survivors were produced. A computing-like process is more likely to succeed in this case. Simple neural-network models suggest that the convergence is a monotonic function of the complexity of the model (e.g., the number of parameters). Monkey babies may develop faster physically than human babies, and they have some learning abilities, but human babies learn much faster — some say with two, one, or zero examples. Human brains are bigger and more complex than monkey brains; hence simple neural-network models would predict a slower learning rate. In contrast, a computing-like process with an architecture designed to learn or adapt can be expected to learn faster than a process with a nonlearning architecture, even if the computing-like process is more complex. Such *learning or adaptive architectures* are expected to include many concepts, from internal model principles to abstractions and other symbol-manipulation capabilities.

Saddle points are important points between stable and unstable manifolds. These points are the simplest type of exchange locations, places where the bifurcation phenomena occur. *Sets of saddle points* could be seen as having a hierarchy similar to stable manifolds (i.e., from the simple saddle point all the way up to complex submanifolds that are foliated surfaces). Such issues are part of the work still in progress.

7.3.2 Coupling of Chaotic Systems as a Synchronization Process

One of the challenges is that DNA-related dynamic systems require that *both continuous and discrete* aspects need to be modeled (i.e., we need to consider biological systems as nonlinear *hybrid dynamic stochastic systems*. An important subset is represented by *chaotic systems*. They exhibit the kind of coupling phenomena that could be seen at the system level as a synchronization of subsystems, or a global meta-state. Different types of approaches have been used, including phase-synchronization, intermittent-synchronization with lag, intermittent lag, and others.[27] The underlying concept is related to the existence of a separable subspace requirement of the combined overall phase space. We define a *generalized synchronization* of N systems S_1, ... S_n to be a decomposition of the tensor product of the phase spaces $Ph(s_1) \times Ph(S_2) \times \ldots \times Ph(S_n)$ into local phase spaces $Ph(S_1)$, $Ph(s_2)$, ... $Ph(S_n)$ when one of the phase spaces is known. This definition generalizes the definition used by others.[28,29] It can be seen to encompass the quantum states, because there is no limitation imposed in the nature of the phase space. Therefore, the entanglement between two or more chaotic subsystems (quantum included) could be seen as part of the synchronization process. In the quantum case, the decomposition is not necessarily into local independent subspaces but into the realizable quantum states (e.g., $Ph(S_1) \times Ph(s_2)$, $Ph(S_1) \times Ph(S_3)$, ...), either local or nonlocal.[T45] The nonlocal case is important in DNA, because two far-apart information units are spatially coupled due to the folding process in the chromosomes. Quantum systems techniques have been proposed to analyze bifurcation problems in biological systems.[30,T44]

A slightly higher-level example of molecular- or nanoarchitecture and its stability is the structure of tubulin. Fygenson[16] defines an entropic bristle domain (EBD) hypothesis, in which beta-tubulin monomer substructures unfold into an EBD on the interior of the microtubule. This hypothesis attempts to unify diverse results in the microtubule field, from structural to kinetic, including a stability analysis of quasi-one-dimensional polypeptides edge-perturbations, with the goal of developing a model of the "mysterious catastrophe and rescue transitions that characterize microtubule dynamic instability."[16] The two-dimensional arrays of alpha- and beta-tubulin contain electron-states and support mechanical, electronic, and information flows. Arrays of neighboring dipole–dipole (i.e. multipole super-lattice) interactions generate spaces and systems with complex dynamics of the types discussed in this chapter (Figure 7.3). Local nonlinearities can generate chaotic (attractor) behavior, hence computing-like dynamics — providing model support for the observed catastrophes, recoveries, and other dynamic instabilities. The computing-like properties of tubulin and related structures (see below the discussions on the Turing-complete game-of-life neighborhood and finite-state automaton rules) suggest that the molecular or signal-transport functions of webs of tubulins can include the kind of functions found in electronic switching and routing systems.

7.3.3 Topological Aspects of Flows as a Network Architecture

Designing complex systems with many parts usually involves information flows at the highest level of abstraction. At a global scale, designing the Internet involves designing a flow as a web of links with nodes for routing and switching. At the molecular level, a stunning view is presented by the movies made of a web of microtubules that transport large molecules, especially the seemingly random switching between riding on one microtubule to riding on another microtubule that intersects with the first one. If this process involves random switching, a diffusion process is generated. Given that the dipole–dipole interaction states on the surface of a microtubule are a finite-state machine that generates the dipole waves for transport, some of these dipole–dipole nearest neighbor interactions may represent an asynchronous finite-state automaton of the type encountered in the game-of-life, a *Turing-complete* computing structure, that is capable of emulating goal-oriented routing mechanisms.[31] Similarly, the Golgi "routing apparatus," an intracellular organelle, is involved in the vesicular or membrane traffic in a host cell.[32]

At a higher system level, neural activities in the brain use all the different communication modes to control peripheral (e.g., fingers) functions, including one-to-one wires, broadcast, and packet-switching. It would not be too surprising if all these modes were also used at the molecular or nanoscale level. Figure 7.3 gives a simple example of a nanoarchitecture with an asymmetric flow, a super-lattice form of symmetry-breaking. The figure shows how a two-dimensional array of locally homogeneous flows can create a *global asymmetry*, a slower North–South flow.

Flow examples as in Figure 7.3 originally came out of research on lattice gas architectures.[3,T37] Such lattice architectures have been used in many disciplines, from oil exploration in geophysics at MIT, to biological modeling of early life at NASA/Ames, to modeling traffic flow. The particular *slow North–South flow* in Figure 7.3 was discovered while studying slow traffic flows, but this discovery was recognized as a fundamental information flow and a topological architectural design example that is applicable to many problem domains at all scales.

The *fast East–West flow* traffic models the high lateral diffusion in a cell wall (more like a two-sided ocean-surface model). The *slow North–South flow* may model out-diffusion through the cell wall. In networking, the slow North–South flow can be used to keep out unwanted traffic as spam messaging. All that is needed to achieve this asymmetry is a two-dimensional array of local traffic circles, or vortices. In a semiconductor material, such a super-lattice structure, or nanoarchitecture, would implement a topological diode circuit, a two-dimensional version of the recently discovered one-dimensional nanotubes with kinks that act as very low-power diodes.

7.3.4 Chiral Topology Aspects as Fundamental Flows: The DNA Connector

The *slow North–South flow* is a complex multidirectional traffic flow. Can such a network communicate on even simpler flows? Are there any topological properties that are embedded in such complicated flows, and can they be used as stepping-stones or design principles in a hierarchy of flows? The short answer is, yes.

To gain insights, consider two points, A and B, in a plane. We want to connect the two points through a communication network that has different routing possibilities, so that multiple paths can be used in getting from A to B. The ability to use multiple paths avoids the auto-blocking problem (as in freeway on-ramp traffic lights) or traffic jams (hot-spots) that can occur on a single path. Moreover, in a multiple-path network there is no reason to consider *a priori* that there are special directions (turning privileges, or bifurcation probabilities). In this case, a classical open disk topology (chiral, but not including the bending type) is sufficient to cover the plane that supports the network. The Euclidean plane has the basic property that the shortest path between two points is the straight line that connects them.

The shortest path principle leads to the well-known triangle inequality in the Euclidean plane P. For all points A, B, and C in P, we have a distance $d(.,.)$ inequality (triangle inequality):

$$d(A, B) \leq d(A, C) + d(C, B)$$

This inequality is characteristic to the Archimedean metric. Consider now that our network system is formed by a network of circular disks, in which the information on the disk boundaries flows in a given direction, like on a chessboard full of traffic circles. Figure 7.4 illustrates such a non-Archimedean point-to-point AB-flow routing space within a double "light cone" convex hull that is reminiscent of a Fresnel zone in optics and antenna theory. We refer to these disks as *network chiral cells,* because they have a left (counterclockwise) or right (clockwise) flow of information. Assume that the disks are chosen in such a way that they cover the space between the two points for input and output, A and B, with a finite number of disks. Every point of contact between two network chiral cells is considered a point of conditional exchange, a local permutation, or simply a *switch.* In this type of topography there are two main types of switches, a *unidirectional switch* (to the right or left), and a *two-directional switch* (bidirectional, or counter flows), as shown in Figure 7.5.

Flows in the two-directional switches can generate *loops* (feedback, echoes, reverberations, standing waves, oscillations), and they are known to be of the *Lie type* in topology and algebra. The unidirectional switch is more difficult because it gives only a single direction, but with a parallel split of paths (multiple parallel lanes, multichannel). This type of switch requires more information about which direction to follow and which lane to select.

Again, there is an analogy to Internet networking, with the one-way switch serving as the element of decision on the most probable path to choose. The address look-up (of forwarding) tables in Internet routers are dynamically generated in every router, given the router's prediction of the best route. Hence, the expectation of the most probable path to follow in order to arrive at output B is given by the router's look-up table.

From a scattering viewpoint, switches in unidirectional flows decide to follow the best path or (if unavailable) get *deflected* on to a close, available path toward the

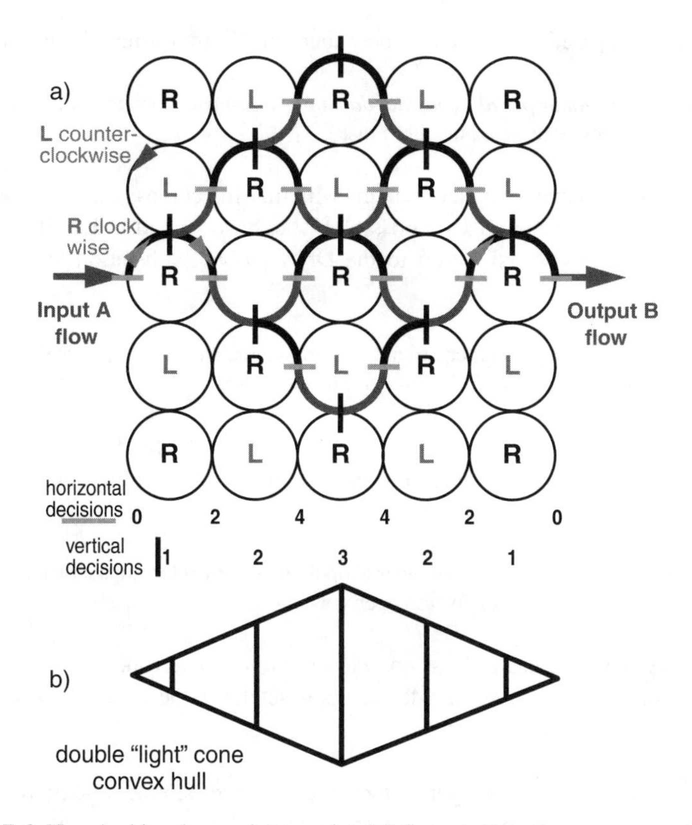

FIGURE 7.4 Non-Archimedean point-to-point AB-flow routing space.

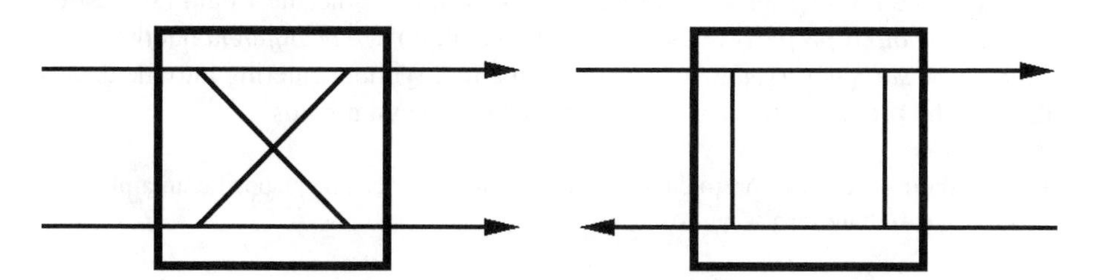

FIGURE 7.5 Unidirectional and two-directional switches.

desired subspace with goal B. If this subspace is not reachable by any path, a two-directional switch is used to *reflect* the flow back into the subspace containing the source A from which it came. On the Internet, if such a flow fails, the packet flow is discarded. *Reflection* preserves flow mass and bits, and it acts as a backpressure flow control. Hackers and spammers would not like that, since they would get flooded or effectively disconnected. A *return-to-sender* response is built in, because the message has to be accepted or returned by B; discarding the message is not an option.

With this preparation, we state a new theorem,[T45] after a few definitions:

Definition 1*:* A *planar pseudo-compact domain* is a compact domain with a countable set of missing points (e.g., a punctured disk).

Extensions to planar compact domains with missing curves, uncountable number of points, and fractal structures are possible[T45] but have second-order effects that are not relevant in first approximation to the DNA global explanation and model we develop.

Definition 2: For every planar curve, a direction to circulate along the curve is defined. This direction is called the *chiral direction* of the curve.

For vector fields, the chiral direction is a result of the field properties and can be calculated. In our sense, the definition is more general and applies to all types of structures.

Definition 3: A *planar chiral topology* is a topology on a pseudocompact planar domain with the following properties for the open sets:

1. Every open set has at least one closed curve inside (one or more loops).
2. All the closed curves inside an open set have the same direction (no counter flow) (Figure 7.6).

This definition is new and generalizes the notion of simple, classic topology. A homology and cohomology are associated with it.[T45]

An *open set* is a disk without a border. It is pseudocompact because some interior points can be missing (piercing through a cell with a thin needle would not destroy it). The *homotopy property* (existence of a global flow) *will be different* but the *chiral behavior is still conserved* to a first approximation. Without entering into the details that will be presented by Barbu,[T45] the following theorem holds:

Theorem: A network constructed with a chiral topology can be mapped onto a plane in a non-Archimedean way.

For example, consider the network in Figure 7.4. All switches at the tangent points C of two touching circles are equidistant from the points A and B if they are situated on the same vertical line relative to the Euclidean line that conceptually connects A and B in the bottom diamond (Figure 7.4b). If this chessboard arrangement of chiral topology induces a non-Archimedean space on top of the Euclidean space represented by plane P, this is equivalent to all points A, B, C in P and C on a vertical line inside the convex hull, and we have a distance metric $d(.,.)$ with the equality:

$$d(A, B) = d(A, C) + d(C, B)$$

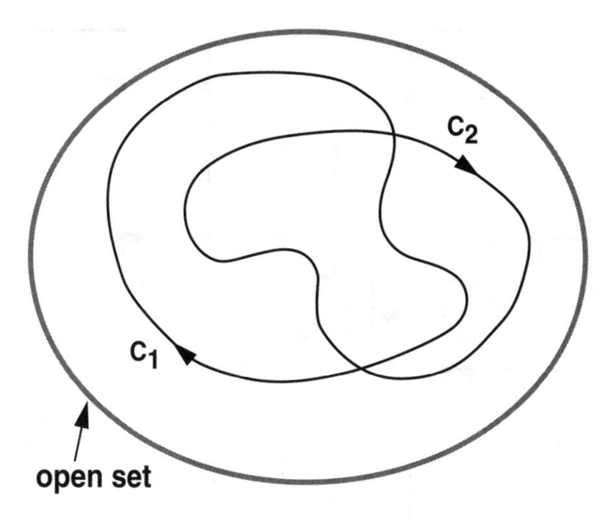

FIGURE 7.6 Open set with clockwise chiral flow direction.

This is a typical *Manhattan* (city-block) distance, an example of a non-Archimedean metric as used in networking and chip design. In addition, the chiral topology could create non-Archimedean spaces, which in this case could be considered projections of the circles situated on the two cones of light passing in three dimensions through A and B (Figure 7.4b). The link to three-dimensional Riemannian geometry is straightforward for these networks.

The theorem has the following corollary:

> **Corollary:** If the chiral topology arrangement is preserved while scaling only the size and number of cells, a rescaled network is created with different densities but without altering the properties of the space.*

The size or number of cells does not change the non-Archimedean property of this space. This network has the ability to create many further properties for the network with a number of switches.

If the cells are open *squares* (compressed circular cells), then a tangent point becomes a tangent side at a traffic avenue (one-way or two-way), as shown in Figure 7.7. This results in an intersection with saddle-point (mountain pass, or X-gate) type of flow. The switches will be now more complex because they are situated at the intersection of four avenues.

If the boundary is separated with two-way avenues at the crossing, then this topology is equivalent to one of the array topologies described elsewhere,[3,T17] more recently rediscovered and described as a two-dimensional *indirect mesh.* [33] In this case, the switch is a four-way bidirectional switch (Figure 7.8).

---

* This can be seen as a generalization of *Moore's law* for chip manufacturing, and an explanation of infants growing mainly in size to adults without major structural changes.

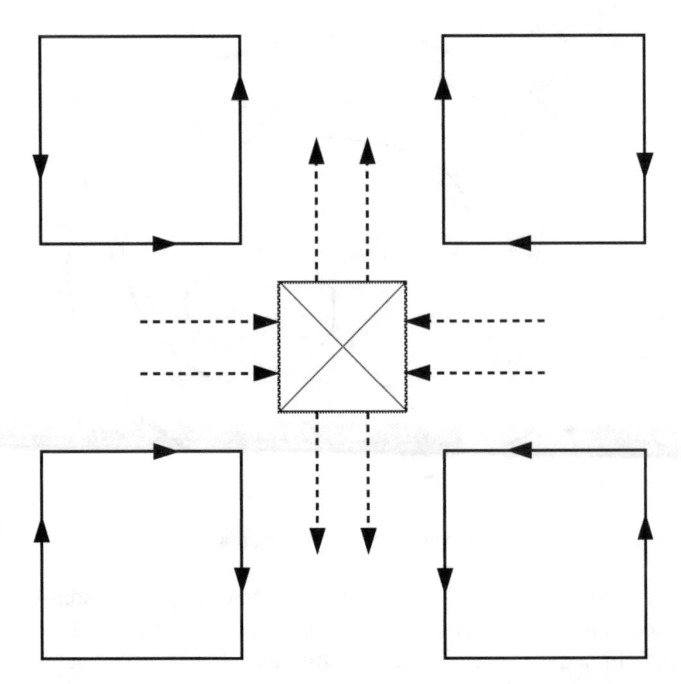

FIGURE 7.7 Open-disk to open-square transform.

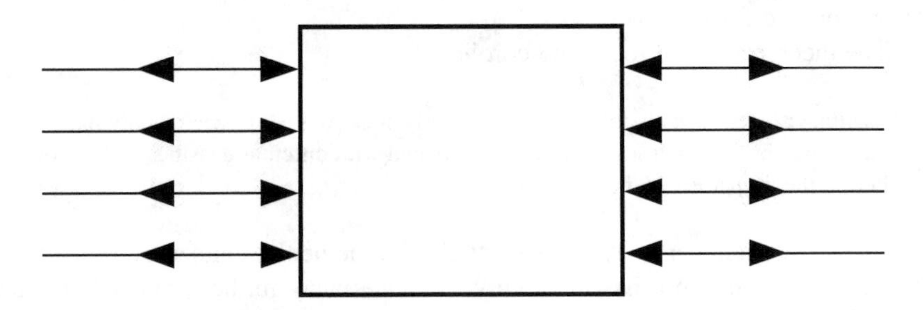

FIGURE 7.8 Four-way bidirectional switch.

We have developed optical switching mirrors, or X-gates,[3,T40,T41,T42] and a patent[34] that correspond to the diagonals in the switch box. In biology, this type of switching function is found in membrane sodium-calcium pumps. Compared to the other array architectures for rotations, this one corresponds to a hyperbolic rotation. A. Huang, the inventor of many sorting networks, noted their connections to fast Fourier transforms (FFTs), which are unitary or orthogonal transforms — complex rotations, or binary switches if the values are constrained to two levels.

The next operation that is equivalent to creating a unique flow in a membrane is given by a topology that has the merged-squares boundary flow, single-line saddle-point flow characteristics shown in Figure 7.9.

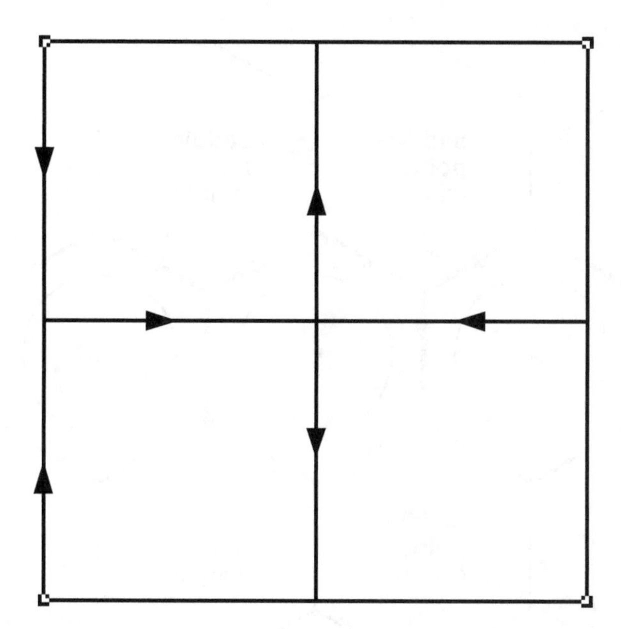

FIGURE 7.9 Merged-squares boundary flow, single-line saddle-point flow.

This topology is obtained from the preceding one by considering only a single boundary mesh. This shows that the same triple connections could be obtained with only four input/output lines if every line is bidirectional (full-duplex). Again, sodium-calcium pumps are examples of bidirectional flow.

The flow in DNA is created by *p*-electrons flowing only in one direction. A bidirectional network was obtained by *local looping* or *complementary networking* topologies.

In the squares, the nonexclusionary property is conserved. For increased complexity, consider hexagons (Figure 7.10). This D3 triangle-group object has the distinctive property of a *fast-North and slow-South flow* that is inherent with the D3 groups. As we have seen in Figure 7.3, the choice of the sense of rotation in the hexagon can destroy the chiral aspect. It is easy to demonstrate that a hexagonal topology has no inherent chiral flow, only complex chiral hexagonal chains exist.

This hexagonal chiral topology introduces a mixture of chiral and nonchiral cells. A fixed-lattice gas model would have only local chiral chains. However, by randomizing the left and right rotations (akin to a mean-field approximation), the required symmetry for a gas model is obtained. Control cells* are necessary in these

* These control cells serve as *controllers* to the circular cells acting as *data-paths*, a topological equivalent to a *Harvard-style* computer architecture, for example, which is popular in DSPs — Digital Signal Processors. A similar distinction can be made between *eXchange*-type and *Rotation*-type gates.

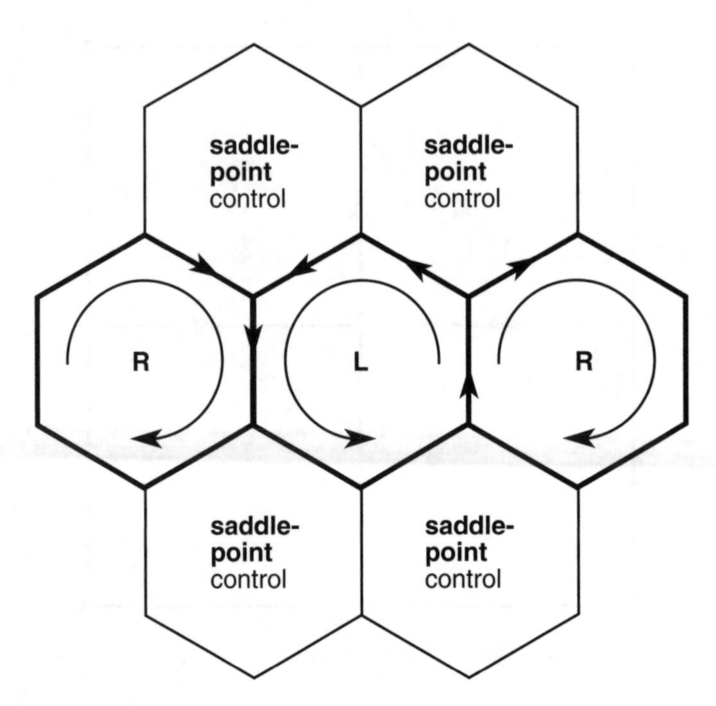

FIGURE 7.10 Chiral hexagonal chain.

complex flows. The properties of the hexagonal arrangements are topological. Observe the similarity between Figure 7.3 and Figure 7.10. The last is a direct consequence of the chiral topology and confirms the practical realizations related to Figure 7.3.

The non-Archimedean structure induced by the chiral topology was one of the fundamental features of secure systems and cryptography. This is a very recent subject in communications research. We present only a brief introduction in order to outline the applications in networking and the DNA structure.

This overview on chiral topology shows that *the DNA connectors are unidirectional (free-ways) and only the information units carry out the actual routing. In complex spatial organization, complex flows could occur and information units will be used as control units (routers).* This is the main conclusion from this section on topology and DNA structure. For details, see Barbu.[T45]

Traditionally, the topology has been treated as an independent aspect of system design and manufacturing. As the science and technology are moving to the nanoscale level, we will need to consider the interdependency between the topology and data representation.

7.4 DATA REPRESENTATION, FLOW MODELING, AND THE DYNAMIC IN BIOSYSTEMS

The dependence of biosystems on event histories has already been established (see chapter by You Jacobs in this book and early work on bone growth by D. Carter et al.). Many bioregulatory process diagrams actually use mixed-level representations in electronic design, including Boolean signals, AND gates for state information, delay-line memories for event histories, and multiplexers or switches to complete the circuits.

7.4.1 SIGNAL CODING: DATA REPRESENTATION AND RICCATI EQUATIONS IN FLOW MODELS

Signals are either binary (enabling, disabling) or ternary (enabling, disabling, not-connected), as in communication systems circuits (e.g., return-to-zero, RZ, versus not-return-to-zero, NRZ) or multilevel digital (number of molecules) or analog deterministic or stochastic parameters. Biosystem design needs at least extensions of the tools used in electronic circuit design. Extensions tend to be not unique (i.e., they offer several different and hopefully complementary choices). This leads to the second frontier issue: a new *hybrid* model of globally mixed discrete-continuous systems supporting all the data representations may be required as a complement to locally mixed discrete-continuous systems; see the hybrid approach by Tomlin.[35] [GT'73] These locally mixed hybrid systems exhibit nonunique Zeno states that may be due to *over-abstraction*. Globally defined, possibly distributed states, defined by attractors of nonlinear (e.g., Riccati) equations, or the global configuration spaces associated with the local hybrid discrete and continuous systems, can be used as complementary representations, or the representations may be combined. Such global-attractor-based models also appear in nonergodic and chaotic flow and are reminiscent of quantum-mechanical models (e.g., using stochastic control theory).

As a simple example, consider an approach based on second-order, nonlinear Riccati equations. Such an example can be derived by realizing that probabilities form a straightforward way to get AnaLogic (Analog and/or Logic) behavior: map the logical True/False to One/Zero (the OZ mapping); alternatively, for biocircuits involving the reversible chemical logic, map True/Nil to One/Zero. The Riccati equation arises, for example, by constraining a Boolean variable, x, to One or Zero, as in $(x - One)(x - Zero) = x^2 - x = 0$, and from implementing the nonlinear (for example, sigmoid) functions as implicit solutions to Riccati equations. Given that Riccati-equation solutions are rational expressions of exponential functions, this is not much of a constraint. For example, the popular sigmoid function, *arctan()*, has a simple difference or differential (Riccati) equation that converges to the desired *arctan(.)* value as fast as needed (faster than the fastest biocircuit time-constant). Solutions to these global Riccati equations can be found in many ways. There are fast-converging doubling and eigen-decomposition algorithms that can take advantage of the sparseness of the matrices involved.[3,T7,T25,T26] As in communication theory and artificial neural networks, an alternative to mapping of Boolean functions into algebraic functions is the mapping of True/False/Nil to Minus-One/One/Zero (the MOZ mapping). This {True, False, Nil} k=3-set is compatible with V. Pratt's work.[41]

7.4.2 From Xmoz Gates to the Game of Life

Standard binary logic supports Boolean functions using two values (1 and 0, True and False, or True and Nil). The use of multiple levels (e.g., strength of bonds) is more efficient but possibly less robust and noisier. It is again useful to introduce the concepts of *Nil, undefined,* or *not-connected* for logic or connectivity reasons. Logically, such terms help avoid singular situations, well-known logic paradoxes, or situations in which subsystems can be topologically either connected or isolated. From a switching viewpoint, pass-gates can serve such functions, keeping unwanted information or objects out; electrically, a series switch makes either a low-energy-loss/low-impedance connection, or a high-impedance (high-Z) disconnect, or high-energy barrier (a form of Nil state). Any physical realization is required to satisfy all physical conservation and dynamics laws, such as mass, charge, momentum, energy, and more general flow invariants. A good example of such realizations are X-type gates.

X-type gates have at least two types of properties: they are logically complete, and they satisfy at least one conservation law. Binary X-gates are logically complete, are invertible, and preserve the Hamming weight in the output. The Hamming weight is indicated in the first column in the three tables in Figure 7.11. In the two-level cases, the Xoz-gate map is ({True, False}={One, Zero}) and the Xmo-gate map is ({True, False}={Minus-one, One}). In the third (left-most) example, the three-level Xmoz-gate map is ({True, False, Nil}={Minus-one, One, Zero}). The three levels per input or output connection in the Xmoz gate have more information capacity than the two-level gates: $3^2 = 9$ states, compared to $2^3 = 8$ states of the two-level gates. Given the new (3rd) level, Nil, the two- and three-level smallest (primitive) gates can be mapped into each other by adding an all-Nil state to the set of eight binary states in order to get a complete (1:1) mapping to the nine ternary-state set,

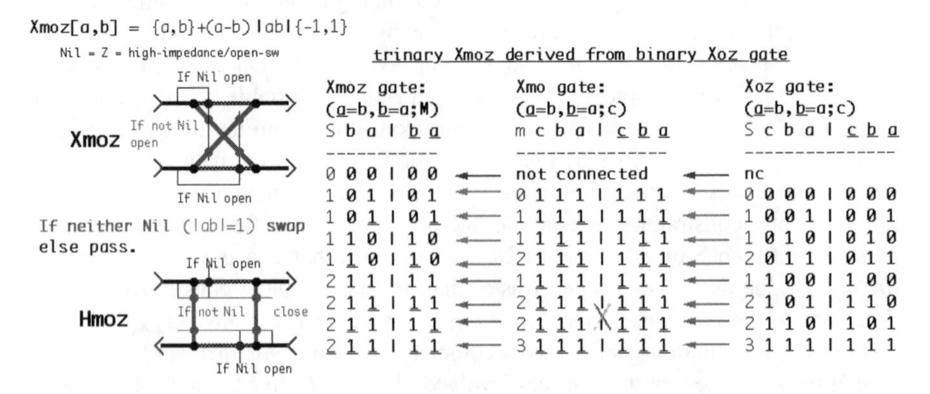

FIGURE 7.11 Xmoz gates.

a proof that two-input Xmoz gates are equivalent over the Boolean functions space. The three-input/output Xoz and Xmo gates are logically complete with respect to Boolean functions because they contain two two-way switches, which are logically complete (through Shannon decompositions or binary decision diagrams). To summarize, any one of the X-type gates, by itself, is logically complete; hence it can implement a computer, such as a Turing machine (Figure 7.12).

Depending on the precise physical properties of a biological subsystem, a well-matched logic primitive, such as an X-type gate, can implement a universal computing process. All that is needed is *interconnect-completeness* (e.g., who gets to talk to whom, the small-world model, the dating-game). Good examples of such an X-type gate in biochemistry are sodium pumps, or more generally cell-wall functions that have connect and/or disconnect flow functions.

A more elaborate example is again the *game-of-life*, which demonstrates that a simple Xmoz-like three-level interaction rule and nearest-neighbor connections are sufficient to be Turing complete. The objection often cited in biology against the game-of-life — that it is too simple to be meaningful for biological modeling — misses a much more important point: Simple neighbor rules, such as the game-of-life uses, are *Turing complete*. That is, they are capable of *emulating* any other computable biological behavior, and thus one can compile any complex biological behavior onto a simple game-of-life hardware device or a software process.

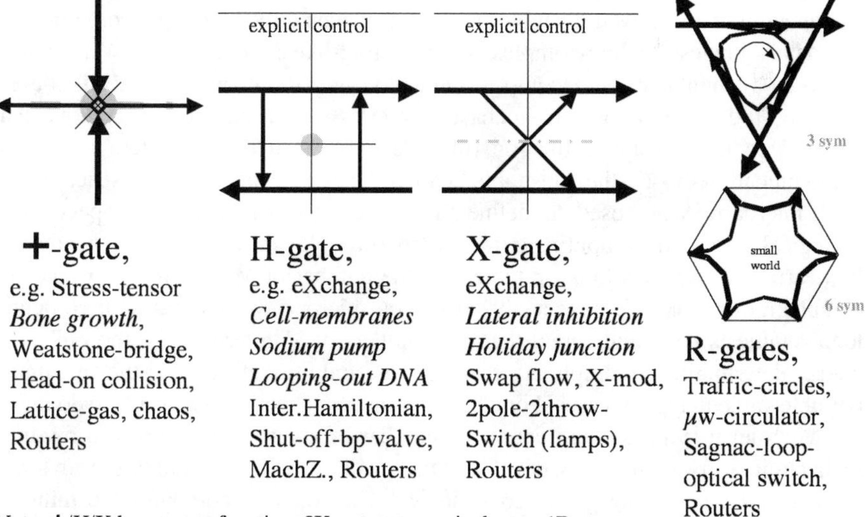

FIGURE 7.12 Common eXchange-type gates.

7.4.3 Biologically Robust Design

A third issue pushing the frontier is *robustness*. Today, computer systems encapsulate robustness and reliability in the hardware, with the result that software and human operators dominate system uptime failures [see http://roc.cs.berkeley.edu/]. Current concepts in robust biological circuits and systems include *cascading* to reliably turn processes off; *parallel paths* to reliably turn processes on; and replication or competition or cooperation to increase chances of success. What is missing are ways to aggregate unreliable components or subsystems to enhance reliability, *without having single points of failure*. We have developed nonscalar primitives (X- and R-type gates with at least three inputs and three outputs) that can be reliably aggregated. It is conceptually based on the ancient principle of simultaneous exchange (trade, barter). These primitives show up in many contexts, including reversible computing (chemical logic), rotations, scattering theory, sorting, routing, coding, code breaking, fluid-flow, and chaos, and in biocircuits such as sodium pumps. Some of these concepts are already implemented in reliable biosystems and simply need to be recognized. To identify and to recognize such biosystem functions, a good method is to study electronic systems from a biological perspective. Electronic system designers tend to encapsulate different aspects of a system (e.g., loops — states), whereas biological systems implement many local and global interactions (loops), which may consist of lateral inhibition of neighbor cells (logical negation or inverters) that are coupled across cell boundaries to achieve the given functionality when engineers encapsulate every process inside a controllable cell.

A fourth issue already mentioned is *complexity*. Complexity in data-representation is related to the complexity of the Boolean functions' form. The *complexity of Boolean functions* derived from biological data can be quite low (e.g., from microarray data), suggesting low-complexity computing-like models. M.E. Wall et al.[36] proposed a singular-value decomposition and principal-component analysis (developed to study correlations in stochastic systems) as an algebraic technique that attempts to find an approximation (the "desirable" part) of a system model. In communication theory, the "desirable" part is the size of the signal subspace. However, microarrays are used to define the interconnect or switching aspects of a biological system. In computing terms, microarrays play the role of a programmable-logic array (PLA), or a look-up table, and the number of Min-terms determines its complexity or rank. We have analyzed several microarray data sets with computer logic-synthesis tools and found that the numbers of Min-terms, like the rank of a matrix determining a complexity measure, are only about 10% of the maximum possible value in typical cases. For a better complexity metric, one needs to do logic- or switch-level synthesis and count the number of gates or switches in a minimal realization. Software has been developed that synthesizes Boolean functions in terms of X and R gates using group concepts.[3,T37] The software computes the minimal number of sodium pumps or equivalent biosystem routing elements required to realize any Boolean function.[3,T37,T41] (See X-gate/modulator patent.[34]) The group concepts can also be used to generate nanoarchitectures derived from group graphs.

Special cases of groups[6] are either related to structure, such as hyper-cubes[37] and icosahedrals,[38] or related to scale (more generally an *affine transformation* that combines shift, rotate, and scale), such as fractal DNA assembly.[8,9] More generally, the above-mentioned groups are a subset of the groups used in SUSY.[1] We expect that, due to the complexity found in nanobiology, most if not all of the groups found in SUSY will apply, because one of the largest groups used today is the monster group, for which its researchers were awarded a Field medal. A majority of the groups used in SUSY are Abelian; their classification is essentially accomplished. However, the usage and classification of the large non-Abelian and/or Lie groups is still an open topic.

Preliminary experiments on Boolean functions from biological data using computer-aided design (CAD) Boolean-function synthesis have indicated that the complexity can be quite low; the actual number of Min-terms in an optimized PLA realization was only 10 to 20% of the maximum value. The internal representations of many CAD tools use binary-decision diagrams (BDDs), which are based on switch-level primitives, also equivalent to the Shannon-decomposition in communication theory. Since switch-level models are closer to physical realization due to the implicit conditional exchange, they are better matched to biocircuits. The X-gates (sodium pumps), are switch-level representations that should be used in biocircuits. For example, cascades in biocircuits are simply cascaded switches, parallel paths are parallel switches, and enabling and inhibiting are reminiscent of P/N (positive/negative, donor/acceptor) switches. The traditional 3′ and 5′ ends of DNA are actually donors/acceptors. Some molecular processes are related to semiconductors. This is a macro-level view of cells. The algebraic mapping of Boolean functions is consistent not only with probabilities but also with physical concepts such as rotation (reversible logic, quantum computing, and unitary logic).

This is not too surprising from a biochemistry viewpoint. We have some evidence that algebraic representations often have reduced complexity compared with Boolean representations (factors of 2 to 4 less have been observed). X-type gates (such as sodium pumps) relate to orthogonal functions and rotations in this domain, hinting again at energy conservation or designs driven by free energy. The appearance of rotations is also not surprising, because rotations relate to state transitions that preserve the length of the state-vector in the state-space. That means that the free energy is conserved, and adiabatic processes are involved from a global perspective. Locally, large rotations in the state-space are decomposed into elementary rotations, and therefore related to local energy and momentum flows, that from a scattering point-of-view determine the local topology of a system. As a remainder, local flows can contain singularities, such as saddle points, that can (1) determine the complexity or information flow of a system, (2) contribute to robustness, error correction, and help trade-offs, or (3) enhance sensitivity and robustness of a system. After discussing synchronization aspects above, we next focus on the clock system itself. This example uses many of the key concepts already mentioned.

Electronic-systems design has found that *toroidal clock networks* are much more accurate and have much less phase jitter then the same individual clock. Recent research[39] has found that *brain clock* mechanisms have *double ring* structures. These two findings are related. Advances in electronic-circuit speeds into the GHz level have reintroduced the issue of clock accuracy and phase jitter. Some of the ingenious solutions involving feedback loops in one or multiple dimensions (again not typically accessible) seem to already exist in biocircuits. There are interesting physics connections to whisper-gallery modes in photonics that again seem to be reflected in biocircuits. This may explain why certain biological clocks are surprisingly accurate. This also applies to the space dimension and inertial navigation based on photonic loops (based on the Sagnac effect, the E-B-field dual to the Aharonov-Boem effect; both are forms of Hopf invariants[40]). The fluid flow version of these effects also exists as sonically excited toroidal fluid-flow resonance implementations, so it is quite possible that such structures are already implemented in biocircuits. Based on these insights we propose the following *conductor architecture* model of DNA. Every *p*-electron transition is a clock delay globally averaged on the total dual-strand DNA length and locally averaged through the toroidal averaging in the local loops and correspondent networks.

Other coding principles used in electronic systems designs, such as robust Gray-like codes, are used as address decoding in biological cell-wall locations (Figure 7.13). Neighboring codes differ by only one bit, and nothing else changes due to this architectural choice. This robust coding principle can be recognized, for example, in the two-dimensional expression patterns on the wing of a fly, where the different genes are expressed as different patterns that typically have some overlap. In any area, the overlap is such that typically only one gene is different at any subarea border. This type of architecture exists in electronic memory-array designs, where each subarea (e.g., a memory column) is addressed separately.

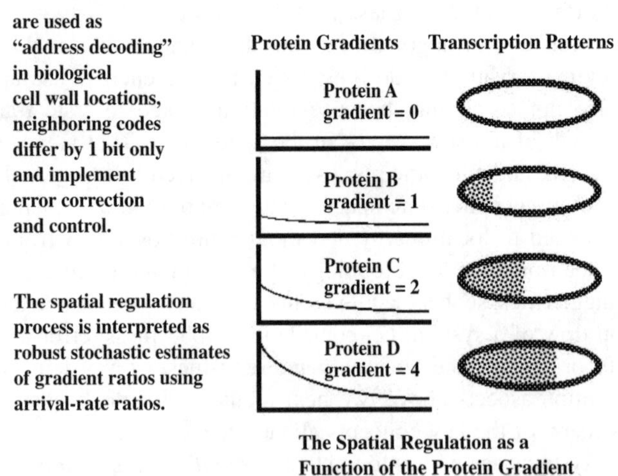

FIGURE 7.13 Robust Gray-like codes.

The extra robustness of Gray coding is not used in today's electronic memory, but if feature sizes approach the nanoscale such robust coding techniques are expected to be reintroduced. Unfortunately, the achievement of such robustness via simple coding is limited by reliable decoding. In most cases, this involves the use of linearity or at least superposition, typical using bulk effects, such as *mean-field theories*, that do not rely on nontrivial (e.g., nonscalar) nanoscale structures. A good spatial counter-example occurs in magnetic materials, in which the macroscopically stable (nonlinear) hysteresis is achieved by an aggregation of magnetic grains in micro- or nanoarchitecture, a stochastic topological arrangement involving loops and saddle points of magnetic dipoles. A visual inspection reveals vortex flow patterns, using Sommerfeld's analogy of velocity~E-field and vorticity~B-field, reminiscent of watersheds and, more globally, continental divides. Similar biological examples include tissues with webs of sodium pumps, mentioned above, or cascaded counter-flow (loop) architectures found in organs that extract, for example, urine. There are many more such examples. We have found that experts familiar with a concept in one field have no problem recognizing an implementation of the concept in a different field after they have been introduced to a suitable Rosetta-stone translation table. The existence of computer language and graphic tools has enabled the mechanization of such translations, although such tools still tend to be rather specialized. For example, text processing was long limited to Latin-based languages. Symbolic computer languages have aided in the manipulation of mathematical expressions for some time now, but advanced mathematical (especially topological) concepts such as Chu spaces[41] are still waiting to be exploited in modeling and computing.

7.5 CONCLUSIONS AND OUTLOOK

The DNA model proposed here promises to give a unified explanation of communication systems at the DNA level. It has scaling capabilities, because it can incorporate the quantum aspects as well as the macromolecular aspects at a higher scale using the same mechanism of gauges in rescaling that is known in the M-theory and dynamic systems. The renormalization properties are due to the unidimensional shift (in the connector) or three-dimensional shifts (in the chromosomes). The rotations are essential to the DNA folding. The recombination as splicing and litigation is important because it implements the conditional exchange — the very nature of evolution in dynamical systems. Further work is necessary to validate and take advantage of this model.

We believe the system-architecture design principles and technologies for biological and electronic systems will meet at the nanoscale, requiring new concepts and new tools, and will create a large field of scientific and ultimately industrial advances. Consider the expanded set of component disciplines — *chaotic attractors* and *global states*; complex, nonscalar, nonlinear, multiscale, stochastic dynamic, computing, and communicating systems; new industrial DNA consumer products; new complex procedures; and new medical treatments — as we learn how to speak the new language of the DNA and other nanotechnologies.

NOTE

The first author was not affiliated with Maxim while working on this chapter. An updated version of the chapter will be available from the second author.

REFERENCES

1. Mead, C., and L. Conway, *Introduction to VLSI Systems*, Addison-Wesley, Reading, MA, 1980.
2. Abbot, L.F., and T.B. Kepler, Model neurons: from Hodgkin-Huxley to Hopfield, in *Statistical Mechanics of Neural Networks*, L. Garrido, Ed., Springer, Berlin, 5–18, 1990.
3. Morf, M. et al., Scattering theory, systems modeling and computing (see theses below).
4. Carbone, A., and M. Gromov, Mathematical slices of molecular biology, *La Gazette des Mathematiciens*, Numéro spécial 88, 11–80, Société Mathématique de France, 2001.
5. Carbone, A., and M. Gromov, Functional labels and syntactic entropy on DNA strings and proteins, *Theor. Comput. Sci.*, 303, 35–51, 2003.
6. Carbone, A., and N.C. Seeman, Circuits and programmable self-assembling DNA structures, *Proc. Natl. Acad. Sci. U.S.A.*, 99, 12577–12582, 2002.
7. Carbone, A., and N.C. Seeman, A root to fractal DNA assembly, *Nat. Comput.*, 1, 469–480, 2002.
8. Carbone, A. and N.C. Seeman, Coding and geometrical shapes in nanostructures: a fractal DNA-assembly, *Nat. Comput.*, 2, 133–151, 2003.
9. Carbone, A., and N.C. Seeman, Molecular tiling and DNA self-assembly, Special volume, Tom Head Festschrift, N. Jonoska, G. Paun, and G. Rozenberg, Eds., *Lecture Notes in Computer Science 2950*, Springer, 2003.
10. Benenson, Y., T. Paz-Elizur, R. Adar, E. Keinan, Z. Livneh, and E. Shapiro, Programmable and autonomous computing machine made of biomolecules, *Nature*, 414, 430–434, 2001.
11. Regev, A., and E. Shapiro, Cellular abstractions and computation, *Nature*, 419, 343, 2002.
12. Bruckstein, A.M., M. Morf, et al. Demodulation methods for an adaptive neural encoder model, *Biol. Cybern.,* 49, 45–53, 1983.
13. Powell, J., M. Morf, et al. Vertical cavity X-modulators for reconfigurable optical inter-connection and routing, *International Conference on Massively Parallel Processing Using Optical Interconnections 1996*, Maui, Hawaii, October 1996.
14. Huang, A., U.S. patent # 4943909.
15. Cover, T., and M. Morf, MURI-ARO project on information-physics and quantum computing, Project reports, Stanford University, 2001–2004.
16. Fygenson, D.K., A unifying hypothesis for the conformal change of tubulin, preprint, Physics Dep., University of California, Santa Barbara, 2001.
17. Hammele, M., and W. Zimmermann, Modeling oscillatory microtubule — polymerization, *Theor. Phys.*, University of the Saarland, Germany, October 8, 2002.
18. Hud, N., and K. Downing, Cryoelectron microscopy of λ phage DNA condensates in vitreous ice: the fine structure of the DNA toroids, *Proc. Natl. Acad. Sci. U.S.A.*, 98, 14925–14930, 2001.

19. Broer, H.W., KAM theory: the legacy of Kolmogorov's 1954 Paper, Department of Mathematics and Computer Science, Groningen University, NL, or www.math.rug.nl/~broer/pdf/kolmo100.pdf.
20. Fenichel, N., Ph.D. thesis, New York University, 1970.
21. Fenichel, N., Persistence and smoothness of invariant manifolds for flows, *Ind. Univ. Math. J.*, 21, 1971.
22. Fenichel, N., Geometric singular perturbation theory for ordinary differential equations, *J. Diff. Eqns.*, 31, 53–98,1979.
23. Irwin, M., On the stable manifold theorem, *Bull. London Math. Soc.*, 2, 196–198, 1970.
24. Irwin, M., A new proof of the pseudostable manifold theorem, *J. London Math. Soc.*, 21, 557–566, 1980.
25. Wiggins, S., *Normally Hyperbolic Invariant Manifolds in Dynamical Systems,* Springer, 1994.
26. Hirsh, M., *Differential Topology*, Springer, 1997.
27. Pikovsky, A., M. Rosenblum, and J. Kurths, *Synchronization: A Universal Concept in Nonlinear Sciences*, Cambridge University Press, 2001.
28. Pastur, L., S. Boccaletti, and P.L. Rarnazza, Detecting local synchronization in coupled chaotic systems, preprint, August 26, 2003.
29. Boccaletti, S., J. Kurths, G. Osipov, D.Valladares, and C. Zhou, *Phys. Rep.* 366, 1, 2002.
30. Patel, A., Quantum Algorithms and the Genetic Code, CTS and SERC, Indian Institute of Science, Bangalore-560012, arXiv:quant-ph/0002037 v3 6 Feb 2001.
31. Muench, D., preprint 2004, www.science.ucalgary.ca/research/Muench.pdf.
32. Chlamydiae, Chlamydial re-routing of host cell membrane traffic, preprint, 2004, www.chlamydiae.com/docs/biology/biol_membtraffick.htm.
33. Kumar, S., A. Jantsch, J. Soininen, M. Forsell, M. Millberg, J., Oberg, K., Tiensrj, and A. Hemanji, A network on chip architecture and design methodology, Proc. IEEE Computer Society Annual Symposium on VLSI, April 2002, pp. 105–112.
34. X-gate U.S. patent: US5909303, *Optical modulator and optical modulator array,* J.A. Trezza, M. Morf, J.S. Harris, Stanford, June 1, 1999–January 3, 1997.
35. Ghosh, R., and C. Tomlin, Lateral inhibition through delta-notch signaling: a piecewise affine hybrid model, vol. 2034, LNCS series, Springer-Verlag, 2001.
36. Wall, M.E., A. Rechtsteiner, and L. M. Rocha, Singular value decomposition and principal component analysis, in *A Practical Approach to Microarray Data Analysis*, D.P. Berrar, W. Dubitzky, M. Granzow, Eds. Norwell, MA, Kluwer, 91–109, 2003.
37. Jimenez-Montano, M.A., C.R. de la Mora-Basanez, and T. Poeschel, On the hypercube structure of the genetic code, in *Proc. 3. Int. Conf. Bioinformatics and Genome Research*, H.A. Lim and C.A. Cantor, Eds., World Science, 1994.
38. Johnson, J.E., Functional implications of protein-protein interactions in icosahedral viruses, *Proc. Nalt. Acad. Sci. U.S.A.*, 93, 27–33, 1996.
39. Xie, X., R.H.R. Hanloser, and H.S. Seung, Double-ring network model of the head-direction system, *Phys. Rev. E*, 66, 041902, 2002.
40. Marsh, G.E., *Force-free Magnetic Fields*, World Science, 1996.
41. Pratt, V.R., Event-state duality: the enriched case, Proc. CONCUR'02, Brno, Aug. 2002; see also http://chu.stanford.edu/ and http://boole.stanford.edu/.

SELECTED THESES (ALSO THE REFERENCES THEREIN)

T1. Friedlander, B., Scattering Theory & Lin. Least Squares Estimation, Ph.D. thesis, Stanford, 1976.
T2. Vieira, A., Orth. Matrix Polynomials, Modeling & Ladder Forms, Ph.D. thesis, Stanford, 1977.
T3. Kung, S.-Y., Multivariable and Multidimensional Systems Analysis, Ph.D. thesis, Stanford, 1977.
T4. Wood, S.L., A System Theoretic Approach to Image Reconstruction, Ph.D. thesis, Stanford, 1978.
T5. Mathews, R.G., Sound Processing for an Auditory Prosthesis, Ph.D. thesis, Stanford, 1978.
T6. Capitant de Villebonne, P., Distortion-Free Compres. Techniques, Ph.D. thesis, Stanford, 1978.
T7. Newkirk, J., Computational Issues In Estimation and Control, Ph.D. thesis, Stanford, 1979.
T8. Godfrey, B., Deconvolution of Seismic Data, Ph.D. thesis Stanford, 1979.
T9. Lee, D.T.L., LadderForms & Fast Estimation Algorithms, Ph.D. thesis, 1980.
T10. Fortes, J., Estim. 3-D Reconstr., Counting Stat., Med. Imaging, Ph.D. thesis, Stanford, 1980.
T11. Nunes, P.R., Estimation Algorithms for Medical Imaging, Ph.D. thesis, Stanford, 1980.
T12. Porat, B., Ladderform Estimation Algorithms, Ph.D. thesis, Stanford, 1981.
T13. Schmidt, R., Signal Subspace M. Emitter Location & Spectral Est., Ph.D. thesis, Stanford, 1981.
T14. Levy, B., Algebraic Approach to Multi-Dimensional Systems., Ph.D. thesis, Stanford, 1981.
T15. Stirling, W., Jump Exitation Modeling & System Parameter Est., Ph.D. thesis, Stanford, 1982.
T16. Muravchik, C., Arrays for Finite Rank Process Ladder Models, Ph.D. thesis, Stanford, 1982.
T17. Delosme, J.-M., Alg. & Arch. for Finite Shift-Rank Processes, Ph.D. thesis, Stanford, 1982.
T18. Ahmed, H. M., Signal Processing Algorithms and Architectures, Ph.D. thesis, Stanford, 1982.
T19. Roberts, M., Info. Thy. Ap. to FS Modeling of Large Scale Systems, Ph.D. thesis, Stanford, 1982.
T20. Roth, E., Estimation Approach to Medical Imaging, Ph.D. thesis, Stanford, 1982.
T21. Su, G., Subspace Alg. for Location and MD Spectral Estimation, Ph.D. thesis, Stanford, 1983.
T22. Hadidi, M.T., Modeling of Multichannel Autoregressive Processes, Ph.D. thesis, 1983.
T23. Nehorai, A., Algorithms for System ID and Source Location, Ph.D. thesis, Stanford, 1983.
T24. Smith, J.O., DSP and Systems Identification for the Violin, Ph.D. thesis, Stanford, 1983.
T25. Ang, P.-H., Fast Algorithms and Specialized Architectures, Ph.D. thesis, Stanford, 1984.
T26. Krishnakumar, A.S., Divide & Conquer Eigenproblem Alg. & Arch., Ph.D. thesis, Stanford, 1984.

T27. Bruckstein, A., Scattering Models in Signal Processing, Ph.D. thesis, Stanford, 1984.
T28. Wax, M., Source Location Algorithms, Ph.D. thesis, Stanford, 1984.
T29. Van Campernolle, D., Speech Processing for Chochlear Prosthesis, Ph.D. thesis, Stanford, 1985.
T30. Wagner, B.R., Cipon: A Model for Distributed Systems, Ph.D. thesis, ETH Zurich, 1986.
T31. Muller, P., A Prolog Interpreter Without Cl. Backtracking, Ph.D. thesis, ETH Zurich, 1986.
T32. Guzzella, L., Robustness Regulators with Variable Structure, Ph.D. thesis, ETH Zurich, 1986.
T33. Wong, Y., Algorithms for Systolic Array Synthesis, Ph.D. thesis, Yale University, 1988.
T34. Nussbaum, M., Delayed Evaluation in Logic Programming, Ph.D. thesis, ETH Zurich, 1988.
T35. Heeb, H.R., Rule-based System for VLSI Polycell Generation, Ph.D. thesis, ETH Zurich, 1988.
T36. Lam, J., An Efficient Simulated Annealing Schedule, Ph.D. thesis, Yale, 1988.
T37. Lee, F.F., Scalable Comp. Arch. for Lattice Gas Sim., Ph.D. thesis, Stanford, 1993.
T38. Flachs, B., Sparse Adaptive Memory, Ph.D. thesis, Stanford, 1994.
T39. Ho, M., Fast Multi-Dim. Adaptive Alg. for Digital Com. Ap., ADSL, Ph.D. thesis, Stanford, 1994.
T40. Trezza, J., Creat. of Efficient Quantum Well Optoelectronic Switches, Ph.D. thesis, Stanford, 1995.
T41. Powell, J., Application of X-Modulators, Ph.D. thesis, Stanford, 1996.
T42. Levine, E.I., State Feedback for Neural Networks, Ph.D. thesis, Stanford, 1996.
T43. Mencer, O., Adaptive Architectures & Reconfigurable Arithmetic, Ph.D. thesis, Stanford, 1999.
T44. Yard, J., Quantum-Information & Channel-Capacity, Ph.D. thesis, Stanford, exp. 2004.
T45. Barbu, S., Nonlinear Dynamical Systems in Information-Theory, Quantum Systems and Genomics, Ph.D. thesis, exp. 2004.

8 Single-Molecule Optical Trap Studies and the Myosin Family of Motors

David Altman and James A. Spudich

CONTENTS

0-8493-1940-4/05/$0.00+$1.50

8.1 INTRODUCTION

Since 1994, optical traps have been used to probe the detailed mechanism by which myosins convert chemical energy of the cell into directed motion. In this chapter, we discuss the optical trap transducer and its role in single-molecule studies of these molecular motors. We focus on how this tool has developed as the questions of the myosin field have changed. Though myosin studies go back to the first half of the twentieth century, the last ten years have been witness to exciting advances in our understanding of the fundamental acto-myosin interaction. Optical traps played no small role in this process, and we attempt to present here a synopsis of how this tool has been utilized.

We begin our discussion with an introduction to myosins, focusing on myosins II, V, and VI. These motors have been studied rigorously and serve as our case studies for discussing optical trap experiments. We then briefly describe early myosin studies, previous to work done on the single-motor level. We continue by discussing the benefit of single-molecule experiments in general and describe standard single-molecule assays that utilize optical traps. Finally, we describe the optical trap studies of myosins II, V, and VI and consider the insights gained from these studies.

8.2 MYOSINS

8.2.1 Overview of the Myosin Family

Myosins are a diverse family of molecular motors that convert chemical energy to directed mechanical motion through actin-activated ATP hydrolysis. There are more than 18 distinct classes, and in humans alone, there may be as many as 40 expressed myosins.[1] Myosins are defined by a highly conserved ~80-kDa catalytic domain, denoted the *head*, which contains the actin- and nucleotide-binding sites. Following the head is a *neck* region consisting of an α-helical strand stabilized by the binding of one or more light-chain subunits. The *tail* domain, whose sequence varies the most among different myosin classes, is the third and most C-terminal domain. It serves diverse purposes such as allowing myosins to self-associate, anchoring a myosin so it can move relative to actin, and allowing a myosin to bind specific cargo.[2] Taken together, these three domains make up the myosin heavy chain.

Myosin ATPase activity is coupled to its binding and release of actin such that the motor is able to do mechanical work. Figure 8.1B describes the nucleotide- and actin-bound states of the motor, with **M** representing the motor, **A•M** representing the myosin in an actin bound state, and **ATP, ADP,** and **ADP•P$_i$** representing the nucleotide states of the motor. Highlighted in gray is the motor's predominant kinetic pathway.[3] Pictured in Figure 8.1A are the global structural changes associated with the acto-myosin chemomechanical cycle.

We describe the motor's stepping pathway starting from the left-most state in Figure 8.1A and moving around the figure clockwise. This is the same as starting from the bottom-left of Figure 8.1B and following the highlighted path. The actin-bound myosin in its nucleotide-free state (the so-called *rigor* state) binds ATP and dissociates from actin. It hydrolyzes its ATP and enters what is denoted the *pre-stroke* state, in which it is poised structurally for a mechanical stroke. Upon rebinding to actin, the motor releases its phosphate, undergoes a "working stroke," and enters its *post-stroke* state. It then releases ADP and returns to the *rigor* state.

A popular paradigm to describe the link between the biochemistry of the stepping cycle and the global structural changes in the motor is denoted the *lever-arm*

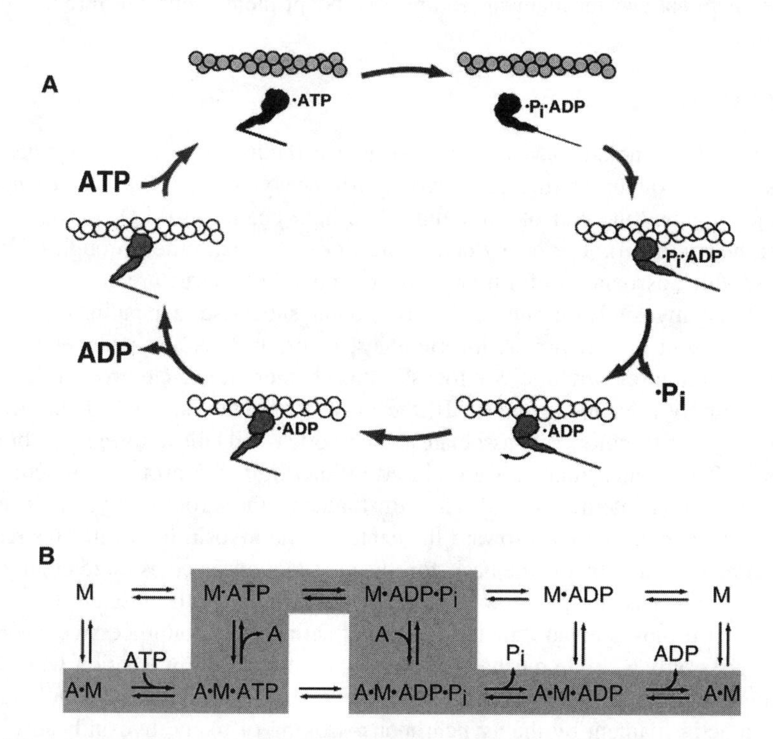

FIGURE 8.1 Myosin chemomechanical cycle. (A) Global structural changes of myosin associated with its actin-activated ATPase cycle. (B) Actin- and nucleotide-bound states of myosin. Myosin is represented by **M**, and actin-bound myosin is represented by **A•M**. The predominant kinetic pathway of the actin-activated myosin ATPase cycle is highlighted in grey. (Adapted from Murphy, C.T. et al., A myosin II mutation uncouples ATPase activity from motility and shortens step size, *Nature Cell Biol.*, 2001. With permission.)

model.[4–6] According to this model, small nucleotide-dependent structural changes in the myosin head are amplified by a relatively rigid lever arm, resulting in a large directed motion of the motor. The light-chain stabilized neck region is often envisioned to serve this role as the lever arm. This hypothesis has been a contentious point in myosin studies and is discussed in detail later in the chapter.

Though the chemomechanical cycle described here is common to the different myosin classes, differences in the rate constants describing the individual transitions can dictate differing function. For example, muscle myosin II (discussed in detail later) works in large ensembles of motors. For such a motor, it is advantageous to spend most of the chemomechanical cycle off of the actin so as to not hinder the stroking of neighboring motors. For myosin II, this is achieved by "tuning" the kinetic cycle such that its rate-limiting transition occurs while it is dissociated from actin. Different myosins are also able to use the cycle described in Figure 8.1 to achieve disparate types of motion. For example, myosin VI (discussed in detail later) is able to move along actin in the opposite direction from all other characterized myosins.

The three myosins we discuss in this chapter all exhibit stepping cycles with unique chemical and mechanical features that adapt them to their different cellular functions.

8.2.2 Myosin II

Myosin II, also denoted *conventional* myosin, was the first myosin to be studied and consists of a hexameric structure of two myosin heavy chains (~200 kDa each) and four light chains (one pair of calmodulin-like light chains, ~20 kDa each, binding to each heavy chain). The heavy chains are linked to each other through their tail regions, which associate to form an elongated coiled-coil structure.

Different myosin II variants can be divided into subclasses depending on whether they are present in vertebrates, invertebrates, or protozoans and whether they are muscle or nonmuscle motors.[2] We focus in this chapter on muscle myosin II, which we will simply refer to as myosin II, the most extensively studied of the myosin motors. Myosin II motors self-associate at their coiled-coil tails to form large, bipolar filaments. These thick filaments are placed in interdigitating arrays with actin filaments, and upon interacting with an actin filament, the stroking myosin II heads tend to move actin filaments toward the center of the myosin filament. This results in a contractile force that is the basis for diverse functions such as rapid contraction of insect flight muscles and slow contractions of tonic smooth muscle.[7]

Myosin II moves toward the barbed end of actin and is nonprocessive, meaning it undergoes only a single mechanical translocation for each interaction with actin. The numerous motors in a myosin II thick filament, however, are able to continuously move an actin filament by the asynchronous stroking of many myosin heads.

Through controlled proteolysis, myosin II can be cut into smaller fragments. Through papain digestion, the myosin subfragment 1 (S1) is liberated from the tail domain. The S1 fragment consists of the catalytic domain and the neck region with its two bound light chains. S1 heavy chains cannot dimerize like the wild-type motor. Through chymotrypsin digestion, the motor is cleaved to form heavy meromyosin

(HMM) and light meromyosin (LMM). HMM consists of the catalytic domain, the light chains, and a portion of the tail and is able to dimerize. Experiments on myosin II often utilize the HMM and S1 fragments, and for studies of other, unconventional myosins, HMM-like and S1-like fragments are also used. As with the myosin II proteolytic fragments, unconventional HMM-like fragments typically dimerize, and S1-like fragments do not.

The myosin II motor is illustrated in Figure 8.2.

8.2.3 Myosin V

Myosin V, the most thoroughly studied unconventional myosin, also consists of two myosin heavy chains that dimerize by the formation of a coiled coil at their tail regions. A globular domain is located at the C-terminal end of the tail and is hypothesized to bind cargo or specify subcellular localization.[8–10] Each heavy chain binds six light chains, many of which are bona fide calmodulin molecules, along its neck, resulting in a relatively long lever arm. Myosin V does not self-associate like myosin II, and a single motor is capable of transporting organelles in various *in vivo* systems. The motor is implicated in melanosome transport,[8] possibly involving an interaction with the microtubule motor kinesin,[11] as well as in membrane trafficking[12] and movement of vesicles in the brain.[13,14]

Like myosin II, myosin V moves predominantly toward the barbed end of actin. The motor interacts processively with a filament, meaning it undergoes numerous mechanical transitions for each encounter with actin, thus allowing it to carry cargo along actin. In this chapter, we will discuss in detail the experiments that have elucidated the mechanism of myosin V processivity.

The myosin V motor is illustrated in Figure 8.2.

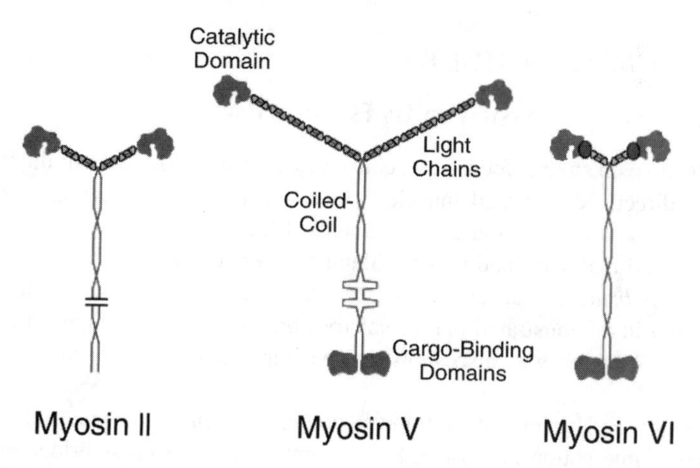

FIGURE 8.2 Depictions of Myosins II, V, and VI. Each motor consists of two identical heavy chains joined at their tail regions by the formation of a coiled-coil rod. Light chains bind at the neck region between the catalytic domains and the coiled-coil. The oval between the myosin VI catalytic head and light chain is a 53-amino acid insert unique to this motor,[22] which is known to bind a calmodulin light-chain.[16]

8.2.4 Myosin VI

Myosin VI, a more recently studied unconventional myosin, has proven to be an intriguing and unique motor. It has surprised many by defying the paradigm of motility resulting from myosin II and myosin V studies, a point we explore in depth later. Like myosins V and II, myosin VI is hypothesized to consist of two heavy chains that are linked at their tail region through a coiled-coil structure, and at the end of this tail region is a C-terminal globular domain hypothesized to be involved in cargo binding. Each heavy chain binds two calmodulin molecules,[15,16] and like myosin V, myosin VI motors do not self-associate.

Myosin VI is hypothesized to perform two distinct roles *in vivo*. The motor is hypothesized to function in some systems as a transporter. For example, it is implicated in the movement of endocytic cargo into a cell[17] and in the transport of proteins to the leading edge of a migrating cell.[18] In other systems, the motor functions as an anchor, linking whatever is bound at its tail to an actin filament. Hypothesized anchoring function includes linking actin-regulatory proteins to an actin-complex during Drosophila spermatogenesis[19,20] and mooring stereocilia to the hair cells of inner ear sensory epithelia.[21]

Unlike myosins V and II, myosin VI moves predominantly toward the pointed end of actin[22] and is the only characterized myosin to exhibit such behavior. The ability to transport cargo suggests myosin VI is also a processive motor like myosin V, but its ability to perform two functions also suggests that myosin VI has a mechanism of switching its kinetics from that of a processive transporter to that of an anchor. We discuss in this chapter experiments that have probed and elucidated this mechanism.

The myosin VI motor is illustrated in Figure 8.2.

8.3 EARLY MYOSIN STUDIES

8.3.1 Reducing the System to Its Essential Parts

Early acto-myosin studies focused on conventional myosin II. Some of the first work was done directly on striated muscle, which consists of interdigitating arrays of myosin II thick filaments and actin filaments. Electron microscopy on sections of muscle showed that actin and myosin filaments were connected through structures, denoted *cross-bridges,* that extended from the myosin filaments.[23,24] Studies with purified myosin demonstrated that these cross-bridges were likely part of the HMM subfragment,[25] which was also shown to be implicated in actin- and nucleotide-binding.[26]

In 1969, H.E. Huxley put forth the *swinging-cross bridge model* to describe the actin-myosin interaction in muscle. By combining previous cross-bridge work with detailed X-ray diffraction studies,[27] he hypothesized that actin and myosin filaments interact through "repetitive interaction of the cross-bridges with the actin filaments."[25] According to this model, for each ATP hydrolysis by a myosin, the cross-bridge goes through a single cycle of its tension-generating interaction with actin.

This interaction was hypothesized to involve a change in the angle of the cross-bridge attachment to the actin filament followed by detachment from actin.

A picture of this interaction was refined by further biochemical studies using purified actin and myosin, as well as by mechanical studies of intact muscle. Three-dimensional structures from electron micrographs yielded low-resolution structures of the acto-myosin complex[28] and kinetic studies revealed the nature of the ATPase cycle of the motor.[3,29–31] The dynamics of the cross-bridge was elucidated by time-resolved X-ray diffraction studies of muscle that revealed changes in cross-bridge structure corresponding to tension development.[32]

Though these early studies could probe the chemical and mechanical properties of the cross-bridge, a means of directly studying the fundamental interaction between actin and myosin was still lacking. In 1983, Sheetz and Spudich developed such an assay when they observed myosin moving over a surface of actin filaments from the alga *Nitella*.[33] In these studies, *Nitella* cells were cut open and spread out, and the cytoplasmic contents were washed away, leaving behind bundles of actin filaments. Beads coated with HMM were applied to the cell surface, and in the presence of ATP, the beads moved along the actin filaments directionally.

The *Nitella* assay provided a means of observing motility of myosin along actin in a reproducible and quantifiable manner. Using this assay, Sheetz and Spudich confirmed that the sliding of myosin and actin filaments in muscle arose from a fundamental acto-myosin interaction. These experiments also suggested that the basis for motility of unconventional myosins could be analogous to the basis of muscle contraction.

This assay was further improved by establishing systems in which actin filament geometry was better characterized. This was first achieved by attaching filaments to a surface using severin, a protein that binds specifically to the barbed-end of actin. The filaments on the surface were then oriented by flow, and myosin-coated beads were seen to move directionally along the filaments, against the flow of buffer.[34]

In 1984, Yanagida et al. observed fluorescently labeled actin in solution using light microscopy,[35] a feat which allowed for a final sophistication of this acto-myosin assay. In the presence of ATP, Kron and Spudich observed such fluorescently labeled filaments moving directionally over a glass-surface coated with myosin.[36] This form of the assay, in which motor on a surface moves actin in solution, provides a means of directly observing and quantifying myosin-driven actin motility while requiring a relatively low number of myosin motors. We refer to this experiment as the *in vitro motility* or *gliding filament* assay.

8.3.2 *In Vitro* Motility Assays: Insights and Limitations

Studies using the *in vitro* motility assay helped bolster the cross-bridge model by demonstrating that myosin II maintains a sliding motion of actin, and that this movement is generated by a cyclical interaction between myosin and actin that is coupled to ATP hydrolysis (a discussion of *in vitro* motility assays and the cross-bridge model is found in a review by H.E. Huxley[37]). The single-headed myosin II fragment S1 was shown to be sufficient to establish tension on actin[38] and to drive sliding motion of filaments,[39] indicating that acto-myosin sliding is occurring through this N-terminal region of the motor. Also, Kron and Spudich observed that the velocity of actin motility is limited not by the power output due to ATP hydrolysis, but instead by a kinetic limitation in the sliding interaction. They hypothesized that this limiting factor is likely the time required for the motor to stroke and detach from actin.[36]

Though these motility assays answered many questions about the cross-bridge interaction, inevitably they also raised questions that required more refined experiments. Two fundamental characteristics of stepping, coupling and step size, could only be indirectly inferred from *in vitro* motility experiments, and divergent values arose from different studies. *Coupling* of a motor refers to the correlation between ATP hydrolysis and mechanical translocation; a tightly coupled mechanism predicts a single ATP hydrolysis for each mechanical transition, while a loosely coupled motor undergoes a variable number of translocations for each ATP hydrolyzed. Because a cross-bridge is ~20 nm long, an upper limit to the step size predicted by a tightly coupled model is 40 nm. While some labs concluded that myosin II translocates actin 100–300 nm for each ATP hydrolyzed, presumably through a loosely coupled mechanism,[40,41] others suggested a step of 5–30 nm, possibly through a tightly coupled mechanism.[42–44]

These inconsistencies arose largely because the step size of a motor can only be indirectly inferred from the measured actin sliding velocity. In particular, there was considerable disagreement over the number of myosins interacting with an actin filament for a given experiment. Models estimating the number of heads interacting with a given length of actin filament were devised,[41,44] but these different models made assumptions which yielded differing predictions. Resolving these uncertainties required experiments that could probe the acto-myosin interaction on the single-molecule level, looking directly at a single myosin interacting with an actin filament.

This was not possible with the *in vitro* motility assay because a single myosin II cannot maintain continuous motion of an actin filament. The motor spends most of its stepping cycle detached from actin, and so a filament tends to diffuse away from the myosin surface. Attempts were made to observe fewer myosins interacting with an actin filament through refinement of the assay. The viscosity of the experimental buffer was increased to slow actin diffusion,[43] an actin filament was attached to a microneedle,[38,45] and an actin filament was fastened to a latex bead controlled by an optical trap.[46] Though these experiments improved upon the original assay, it was the next generation of experiments, those directly observing stepping of a single motor, which would help settle the conflicting results from different myosin studies.

8.4 SINGLE-MOTOR OPTICAL TRAP ASSAYS

8.4.1 Single-Molecule Experiments

A weakness of *in vitro* motility assays, as well as other bulk myosin experiments, is that they rely on observations of numerous motors asynchronously cycling through their chemomechanical cycle. Such results represent ensemble averaging of the states of many motors and predominantly yield information about the slowest transition in the cycle (the transition which most motors are waiting to undergo at any time).

Information about other, more rapid transitions is "blurred" by averaging. In single-molecule experiments, however, individual transitions within the motor's stepping cycle are observed, and their detection is limited only by experimental resolution.

Various methods have been used to observe single myosin activity (a discussion of single-motor studies of myosin II is presented by Molloy and Knight[47]). Common to many of these is the use of a *transducer*, a detection system that converts the force and movement of the motor to an electrical signal.[47] In this chapter, we will focus on the use of optical tweezers as a transducer in single-protein myosin studies. Though other tools, such as glass microneedles[45] and atomic force microscopes,[48] have also been successful in probing these motors, our goal here is to relate the development and evolution of a particular single-molecule detection system. We also explore the equally important and challenging task of developing appropriate tools to analyze data collected from these assays. To this end, we discuss insight gained into the mechanisms of *conventional* myosin II as well as *unconventional* myosins V and VI through optical trap assays.

8.4.2 The Optical Trap

An optical trap consists of a laser beam that is strongly focused in an aqueous solution with a short focal length lens, usually a high numerical aperture microscope objective.[49] When the laser passes through an object near its focus that has an index of refraction higher than the surrounding medium, the object refracts light away from the laser's focal point. Because of the light's momentum, the refracted laser light imparts a force on the object, thus trapping it at the focus. Depending on the size and refractive index of the particle and the power and wavelength of the trapping laser, the trap can impart a force on the order of 1–100 pN.[50]

Simmons et al. demonstrated that, for a 1 μm-diameter glass bead (about the size of beads commonly used in single-molecule optical trap assays), the distance between the bead and trap is linearly related to the trap's restoring force for displacements from the trap of up to ~200 nm.[50] Thus, the trap behaves as a linear spring over this range of distances, with the apparent spring constant denoted the *trap stiffness*. By measuring this stiffness, a given distance between the bead and trap can be converted to a known force on the bead.[51]

In addition, Simmons et al. demonstrated that a micron-sized bead in a trap is capable of transducing forces on the order of piconewtons on a time scale of milliseconds.[50] This is the same force and time scale of the acto-myosin interaction, making an optical trap an ideal transducer for these single-motor studies.

8.4.3 The Three-Bead Assay

We denote the optical trap experimental geometry most frequently used for myosin studies the *three-bead assay*. A detailed description of how to build, calibrate, and use an optical trap system for this assay is presented by Sterba and Sheetz[52] and by Rice et al.[53] We go through key aspects of the assay below.

For this assay, an actin "dumbbell" is made in solution by attaching ~1 μm-diameter glass or plastic beads to both ends of an actin filament. These beads serve as "handles," and each is held in an independently controlled optical trap. Myosin is sparsely adsorbed on a surface that has been coated with beads. These surface beads serve as platforms, lifting the motor above the experimental surface. The beads on the ends of the actin dumbbell are two beads of the three-bead assay while the surface bead is the third.

The actin is stretched taut using the two optical traps holding the dumbbell beads, and the dumbbell is moved over a platform bead. When a motor interacts with the actin filament, the dumbbell is translocated, resulting in displacement of the dumbbell beads relative to their traps. Thus, detection of the trapped beads' positions is the empirical readout of the assay. Detection of the beads is usually achieved by projecting their images onto quadrant-photodiodes.

This experimental setup is depicted in Figure 8.3A.

8.4.4 The Single-Bead Assay

Another useful single-motor optical trap experiment, denoted the *single-bead assay*, looks similar to the three-bead assay, only turned upside-down. In this assay, actin filaments are attached to the experimental surface. A bead in solution, approximately one micron in diameter, is sparsely coated with motor. The bead is held in an optical trap and lowered to the actin surface. As the motor binds and translocates along actin, the bead is moved out of its trap. Thus, detection of the bead position is again the means of observing the acto-myosin interaction.

This experimental setup is depicted in Figure 8.3B.

8.4.5 Experimental Considerations for the Optical Trap

The stiffness of an optical trap has a strong influence on the acto-myosin interaction and, thus, must be adjusted appropriately. This is true for any experiment involving the transduction of displacements and forces on the small scales of a single motor assay. When a myosin interacts with actin in either the single- or three-bead assay, the bead that is displaced from its trap experiences a restoring force from the laser. The force then affects the acto-myosin interaction itself. Thus, the trap stiffness must be kept sufficiently low so that effects of the trap do not dominate the interaction to be observed. In general, the transducer compliance must be much less than the stiffness of the cross-bridge.

At the same time, increased trap-stiffness will quash Brownian noise and thus increase signal-to-noise. Single-myosin studies tend to use transducers with stiffnesses of ~0.05 pN/nm.[47] For such an experiment, the stiffness is much lower than the acto-myosin stiffness, which has been measured to be ~0.2–0.7 pN/nm for myosin II,[54–58] and so the transducer is not significantly hindering the motor's mechanical transition.

The lasers used for trapping must also be chosen so as to minimize photodamage to the biological sample. For this reason, trapping lasers in myosin experiments are typically in the near-infrared range, with wavelengths around 1 μm. Solid-state lasers such as Nd:YAG, Nd:YVO_4 (both 1064 nm) and Nd:YLF (1047 nm) are typically used.[53]

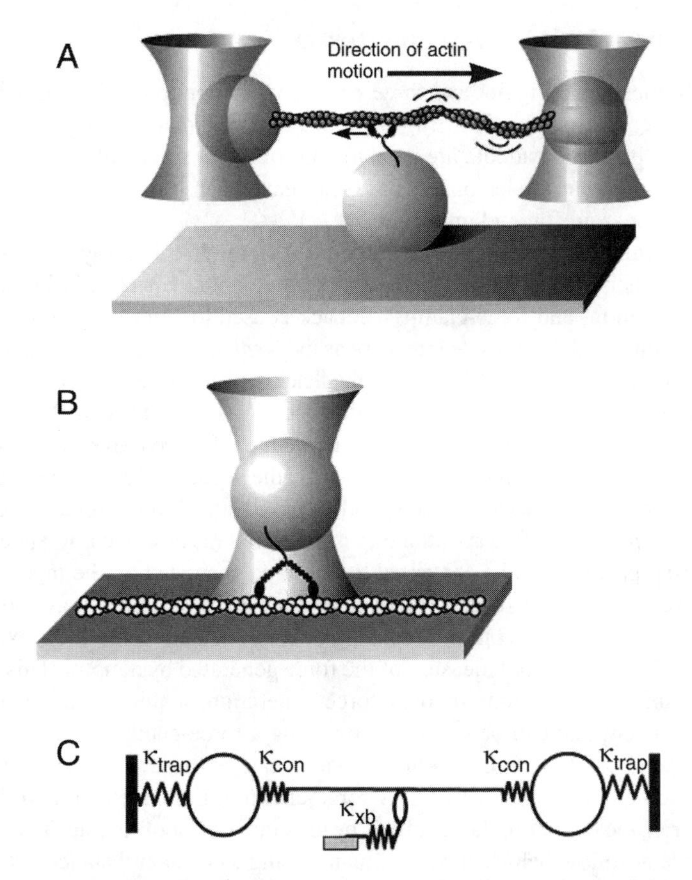

FIGURE 8.3 Single-molecule optical trap assays. (A) The three-bead assay with myosin VI. This assay is described in detail in the text. Two independently controlled optical traps (represented by hourglasses) hold an "actin dumbbell," consisting of micron-diameter beads (represented by spheres) attached to the ends of an actin filament. The myosin is held above the surface by a 1.5-μm-diameter bead. In the assay represented here, the actin is kept just slack between the dumbbell beads. When the motor translocates the actin, one bead is pulled from its trap (the bead on the left), while the other remains in the center of its trap (the bead on the right). (Adapted from Rock, R.S., et al. (2001). Myosin VI is a processive motor with a large step size. *Proc. Natl. Acad. Sci. U.S.A.,* 98, 13655–13659.) (B) Single-bead assay with myosin V. This assay is described in detail in the text. Actin is adsorbed to the surface, and a micron-diameter bead with a motor attached to it is held in an optical trap (represented by the hourglass). The bead is brought down to the surface to allow myosin and actin to interact. (Adapted from Rief, M., et al. (2000). Myosin-V stepping kinetics: a molecular model for processivity. *Proc. Natl. Acad. Sci. U.S.A.,* 97, 9482–9486.) (C) Cartoon of the chief compliant elements of the three-bead assay. The traps are represented by springs with stiffness κ_{trap}, the bead-actin linkages are represented by springs with stiffness κ_{con}, and the motor's attachment to the actin-filament is represented by a spring with stiffness κ_{xb}. The actin filament is drawn as a rigid rod. (Adapted from Veigel, C., et al. (1998). The stiffness of rabbit skeletal actomyosin cross-bridges determined with an optical tweezers transducer. *Biophys. J.,* 75, 1424–1438.)

8.4.6 Optical Trap Assays with Feedback

The capabilities of the single- and three-bead assays are improved by implementing feedback to control the optical trap position. In experiments using feedback, the trapped-bead position readouts are used to control the positions of the optical traps. The two predominant modes of feedback that have been utilized in myosin studies are the position- and force-clamp.

We first consider feedback in the three-bead assay. In this assay, one of the two dumbbell beads is pulled out of its trap by the motor (the left-bead in Figure 8.3A). For both a position- and force-clamp, feedback is used to control the position of the trap holding this bead. We thus denote these as the *feedback trap* and the *feedback bead.*

To implement a position-clamp, the feedback trap is moved so as to maintain a fixed position of the feedback bead. To do this, the trap is moved to exactly compensate for actin translocation by myosin, resulting in a large effective trap stiffness and near-isometric behavior of the motor. In doing this, the feedback trap is moved until it applies a restoring force to its trapped bead that is identical to the force generated by the motor. Thus, if the stiffness of the trap is measured (as described by Sheetz[51]), the feedback trap position can be converted to the force generated by the myosin.

However, as we will discuss later, rethinking of myosin II assays revealed a problem with the position-clamp. Stretching of compliant elements in the experiment results in an underestimated measure of the force generated by a motor. This problem limits the information about myosin force-generation acquired from a position-clamp, a problem that can be overcome by using a force-clamp instead.

The limitation of a force-clamp is that it can be used only with processive myosins such as myosins V and VI. A force-clamp is implemented as follows: As a myosin translocates actin, the feedback bead is moved out of its trap. The feedback trap is made to follow behind the bead, maintaining a constant distance between the two. Thus the restoring force of the trap is kept constant throughout the processive motor's stepping (we observe *isotonic* behavior of the motor). Since the trap acts approximately as a linear spring, when the stiffness of the trap is measured (as described by Sheetz[51]), the distance between the bead and the center of the trap can be converted to a known applied force. The force-clamp described here is only capable of applying forces against a motor's stepping. Force-clamps utilizing more sophisticated feedback are capable of maintaining constant forces in other directions relative to stepping.[59] An example of time traces for the feedback bead position and the feedback trap position for a force-clamp is shown in Figure 8.4C.

Position- and force-clamps are implemented identically with the single-bead assay. The only difference is that the feedback bead and feedback trap are the only bead and trap in the assay.

8.5 MYOSIN II

8.5.1 Working Stroke Measurements

In 1994, the first measurements of the forces and displacements of single myosin molecules interacting with actin were made using the three-bead assay to study

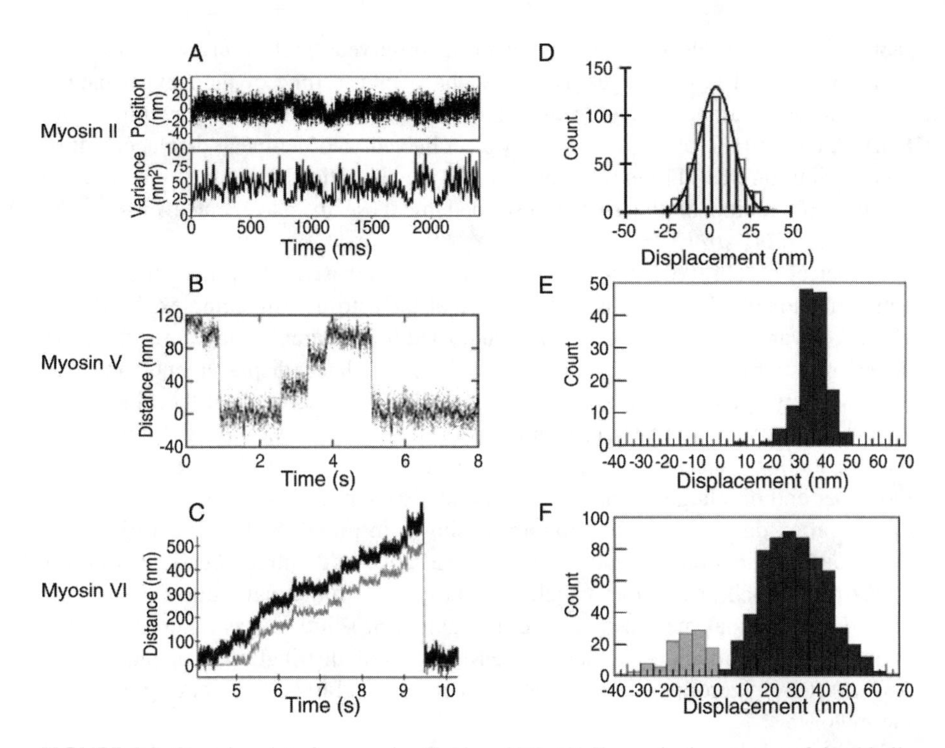

FIGURE 8.4 Stepping data for myosins II, V, and VI. (A) Example time traces of dumbbell-bead position (on top) and variance of dumbbell-bead position (on bottom) for myosin II in the three-bead assay. Variance drops during displacement events due to the increased stiffness of the system as myosin binds the actin-dumbbell. ATP concentration is 5 μM. (Adapted from Mehta, A.D., J.T. Finer, and J.A. Spudich (1997). Detection of single-molecule interactions using correlated thermal diffusion. *Proc. Natl. Acad. Sci. U.S.A.*, 94, 7927–7931.) (B) Example time trace of dumbbell-bead position for myosin V in the three-bead assay. ATP concentration is 10 μM. (Adapted from Mehta, A.D., et al. (1999). Myosin-V is a processive actin-based motor. *Nature*, 400, 590–593.) (C) Example time trace of dumbbell-bead position (black) and trap position (grey) for myosin VI in the three-bead assay with force-clamp. Feedback maintains a constant distance between the feedback bead and its trap, thus imposing a constant force against stepping. ATP concentration is 2 mM, and backward force on the motor is 1 pN. (Adapted from Altman, D., H.L. Sweeney, and J.A. Spudich (2004). The mechanism of myosin VI translocation and its load-induced anchoring. *Cell,* 116, 737–749.) (D) Histogram of myosin II displacements from the three-bead assay. Mean displacement is ~5 nm (n = 666). ATP concentration is 3 μM. (Adapted from Veigel, C., et al. (1998). The stiffness of rabbit skeletal actomyosin cross-bridges determined with an optical tweezers transducer. *Biophys. J.,* 75, 1424–1438.) (E) Histogram of myosin V steps from the three-bead assay with force-clamp. Mean displacement is 35 ± 6 nm (mean ± SD, n = 131). ATP concentration is 2 mM, and backward force is 1 pN. (Adapted from Rock, R.S., et al. (2001). Myosin VI is a processive motor with a large step size. *Proc. Natl. Acad. Sci. U.S.A.*, 98, 13655–13659.) (F) Histogram of myosin VI steps from the three-bead assay with force-clamp. Mean forward displacement is 30 ± 12 nm (mean ± SD, n = 615) and mean backward displacement is 13 ± 8 nm (mean ± SD, n = 114). ATP concentration is 2 mM, and backward force is 1.7 pN. (Adapted from Rock, R.S., et al. (2001). Myosin VI is a processive motor with a large step size. *Proc. Natl. Acad. Sci. U.S.A.*, 98, 13655–13659.)

rabbit skeletal muscle myosin.[60] Finer et al. observed single translocation events with leading and falling edges as fast as the response time of their experimental apparatus. An example of data from a myosin II three-bead assay is shown in Figure 8.4A. A distribution of displacements was collected, and the mean displacement of this distribution was ~11 nm over various ATP concentrations. We denote the directed mechanical motion of the motor described by these displacement events as the motor's *working stroke*.

Finer et al. observed that Brownian motion decreased when myosin bound to actin, presumably due to the added stiffness of the acto-myosin complex. Molloy et al. took advantage of this phenomenon and used the apparent change in stiffness to detect myosin binding events.[54] This allowed them to detect displacements too small to be seen by eye and thus improved upon the spatial resolution of the Finer et al. studies. By collecting even very small displacement events, Molloy et al. found that the distribution collected by Finer et al., extending from 5–20 nm, described only the upper end of a larger Gaussian distribution extending through zero (an example of this broad displacement distribution is shown in Figure 8.4D). The variance of this broad distribution was identical to the variance in dumbbell bead position for an actin dumbbell in solution, which only moves due to thermal motion. Thus, they argued that thermal motion of the actin filament in solution was randomizing the location of myosin binding to actin. The Molloy et al. distribution of displacements was offset from zero by ~4 nm, which they interpreted to be the working stroke of the motor.[54]

Methods for detection of displacement events in data have been further refined by improvement of variance-based analysis methods[61] as well as through use of correlation between the two dumbbell beads.[57] In an attempt to remove subjectivity from data analysis, Guilford et al. utilized a method called mean-variance analysis, originally developed to analyze ion-channel current recordings, to automate analysis of three-bead assay data.[62]

Contention still exists as to what the true myosin II working stroke is. Though assays utilizing either the variance or bead-correlation analysis methods both arrived at a stroke of ~5 nm,[54,57] Guilford et al., through their mean-variance analysis, measured displacement distributions with peaks at 11 and −11 nm and with very few displacements near zero.[62] Because these two peaks are much more narrow than the Molloy et al. distributions, Guilford et al. argued that perhaps, in their hands, actin and myosin were adopting a rigid geometry that only allowed the myosin to bind to specific sites on the actin.[62] Other arguments, reviewed in detail by Mehta and Spudich,[63] have been made to explain the difference in displacement measurements, but none has adequately resolved the discrepancy.

Experiments performed by Tanaka et al. exploring the dependence of the measured displacement on the relative angle between myosin and actin have suggested that 5 nm may actually be an underestimate of the working stroke.[64] At the optimal angle between the two (5–10°), they measured a 10-nm mean displacement, which decreased to near-zero as they approached 90°. The displacement then increased to 5 nm again at angles of 150–170°. The maximum value of the working stroke observed by Tanaka et al. is considerably larger than 5 nm, and they suggested that it could possibly be as large as 15 nm if they correct for extension of compliant elements in the experimental system. They argued that earlier experiments yielded a lower step-size due to averaging over all the possible myosin-actin orientations.[64] This argument is not entirely satisfactory, however, since Molloy et al. observed displacement distributions whose width corresponded to the thermal motion of the actin dumbbell.[54] According to the Tanaka et al. conjecture, the expected spread in displacements should correspond to a convolution of the spread due to the dumbbell's thermal motion and the spread due to sampling random angle orientations between the actin and myosin. Thus, a much broader displacement distribution would be expected.

To further confuse the issue, Ishijima et al. published results from microneedle studies that indicated a displacement size closer to ~20 nm.[65,66] A thorough discussion concerning these different stroke size measurements is presented by Mehta and Spudich.[63]

8.5.2 Measuring the Force of the Stroke: Rethinking the Position-Clamp

Initial measurements of the force of the myosin working stroke were made using a position-clamp, yielding measurements of 2–4 pN.[54,60,62] However, it was not long until it was recognized that this assay suffers from a serious flaw. As described earlier, the force generated by a myosin is measured by converting the feedback trap's position to a force using the trap stiffness. The restoring force of the trap can then be equated to the force generated by the motor. This approach assumes, however, that the stiffness of the experimental transducer is much greater than the acto-myosin cross-bridge stiffness.[56]

The trap itself is made very stiff, on the order of 10 pN/nm, through implementation of feedback. But the link between the dumbbell bead and actin can be quite compliant, and this compliance can be variable for different dumbbells.[56,57,67] When the feedback trap applies a restoring force to the feedback bead, this compliant actin-bead linkage is able to extend. Thus, the feedback trap does not need to move as far to maintain the constant position of the bead. This results in an underestimated force measurement.

To characterize this experimental problem, Veigel et al. described and measured the stiffnesses inherent to the trapping assay.[56] A cartoon showing the three-bead assay reduced to its principal compliant elements is presented in Figure 8.3C. If the actin filament is assumed to have a very high stiffness, measured to be ~8 pN/nm by Kojima et al.,[68] then the three stiffnesses that must be considered are the stiffness of the traps, κ_{trap}; the stiffness of the actin-bead linkages, κ_{con}; and the stiffness of the cross-bridge, κ_{xb}. κ_{trap} can be measured as described.[51] By driving one of the beads of an actin dumbbell with a sinusoidal wave, Veigel et al. cleverly used the response in the passive bead, both when the dumbbell was attached and detached from a motor, to measure κ_{con} and κ_{xb}. Their results indicated that, though the compliance between the actin and beads can vary, for most trapping assays, $\kappa_{con} >> \kappa_{trap}$ and $\kappa_{con} \approx \kappa_{xb}$.

These compliance relations provided two key facts about these assays. First, they verified that the stiffness of the trap is much less than the stiffness of the cross-bridge, a condition that must be met to assume that the displacement of a dumbbell bead correlates to the distance of the working stroke for an assay without feedback. Nonetheless, compliance corrections for displacement measurements can be made,[56,57] and this error is expected to be quite small. In general, the three-bead assay may yield a measured working stroke which underestimates the actual working stroke by about 10%, or ~0.5 nm.[56]

A more striking insight, however, is that a position-clamp experiment results in a significantly underestimated measurement of the force of the working stroke. Using the values for the compliances measured by Veigel et al. ($\kappa_{con} \approx 0.3$ and $\kappa_{xb} \approx 0.7$) and calculating the apparent stiffness of the system when trap stiffness is large and the dumbbell is attached to a myosin, $\kappa_{system} = \kappa_{xb} \times \kappa_{con}/(\kappa_{con} + \kappa_{xb})$,[56] we calculate a system stiffness of ~0.2 pN/nm. The original position-clamp experiments, however, assumed that the stiffness of the system corresponds only to the stiffness of the acto-myosin complex. Thus, the system stiffness was assumed to correspond to the ~0.7 pN/nm cross-bridge stiffness, resulting in a force measurement that may be more than three-fold underestimated. The actual force produced may be closer to 6–12 pN.

Using very stiff microneedles, Ishijima et al. measured myosin stroke forces at near-isometric conditions of ~6 pN,[65,66] consistent with the above estimate. The force is also likely bounded above by ~9 pN, which was measured to be the force required to rapidly force dissociation of myosin from actin in the absence of ATP.[58]

8.5.3 Displacement Durations and the Kinetics of the Acto-Myosin Interaction

Kinetic information about the myosin II chemomechanical cycle can be inferred from the durations of acto-myosin interactions. The duration of a displacement corresponds to the time the motor spends undergoing the transitions following actin binding and leading to detachment from actin. We denote this as the strongly bound time of the motor, t_s. According to the scheme in Figure 8.1, this corresponds to the duration of the actin-bound states after the motor has hydrolyzed its ATP and before it binds another.

At high (2 mM) ATP concentrations, Finer et al. observed durations close to the resolution of their experiment, on the order of 7 ms or less.[60] Their resolution was predominantly limited by the Brownian motion of the trapped beads, which, as discussed earlier, is quite large due to the low stiffness of the optical trap transducers. For a bead in a linear, spring-like potential, the frequency spectrum of the bead's thermal motion has a roll-off frequency, f_c, which is related to the bead radius, r, the viscosity of the water, η, and the stiffness of the system, κ_x, through the relation:

$$f_c = \frac{1}{2\pi} \frac{\kappa_x}{6\pi\eta r} \tag{8.1}$$

For optical trap assays, f_c is on the order of 100 Hz,[61] and so for an event to be resolved from noise, it must be sufficiently longer than $1/f_c$ or ~10 ms. Thus, events on the order of 7 ms or less cannot be well-resolved from thermal noise in these assays. One way in which the time resolution of the system can be improved is by imposing a low-amplitude, high-frequency oscillation to one of the traps. This modified assay allows for resolution of binding events on the order of a millisecond in duration.[54]

More frequently, however, the problem of time resolution is overcome by lowering the ATP concentration in order to extend displacement events. As Finer et al. lowered the ATP concentration in their experiments to 10 μM and then again to 1 μM, the mean duration fell within the detection range of their apparatus and increased as ATP concentration was lowered. This suggested that, at such low ATP concentrations, ATP binding had become slow relative to the other kinetic events leading to detachment.[60] By this reasoning, at sufficiently low ATP, ATP binding should dominate the displacement kinetics, and the mean duration should be approximately the inverse of the rate of ATP binding. At 1 μM ATP, Finer et al. measured a mean duration of 0.26 s. Taking the inverse of this value, they estimated the second-order ATP binding-rate for myosin II to be ~4 $\mu M^{-1}s^{-1}$,[60] consistent with measurements of the acto-myosin ATPase measured in solution.[69]

Further kinetic information is gained if the measured durations are not averaged, but instead are binned as a histogram. It was observed that the distribution of myosin II durations collected at low ATP has the appearance of a decaying exponential (Figure 8.5A),[55,64] suggesting that a single kinetic step limits the detachment process. The resulting duration distribution obeys the normalized probability distribution:

$$P(t;k) = k\exp(-kt) \tag{8.2}$$

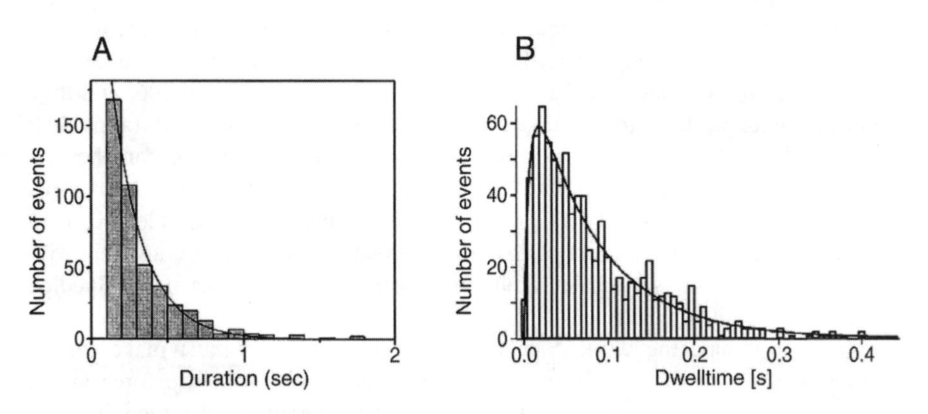

FIGURE 8.5 Displacement duration data for myosin II and dwell time data for myosin V. (A) Histogram of durations for myosin II displacements at 1 μM ATP. The data was fit to a single exponential (the solid line), yielding a time constant of 0.22 s. (Adapted from Tanaka, H., et al. (1998). Orientation dependence of displacements by a single one-headed myosin relative to the actin filament. *Biophys. J.*, 75, 1886–1894.) (B) Histogram of dwell times for myosin V stepping against 1 pN of load in the presence of 2 mM ATP. The data was fit to Equation 8.6 the (solid line), yielding $k_1 = 150\ s^{-1}$ and $k_2 = 12.5\ s^{-1}$. (Adapted from Rief, M., et al. (2000). Myosin-V stepping kinetics: a molecular model for processivity. *Proc. Natl. Acad. Sci. U.S.A.*, 97, 9482–9486.)

where k is the rate of the limiting transition event. Assuming that ATP binding is sufficiently slowed, fitting this exponential to the distribution of durations yields the apparent rate of ATP binding. Tanaka et al. collected displacement durations at 1 μM of ATP and fit the distribution to Equation 8.2, yielding a rate of ~4 s^{-1},[64] consistent with the ATP-binding rate calculated by Finer et al.

8.5.4 Coupling and Efficiency

We now attempt to relate the empirical results discussed above to the coupling and efficiency of the motor. In doing this, we will compare results from *in vitro* motility and trapping assays to other bulk motor experiments, allowing us to verify our assumption that these studies are probing an interaction that is the basis for more macroscopic phenomena.

As Huxley described in his review of myosin II studies,[37] previous muscle studies[70–72] are consistent with a tightly coupled stepping mechanism for the range of measured working strokes of 10–20 nm. Structures of chicken skeletal muscle myosin, however, suggested that the myosin II neck is an ~8.5 nm helix,[73] a result that is hard to reconcile with a displacement as large as 20 nm. We thus estimate that the working stroke is on the order of 10 nm, consistent with many of the trapping experiments discussed.

Optical trap studies suggested that the strongly bound time of the motor at high ATP concentrations is less than 7 ms. A better estimate for this value is inferred from *in vitro* motility studies done by Uyeda et al., who observed actin sliding driven by rabbit muscle myosin in the presence of methylcellulose to reduce filament diffusion from the experimental surface.[44] They measured the duty ratio f of the motor, defined as the fraction of time the motor spends attached to actin relative to its full stepping cycle. For a nonprocessive motor, such as myosin II, f is expected to be small. Uyeda et al. measured f to be 0.050 ± 0.006, and they measured an upper limit to the full cycle time of ATP hydrolysis at saturating ATP for a single motor, t_c, of 77 ± 5 ms. The strongly bound time t_s can then be calculated by the relation $t_s = f \times t_c$, resulting in an upper limit to t_s of 3.9 ± 0.5 ms.

These values suggest that the velocity of actin motility by an ensemble of myosin II motors should be on the order of 2.6 μm/s or greater. This value is consistent with the 3–4 μm/s velocities of skeletal muscle myosin measured by Kron and Spudich using *in vitro* motility assays.[36]

To approximate the work done by a myosin stroke, we must make certain assumptions. If we assume the myosin applies its ~6 pN isometric force at the beginning of the stroke and maintains this force over the entire stroke, then the work done by a 10 nm stroke is ~60 pN nm. If, on the other hand, the stroke is due to energy release by a perfectly elastic element, we expect half this, or about 30 pN nm. From muscle studies, the work output from a single myosin hydrolysis was estimated to be about 50 pN nm.[63] This agrees with the values above, and is consistent with the hypothesis that a single ATP hydrolysis is tightly coupled to a myosin stroke. Since the energy liberated from one ATP hydrolysis is about 100 pN nm, these values indicate that the motor may be quite efficient.

To directly detect the motor's coupling of ATP hydrolysis to force generation, Ishijima et al. combined three-bead assay studies of a single-headed myosin with total internal reflection fluorescence (TIRF) microscopy to visualize a fluorescent ATP analog, Cy3-ATP.[74] They observed a fluorescent signal when the head bound the nucleotide and no signal when the head released its hydrolysis product, Cy3-ADP. For many of the motors they observed, the release of bound nucleotide coincided with force generation, consistent with a tightly coupled relation between mechanical motion and hydrolysis. However, they observed a large mean displacement of 15 nm, which they hypothesized is indicative of numerous mechanical transitions for each ATP hydrolysis. They also observed some myosins produce force several milliseconds after product release. From this result, they conceived of a scenario in which the energy of hydrolysis is stored in the molecule, allowing it to bind actin and generate its force several milliseconds after product release.[74]

Though it is apparent that contention and unanswered questions still surround these optical trap studies, they have nonetheless been important in elucidating the fundamental acto-myosin interaction. And as we will see in our discussions of myosin V and VI, the single-molecule work with myosin II helped develop the sophisticated tools that would be further utilized and refined for future myosin studies.

8.6 MYOSIN V

8.6.1 Demonstrating Processivity: A Return to Gliding Filament Assays

Because myosin V is processive, meaning it can undergo numerous mechanical steps for each interaction with actin, single-molecule studies of this motor differ slightly from those of myosin II. Processivity is a trait also shared by the microtubule-based motor kinesin, a motor that had already been rigorously studied on the single-molecule level[75,76] while myosin V studies were in their early stages. Thus, many myosin V studies parallel earlier kinesin studies.

Activity of single myosin V motors was first observed by Mehta et al. using the *in vitro* motility assay to study tissue-purified myosin V from chick brain.[77] As discussed earlier for myosin II, these motility assays could not be used for single-molecule studies since a single myosin II cannot maintain continuous motion of actin. For a processive myosin, however, a single motor is capable of maintaining actin motility, and so *in vitro* motility can be utilized at the single-molecule level.

Mehta et al. used these assays to demonstrate the processivity of myosin V.[77] They looked at myosin V–driven actin motility at decreasing motor surface densities and observed that filament velocity remains unaffected. At low motor densities, they observed filaments swiveling about single-point connections to the surface, behavior also seen with processive kinesins.[78] To demonstrate that single motors led to this observed behavior, Mehta et al. performed two assays that are essential to verifying processivity. First, they measured the rate at which actin filaments bound to a myosin-coated surface at varying motor surface densities. They observed a first-power dependence of landing rate on surface density, consistent with studies of other processive motors, such as kinesin.[78,79] They then tabulated the fraction of landing filaments that moved longer than their length before detaching (F) as a function of motor surface density (ρ). They found that their results conformed to the expected relation for a processive motor:

$$F(\rho;\rho_0)=1-\frac{(\rho/\rho_0)\exp(\rho/\rho_0)}{1-\exp(\rho/\rho_0)} \tag{8.3}$$

where ρ_o is a fit parameter.[78]

Using *in vitro* motility assays, Moore et al. measured the myosin V duty ratio.[80] For myosin V, the duty ratio is expected to be large, since the motor processively interacts with actin. Moore et al. measured actin velocities at various surface densities and plotted the velocity, V_{actin}, as a function of the number of myosin heads interacting with a filament, N. They fit this plot to the relation:

$$V_{actin}=V_{\max}(1-(1-f)^N) \tag{8.4}$$

where $V_{\max}$ is the maximum filament velocity, and f is the duty ratio.[41,44] Though their fit yielded a value of 0.51, this method has poor resolution between 0.5 and 1. Thus they concluded that myosin V has a duty ratio >0.5, consistent with bulk biochemical assays of the motor that measured a duty ratio of ~0.7[81] and consistent with the expected high duty ratio anticipated for a processive motor.

8.6.2 Demonstrating Processivity with Optical Traps

Optical trap assays provide another, more direct means of demonstrating processivity. Care must be taken, however, to ensure that observed activity in an optical trap experiment is not the result of multiple motors working together. This is especially important because, in a single- or three-bead assay, a single processive motor yields data that is often similar in appearance to data from an ensemble of nonprocessive motors. For either scenario, the time trace of the detection bead-position has the appearance of a staircase (Figure 8.4B and Figure 8.4C show staircase data from processive motor studies). Each staircase consists of rapid translocations of the bead interspersed with durations during which bead-position remains relatively fixed.

To ensure that staircases in a three-bead assay result from a single processive motor and not an ensemble of motors, numerous surface platforms must be tested for motor activity. Assuming a homogeneous distribution of motors over the experimental surface, the number of motors on a surface platform is dictated by a Poisson probability distribution function. According to this distribution, when motor activity is observed on 10% of the platforms, ~95% of the platforms with motor activity contain only a single motor.[82] Thus, if staircases are observed in an experiment in which activity is seen on ~10% or less of the surface platforms, it is very likely the staircases are the result of a single motor. Similarly, for a one-bead assay, numerous motor-coated beads are tested for motor activity, and for experiments in which motor activity is detected on ~10% or less of the motor-coated beads, staircases likely result from a single motor.

Using this test for single-motor activity, Mehta et al. used a three-bead assay to demonstrate that individual myosin V motors are able to maintain continuous motion of an actin filament and produce staircase time traces,[77] thus providing direct evidence for myosin V processivity.

8.6.3 Of Strokes and Step Sizes

For a processive motor, the two experimental observables of optical trap assays are denoted dwell time (τ) and step size (d). Looking at a staircase (Figure 8.4B and Figure 8.4C), a dwell time consists of the time duration between consecutive translocations of the bead, and a step size consists of the distance between consecutive dwells. Care must be taken to differentiate *dwell time* and *step size* from *strongly bound time* and *working stroke*, the two observables of the optical traps assays with nonprocessive myosin II. We first discuss the difference between step size and working stroke.

A step size is defined as the total distance traveled by a motor through a full cycle of its stepping pathway. On the other hand, a working stroke describes the distance traversed by the motor due to its force-generating mechanical transition. To elucidate the difference between these two observables, we discuss the hypothesized myosin V stepping cycle. The experiments that have led to this specific model are described in more detail later.

The pathway is shown in Figure 8.6A. Myosin V spends most of its cycle with both heads strongly bound to actin, each with a bound ADP (the left-most state in Figure 8.6A). In this state, the front arm is strained in the direction of the motor's motion. The rear head releases ADP (a in Figure 8.6A), binds ATP (b), and dissociates from actin. This allows for dissipation of strain in the front arm, resulting in an 11-nm translocation of the actin-dissociated head in the direction of stepping. This is rapidly followed by a structural transition in the front head that is amplified by the myosin V neck region to a 25-nm force-generating stroke. The dissipation of strain combined with this working stroke moves the actin-dissociated head a total of 36 nm along the actin, positioning it as the new front head (c). This front head then hydrolyzes its ATP, binds to actin (d), and releases phosphate (e). Upon releasing its phosphate, the front head attempts to carry out its working stroke, but cannot because the motor's two heads are constrained along the length of the actin. This results in the strained state of the front arm. Thus, the motor returns to its original state after having completed a single stepping cycle. Myosin V thus has a force-generating *working stroke* of 25 nm but a total *step size* of 36 nm.

From staircase data of myosin V, Mehta et al.[77] and Rief et al.[83] both measured a step size of 34–40 nm using tissue-purified chick brain myosin V, though they used different methods. Mehta et al. used the three-bead assay. A problem with this assay is that each time the detection bead is pulled from its trap, compliant elements in the experimental system extend. To deal with this problem, they

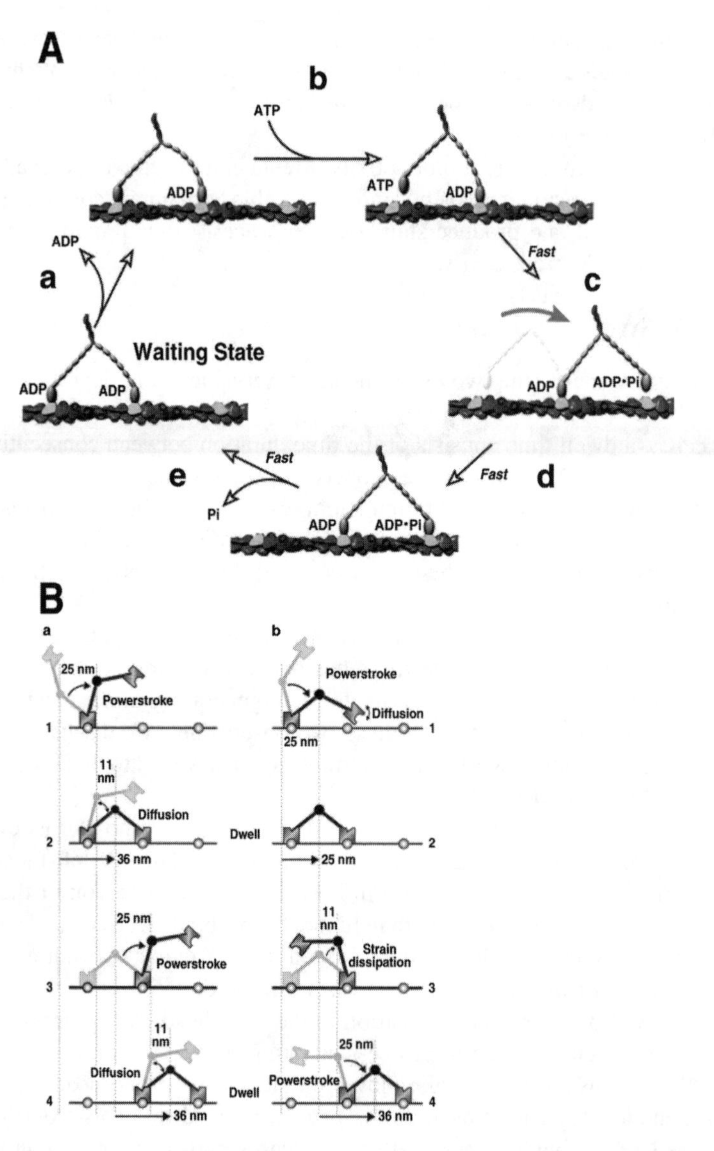

FIGURE 8.6 Representations of the myosin V stepping cycle. (A) Depiction of the processive stepping of myosin V. This model is described in detail in the text. (Adapted from Vale, R.D. (2003). Myosin V motor proteins: marching stepwise towards a mechanism. *J. Cell Biol.,* 163, 445–450.) (B) Two models for the mechanical transitions associated with myosin V stepping. These are discussed in the text. Briefly, according to model (a), one head attaches to actin and undergoes a 25-nm working stroke (1). The free head diffuses 11 nm to its next binding site (2), resulting in a 36-nm first step. When the rear head dissociates from actin, after binding ATP, stepping proceeds identically. According to model (b), the working stroke of the bound myosin head in (1) positions the free head near its next actin-binding site. Binding of this free head (2) then results in a 25-nm first step. When the rear head dissociates from actin, dissipation of strain in the molecule yields an 11-nm displacement followed rapidly by another 25-nm working stroke. Thus, all steps after the first are 36 nm. (Adapted from Spudich, J.A. and R.S. Rock (2002). A crossbridge too far. *Nat. Cell Biol.,* 4, E8–10.)

applied a large-amplitude oscillation to the detection bead's optical trap. If no motor was bound to the actin, the bead position oscillated with the trap. When a motor bounded, however, the bead followed the trap until the actin-myosin linkage pulled taut, resulting in clipping of the bead's oscillation. When the motor translocated the actin, the position of clipping changed, and the distance between the consecutive clipping positions was equated to a step size. This method overcomes the problem of extension of compliant elements because, when the oscillation is clipped, all compliant linkages are fully and equivalently extended.[77] Rief et al. on the other hand, used the single-bead assay and implemented a force-clamp feedback.[83] Because a constant force was maintained against the motor's stepping, compliant linkages were equally extended throughout a staircase.

Stepping distributions for myosin V studies were much more narrow than myosin II displacement distributions. This is not surprising, however, since a myosin II displacement distribution has a large spread due to the myosin binding to various sites along the actin as the filament undergoes random thermal motion. Myosin V step size distributions, however, are not affected by this thermal motion, since the motor remains bound to the actin throughout its stepping. Thus, the spread in the myosin V step distribution is only indicative of the motor's propensity to bind to monomers surrounding its most frequent binding site. This variability is presumably due to the actin-dissociated head undergoing a diffusive search to find its actin-binding site after the actin-bound head has undergone its working stroke (Figure 8.6A (c)).[83]

A striking feature of the myosin V step size is that it is about the same as the 36-nm pseudo-repeat of an actin filament.[84] Thus, the motor should be capable of walking along the top of actin without spiraling around the filament, a desirable trait for a motor that processively transports cargo. The Rief et al. single-bead experiment confirmed this by showing that myosin V can walk along a filament that is tightly adsorbed to a surface. If the motor is not hindered from rotating around the actin filament, however, as is the case in the single-bead assay, the motor tends to take a step that is just short of the pseudo-repeat, thus resulting in a slow left-hand spiral about actin.[85]

The force-generating working stroke of myosin V was observed through the use of an HMM-like murine myosin V construct with impaired processivity. Using the three-bead optical trap assay, Moore et al. observed that, at low surface densities, this motor predominantly exhibited single displacement time traces, similar in appearance to myosin II data.[80] They observed a mean displacement of ~23 nm and hypothesized that this distance corresponds to the working stroke of the motor. Presumably, this nonprocessive myosin V construct could not translocate the remaining distance required for the full 36-nm step size.

The length of the working stroke was directly measured by Purcell et al., using an S1-like chicken myosin V construct. Similar to the Moore et al. measurement, the S1 construct yielded nonprocessive displacement data with a mean displacement of ~20 nm.[4] This led to two models of myosin V stepping, both involving an ~25 nm stroke size and an ~11 nm displacement achieved by some other mechanism. Shown in Figure 8.6B, these two models are similar but can be empirically distinguished. According to *Model a*, the actin-bound myosin V head undergoes a 25-nm working stroke. The actin-dissociated head then undergoes an 11-nm diffusional search to find its actin-binding site. This cycle is then repeated identically. *Model b*, on the other hand, proposes that the 25-nm working stroke places the actin-dissociated head close to its actin-binding site. Thus, this head diffuses only a short distance to find its actin-binding site. It then binds to the actin, as the new lead head, and enters its strained state. When the rear head releases from the actin, the strain is dissipated, resulting in the 11-nm displacement. *Model a* thus predicts that the 11-nm distance is the result of a diffusional search, while *Model b* predicts it is the result of dissipation of strain.

As Figure 8.6B illustrates, though both models predict a step size of 36 nm, the first model predicts that the first step in a staircase should be 36 nm, while the second model predicts a first step of 25 nm. Veigel et al. demonstrated, using tissue-purified mouse myosin V, that the motor exhibits a shortened 25 nm initial step followed by consecutive 36 nm steps,[86] supporting *Model b.*

8.6.4 Dwell Time Measurements

While the *duration* measurements of nonprocessive myosin II assays yield information about transitions that occur while the motor is bound to actin, a *dwell time* measurement describes the time it takes the motor to undergo an entire cycle of its stepping pathway. We consider a single cycle of a generalized stepping pathway of N transitions:

$$1 \overset{K_1}{\leftrightarrow} 2 \overset{K_2}{\leftrightarrow} \dots \overset{K_{N-1}}{\leftrightarrow} N \overset{K_N}{\leftrightarrow} 1' \tag{8.5}$$

where the motor goes from its initial state **1** to the identical state **1′** after translocating a single step. A histogram of empirical dwell times, when normalized, yields the probability density, $P(t)$, for a motor starting at its initial state to complete a full stepping cycle in the interval of time between t and $t+dt$. We can calculate this theoretical dwell time distribution for a kinetic scheme (Equation 8.5) by solving the kinetic equations, assuming a motor starts at state **1** and ends at state **1**. The normalized probability distribution $P(t;k_{\pm1},k_{\pm2},\dots k_{\pm N})$, where $k_{\pm i}$ are the forward and reverse rates corresponding to K_i, is calculated by imposing the initial conditions [**1**] = **1** and [**2**] = … = [N] = [**1′**] = 0 and solving for d[**1′**]/dt.

When Rief et al. binned dwell times collected at saturating ATP conditions, they observed that, at short dwell times, the distribution rose rapidly with increasing dwell time while, at higher times, it fell off gradually with increasing dwell time, as shown in Figure 8.5B.[83] This observed dwell distribution was consistent with a stepping cycle approximated by two irreversible transitions. The probability density for such a model is calculated to be:

$$P(t,k_1,k_2) = \frac{k_1 k_2}{k_1 - k_2}\left[\exp(-k_2 t) - \exp(-k_1 t)\right] \tag{8.6}$$

At saturating (2 mM) ATP, the dwell distribution, when fit to this relation, yielded a very rapid rate of 150 s^{-1} and a slower rate of 13 s^{-1}. This slower rate is consistent with ADP release rates as measured in bulk motor studies.[81,87,88] Before these studies, however, it was still a point of debate as to whether this transition was the rate-limiting step in the motor's chemomechanical cycle. The optical trap study thus presented evidence that ADP-release is rate limiting, and that at saturating ATP, it is more than ten-fold slower than all other transitions in the stepping cycle.

At very low ATP concentration (2 μM), the dwell time distribution could be well fit by Equation 8.6 with rates of 14 s^{-1} and 3 s^{-1}. Presumably, the fast rate corresponds to ADP release, and the other rate now corresponds to ATP binding, which has been slowed by reducing the ATP concentration. In support of this model, the assay was repeated at concentrations ranging from 2 to 20 μM ATP. When fitting Equation 8.6 to each of these distributions, one rate was observed to be constant over all ATP concentrations, at ~13 s^{-1}, while the other increased linearly with increasing ATP concentration, consistent with a second-order rate of ATP binding.

Rief et al. argued that their results supported a tightly coupled model in which a single mechanical transition involves a single ATP hydrolysis. They claimed that, if two ATP molecules are hydrolyzed for each translocation, then, at sufficiently low ATP, the dwell distribution should describe three rates, one corresponding to ADP release and two identical rates describing binding of two separate ATP molecules. However, as they point out, this argument is true only if the two ATP binding rates are identical. For example, one can envision a model in which binding of two ATP molecules is highly cooperative; such a mechanism can also be explained by the Rief et al. results.

8.6.5 Settling the Lever Arm Debate?

As discussed earlier, a popular and contentious model of myosin stepping is the lever-arm hypothesis. According to this model, small nucleotide-dependent changes in the catalytic head of the motor are amplified to large, force-generating displacements through a relatively rigid lever arm, thought to be the light-chain bound neck region. *In vitro* motility studies utilizing *Dictyostelium* myosin II constructs with light-chain binding domains of varying lengths supported this model.[89] Uyeda et al. observed that, for constructs containing zero, one, two, or three light-chain binding domains, the sliding velocity of actin filaments was directly proportional to the number of bound light chains (Figure 8.7A). The average sliding velocity (v) of the actin is primarily determined by the stroke size (x_{stroke}) and the duration of the strongly actin-bound state of the motor (t_s) through the relation $v = x_{stroke}/t_s$.[89] The Uyeda et al. results could be explained by a stroke size that changes with neck length, consistent with the lever arm hypothesis. However, these studies could not rule out the possibility that the changing velocity is due to a change in t_s for their myosin II constructs.

Myosin V proved to be an ideal motor to test the lever arm model because of its very long putative lever arm, which consists of six light-chain binding domains. When Moore et al. plotted the working stroke as a function of the number of light-chain binding domains in the neck domain for a series of myosin II–HMM neck length mutants and included their myosin V HMM-like construct with six light

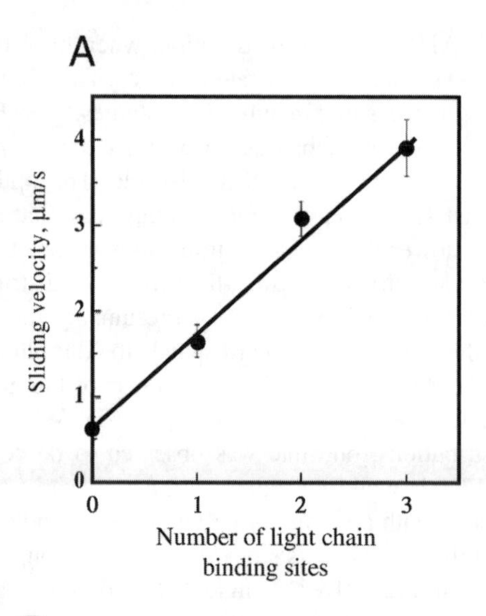

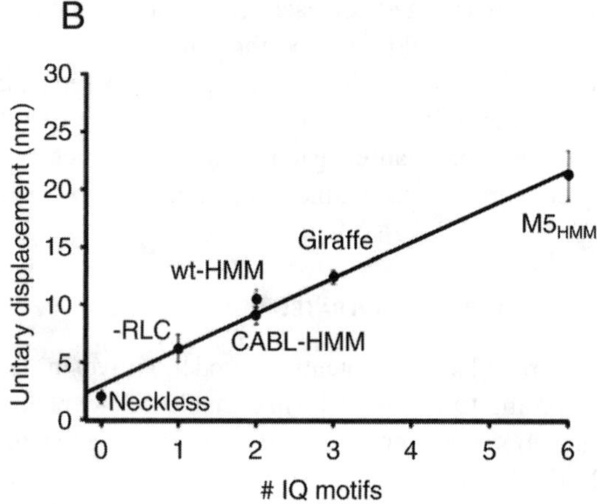

FIGURE 8.7 Dependence of sliding velocity and working stroke on lever arm length. (A) Actin sliding velocities of *Dictyostelium* myosin II constructs in the gliding filament assay. The zero light-chain measurement utilized a mutant lacking both light-chain binding sites; the one light-chain measurement utilized a mutant lacking a regulatory light-chain binding site; the two light-chain measurement utilized a construct with both light-chain binding sites; and the three light-chain measurement utilized a myosin construct with an additional essential light-chain binding site. (Adapted from Uyeda, T.Q., P.D. Abramson, and J.A. Spudich (1996). The neck region of the myosin motor domain acts as a lever arm to generate movement. *Proc. Natl. Acad. Sci. U.S.A.*, 93, 4459–4464.) (B) The working stroke (denoted here as the "unitary displacement") of smooth muscle myosin II–HMM neck length mutants and a nonprocessive myosin V HMM. These data were obtained from Warshaw et al.,[104] except for the Myosin V HMM ($M5_{HMM}$)data, which is from the Moore et al. studies.[80] wt-HMM and CABL-HMM contain two IQ motifs. Zero, one, three, and six IQ motifs were assigned to Neckless, -RLC, Giraffe, and the $M5_{HMM}$, respectively. The linear fit does not pass through zero but intersects the y-axis, suggesting that the lever may extend into the motor domain. (From Moore, J.R., et al. (2001). Myosin V exhibits a high duty cycle and large unitary displacement. *J. Cell Biol.*, 155, 625–635. With permission.)

chains (Figure 8.7B), they observed a strikingly linear relation.[80] This is expected for the lever-arm hypothesis, which predicts that the working stroke is linearly dependent on the length of the neck.

Purcell et al. and Sakamoto et al. were able to verify this relation through trapping assays using myosin V neck-length mutants. Myosin V constructs were made with neck regions consisting of 1, 2, 4, 6, and 8 light-chain binding domains. These constructs yielded a working stroke that was linearly related to neck length, ranging from 7 to 30 nm.[4,5] For myosin V HMM-like constructs, constructs that had 4–8 light chains were observed to be processive while the 1 and 2 light-chain constructs were not. As with the working stroke, step size is proportional to neck length for the processive 4 and 6 light-chain constructs.[4] These results strongly favor the lever-arm model as the myosin V stepping mechanism.

Nevertheless, these experiments did not end the debate over the lever-arm model. Previous to these studies, Tanaka et al. used an assay similar to the three-bead assay to examine a myosin chimaera consisting of the myosin V catalytic head followed by only one of the motor's six light chains. The myosin V coiled-coil tail domain was then replaced with a portion of the tail region of chicken smooth-muscle myosin II.[90] Thus, this motor is similar to the single light-chain constructs studied by Purcell et al. except for the different tail domains. Strikingly, Tanaka et al. found that their construct stepped processively, with steps somewhat shorter than but similar to those of a wild-type myosin V. This led them to formulate an alternate model of stepping in which a myosin V head, upon binding to actin, alters the actin filament so as to create a "potential slope."[90] The head slides down this potential, along an actin proto-filament, for 36 nm. The second myosin V head then crosses over and binds to the other proto-filament and continues the stepping cycle. This crossing over is necessary to explain the ability of myosin V to step continuously along an actin that has been adsorbed to a surface. This proposed model accounts for a step size that is apparently impervious to the motor's neck length. Another possible explanation for the Tanaka et al. results is that the smooth-muscle myosin II coiled-coil can unravel and allow for a large, diffusive ~30-nm step[91] (a mechanism similar to myosin VI stepping — see the discussion below).

The lever-arm model may account for the stepping behavior of myosin V, and, as we saw in Figure 8.7A and 8.7B, it may also explain the stepping of myosin II. From these results, it appeared (to the optimistic) that this model might explain in general the stepping of the different classes of myosins. As we will see, however, early on in single-motor studies of myosin VI, it became apparent that this unique motor was not going to fit so easily into the lever-arm paradigm. Nonetheless, our current model of myosin VI stepping, described below, fits well with the concepts already developed in studies of myosins II and V.

8.7 MYOSIN VI

8.7.1 The Lever-Arm Model Falls Short

Because the neck region of myosin VI contains two light-chain binding domains, the step size of this motor, based on the lever-arm model, is predicted to be around

10 nm (~1/3 of the myosin V step and similar to that of myosin II). This is consistent with the 12-nm observed working stroke of a single-headed myosin VI construct (R. Rock, A. Dunn, unpublished observations). Studies of a porcine myosin VI using a three-bead optical trap assay with force-clamp, however, showed a mean step size of ~30 nm for a myosin VI stepping against 1.7 pN load[92] and ~36 nm at lower loads,[93] too large to be explained simply by the lever-arm model. The stepping distribution of myosin VI (Figure 8.4F) also shows two striking features: (a) the stepping distribution of the motor is much broader than a myosin V stepping distribution, and (b) myosin VI takes frequent backward steps, behavior not often seen with myosin V.

Figure 8.8 depicts the stepping of myosin VI and myosin V on an actin filament. Painted on the actin are the most frequent binding sites of myosin V in green and the most frequent binding sites of myosin VI (stepping against 1.7-pN load) in red, as determined by their stepping distributions (Figure 8.4E and 8.4F) (see color insert following page 204). As the cartoon illustrates, myosin V is able to walk along the pseudo-repeat of the actin filament and does not bind to actin monomers promiscuously, predominantly finding either its most frequent binding site or a binding site directly adjacent to it. Myosin VI, on the other hand, binds to a wide range of actin monomers and finds binding sites toward both the pointed and barbed ends of the filament. From this illustration, it appears that the myosin VI binding sites lie along a surface of the actin filament. Looking down the actin facing the pointed-end, this surface is rotated to the left relative to the location of the bound myosin head.

A likely explanation for this stepping behavior is that the actin-dissociated myosin VI head undergoes a diffusive search for its binding site at the end of an extended and flexible part of the motor. Through such a diffusive search, the head could interact with and bind to various actin monomers surrounding its most frequent binding site, resulting in a wide step distribution. Also, a diffusing head would occasionally find binding sites toward the barbed end of the motor, explaining the frequent backward stepping.

The region of the motor hypothesized to be in this extended and flexible conformation consists of the first N-terminal ~70 amino acids of the myosin VI tail, adjacent to the two light chains, denoted the *proximal tail domain*. Using the Paircoil program,[94] which relates the relative propensity for an amino acid sequence to form a coiled-coil, this region shows a low propensity to form a coiled-coil for eight myosin VI variants from seven organisms. This region has also been observed to be susceptible to proteolysis, suggesting it exists in a relatively unstructured conformation (B. Rami, B. Spink, unpublished observations).

Motivated by this model and for reasons discussed in detail below, Altman et al. explored the effects of applied forces on the stepping of myosin VI.[93]

8.7.2 Load-Affected Stepping Behavior

Using force-clamp feedback to control optical trap position, we can observe the stepping of a motor against a constant force. Thus, by studying the motor's stepping against different loads, we can probe its force-dependent behavior. Understanding this behavior then requires a theoretical framework to interpret such force-dependent

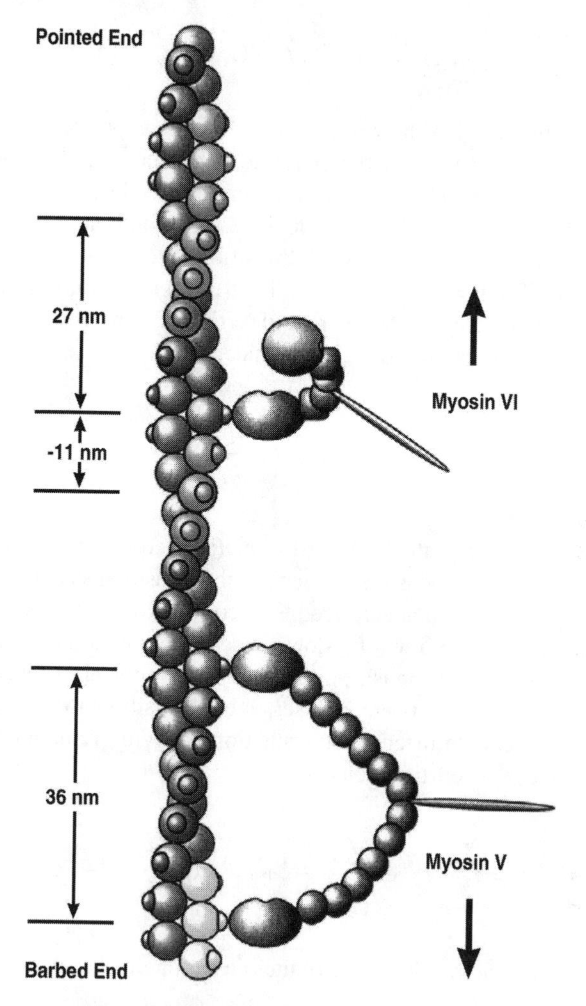

FIGURE 8.8 Model of myosins V and VI stepping on actin. (See color insert following page 204.) Knobs on the actin indicate stereospecific myosin-binding sites (although the actual binding site is located between two actin-monomers, binding is shown on only one monomer for simplicity). Both motors are bound to actin subunits facing directly right. The actin subunits in green indicate the preferred binding sites of the lead myosin V head while it is searching for its binding site. The actin subunits in red indicate the preferred binding sites for the actin-dissociated myosin VI head, with the spread from red to blue reflecting the width of the empirical step size histogram. (From Rock, R.S., et al. (2001). Myosin VI is a processive motor with a large step size. *Proc. Natl. Acad. Sci. U.S.A.*, 98, 13655–13659. With permission.)

effects. We consider again the kinetic scheme represented by Equation 8.5. We have already discussed how to calculate the theoretical probability density $P(t;k_{\pm1},k_{\pm2},\ldots k_{\pm N})$ for a general kinetic scheme. We now focus on another property of a kinetic stepping model: the theoretical mean dwell time, $\bar{\tau}(k_{\pm1},k_{\pm2},\ldots k_{\pm N})$. The expected mean dwell can be calculated from the probability density through the relation

$$\overline{\tau}(k_{\pm 1}, k_{\pm 2}, \ldots k_{\pm N}) = \int_0^{+\infty} P(t; k_{\pm 1}, k_{\pm 2}, \ldots k_{\pm N}) t dt \tag{8.7}$$

We now ask the question: how does load affect the expected mean dwell time?

The effect of a load is to make the motor do either more or less work to undergo its mechanical transitions. Thus, load has the effect of selectively perturbing these transitions in a stepping cycle. For example, if a 2-pN load is applied in the direction opposite to that of a 2-nm mechanical transition, the transition will require an additional ~2 pN × 2 nm = 4 × 10^{-21} J ≈ $k_B T$ (at room temperature) of work. Using an Arrhenius-Eyring formulation, a rate describing a mechanical transition is expected to have the following load dependence:

$$k = k(0)\exp\left(-\frac{F\delta}{k_B T}\right) \tag{8.8}$$

where F is the magnitude of the backward force (applied in the direction opposite to that of the mechanical step), δ is the distance to the transition state of the mechanical transition, and *k(0)* is the rate at zero load.[95] According to this relation, a mechanical rate with a positive/negative δ will be slowed/accelerated by a backward load.

We consider mechanical transitions that are approximately irreversible. For the class of kinetic schemes consisting of a series of reversible biochemical transitions and a series of irreversible mechanical transitions, solving Equation 8.7 yields the load-dependent mean dwell time relation

$$\tau = \tau_b + \tau_m^{(1)} \exp\left(\frac{F\delta_1}{k_B T}\right) + \tau_m^{(2)} \exp\left(\frac{F\delta_2}{k_B T}\right) + \ldots \tag{8.9}$$

where the ith mechanical transition contributes an exponential term with characteristic distance δ_i . $\tau_b / (\tau_b + \tau_m^{(1)} + \tau_m^{(2)} + \ldots)$ and $\tau_m^{(i)} / (\tau_b + \tau_m^{(1)} + \tau_m^{(2)} + \ldots)$ represent the fraction of the motor's chemomechanical cycle that it spends undergoing its biochemical and its ith mechanical transition, respectively, and δ_i represents the distance corresponding to the ith mechanical transition. If the stepping cycle consists of a single mechanical transition, or if one of the mechanical transitions is either much slower than the others (it has a large $\tau_m^{(i)}$) or is much more sensitive to load (it has a large δ_i), Equation 8.9 reduces to the simpler form:

$$\overline{\tau} = \tau_b + \tau_m \exp\left(\frac{F\delta}{k_B T}\right) \tag{8.10}$$

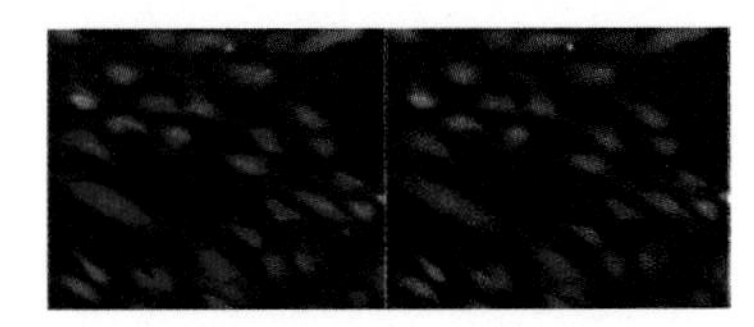

Figure 6.10 Image of the intracellular Ca^{2+} level in cells before (left) and after (right) flow exposure. The color reflects the $[Ca^{2+}]_i$ with the red indicating a high level and the green for a low level.

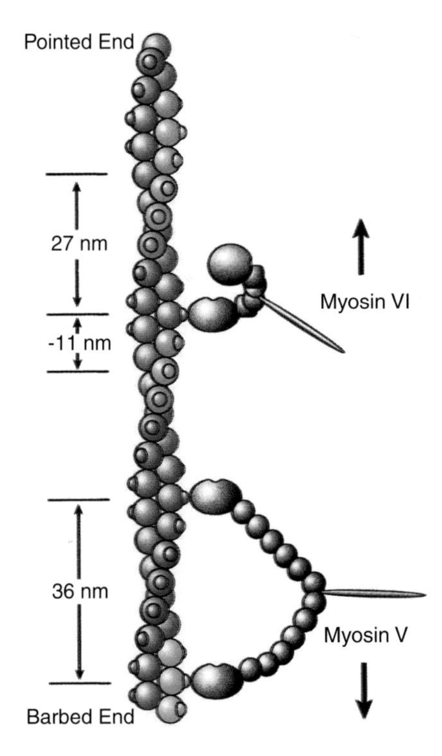

Figure 8.8 Model of myosins V and VI stepping on actin. Knobs on the actin indicate stereospecific myosin-binding sites (although the actual binding site is located between two actin-monomers, binding is shown on only one monomer for simplicity). Both motors are bound to actin subunits facing directly right. The actin subunits in green indicate the preferred binding sites of the lead myosin V head while it is searching for its binding site. The actin subunits in red indicate the preferred binding sites for the actin-dissociated myosin VI head, with the spread from red to blue reflecting the width of the empirical step size histogram.

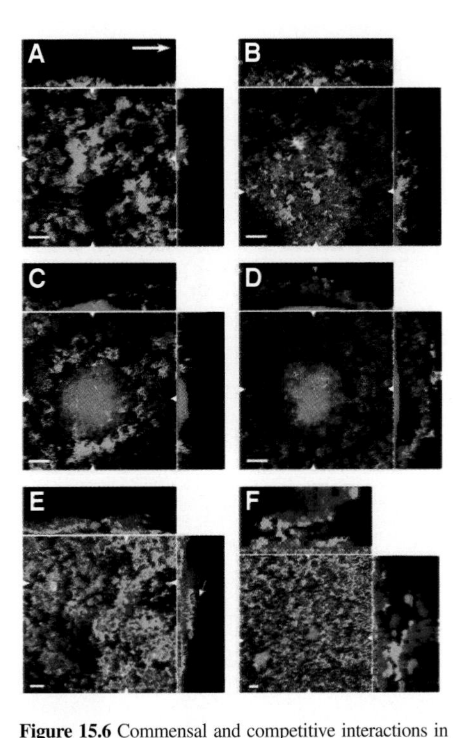

Figure 15.6 Commensal and competitive interactions in a two-member biofilm consortium. SCLM micrographs show the dynamically changing structural relationships between *Acinetobacter* strain C6 and *P. putida* R1 cells with differential growth activities in a benzyl alcohol containing biofilm. *P. putida* R1 and *Acinetobacter* strain C6 were hybridized with PP986 labeled with CY5 (blue) and ACN449 labeled with CY3 (red), respectively. The actively-growing *P. putida* R1 cells can be detected as cells emitting green fluorescence due to the *rrnBP*1-*gfp*^{AGA} fusion inserted in the chromosome of *P. putida* R1. These cells appear as cyan due to the combination of green (GFP) and blue (hybridization). Panels A and B are representative of the biofilm structures observed at day 1 and 2. Panels C and D are examples of the large *Acinetobacter* strain C6 microcolony that developed after 3 days (C) and which later was overgrown by *P. putida* R1 (D). Panels E and F are examples of *Acinetobacter* strain C6 microcolonies (arrow) that were established in the upper part of *P. putida* R1 cell clusters after 2 to 3 days (E), which resulted in production of large macrostructures of associated *P. putida* R1 and *Acinetobacter* strain C6 cells (F). Shown to the right and above the x-y plots are vertical sections through the biofilm collected at the positions indicated by the white triangles. The arrow indicates the direction of flow. Bars, 20 μm. (From Christensen, B.B. et al., *Appl. Environ. Microbiol.*, 2002, 68(5), 2495–2502. With permission.)

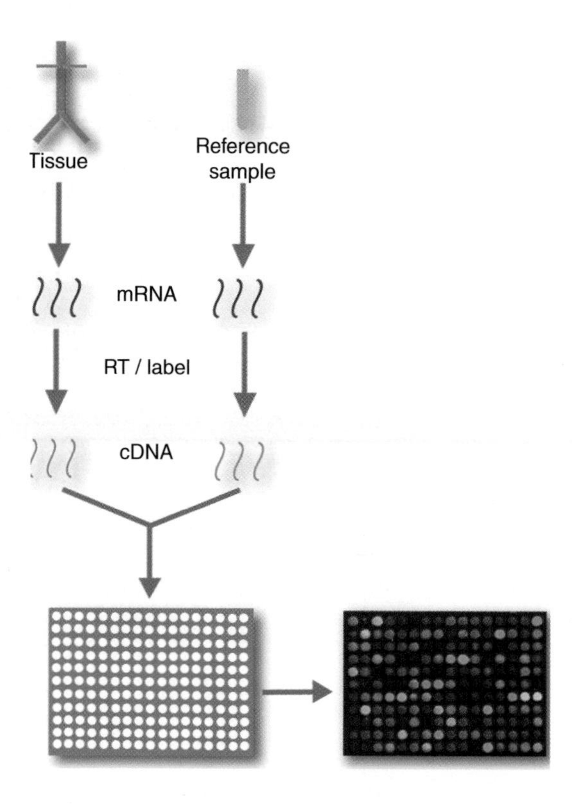

Figure 17.1. Vascular gene expression analysis using a DNA microarray. Total mRNA is first extracted from the tissue of interest (aorta in this example) and then reverse transcribed (RT) in the presence of red fluorescently labeled nucleotide precursors to cDNA. A reference sample is prepared similarly with a green fluorescent tag. The two fluorescently labeled cDNA populations are then mixed and competitively hybridized with a DNA microarray where each vascular gene is represented as a distinct spot on the microarray. The fluorescently labeled cDNA sequences representing an expressed transcript hybridize with their corresponding gene sequence target on the microarray. Relative abundance of transcripts is then determined as the ratio of "red" to "green" fluorescence measured for each gene on the array. In this manner, massively parallel quantitative evaluation of gene expression of a tissue of interest can be performed.

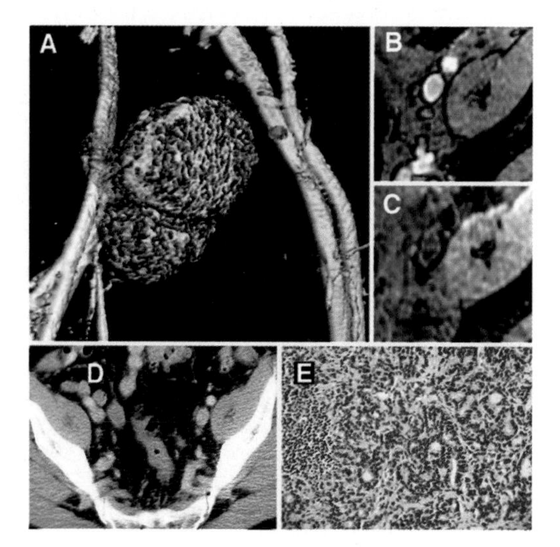

Figure 18.5. Three-dimensional reconstruction of pelvic lymph nodes (A), conventional MRI (B), MRI with lymphotrpoic superparamagnetic nanoparticles (C), abdominal CT (D), and histology (E) of lymph node metastases with prostate cancer. Panel A shows 3-D reconstruction of prostate, iliac vessels and metastatic (red) and normal (green) lymph nodes. Panel B shows that conventional MRI signal intensity is identical in positive and negative lymph nodes. Panel C shows increased activity in the metastatic nodes (arrow). Panel D shows that CT cannot discriminate between positive and normal lymph nodes. Panel E shows the histology of prostate cancer that has metastasized to a lymph node. (Reprinted from Harisingham MG and others. Noninvasive detection of clinically occult lymph-node metastases in prostate cancer. *N Eng J Med* 2003; 348,: p. 2497, Figure 3. With permission.)

Wang et al. showed that Equation 8.10 also describes the load-affected mean dwell for two other classes of kinetic models.[96] One class involves kinetic schemes consisting of reversible transitions in which one of the states in the cycle is a composite of two sequential substates in rapid and load-dependent equilibrium. The other class involves stepping pathways consisting of reversible transitions in which there exists a load-dependent transition to an off-pathway state. For both these schemes, the load-affected mean dwell is expected to obey Equation 8.10, though the meaning of the δ term is slightly changed. For the first class, δ signifies the distance between the two substates, and for the latter, it is the physical distance between the on-pathway and off-pathway states.

For myosin V, Mehta et al. observed that mean dwell shows little dependence on load at low ATP and obeys Equation 8.10 at saturating ATP (Figure 8.9A).[77] They hypothesized that, at saturating ATP, a transition that is slower than ATP binding is associated with a load-dependent mechanical transition. At low ATP, this mechanical transition is much more rapid than the slowed ATP-binding rate, which thus obscures the load affected behavior. Fitting Equation 8.10 to the saturating ATP mean dwell data yields $\tau_b/\tau_m{>}100$, indicating that the mechanical transition is very rapid relative to the rest of the chemomechanical cycle, and a value for δ of 10 to 15 nm, which may correspond to the 20- to 25-nm stroke of the motor (or perhaps a portion of it).

For myosin VI, we hypothesized that the stepping cycle could involve a diffusive search of the actin-dissociated head at the end of a flexible, extended domain. We thus predict that effects of load on myosin VI stepping may differentiate it from myosin V, which steps with a rapid stroking of a relatively rigid lever arm.

8.7.3 Effects of Load on Myosin VI: Formulating a Model of Stepping

Using a three-bead optical trap assay with force-clamp feedback, the mean step size and the dwell time of myosin VI were measured at various forces against its stepping.[93] The motor is able to step against increasing loads until forces of ~2.1 pN, denoted its *stall force*. The motor's step size in both the forward and backward direction is largely unaffected by load, though the forward step gradually decreases ~8 nm as load increases to stall. The mean dwell time of the motor's stepping remains fairly impervious to load for forces up to 2 pN, and at loads greater than 2 pN, it increases dramatically with increasing force (Figure 8.9B).[93]

Can these results be reconciled with a diffusive stepping model? The duration of a diffusive search is sensitive to load,[97] so an important test is to see whether the model can account for a mean dwell time that is load independent for forces up to 2 pN. The time required for a spherical object to diffuse a distance x_0 in one dimension is given by the first-passage relation:

$$t = 2 \times \frac{x_o^{\,2}}{2D} \times \left(\frac{k_B T}{F x_o} \right)^2 \times \left\{ \exp\left(\frac{F x_o}{k_B T} \right) - 1 - \frac{F x_o}{k_B T} \right\} \tag{8.11}$$

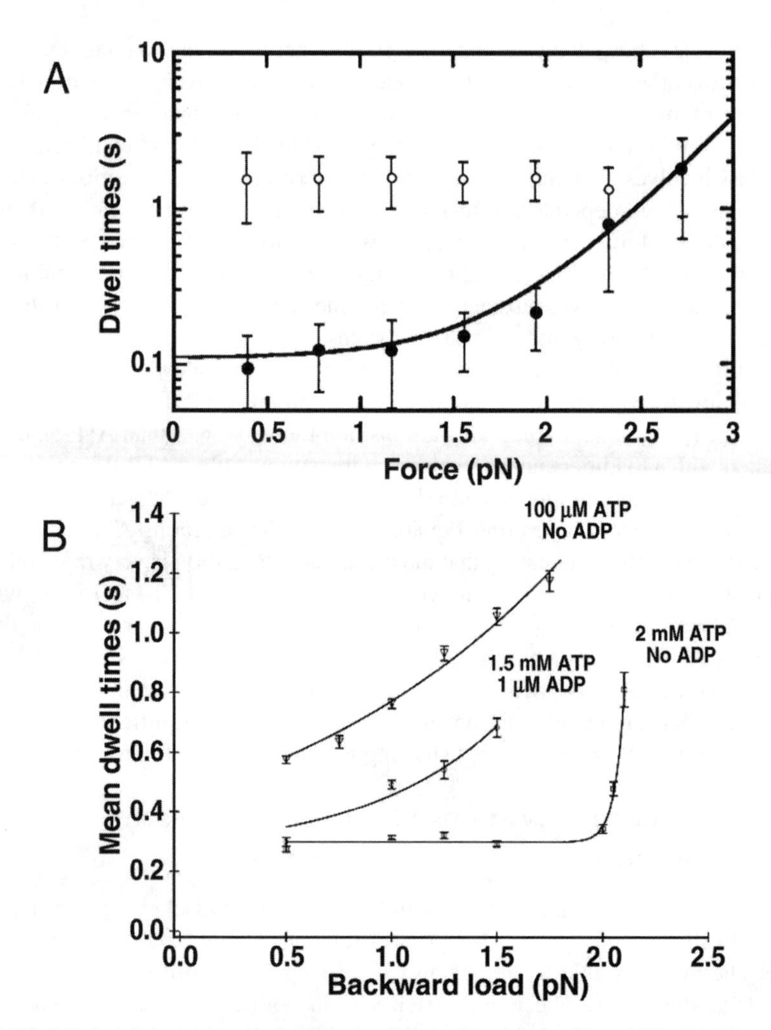

FIGURE 8.9 Load-dependence of mean dwell time for myosins V and VI. (A) Mean dwell as a function of load for myosin V in the presence of 2 mM ATP (filled circles) and 1 μM ATP (open circles). Mean dwells and errors on these values are calculated by the maximum likelihood method, as described by Colquhoun and Sigworth.[101] The 2 mM ATP data is fit to Equation 8.10 (solid line), yielding $\tau_b/\tau_m > 100$ and $\delta \sim 10$ to 15 nm. (Adapted from Mehta, A.D., et al. (1999). Myosin-V is a processive actin-based motor. *Nature*, 400, 590–593.) (B) Mean dwell as a function of load for myosin VI in the presence of 2 mM ATP (squares), 1.5 mM ATP and 1 μM ADP (circles), and 100 μM ATP (triangles). Errors on the mean dwell are calculated using a bootstrap method, as described by Altman et al.[93] Equation 8.10 was fit to the three data sets (solid lines), yielding $\tau_b = 0.30$ s, $\tau_m = 0.28 * 10^{-20}$ s, and $\delta = 91$ nm for the 2 mM ATP data, $\tau_b = 0.16$ s, $\tau_m = 0.30$ s, and $\delta = 2.9$ nm for the 100 μM ATP data, and $\tau_b = 0.26$ s, $\tau_m = 0.04$ s, and $\delta = 6.3$ nm for the 2 mM ATP and 1 μM ADP data. (Adapted from Altman, D., H.L. Sweeney, and J.A. Spudich (2004). The mechanism of myosin VI translocation and its load-induced anchoring. *Cell,* 116, 737–749.)

where D is the diffusion constant for the object, which depends on the radius of the object and the viscosity of the medium in which the diffusion occurs, F is the magnitude of the external force, and k_BT is the thermal energy.[97]

In the following calculations, it is assumed that the radius of the myosin VI head that is diffusively finding its next binding site is comparable to that of a myosin V head (~40 Å), and that, in the optical trap assays, it is diffusing in water at room temperature. As described above, the working stroke is ~12 nm while the step size is ~30 nm. The motor thus must undergo a diffusive search of mean distance ~20 nm (though, perhaps as large as 30–35 nm to account for the broad stepping distribution). At the same time, this diffusive search is expected to occur against loads of up to 2 pN without a slowing of the stepping kinetics. Equation 8.11 predicts that a 20-nm diffusive search occurs in 3 μs for the unloaded motor and in 1 ms for a motor experiencing a 2-pN load. Because these durations are much more rapid than the 0.3-s mean dwell of the motor, a load-sensitive diffusive search need not be inconsistent with the observed load-insensitive mean dwell if the stepping cycle involves a 20-nm diffusion.

The model does not describe the data so well, however, when considering the larger observed steps of the motor that require longer diffusive searches. For example, a 30-nm diffusive search takes 6 μs for the unloaded motor, but is slowed to 130 ms for a diffusing head experiencing a 2-pN load. Thus, we would predict that a motor experiencing 2 pN of load should either take fewer 40-nm steps or exhibit a noticeably increased mean dwell time, neither of which is observed.

The stepping model must also account for the load dependence of the mean dwell data. Fitting Equation 8.10 to this data yields $\tau_b > \tau_m$ and $\delta \sim 90$ nm (Figure 8.9B). Though a value of δ larger than the step size seems unphysical, Altman et al. conjecture that the value is only approximate, due to the fact that only two data points describe the rapid rise in dwell at high loads.[93] What is most significant, however, is that the fit values of τ_b and τ_m indicate a mechanical transition that is extremely rapid relative to the rest of the chemomechanical cycle, likely more rapid than a diffusive search.

To account for this behavior, Altman et al. hypothesized that, while the actin-dissociated head is undergoing its diffusive search, it is shielded from load, presumably through the actin-bound head adopting a rigid conformation.[93] This modified stepping model thus predicts that a myosin VI arm undergoes a conformational change between a flexible and elongated state (when the head is searching for its next binding site) and a rigid conformation (when the head is bound to the actin). Again, the proximal tail domain is hypothesized to be in an extended and flexible conformation to account for the large, diffusive stepping. According to the modified stepping scheme, however, this region is also expected to adopt a rigid conformation, perhaps by folding itself into a rigid structure or by binding to another part of the motor or the actin.

An example of a possible stepping model is shown in Figure 8.10. In this cartoon, the proximal tail domain alternates between an extended, flexible conformation and a rigid conformation. When it is in its flexible conformation, this region is drawn as a thin, curved line, and when it is rigid, it is not shown in the cartoon. The simpler

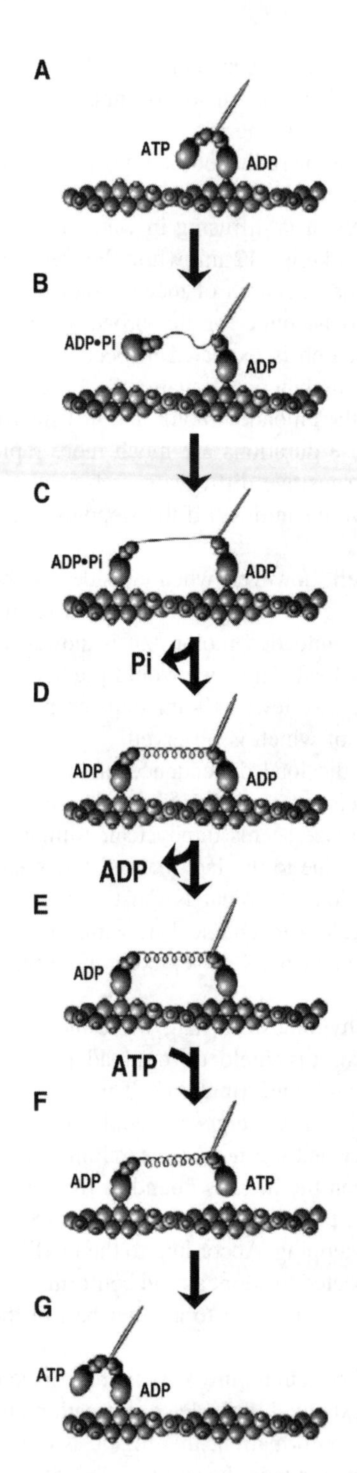

FIGURE 8.10 Model of myosin VI stepping. This stepping model is discussed in detail in the text. In this cartoon representation, the first ~70 residues of the tail just adjacent to the light chain binding domains (referred to as the proximal tail domain) alternate between a rigid and flexible conformation depending on the state of the catalytic head. Each myosin VI monomer within the homodimer is represented as an oval, describing the catalytic domain, a square, representing the ~50-residue insert unique to this myosin, and a circle, representing the light chain. The coiled-coil tail region joining the two heads is represented by a rod. When the proximal tail domain is flexible and extended, this region is represented by a thin line. When this domain is rigid, which is hypothesized to result either from the domain folding itself into a compact structure or docking to part of the motor or actin, it is not shown. Knobs on the actin filament indicate stereospecific myosin-binding sites. (From Altman, D., H.L. Sweeney, and J.A. Spudich (2004). The mechanism of myosin VI translocation and its load-induced anchoring. *Cell,* 116, 737–749. With permission.)

putative stepping scheme, in which this region of the tail remains in a flexible conformation throughout stepping, can be represented by this same cartoon, except the proximal tail domain for each heavy chain remains flexible and extended throughout.

In this stepping scheme, the motor starts in state (A) with one head bound to actin with a bound ADP and the other head dissociated from the filament with bound ATP. The proximal tail domain of the actin-dissociated head must become flexible and elongated to allow for the large diffusive step, while the other proximal tail domain adopts its rigid conformation to shield the diffusing head from load. The actin-dissociated head hydrolyzes its ATP (B) and finds its next actin-binding site (C), becoming the new lead head. Because the proximal domain of a myosin VI head that is actin- and ADP-bound is hypothesized to be in its rigid conformation, the proximal domain of the lead head attempts to transition back to its rigid conformation upon releasing phosphate (D). It cannot, however, since the motor is constrained across the length of the filament. It is thus represented by a taut spring, and cannot adopt its rigid conformation until the rear head releases from the actin.

The next transition, the release of ADP from the rear head, is the rate-limiting transition in the stepping cycle.[93,98] Thus, the motor predominantly dwells in state (D), waiting for ADP to dissociate. The rear head then releases its nucleotide (E) and binds an ATP (F), causing it to dissociate from actin. At this point, the proximal tail domain of the lead head can adopt its rigid conformation (G). This transition, from the flexible to rigid conformation, is predicted to result in a rapid 30 nm translocation of the motor. The motor has thus returned to its original state, having translocated ~30 nm along the actin.

8.7.4 Load Effects on Nucleotide Binding: Formulating a Model of Anchoring

Altman et al. also explored the effect of load on mean dwell at subsaturating ATP conditions and at saturating ATP conditions in the presence of 1 μM ADP (Figure 8.9B).[93] At subsaturating ATP, the load dependence of the dwell time differs greatly from saturating ATP conditions. The mean dwell slows with increasing load at much lower forces, and the rise in dwell time is much more gradual relative to the saturating ATP data. Fitting Equation 8.10 to this data yields a smaller δ of ~3 nm and an increase in τ_m / τ_b. These fits reveal that the motor is now spending more of its chemomechanical cycle near a load-dependent mechanical transition, and that this transition is unique from the more load-sensitive transition apparent in the saturating ATP data. Presumably, ATP binding is associated with this mechanical transition. At saturating ATP, ATP binding is very rapid, and load effects on this rate are obscured by the rate-limiting ADP release transition. At subsaturating ATP, however, the binding rate slows, and a load-dependent change in this rate is apparent in the mean dwell data.

In the presence of 1 μM ADP, the dependence of the mean dwell on load is also changed from the behavior observed at saturating ATP. In the presence of 1 μM ADP, the mean dwell increases with increasing load at much lower forces and increases much less rapidly than is seen in the presence of saturating ATP (Figure

8.9B). Again, the motor is apparently spending a greater portion of its stepping cycle near a load-dependent transition, and this mechanical transition is unique from the one that is apparent in the saturating ATP data, this time with an associated δ of ~6 nm. Since this load-affected transition is only observed in the presence of ADP, Altman et al. hypothesized that this transition is associated with rebinding of ADP to the motor.[93] In the absence of ADP, the apparent rate of ADP binding is zero, and load effects on this second-order binding rate are not apparent. In the presence of ADP, on the other hand, load effects on this rate are apparent in the mean dwell time data. In the presence of sufficient load, binding of ADP to the motor is now able to compete with binding of ATP, resulting in slowing of stepping and an increase in the mean dwell.

The load-dependent ATP-binding and ADP-binding rates both have similar associated δ-values, and these values are both on the same order as the 8-nm decrease in the observed forward step size. This led Altman et al. to conjecture that the applied load may be deforming the rear actin-bound myosin VI head.[93] This deformation pulls the head backward ~8 nm, causing the reduction in observed mean step. According to this model, this deformation also distorts the catalytic head, resulting in a change in the nucleotide-binding properties of the motor.

The load-dependence of nucleotide binding may account for *in vivo* anchoring behavior of myosin VI. As load increases, the reduced rate of ATP binding and increased rate of ADP binding both tend to slow the motor's stepping. Sufficient loads should cause the motor to stall while it remains bound to the actin, causing it to anchor to the filament anything that is bound at its tail. Altman et al. demonstrated that, at physiological nucleotide conditions, it is specifically the increased rate of ADP binding that is most relevant to such anchoring function.[93]

From these results, one can imagine myosin VI playing a role as a tension sensor in a cell. For example, in sensory epithelia of inner ear hair cells, the motor may bind to the membrane between stereocilia.[99] It can then walk along actin bundles in the stereocilia, which are positioned with their pointed ends facing toward the stereocilia roots where they are moored to the body of the hair cell.[100] The motor will thus pull the membrane between the stereocilia taut, at which point a load against the motor's stepping will develop and cause the motor to anchor the membrane to the actin. If the membrane becomes slack again (for example, if membrane is added to the cell), the load will be reduced and the motor will resume walking until the backward force is reestablished. In this way, the motor can maintain a particular tension in this membrane, a function that may be essential to maintaining the structural integrity of the stereocilia and the surface membrane topology of the hair cell.[99] Thus, myosin VI serves as a tension sensor in the cell, a property that we think will likely describe the behavior of a whole host of cellular components involved in maintaining the dynamic city plan of a cell.

8.8 CONCLUSIONS

This chapter is not meant to be an extensive discussion of single-motor studies. Such a chapter would have necessarily touched upon studies of F1-ATPase, kinesin, dynein, and RNA polymerase, among others, all of which have played a role in innovation of single-molecule studies. In fact, many of the analysis tools discussed here were developed from studies of ion channels,[101] so these would also have to be included in a thorough discussion of single-motor work. Nor is this chapter meant to comprehensively discuss all single-molecule work on myosins. A complete discussion would have included electron microscopy, crystallization work, and a myriad of elegant fluorescence studies.

Nonetheless, what we have tried to document here is the development of a tool that has been invaluable to understanding myosin motors. Ten years after the first myosin displacement was directly observed by an optical trap transducer, optical trap assays have been innovated and evolved, and now allow us to probe the detailed myosin stepping mechanism. This has been a remarkable feat, especially considering that the rudimentary output of these experiments is the position of a bead tethered to a myosin or to its actin track. Using this sparse information along with well-conceived motor constructs, experimental setups, and analysis methods, much progress in the myosin field has been made.

In the future, optical trap work will continue to be an important part of new and innovative myosin studies. An important question that remains unanswered is how the two heads of processive myosins, such as myosin V and VI, communicate with each other. Some means of communication is necessary to explain the numerous steps these motors can take before dissociating from an actin filament. Optical trap studies of myosin V have begun to elucidate this mechanism, though contending models of this cooperativity exist.

An innovation that may be part of the next generation of single-molecule studies is the development of sophisticated equipment capable of combining optical traps with other single-molecule detection methods. A likely example is the combination of optical trap transduction with fluorescence microscopy. While optical trap studies allow for the observation of large, mechanical translocations of the motor, fluorescence experiments can be used to detect other events, such as binding of a fluorescently labeled ligand to the motor. Ishijima et al. performed this type of experiment using myosin II and fluorescently labeled nucleotide.[74] They were able to temporally correlate binding events to mechanical motion of the motor. However, this experiment required that the optical trap transduction and fluorescence excitation occur at different locations in their experiment, a restriction that severely limits the experimental geometries capable of being explored by this assay.

The difficulty in using both fluorescence and optical traps at the same time in the same location lies in the fact that the trap laser emits a huge flux of infrared photons, many orders of magnitude greater than the flux of a single fluorophore. This can result in both a large background signal as well as a reduced fluorescence lifetime. Recently, however, Lang et al. demonstrated that, with the proper choice of fluorophore and excitation wavelength, this problem can be overcome.[102] Thus, combined technologies such as this may soon yield some of the next significant advances in single-motor studies.

REFERENCES

1. Berg, J.S., B.C. Powell, and R.E. Cheney (2001). A millennial myosin census. *Mol. Biol. Cell.,* 12, 780–794.
2. Sellers, J.R. and H.V. Goodson (1995). Motor proteins 2: myosin. *Protein Profile,* 2, 1323–1423.
3. Lymn, R.W., and Taylor, E. W. (1971). Mechanism of adenosine triphosphate hydrolysis by actomyosin. *Biochemistry,* 10, 4617–4624.
4. Purcell, T.J., et al. (2002). Role of the lever arm in the processive stepping of myosin V. *Proc. Natl. Acad. Sci. U.S.A.,* 99, 14159–14164.
5. Sakamoto, T., et al. (2003). Neck length and processivity of myosin V. *J. Biol. Chem.,* 2003, 278(31), 29201–29207.
6. Spudich, J.A. (2001). The myosin swinging cross-bridge model. *Nat. Rev. Mol. Cell. Biol.,* 2, 387–392.
7. Ruegg, C., et al. (2002). Molecular motors: force and movement generated by single myosin II molecules. *News Physiol. Sci.,* 17, 213–218.
8. Wu, X., et al. (1998). Visualization of melanosome dynamics within wild-type and dilute melanocytes suggests a paradigm for myosin V function *in vivo. J. Cell Biol.,* 143, 1899–1918.
9. Cheney, R.E., et al. (1993). Brain myosin-V is a two-headed unconventional myosin with motor activity. *Cell,* 75, 13–23.
10. Reck-Peterson, S.L., P.J. Novick, and M.S. Mooseker (1999). The tail of a yeast class V myosin, myo2p, functions as a localization domain. *Mol. Biol. Cell,* 10, 1001–1017.
11. Rogers, S.L., and V.I. Gelfand (1998). Myosin cooperates with microtubule motors during organelle transport in melanophores. *Curr. Biol.,* 8, 161–164.
12. Mermall, V., P.L. Post, and M.S. Mooseker (1998). Unconventional myosins in cell movement, membrane traffic, and signal transduction. *Science,* 279, 527–533.
13. Tabb, J.S., et al. (1998). Transport of ER vesicles on actin filaments in neurons by myosin V. *J. Cell Sci.,* 111, 3221–3234.
14. Evans, L.L., et al. (1998). Vesicle-associated brain myosin-V can be activated to catalyze actin-based transport. *J. Cell Sci.,* 111, 2055–2066.
15. Hasson, T. and M.S. Mooseker (1994). Porcine myosin-VI: characterization of a new mammalian unconventional myosin. *J. Cell Biol.,* 127, 425–440.
16. Bahloul, A., et al. (2004). The unique insert in myosin VI is a structural calcium-calmodulin binding site. *Proc. Natl. Acad Sci. U.S.A*., 2004, 6; 101(14), 4787–4792.
17. Buss, F., J.P. Luzio, and J. Kendrick-Jones (2001). Myosin VI, a new force in clathrin mediated endocytosis. *FEBS Lett.,* 508, 295–299.
18. Buss, F., J.P. Luzio, and J. Kendrick-Jones (2002). Myosin VI, an actin motor for membrane traffic and cell migration. *Traffic,* 3, 851–858.
19. Fabrizio, J.J., et al. (1998). Genetic dissection of sperm individualization in *Drosophila melanogaster. Development*, 125, 1833–1843.
20. Rogat, A.D. and K.G. Miller (2002). A role for myosin VI in actin dynamics at sites of membrane remodeling during *Drosophila* spermatogenesis. *J. Cell Sci.,* 115, 4855–4865.
21. Hasson, T., et al. (1997). Unconventional myosins in inner-ear sensory epithelia. *J. Cell Biol.,* 137, 1287–1307.
22. Wells, A.L., Lin, A.W., Chen, L.Q., Safer, D., Cain, S.M., Hasson, T., Carragher, B.O., Milligan, R.A., and Sweeney, H.L. (1999). Myosin VI is an actin-based motor that moves backwards. *Nature,* 401, 505–508.

23. Huxley, H.E. (1957). The double array of filaments in cross-striated muscle. *J. Biophys. Biochem. Cytol.,* 3, 631–648.
24. Huxley, H.E. (1953). Electron microscope studies of the organisation of the filaments in striated muscle. *Biochim. Biophys. Acta,* 12, 387–394.
25. Huxley, H.E. (1969). The mechanism of muscular contraction. *Science,* 164, 1356–1365.
26. Szent-Gyorgyi, A.G. (1953). Meromyosins, the subunits of myosin. *Arch. Biochem. Biophys.,* 42, 305–320.
27. Huxley, H.E. and W. Brown (1967). The low-angle x-ray diagram of vertebrate striated muscle and its behaviour during contraction and rigor. *J. Mol. Biol.,* 30, 383–434.
28. Moore, P.B., H.E. Huxley, and D.J. DeRosier (1970). Three-dimensional reconstruction of F-actin, thin filaments and decorated thin filaments. *J. Mol. Biol.,* 50, 279–295.
29. Yount, R.G., D. Ojala, and D. Babcock (1971). Interaction of P–N–P and P–C–P analogs of adenosine triphosphate with heavy meromyosin, myosin, and actomyosin. *Biochemistry,* 10, 2490–2496.
30. Goody, R.S. and F. Eckstein (1971). Thiophosphate analogs of nucleoside di- and triphosphates. *J. Amer. Chem. Soc.,* 93, 6252–6257.
31. Goody, R.S. and W. Hofmann (1980). Stereochemical aspects of the interaction of myosin and actomyosin with nucleotides. *J. Muscle Res. Cell Motil.,* 1, 101–115.
32. Huxley, H.E., et al. (1982). Time-resolved X-ray diffraction studies of the myosin layer-line reflections during muscle contraction. *J. Mol. Biol.,* 158, 637–684.
33. Sheetz, M.P. and J.A. Spudich (1983). Movement of myosin-coated fluorescent beads on actin cables *in vitro*. *Nature,* 303, 31–35.
34. Spudich, J.A., S.J. Kron, and M.P. Sheetz (1985). Movement of myosin-coated beads on oriented filaments reconstituted from purified actin. *Nature,* 315, 584–586.
35. Yanagida, T., et al. (1984). Direct observation of motion of single F-actin filaments in the presence of myosin. *Nature,* 307, 58–60.
36. Kron, S.J. and J.A. Spudich (1986). Fluorescent actin filaments move on myosin fixed to a glass surface. *Proc. Natl. Acad. Sci. U.S.A.,* 83, 6272–6276.
37. Huxley, H.E. (1990). Sliding filaments and molecular motile systems. *J. Biol. Chem.,* 265, 8347–8350.
38. Kishino, A. and T. Yanagida (1988). Force measurements by micromanipulation of a single actin filament by glass needles. *Nature,* 334, 74–76.
39. Toyoshima, Y.Y., et al. (1987). Myosin subfragment-1 is sufficient to move actin filaments *in vitro*. *Nature,* 328, 536–539.
40. Harada, Y. and T. Yanagida (1988). Direct observation of molecular motility by light microscopy. *Cell Motil. Cytoskeleton,* 10, 71–76.
41. Harada, Y., et al. (1990). Mechanochemical coupling in actomyosin energy transduction studied by *in vitro* movement assay. *J. Mol. Biol.,* 216, 49–68.
42. Toyoshima, Y.Y., S.J. Kron, and J.A. Spudich (1990). The myosin step size: measurement of the unit displacement per ATP hydrolyzed in an *in vitro* assay. *Proc. Natl. Acad. Sci. U.S.A.,* 87, 7130–7134.
43. Uyeda, T.Q., et al. (1991). Quantized velocities at low myosin densities in an *in vitro* motility assay. *Nature,* 352, 307–311.
44. Uyeda, T.P.Q., S.J. Kron, and J.A. Spudich (1990). Myosin step size estimation from slow sliding movement of actin over low densities of heavy meromyosin. *J. Mol. Biol.,* 214, 699–710.
45. Ishijima, A., et al. (1991). Sub-piconewton force fluctuations of actomyosin *in vitro*. *Nature,* 352, 301–306.

46. Simmons, R.M., et al. (1993). Force on single actin filaments in a motility assay measured with an optical trap. *Adv. Exp. Med. Biol.,* 332, 331–336; discussion 336–337.
47. Knight, A.E. and J.E. Molloy (2000). Muscle, myosin and single molecules. *Essays Biochem.,* 35, 43–59.
48. Kitamura, K., et al. (1999). A single myosin head moves along an actin filament with regular steps of 5.3 nanometres. *Nature,* 397, 129–134.
49. Chu, S., et al. (1986). Experimental observation of optically trapped atoms. *Phys. Review Rev. Letters,* 57, 314–317.
50. Simmons, R.M., et al. (1996). Quantitative measurements of force and displacement using an optical trap. *Biophys. J.,* 70, 1813–1822.
51. Sheetz, M., ed. (1998). Laser tweezers in cell biology. *Methods Cell Biol.,* Vol. 55, Elsevier, New York, 228 pp.
52. Sterba, R.E. and M.P. Sheetz (1998). Basic laser tweezers. *Methods Cell Biol.,* 55, 29–41.
53. Rice, S., T.J. Purcell, and J.A. Spudich (2003). Building and using optical traps to study properties of molecular motors. *Methods Enzymol.,* 361, 112–133.
54. Molloy, J.E., et al. (1995). Movement and force produced by a single myosin head. *Nature,* 378, 209–212.
55. Finer, J.T., A.D. Mehta, and J.A. Spudich (1995). Characterization of single actin-myosin interactions. *Biophys. J.,* 68, 291S–296S; discussion 296S–297S.
56. Veigel, C., et al. (1998). The stiffness of rabbit skeletal actomyosin cross-bridges determined with an optical tweezers transducer. *Biophys. J.,* 75, 1424–1438.
57. Mehta, A.D., J.T. Finer, and J.A. Spudich (1997). Detection of single-molecule interactions using correlated thermal diffusion. *Proc. Natl. Acad. Sci. U.S.A.,* 94, 7927–7931.
58. Nishizaka, T., et al. (1995). Unbinding force of a single motor molecule of muscle measured using optical tweezers. *Nature,* 377, 251–254.
59. Block, S.M., et al. (2003). Probing the kinesin reaction cycle with a 2D optical force clamp. *Proc. Natl. Acad. Sci. U.S.A.,* 100, 2351–2356.
60. Finer, J.T., R.M. Simmons, and J.A. Spudich (1994). Single myosin molecule mechanics: piconewton forces and nanometre steps. *Nature,* 368, 113–119.
61. Knight, A.E., et al. (2001). Analysis of single-molecule mechanical recordings: application to acto-myosin interactions. *Prog. Biophys. Mol. Biol.,* 77, 45–72.
62. Guilford, W.H., et al. (1997). Smooth muscle and skeletal muscle myosins produce similar unitary forces and displacements in the laser trap. *Biophys. J.,* 72, 1006–1021.
63. Mehta, A.D. and J.A. Spudich (1999). Single myosin molecule mechanics. *Adv. Struct. Biol.,* 5, 229–270.
64. Tanaka, H., et al. (1998). Orientation dependence of displacements by a single one-headed myosin relative to the actin filament. *Biophys. J.,* 75, 1886–1894.
65. Ishijima, A., et al. (1994). Single-molecule analysis of the actomyosin motor using nano-manipulation. *Biochem. Biophys. Res. Commun.,* 199, 1057–1063.
66. Ishijima, A., et al. (1996). Multiple- and single-molecule analysis of the actomyosin motor by nanometer-piconewton manipulation with a microneedle: unitary steps and forces. *Biophys. J.,* 70, 383–400.
67. Dupuis, D.E., et al. (1997). Actin filament mechanics in the laser trap. *J. Muscle Res. Cell Motil.,* 18, 17–30.
68. Kojima, H., A. Ishijima, and T. Yanagida (1994). Direct measurement of stiffness of single actin filaments with and without tropomyosin by *in vitro* nanomanipulation. *Proc. Natl. Acad. Sci. U.S.A.,* 91, 12962–12966.

69. White, H.D. and E.W. Taylor (1976). Energetics and mechanism of actomyosin adenosine triphosphatase. *Biochemistry,* 15, 5818–5826.
70. Huxley, A.F. and R.M. Simmons (1971). Proposed mechanism of force generation in striated muscle. *Nature,* 233, 533–538.
71. Huxley, H.E. and M. Kress (1985). Crossbridge behaviour during muscle contraction. *J. Muscle Res. Cell Motil.,* 6, 153–161.
72. Goldman, Y.E. (1987). Kinetics of the actomyosin ATPase in muscle fibers. *Annu. Rev. Physiol.,* 49, 637–654.
73. Rayment, I., et al. (1993). Structure of the actin-myosin complex and its implications for muscle contraction. *Science,* 261, 58–65.
74. Ishijima, A., et al. (1998). Simultaneous observation of individual ATPase and mechanical events by a single myosin molecule during interaction with actin. *Cell,* 92, 161–171.
75. Svoboda, K., et al. (1993). Direct observation of kinesin stepping by optical trapping interferometry. *Nature,* 365, 721–727.
76. Svoboda, K. and S.M. Block (1994). Force and velocity measured for single kinesin molecules. *Cell,* 77, 773–784.
77. Mehta, A.D., et al. (1999). Myosin-V is a processive actin-based motor. *Nature*, 400, 590–593.
78. Howard, J., A.J. Hudspeth, and R.D. Vale (1989). Movement of microtubules by single kinesin molecules. *Nature,* 342, 154–158.
79. Hancock, W.O. and J. Howard (1998). Processivity of the motor protein kinesin requires two heads. *J. Cell Biol.,* 140, 1395–1405.
80. Moore, J.R., et al. (2001). Myosin V exhibits a high duty cycle and large unitary displacement. *J. Cell Biol.,* 155, 625–635.
81. De La Cruz, E.M., et al. (1999). The kinetic mechanism of myosin V. *Proc. Natl. Acad. Sci. U.S.A.,* 96, 13726–13731.
82. Block, S.M., L.S. Goldstein, and B.J. Schnapp (1990). Bead movement by single kinesin molecules studied with optical tweezers. *Nature,* 348, 348–352.
83. Rief, M., et al. (2000). Myosin-V stepping kinetics: a molecular model for processivity. *Proc. Natl. Acad. Sci. U.S.A.,* 97, 9482–9486.
84. Sheterline, P. and J.C. Sparrow (1994). Actin. *Protein Profile,* 1, 1–121.
85. Ali, M.Y., et al. (2002). Myosin V is a left-handed spiral motor on the right-handed actin helix. *Nat. Struct. Biol.,* 9, 464–467.
86. Veigel, C., et al. (2002). The gated gait of the processive molecular motor, myosin V. *Nat. Cell Biol.,* 4, 59–65.
87. Trybus, K.M., E. Krementsova, and Y. Freyzon (1999). Kinetic characterization of a monomeric unconventional myosin V construct. *J. Biol. Chem.,* 274, 27448–2756.
88. Wang, F., et al. (2000). Effect of ADP and ionic strength on the kinetic and motile properties of recombinant mouse myosin V. *J. Biol. Chem.,* 275, 4329–4335.
89. Uyeda, T.Q., P.D. Abramson, and J.A. Spudich (1996). The neck region of the myosin motor domain acts as a lever arm to generate movement. *Proc. Natl. Acad. Sci. U.S.A.,* 93, 4459–4464.
90. Tanaka, H., et al. (2002). The motor domain determines the large step of myosin-V. *Nature,* 415, 192–195.
91. Vale, R.D. (2003). Myosin V motor proteins: marching stepwise towards a mechanism. *J. Cell Biol.,* 163, 445–450.
92. Rock, R.S., et al. (2001). Myosin VI is a processive motor with a large step size. *Proc. Natl. Acad. Sci. U.S.A.,* 98, 13655–13659.

93. Altman, D., H.L. Sweeney, and J.A. Spudich (2004). The mechanism of myosin VI translocation and its load-induced anchoring. *Cell,* 116, 737–749.
94. Berger, B., et al. (1995). Predicting coiled coils by use of pairwise residue correlations. *Proc. Natl. Acad. Sci. U.S.A.,* 92, 8259–8263.
95. Fisher, M.E. and A.B. Kolomeisky (1999). The force exerted by a molecular motor. *Proc. Natl. Acad. Sci. U.S.A.,* 96, 6597–6602.
96. Wang, M.D., et al. (1998). Force and velocity measured for single molecules of RNA polymerase. *Science,* 282, 902–907.
97. Howard, J. (2001). *Mechanics of Motor Proteins and the Cytoskeleton,* Sunderland, MA, Sinauer Associates, Inc., 367.
98. De La Cruz, E.M., E.M. Ostap, and H.L. Sweeney (2001). Kinetic mechanism and regulation of myosin VI. *J. Biol. Chem.,* 276, 32373–32381.
99. Cramer, L.P. (2000). Myosin VI. Roles for a minus end-directed actin motor in cells. *J. Cell Biol.,* 150, 121–126.
100. Tilney, L.G., D.J. Derosier, and M.J. Mulroy (1980). The organization of actin filaments in the stereocilia of cochlear hair cells. *J. Cell Biol.,* 86, 244–259.
101. Colquhoun, D., and F.J. Sigworth (1944). Statistical analysis of records, in *Single-Channel Recording*, B. Sakman, and E. Neher, Eds., New York, Plenum Press, 233–263.
102. Lang, M.J., P.M. Fordyce, and S.M. Block (2003). Combined optical trapping and single-molecule fluorescence. *J Biol.,* 2, 6.
103. Spudich, J.A. and R.S. Rock (2002). A crossbridge too far. *Nat. Cell Biol.,* 4, E8–10.
104. Warshaw, D.M., et al. (2000). The light chain binding domain of expressed smooth muscle heavy meromyosin acts as a mechanical lever. *J. Biol. Chem.,* 275, 37167–37172.

9 Biomineralization: Physiochemical and Biological Processes in Nanotechnology

Brent R. Constantz

CONTENTS

9.1 INTRODUCTION

Biomineralization is the process where organisms form minerals. Mineralized skeletons of animals and plants derive their exceptional material properties from the extremely small size of the crystals comprising these nanostructures. Understanding biomineralization, with the potential of forming new nanomaterials via entirely new processes, is extremely attractive because, to date, synthetic processes process have been limited in the ability to reproduce the remarkable properties of mineralized skeletons. Minerals are generally studied by geoscientists and specifically by mineralogists. The processes responsible for mineral formation, in association with life, differ in certain aspects, and to varying degrees, from the processes we understand that control the formation of minerals in the typical geologic and synthetic synthesis realms. Mineralogists are beginning to appreciate that most mineralization processes at low temperatures occur in the presence of some biological effect. Materials scientists are just beginning to appreciate how study of low-temperature mineralization, modulated by biological effects, may lead to breakthroughs in nanotechnology. While the descriptive study of "calcified tissues" by biologists, biochemists, chemists, and medical researchers has a long history, the application of fundamental principles from basic mineralogy is just emerging.

0-8493-1940-4/05/$0.00+$1.50

Robert Hooke (1635–1703) compared the spontaneous generation of microbes with that of the "silver tree," a dentritic structure with plant-like morphologies formed from an amalgam of silver and mercury dissolved in nitric acid, which had been studied by Isaac Newton. Interest in comparing biological structures with artifacts of purely inorganic nature engaged Newton and Hooke. Leduc and Herrera devoted themselves for several decades to the production of lifelike structures from various combinations of crystals and inorganic fluids, as part of the now largely forgotten fields of "synthetic biology" or "plasmogeny." Advocates of complexity theory, which likens the emergence of complex patterns in dynamical systems with biological phenomena, make comparisons between the complicated morphologies of inorganic systems and exquisite biological structures, like mineralized skeletons.

Minerals composing animal and plant skeletons range from forms indistinguishable from their inorganic counterparts to forms only seen in skeletons. Organisms appear to control the composition, crystallographic organization, nucleation and growth, and composite structure of skeletal crystals to varying degrees, reflected in the extent to which they differ from their inorganic counterparts. The mechanisms whereby organisms affect composition, crystallographic orientation, and morphology and exhibit higher-level structural organization could serve as examples for innovative new nanomaterials and devices. Lowenstam[1] termed the continuum of processes he observed as "physiochemically-dominated biomineralization" and "biologically-dominated biomineralization," reflecting the apparent lack of biological involvement in some skeletal mineral formation processes in contrast to the clear biological control observed in others.

In this chapter I present the two main themes in biomineralization, crystal nucleation and crystal growth, with a focus on examples from my own experiences. The first portion of the chapter discusses mechanisms of nucleation in biomineralization and their application to a nanocrystalline coating problem. In the second portion, I discuss the growth of crystals in nanostructures that are formed via biomineralization processes and their application to the development of a nanophasic composite. By way of example, insight about the translation of biomineralization processes and the products they produce into new systems in nanotechnology should develop. There are some general texts on biomineralization that provide overviews of this broad and highly multidisciplinary subject area.[2,3]

9.2 CRYSTAL NUCLEATION

Both homogeneous nucleation and heterogeneous nucleation mechanisms are commonly seem in biomineralization processes. The level of biological "control" over the nucleation event varies widely, from almost pure physiochemical nucleation to highly controlled, oriented, crystal nucleation. Homogeneous nucleation of biominerals occurs when the organism modulates the saturation state of a local solution to a state of supersaturation. Supersaturation is a measure of the extent the solution is out of equilibrium and represents the thermodynamic driving force for inorganic precipitation. Supersaturation's driving inorganic precipitation is often offset by the kinetics constraints of nucleation. During the homogeneous formation of molecular aggregates, the expenditure of interfacial free energy is balanced by the released energy in the formation of bonds in the bulk aggregate, until a whole stable nuclei is attained; otherwise, the aggregates growing against a gradient of free energy required to create a new solid–liquid interface fall apart.

Homogeneous nucleation occurs from the spontaneous formation of nuclei in the bulk of the supersaturated solution that appears to be rare in biological systems due to the ubiquitous presence of organic substrates. Organic compounds in solution may significantly reduce the interfacial energy and hence increase the rate of nucleation. This heterogeneous nucleation may occur at lower supersaturation levels than those required for homogeneous nucleation, because the nuclei are stabilized by molecular interactions on the crystal aggregate interface. In biomineralization processes, heterogeneous nucleation is very common, frequently on an organic substrate or matrix, erected first by the organism, then subsequently mineralized after mineral nucleates on the organic component. A number of charged matrices, especially from mollusks, have been described, sequenced, and manipulated to gain further understanding of their putative role in directing crystal nucleation in some cases of biomineralization. Whether soluble or insoluble, organic moieties clearly induce nucleation in many taxonomic groups.[2,3] Future lines of investigation in this area involve understanding how specific mineral polymorph may be controlled by the organic phase and how an oriented insoluble organic substrate can nucleate crystals with an aligned orientation.

Amorphous precursor phases have been implicated as prevalent in skeletonizing systems from the ubiquitous carbonate minerals of marine invertebrates, to phosphate mineral precursors in bones and teeth, and the primary production of biogenic silica. The formation of amorphous calcium carbonate occurs from highly supersaturated solutions in the presence of other additives such as Mg that inhibit crystalline phases from forming. The concentrations required for amorphous calcium carbonate to form are similar to those found in seawater.[4] Various proteins are able to induce the formation and stabilization of amorphous calcium carbonate in the absence of Mg, which would otherwise form a crystalline phase of calcium carbonate.[5] *In vitro* experiments with proteins extracted from sea urchin larval spicules show that Mg has to be present for these proteins to induce amorphous calcium carbonate formation.[6]

When studying the skeletogenetic process of reef corals, which form calcium carbonate exoskeletons, I learned that the initial phase of mineral is produced intracelluarly and exocytosed to direct the site of extracellular aragonite (calcium carbonate) crystal growth, termed a "center of calcification."[7,8] Analysis of the initial mineral produced demonstrated that it had a highly modulated, partially amorphous structure, that was partially dolomitic calcium carbonate.[9] Reef coral skeletons have been found to have organic moieties intercalated within the mineral phase and especially at these centers of calcification.[10] The material at these "centers of calcification," which was of intracellular origin, demonstrated a remarkable capacity to nucleate a very large number of aragonite crystals at a very high density, compared to surrounding surfaces, even after they were separated from the animal and tested on the bench top in inorganic supersaturated solutions — a physiochemically dominated process.[1,7]

In creating synthetic nanomaterials from aqueous solutions, examples from crystal nucleation in biological systems may provide nucleation mechanisms and processes that could be replicated in synthetic systems. It would be desirable, for instance, to employ some of the biological strategies of attaining consistent critical-size nuclei with high nucleation potentials. Specifically, forming nuclei that are stabile, yet not yet crystalline, but semiamorphous, may afford new possibilities.

By way of example, I led an effort that developed a new nanocrystalline coating for artificial hip and knee prostheses used in total joint arthroplasty. The objective was to provide a coating on the Co–Cr and Ti–alloy porous-surfaced implants that would be more osteoconductive (e.g., conductive to bone ingrowth). The clinical objective was to allow patients to weight-bear early on uncemented implants by increasing the rate of biological fixation via bone ingrowth of the porous coating.

At the time, 1988, calcium phosphate minerals, specifically hydroxyapatite, where well known to be stable and osteoconductive in the body, and methods of plasma-spraying sintered hydroxyapatite, to form a coating, had been attempted. The varying partial melting of the mineral during the plasma-spray process resulted in high variations in crystallinity of the plasma-sprayed coatings, ranging from amorphous glass to highly crystalline, which gave them variable solubility *in vivo*, resulting in delamination of the coating from the metal surface *in vivo*, and catastrophic clinical failures. We predicted that solution-precipitated crystals would have consistent crystallinity and be stabile *in vivo*. We also predicted that a nanocrystalline hydroxyapatite coating would have a high surface area with the potential of binding endogenous proteins and other organic moieties, conducive to bone ongrowth.

The first difficulty encountered was that crystals would not nucleate from solution on the alloys used for prostheses. We developed a method that forms amorphous calcium phosphate nuclei via homogeneous precipitation from a supersaturated solution of calcium acetate and ammonium phosphate that contacted the alloy substrate as a flowing amorphous gel; in the process of stabilization of the amorphous calcium phosphate to hydroxyapatite, a bond to the metal develops.[8,11-13] The organic counterions appear to both help stabilize the nuclei which form, so more nuclei form and persist, and also prevent the nuclei from becoming well crystalline, preserving their high nucleation potential. Not only were we able to get good adhesion of the nuclei to the alloy surfaces, but the nuclei acted like the centers-of-calcification in reef corals and nucleated a very high density of hydroxyapatite crystals, about 10 nm across and 200–400 nm long, distributed in a very high density. When implanted in the body, the surfaces of these implants were not only covered with an osteoconductive coating, but the surface area created by the high density of nucleating crystals of hydroxyapatite, which has highly charged crystal faces, allowed a variety of growth factors and other drugs to be delivered via adsorption to these surfaces. This coating is currently applied to the most commonly used uncemented porous hip and knee prostheses, both in the United States and worldwide, which is manufactured and distributed by Stryker Corporation.

Mechanisms of nucleation in biomineralization may be diverse, ranging from the modulated nuclei of reef corals to oriented insoluble organic matrices of mollusks, and others undoubtedly exist. As I have shown above in the orthopaedic implant nanocrystalline coating example, the novel nucleation processes seen in biomineralization provide wonderful examples for synthetic nanotechnology applications.

9.3 CRYSTAL GROWTH

Organisms forming minerals from solution require the continual addition of ions to the growing mineral surface and their subsequent incorporation into lattice sites. Specific mechanisms of crystal growth, which have been confirmed with atomic force microscopy, vary with degree of supersaturation. Mass transport and diffusion-limited growth and polynucleation of surface growth islands occur at high degrees of supersaturation. Layer-by-layer growth and screw dislocation growth occurs at moderate to low levels of supersaturation. The rate of crystal growth is dependent on the supersaturation level and the number of these active sites of crystal growth. Generally crystal growth is terminated when the supersaturation level falls to the equilibrium level defined by the solubility product of the mineral.

Crystal growth is inhibited in the presence of soluble additives that can become incorporated in the crystal or modify crystal morphology. The morphology of forming crystals results from the relative growth rates of different crystal faces, with the slow-growing surfaces dominating the final form. Equilibrium crystal morphology consists of the symmetry-related faces that give the minimum total surface energy. These crystal habits can be predicted from knowledge of the surface structures and their bonding interactions. Proteins extracted from biominerals induce crystal habit modification by changing the relative growth rates of different crystal faces through molecular-specific interactions with particular surfaces. Electrostatic, stereochemical, and structural matching of crystal faces with low and high molecular weight biological additives can modify the surface energy of a surface and modify its mechanism of growth.

The growth and development of long bones follow two principal processes: endochondral ossification and periosteal bone mineralization. While endochondral ossification is responsible for the extension of long bones' length, periosteal mineralization guides the development of long bones' girth, resulting in cortical (solid) bone. Bone microstructures take several forms and higher-level organizations. Fundamentally, in bone resulting from periosteal mineralization, the mineralized bone is composed of mineralized collagen fibril bundles. From histological observation, its clear that the higher level organization of the collagen fibril bundles occurs before mineralization begins. Osteoid, the organized bone matrix of collagen fibril bundles is produced by osteoblasts. These cells form linear arrays that secrete the insoluble Type I collagen organic matrix of bone according to developmental patterns guided and influenced by mechanical force. This premineralized matrix could be thought of as a sort of scaffolding prior to mineralization, since bone derives nearly all of its mechanical properties from the mineralization of the organic matrix. The organic matrix is 90% Type I collagen, and a variety of other noncollagenous proteins make up the remaining 10%. While some of these moieties could play a role in biomineralization, most appear to simply be bound in the organic matrix prior to and during its mineralization. Due to the highly charged nature of bone mineral, charged soluble moieties bind strongly and preferentially to and become entombed in mineralized bone matrix. That some or any of the noncollagenous proteins may have a putative role in mineral nucleation or regulating growth has yet to be demonstrated.

Collagen fibrils consist of a triple stranded superhelix of three collagen molecules referred to as a tropocollagen filament. The three noncoaxial polypeptides self-assemble to produce an arrangement that maximizes interchain cross-links. Fibril bundles of tropocollagen are organized in a quarter-stagger model, in which tropocollagen filaments are lined up head to tail in rows that are staggered by 64 nm along their long axis, so the filaments are transposed by 64 nm along their lengths. This arrangement results in a regular array of small gaps, or hole zones, between the filaments that are about 40 nm in length and 5 nm in width.[14] A particularly important cross-link occurs between the amino-terminus of one filament and those close to the carboxy-terminus of an adjacent superhelix, giving rise to a mismatch between the superhelices. Structural studies of collagen fibril bundles indicate that adjacent hole zones overlap to produce grooves that are organized in parallel rows along the long axis of the fibril bundle structure.

At the onset of bone mineralization, the hole zones and their associated grooves in the collagen fibril bundle matrix initially confine the plate-shaped carbonated hydroxyapatite crystals. During this phase of mineralization, transmission electron microscopic images of these fibrils, taken perpendicular to the fibril axes and the a and c crystallographic axes of the crystals, show a banded array of mineralized and nonmineralized zones in the collagen fibril bundles due to the distribution of the grooves where the crystals grow. The equilibrium growth habit of carbonated hydroxyapatite with the composition of bone mineral is not distinguishable from that seen in bone mineral. In other words, despite the ubiquitous surroundings of protein, which could potentially modify the crystal growth habit, carbonated hydroxyapatite crystals that look like bone crystal can be homogeneously precipitated from inorganic solutions. This indicates that it is not necessary to invoke any crystal growth modulation process with the noncollagenous proteins present in bones matrix in order to achieve their crystal shapes. The fact that crystals appear in the gaps and grooves initially, however, indicates that the crystals must nucleate in the hole zone because the collagen fibril bundle packing is too restricted to allow already formed or even forming nuclei to be transported into the holes zones from outside.

How the crystals nucleate in the holes zone is a topic of considerable investigation, but remains unclear. If noncollagenous components of bone's organic matrix are present in the hole zone and responsible for the oriented nucleation of the carbonated apatite crystals in the hole zones, then it would be expected to occur across all taxa that form bone. There are several conservative proteins that appear to have such a possible role. The nucleated crystals are initially confined to the hole zone and grow to be about 45 nm in length (crystallographic c axis), 20 nm in width (crystallographic a axis), and only about 3 nm in thickness. The crystals are very thin with a large flat surface (the slowest growing direction) in the plane formed by the a and c crystallographic axes. Individual crystals are oriented with their crystallographic a axis aligned in the direction of the groove and their crystallographic c axis aligned in the direction of the collagen fibril bundle. As crystal growth proceeds, the crystals grow beyond the holes zones and grooves, indicating a lack of any crystal growth termination process seen in many other biomineralization systems.

The interstitial fluids of bone are supersaturated with respect to bone's mineral phase, so crystal dissolution does not occur on a normal basis. For crystal growth to occur within the hole zones and grooves of the collagen fibril bundles, calcium and phosphate ions need to transport through the collagen fibril bundle matrix, which is highly restricted. From histological observation of mineralizing fronts in osteoid, it is apparent that the osteoblastic lining cells which produce osteoid are remote from the mineralizing zones in the osteoid, and given the diffusion distances, it is unlikely that the osteoblasts are "pumping" calcium and phosphate ions to the holes zones for mineral nucleation and growth. One marker of the mineralizing zone of osteoid is the enzyme, alkaline phosphatase, and although its possible role in mineralization is not well understood, its presence indicates that elevated pH occurs in and around the mineralization zone. Although supersaturation levels *in vivo* may involve yet-to-be-proven chemical regulation of the local region through ion transport, complexation-decomplexation, enzymatic regulation, and modifications in ionic strength and water content, it is apparent that the mineral is forming, nucleating, and precipitating in an alkaline subenvironment. The solubility of calcium phosphate, like carbonated hydroxyapatite, which is the equilibrium mineral phase in bone, shows a dramatic solubility decrease with increasing pH,[15] favoring precipitation.

Observing that the mineral phase of bone is difficult to distinguish from native bone mineral and that the noncollagenous proteins in the organic matrix of bone did not appear to be necessary for the growth of the crystal, we predicted that it might be possible to begin with insoluble "nonimmunogenic" Type I collagen, usually extracted from bovine Achilles tendon, and suspend it in a solution supersaturated with respect to carbonated hydroxyapatite to achieve crystal nucleation and growth in the hole zones and grooves of the collagen fibril bundles. At the time, 1990, there was a clinical need for a substitute for autograft bone, typically harvested from the iliac crest of a patient, during orthopaedic reconstructive procedures, and transplanted to the site of bony deficit. Often the morbidity associated with the harvest site exceeded that of the principal operative site. Several cytokines, known to induce the same "osteoinductive" properties of the patient's own autograft bone had been developed and were being produced by recombinant techniques with the intent of developing a substitute for having to take an autograft.

Because the osteoinductive cytokines being developed for the application, most of which were "bone morphogenic proteins" or BMPs, were highly charged, and were described from the noncollagenous protein component of bone, we knew they would bind as well to the mineral phase as they do in native bone. Initial attempts had been made to combine particles of sintered synthetic calcium phosphates, similar to bone mineral in composition, sized as ceramic particles up to a millimeter across, and combined with nonimmunogenic Type I bovine collagen, but the large particles were not stationary, and this movement led to initial cartilage formation, rather than primary new bone formation, as if one compared the healing of an unstable fracture to a stable fracture. It had been necessary to use these large, millimeter-size mineral grains, because small grains caused foreign body responses, overwhelming the entire osteoinductive benefit. We suspected that, if we could reproduce the nanophasic composite relationship of the mineral and the collagen fibril bundles seen in native mineralized bone, we could not only achieve the growth factor binding and delivery capability of the mineral, but also "shield" the extremely small mineral particles inside the hole zones and grooves of the collagen fibril bundles, preventing a foreign body reaction.

We developed a sterile process that involved suspending collagen fibrils in an elevated pH solution to which ionic calcium and ionic phosphate solutions where added at molar ratio appropriate for carbonated apatite formation.[16,17] One early observation we made was that the collagen fibril bundles swell in these solutions proportionately to increase in pH, with the greatest degree of swelling occurring at the most basic pH levels. It appears then, that some of the cross-linking in the collagen fibril bundles is affected by increasing pH, allowing the normal tight bundle packing to loosen up. The swelling of the collagen fibril bundles at high pHs may possibly lower diffusion barriers, allowing ions to transport more freely through the otherwise tightly packed fibril bundles and into the hole zones. This was an interesting possible link to the native mineralization process, because we knew alkaline phosphatase, which has higher activity at basic pH, was present at the zones of mineralization in native bone mineralization and perhaps its role in native mineralization could be that it opened up the local tight packing of the collagen fibril bundles to allow ions to diffuse in for mineralization.

The process "mineralized" the collagen in hours, producing a nanophasic composite ranging from 20 to 30% by mass mineral, determined by conventional ash weight methods. Transmission electron microscopic studies showed the crystals were intimately intercalated in the collagen fibril bundles and showed the characteristic banding seen in native collagen mineralization as well as a less ordered "coating" with mineral. For clinical application, this material could be kept in a sterile hydrated state and packaged in a syringe for injectable delivery. We later found that it could be freeze-dried, and the dried "sponge" had excellent surgical handling properties as well as hemostatic properties. Because of the mineral phase component, charged growth factors could be bound to the materials, and extensive testing, for its osteoinductivity and osteoconductivity, was completed with several BMP and angiogenic factors.[18]

The new nanophasic biomaterial proved to be an excellent alternative to harvested autograft bone. Its first significant clinical use was seen in spinal fusion procedures where autograft bone is routinely harvested from the patient's iliac crest and transplanted to the intervertebral body space to provide a scaffolding for bony fusion of the vertebral bodies. It has since been approved by the FDA for bone void filling in place of autograft bone, a procedure routinely performed in orthopaedic reconstructions and fracture repairs. This product is available without a growth factor today because, when combined with the patient's own marrow aspirate, it exhibits a biological response equivalent to autograft bone without the harvest site morbidity. A version combined with BMP is currently undergoing clinical trials. Both are manufactured and distributed by Johnson and Johnson Inc.

Mechanisms of crystal growth in biomineralization may be diverse, ranging from mass transport and diffusion-limited growth and polynucleation of surface growth islands occur at high degrees of supersaturation to oriented insoluble organic matrices of mollusks, and others undoubtedly exist. As I have shown above in the orthopaedic biomaterial example, the native bone mineralization processes still under substantial investigation with many unsolved problems still provide examples for synthetic nanotechnology applications.

9.4 CONCLUSIONS

In this chapter, I have attempted to provide an overview of crystal nucleation and growth processes in biomineralization. The extrapolation from the nucleation process of reef coral to the orthopaedic implant coating process was only possible after a detailed understanding of crystal nucleation in the skeletogenetic milieu of corals was complete. In the second case, the mineralization process of bone was poorly understood, but insight about the detailed microstructure and biochemistry of periosteal bone enabled a new synthetic bone material to be developed. These two examples show only a small preview of the possibilities of applying biomineralization systems to applied problems.

REFERENCES

1. Lowenstam, H. 1981. Minerals formed by organisms. *Science*, 211, 1126–1131.
2. Lowenstam, H. and Weiner, S. 1989. *On Biomineralization*. New York, Oxford University Press.
3. Mann, S., 2001. *Biomineralization: Principles and Concepts in Bioinorganic Materials Chemistry*. New York, Oxford University Press.
4. Raz, S., Weiner, S, and Addadi, L. 2000. The formation of high magnesium calcite via a transient amorphous colloid phase. *Adv. Mater.* 12, 38–42.
5. Aizenberg, J., Lambert, G., Addadi, L., and Weiner, S., 1996. Stabilization of amorphous calcium carbonate by specialized macromolecules in biological and synthetics precipitates. *Adv. Mater.*, 8, 222–226.
6. Weiner, S. and Dove, P.M. 1983. An overview of biomineralization processes and the problem of the vital effect, in *Reviews in Mineralogy and Geochemistry, Vol. 54, Biomineralization*, Dove, P.M., Deyoreo, J.J., and Weiner, S. Eds., Mineralogical Society of America, Geochemical Society, 1–29.
7. Constantz, B. 1986. Coral skeleton construction: a physiochemically dominated process. *Palaios*, 1, 152–157.
8. Constantz, B. 1989. Skeletal organization in Caribbean *Acropora spp.* (Lamarck), in *Origin, Evolution, and Modern Aspects of Biomineralization in Plants and Animals*. R.E. Crick, Ed., New York, Plenum Press, 1988, pp. 175–199.
9. Constantz, B. and Meike, A. 1989. Calcite centers of calcification in *Mussa angulosa (scleractinia)*, in *Origin, Evolution, and Modern Aspects of Biomineralization in Plants and Animals*. R.E. Crick, Ed., Plenum Press, New York, 1989, pp. 201–207.
10. Constantz, B. and Weiner, S. 1989. Acidic macromolecules associated with the mineral phase of scleractinian coral skeletons. *J. Exp. Zool.*, 248, 253–258.
11. Constantz, B. Apparatus for hydroxyapatite coatings of substrates. United States Patent Number: 5,188,670. February 23, 1993
12. Constantz, B. and Osaka, G. Hydroxyapatite prosthesis coatings. United States Patent Number: 5,164,187. November 17, 1992.
13. Constantz, B.R. and G.C. Osaka. Hydroxyapatite prosthesis coatings. United States Patent Number: 5,279,831. January 18,1994.

14. Katz, E. and Li, S.T. 1973. The intramolecular space of reconstituted collagen fibrils. *J. Mol. Biol.*, 21, 149–158.
15. Nancollas, G. 1989. *In vitro* studies of calcium phosphate crystallization, in *Biomineralization: Chemical and Biochemical Perspectives*. Mann, S., Webb, J., and Williams, R.J.P., Eds., pp. 157–187, Weinheim, VCH Verlagsgesellschaft.
16. Constantz, B.R. and S. Gunasekaran. Mineralized collagen. United States Patent Number: 5,231,169. July 27, 1993.
17. Constantz, B.R. and S. Gunasekaran. Mineralized collagen. United States Patent Number: 5,455,231. October 3, 1995.
18. Gunasekaran, S., Gaspodorwitz, D., Barr, P. and B.R. Constantz. 1994. Mineralized collagen, in *Hydroxyapatite and Related Materials,* P.W. Brown and B. R. Constantz, Eds., pp. 215–223, Boca Raton, FL, CRC Press, 1994.

10 Polyelectrolyte Behavior in DNA: Self-Assembling Toroidal Nanoparticles

Mary X. Tang

CONTENTS

10.1 INTRODUCTION

Although DNA is celebrated for its properties of chemical recognition, which make it the perfect vehicle for storage and propagation of genetic instructions, its non-sequence-specific properties are much less familiar. DNA possesses closely spaced backbone phosphates, which define it as a strongly anionic polyelectrolyte. It is this polyelectrolyte character which dominates its non-sequence-specific behavior, and gives rise to some unusual motifs of self-assembly: at very low concentrations and in solutions of very low ionic strength, DNA can form large, disperse and delicate, yet highly ordered "rafts";[1–3] at high concentrations and in the presence of high amounts of salt, DNA assumes various classic forms of liquid crystals;[4–8] and, as will be reviewed in this chapter, at intermediate concentrations and presence of moderate salt, charge-neutralized DNA forms nanometer-scale, toroidal-shaped particles in which the DNA is packed in a dense, crystal-like manner.

The compaction or "condensation" of DNA into toroids is a remarkably efficient process. For example, the increase in DNA density accompanying the collapse of a random coil T4 phage molecule (about 170 kbp in length) into toroid form is at least four orders of magnitude.[9,10] Indeed, this toroid is likely the form in which the phage packages its own DNA.[11] In fact, it was interest in nucleic acid packing by viruses

0-8493-1940-4/05/$0.00+$1.50

and sperm cells that led to initial research into the condensation of DNA by small, multivalent cationic polyamines.[12,13] Toroids formed from combining DNA with polylysine and polyarginine came to be studied in the 1960s and 1970s, in part because these cationic polypeptides were thought to be convenient models for nucleohistones.[14,15] In recent years, interest in toroidal condensation is sparked by the desire to produce stable, self-assembling colloidal particles for gene transfer and delivery, using functional cationic agents.[16–18]

In this chapter, we will review toroidal nanoparticle condensation with the hope that the concepts outlined here may suggest other, new ways in which DNA may be used as a nanostructural material. We propose that the polyelectrolyte nature of DNA facilitates interesting motifs of self-assembly which may complement its better-known capacity for self-recognition. Moreover, as such motifs are general to all semiflexible, strongly charged polymers, they may also suggest new approaches for incorporating other functional polyelectrolytes as nanoscale scaffolding materials. By exploiting these and other properties of polyelectrolytes, researchers may be able to vastly expand their toolkit for building nanostructures.

10.2 TOROIDAL DNA NANOPARTICLES: EXPERIMENTAL OBSERVATIONS

When mixed in solution, compacted, or condensed, DNA results from the electrostatic interaction of the negatively charged phosphates along the DNA backbone with multi- or polyvalent cationic agents. Often, this takes the form of a white clump at the bottom of a test tube (more on this, in Section 10.5), but individual, nanometer scale particles are also observed. These particles are donut-shaped, with a narrow size distribution and average diameter ranging from about 50 to 300 nm in size, depending on the reference.[10,12–28] High-resolution electron micrographs (EMs) show that the DNA strands are aligned and packed in an orderly manner; the DNA is circumferentially wound around the toroid with allowances for topological crossovers, rather like well-behaved thread on a spool.[10,12,13,19,29,30] Circular dichroism (CD) and X-ray diffraction studies have likewise confirmed that such long-range order is present throughout the toroid.[20–22]

Compact toroids were first documented in 1970, in EMs of particles resulting from mixtures of salmon sperm DNA and polylysine.[14] They have since been also observed in EM and AFM imaging of various types of DNA complexed with many other cationic compounds such as histone proteins,[15] spermine and spermidine,[13,19,23] hexamine cobalt (III),[24–27,29,30] conjugated polylysines,[17] as well as various branched polymers,[16] and even for DNA deposited on surfaces with mobile positive charges.[28] Broadly speaking, cations possessing three or more charges are sufficient to yield DNA toroids in aqueous solutions. In general, divalent cations, such as putrescine or Mg^{+2} can compact DNA only in the presence of alcohols in the solution.[31] These, and numerous other studies, show that counterion-condensed DNA forms toroids over a broad range of solution conditions with a wide variety of cationic condensing agents.

The appearance of toroids and conditions under which they occur provide interesting insight into the forces involved in their formation. First, toroids will form only if DNA is above a minimum length of about 80 to 400 bp, due perhaps to length-dependent cooperativity in adjacent DNA segments or the length required to bend the DNA without kinking.[24,32,33,34] Second, above this minimum and for a given set of condensing conditions, the length of DNA does not seem to affect the resulting size or volume of toroid formed because toroids can be multimolecular. Widom and Baldwin, for example, observed toroids that were consistently about 100 nm in diameter, resulting from the spermine-condensation of DNA fragments whose lengths varied from 0.4 to 49 kpb.[32] The corollary is true; it is possible to form a monomolecular toroid from a single, large DNA molecule, such as derived from T7 (38 kbp) or λ (47 kbp).[23,33] This suggests that gross toroid size, as for micelles or nanocrystals, may be dictated by surface energy or other mass-limiting constraint.[35,36] Third, certain types of DNA have effect on mean toroid size. Topologically constrained forms (i.e., supercoiled versus nicked circular DNA) yield smaller toroids than the comparable linear molecule.[19,25,37,38] DNA with AT tracts (which are inherently curved) also forms smaller toroids compared to DNA without these tracts.[29,30,39] And DNA which contains tracts susceptible to B-Z transition will preferentially adopt the more readily bendable Z-form in toroids.[40] Therefore, DNA in which bending is facilitated results in smaller toroids, thus demonstrating that there is a significant energy penalty associated with DNA bending in the toroid. Fourth, the condensing agent and solution conditions have some effect on toroid size, shape, and the efficiency of particle formation.[27,30,39,41–43] For small cations, the critical concentration at which condensation occurs and the size of the resulting toroids are less than for larger cations of equal charge.[27,41,42] The ionic strength and the nature of the salt also influence toroid size.[29,30] Thus, there are chemical and structural contributions that can affect the condensation process as well.

It should be noted that these observations are commonly based on electron microscopy, the results of which cannot be compared between different studies. In preparation for conventional EM analysis, samples are dehydrated in any of a variety of ways, which distorts the appearance and size of toroids from their solution form. Samples observed using cryo-EM, however, are prepared by rapid freezing, thus remaining hydrated, and thereby best represent the actual form of toroids in solution (see Figure 10.1, references 7, 12, 14). Comparison of cryo- and conventional EM photos indicate that although the DNA in toroids possesses a highly organized, liquid-crystalline structure, these toroids in solution are highly hydrated and can become distorted upon drying.

And lest the wrong impression be given that the formation of DNA toroids is a straightforward process, we note first, that counterion condensation under most solution conditions results in precipitated DNA. Many condensing agents, particularly small multivalent cations and polylysine, are especially conducive to precipitation;[25] however, some cationic polymers with small hydrodynamic dimension are quite good at preventing it.[16] Upon closer examination, the DNA precipitates generally do not, as might be expected, form a disordered tangle, but rather possess fine structure and are actually comprised of agglomerated masses of toroids.[16,25,44] Since the toroids maintain distinct identities, their aggregation may be treated as models

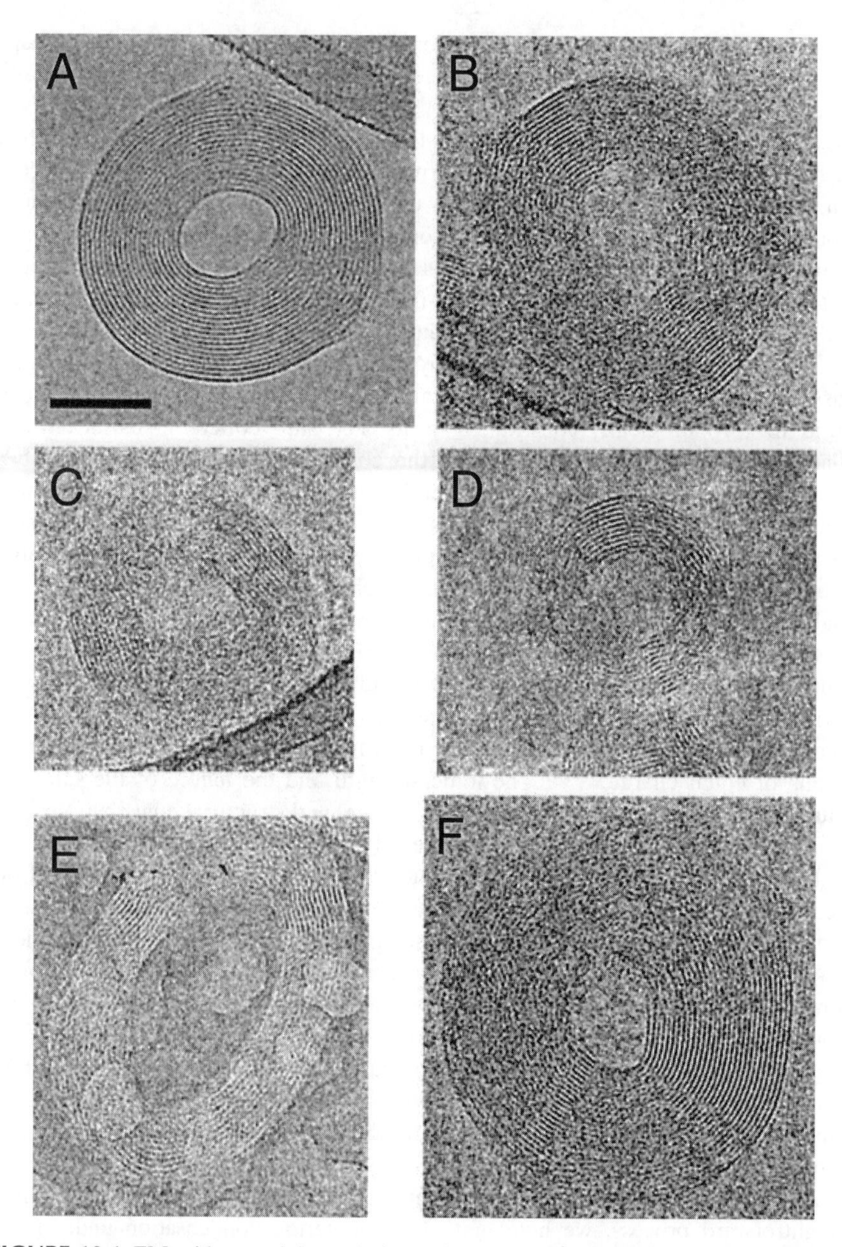

FIGURE 10.1 EM grids containing solutions of condensed λ-phage DNA were vitrified by rapid freezing in liquid ethane and then examined using cryo-transmission electron microscopy. Toroids are viewed from the top, as the plane of the toroids is approximately coplanar with the microscope image plane. Scale bar is 50 nm. All micrographs are shown at the same magnification. (Reprinted with permission from Hud, N.V. and Downing, K.H., Cryoelectron microscopy of λ phage DNA condensates in vitreous ice: the fine structure of DNA toroids, *Proc. Natl. Acad. Sci. U.S.A.*, *98*, 14925, 2001. Copyright 2001, National Academy of Sciences, U.S.A.)

of colloid stability or complex coacervation.[45,46] These approaches and their applicability to some experimental observations are described in Section 10.5. We also note that structures other than toroids are also sometimes observed under condensing conditions, most commonly, thick rods or fibers and spheres. Their appearance may, among other possibilities, be due to hydrophobicity or charge density of the condensing agent used,[27,37,38,43] dehydrating conditions in the presence of alcohols,[47] method of sample treatment for electron microscopy,[23] the use of very large DNA molecules,[37,38] or kinetic trapping during the condensation process.[43]

Toroidal DNA generally retains biochemical activity, although it may be inhibited due to the relative inaccessibility of DNA segments in the interior of the particle. For enzymes acting on curved DNA (such as polymerase) or overlapping DNA segments (such as gyrase), activity is dramatically increased just at the onset of DNA condensation, and then decreases as DNA condenses further and/or aggregates.[48–51] Marx and Reynolds used micrococcal nuclease to enzymatically "slice" through a toroid to yield a series of fragments with lengths corresponding to an integral number of circuits around the toroid.[33,52] Although enzymatic activity on condensed DNA was considerably reduced, in order to yield "sliced" segments, the nuclease would have to have a long residence time and low diffusivity on the toroid surface. Thus, the affinity of some enzymes for DNA, even in toroidal form, may remain high.

The compaction process is reversible. Small, multivalent cations are highly mobile in counterion-condensed DNA and can be readily displaced to allow DNA to return to the random coil form. This can easily be achieved by increasing the salt concentration of the solution (shielding by ions reduces electrostatic interaction),[53–55] applying an external electric field (electrophoretic separation dominates over electrostatic interaction),[56,57] and simple dilution to below the critical condensing concentration.[32] The DNA can then be rinsed and retrieved through standard means (i.e., ethanol precipitation, dialysis, separation/spin column). Toroids formed from counterion condensation by cationic polymers may be decondensed through shielding by the addition of salt; however, the electrostatic interaction is much greater, so that electrophoresis or simple dilution alone do not generally decondense DNA.[16] Because of the larger molecular weight of cationic polymers, separation of salt-decondensed DNA requires more complex means (generally, electrophoresis or size-exclusion chromatography) (personal observations).

Finally, it should be noted that there are solution conditions other than the presence of multi- and polyvalent cations that can also lead to collapse of DNA from an extended state in solution. The presence of alcohols or neutral or anionic polymers in the presence of monovalent salt is thought to cause unfavorable DNA/solution interactions or perhaps even induce dehydration or osmotic stress in the DNA, thus leading to collapse without actually becoming incorporated into the structures.[10,20,47] In fact, early observation of this led to the coining of the term "ψ-DNA" to describe these "polymer and salt-induced" DNA structures, a term which has since sometimes been used interchangeably in the literature to describe structures condensed without neutral or anionic polymers as well.

10.3 "COUNTERION-CONDENSATION" VS. "CONDENSATION"

These terms may seem to be used interchangeably in the literature, but there is a clear distinction. "Counterion condensation" describes the process by which DNA is charge-neutralized, which then facilitates the abrupt collapse from a random coil in solution to a densely packed particle. This process of collapse is also referred to as "condensation," because this term has been long used to more generally describe those biological states in which DNA is compressed (such as in the prophase in mitosis, where chromatin "condenses" into chromosomes).

The Manning-Oosawa "counterion condensation"[58,59] is distinct from the "condensation" used to describe collapse because the former specifically refers to the interaction of the polyelectrolyte with the surrounding solution. Polyelectrolytes can be viewed as having a thick surface coating of mobile counterions. These counterions will have varying degrees of association with the polyelectrolyte, depending on their interactions with each other and the surrounding environment. These interactions have been modeled using the Poisson-Boltzmann equation to examine the electrostatic potential of the polyelectrolyte system under ideal conditions (where the polyelectrolyte is treated as a charged rod of infinite length in a solution of zero salt concentration). This treatment yields a unitless expression that is now commonly referred to as the Manning parameter:[60,61]

$$\xi = e^2/\varepsilon k \mathrm{T} b$$

where e is the value of an electronic charge, ε is the dielectric constant of the surrounding fluid, k is the Boltzmann constant, T is the absolute temperature, and b is the linear distance between charges on the polymer. Conceptually, ξ can be regarded as the ratio of the charge density of a polyelectrolyte ($1/b$) to ability of the fluid to accommodate the charge ($\varepsilon k\mathrm{T}/e^2$, or the inverse of the Bjerrum length).

It is impossible for ξ to be greater than unity (clearly, the polyelectrolyte charge cannot exceed the capacity of the surrounding solution to contain it). What this means physically is that counterions in solution become closely associated with and thereby partially charge-neutralize the polyelectrolyte, thus reducing the effective charge density of the polymer and keeping ξ to below unity. These counterions are not "bound" in the traditional chemical sense, but are associated to a higher degree (i.e., "condensed") with the polyelectrolyte than are the other ions in solution. This close, electrostatic association of counterions with the polyelectrolyte is referred to as "counterion condensation."[60,61]

When a multivalent counterion is present, it will displace mono- (or other lesser) valent counterions condensing on the polyelectrolyte, as there will be a gain in entropy as multiple monovalent ions are released into the solution. G.S. Manning described the efficiency of charge neutralization, f, by multivalent counterions of valence, N, in the following expression:

$$f = 1 - 1/(\mathrm{N}\xi)$$

Higher N results in more efficient charge neutralization (in DNA, f = 76, 88, 92, and 94% for N= 1, 2, 3, and 4).[53–55] It has been observed experimentally that DNA compaction occurs at approximately f = 90% for a broad range of condensing conditions: solutions containing cations of different valence; solutions in which monovalent cation concentration is varied; solutions in which the dielectric constant was varied by the addition of alcohol.[24,27,31]

Monovalent and divalent cations will counterion-condense onto DNA to a degree that should satisfy $\xi < 1$, but do not induce compaction of DNA (unless noninteracting polymers or alcohols are added). At low concentrations, multivalent cations that can induce compaction will counterion-condense onto DNA without causing collapse. Moreover, there is little in counterion condensation theory to describe why such collapse should occur specifically at 90% charge neutralization for DNA. Thus, "counterion condensation" describes only the process of charge neutralization; the abrupt transition, or "condensation" of random coil DNA into a compact toroidal form involves many other forces, in addition to electrostatics, as will be briefly reviewed in the next section.

It should also be noted that Manning proposed the counterion condensation approach might more generally be useful for defining the "limiting laws" by which nonideal polyelectrolyte systems may be examined. Although frequently described as overly simple, the counterion approach has proven to be remarkably effective in explaining experimental findings and stands today as an effective approach to describing polyelectrolyte systems.[62]

10.4 WHAT ARE "SEMIFLEXIBLE POLYELECTROLYTES," AND HOW DO THEY FORM TOROIDS?

It is surprising that toroids, which are highly structured, form spontaneously and abruptly from random coil DNA in solution. The entropic cost of forming these crystal-like structures is high, yet must be clearly more than compensated by the forces that encourage this self-organizing process. This apparently spontaneous formation of order from disorder presents an enticing puzzle for theoreticians and is thus a lively topic of debate. What follows here is, at best, only a perfunctory overview; for a more definitive review of the prevailing thinking on toroid formation, the reader is referred to Bloomfield's review.[63]

In polymer statistics, the "coil-to-globule" transition describes the collapse of a single, highly flexible, random coil polymer chain in a solution in which it is dilute and sparingly soluble. The globule is amorphous, and the transition is gradual. But the transition for DNA collapse is abrupt; moreover, toroid particles possess nearly crystalline structure and may be multimolecular. However, by treating the random coil polymer as a stiff, highly solvated polyelectrolyte, the "coil-to-globule" approach can be adapted to make predictions, which are largely born out by experimental data.[64–67]

So what defines a semiflexible, highly solvated polymer? In polymer statistics, linear polymers in solution are "worm-like" chains, whose segments slip in and around each other in a dynamic manner. This behavior can be described by three structural parameters, which are specific for any given polymer: the contour length, the persistence length, and the excluded volume. These three parameters can also be used to describe the volume occupied by a polymer in solution.[63,64]

The contour length is simply the length of the extended polymer chain. Generally, this is regarded as a statistically weighted average of polydisperse distribution of polymer chain lengths. However, DNA derived from a well-defined source (i.e., plasmids, viral genomes, clonal restriction fragments) is monodisperse, so its contour length is a single value. (The ready availability of well-sized samples confers another degree of experimental control and is yet another reason why DNA is favored for characterization by polymer scientists.) The persistence length, P, describes the rigidity of a polymer. P is commonly described as the length over which a polymer can be regarded as a stiff rod. However, DNA of characteristic length P can form covalently closed minicircles; it is also the length of DNA that wraps around a nucleosome particle. More precisely, P is the length of polymer for which the energy required to bend it into a circle is "kT." For DNA, P is about 50 nm, which makes it a "semiflexible" polymer.[70] (By comparison, P for random coil polypeptides is about 0.5 nm, and about 500 nm for their α-helix counterparts.) The exclusion volume may be viewed as the effective width of a segment of polymer, or alternatively, the space in which two polymer segments will not overlap. Ionized groups are highly solvated in polar solvents, so polyelectrolytes typically possess large exclusion volumes. Because the ionized phosphates along its double backbone also engage in extended hydrogen bonding, random coil DNA in aqueous solutions possesses a particularly large exclusion volume.

The volume occupied by a polymer chain in solution can be described by its radius of gyration, R_g, a geometrically weighted parameter, which can be measured directly. Using static light scattering, Sobel and Harpst determined the R_g of T7 phage DNA as the solution salt concentration was increased from 10 mM to 1 M. Typically, at 1 M salt concentration, electrostatic interactions can be considered strongly shielded. They observed that the R_g decreased by a relatively modest 28%.[70] Thus, this and other studies demonstrated that electrostatic repulsion between phosphate groups to both P and excluded volume has only a modest contribution.[71,72]

There are two important concepts here. First, the stiffness of DNA is inherent to its structure, not its charge—which is not surprising, given that base pairing structure of double-stranded DNA allows for little twisting and no free rotation along its chain. Thus, even when charge-neutralized by the process of counterion-condensation, considerable energy must still be expended to bend the DNA into a toroid with a diameter of 50–300 nm.[34,39] Second, hydration dominates exclusion volume. Thus, the massive volume change the DNA undergoes during collapse into a toroid is due less to a reduction in electrostatic repulsion upon counterion-condensation than to dehydration of the water shell surrounding the polymer chains.[71,73–75]

From another perspective, the "coil-to-globule" transition as applied to semiflexible polyelectrolytes describes a set of solution conditions in which polymer segment–polymer segment interactions are favored over polymer segment–solvent interactions.[65,66] These favorable polymer–polymer interactions are described as "correlated fluctuations" or van der Waals-type forces.[34] Such interactions would clearly be length dependent, which would suggest a minimum length of polymer for which toroids can form, and also highly cooperative, which would suggest that the transition would be abrupt.

These and other additional forces contributing to the formation of toroids are examined extensively in the literature. Of particular note, are the following. Binding between DNA segments in toroids has been attributed to physical bridging by the condensing cations, as shown by X-ray data correlation between spacing of DNA helices and the length of the condensing agent used.[75] The elasticity of DNA, which increases upon condensation, is thought to allow for looping of DNA coils, which facilitates dynamic toroidal wrapping.[76,77] The average size of toroids may be determined by the number and type of stacking defects (i.e., topologically induced crossovers) that can be accomodated when the DNA is toroidally wrapped.[35,36] Alternatively, toroid size and thickness are also thought to be established by the kinetics and dynamics of the nucleation of the initial loop of DNA, which in turn, are affected by the solution conditions, the nature of the condensing agent, and the type of DNA.[29,30]

Again, it is important to note that nothing in the coil-to-globule treatment is specific for DNA; indeed, this phenomenon should be general to all semiflexible polyelectrolytes. In fact, toroids have been observed for counterion-condensation of non-DNA polymers: such as actin,[78] acetan and xanthan,[79] and α-helix forms of polyglutamic acid and polylysine (data to be published).

10.5 PARTICLE AGGREGATION

As commonly observed, precipitates generally result from the compaction of DNA, especially by small, multivalent cations, and in solutions containing any appreciable salt.[25,31,44,48,64,65] However, toroids can exist stably in solution under certain conditions, and for purposes of nanoscale manipulation, it is important to understand the forces that influence solution stability against aggregation.

The fine line between particle stabilization and aggregation lies in the balance between repulsive and attractive forces between particles. This balance of forces is neatly circumscribed in the DLVO theory, which is schematically depicted in Figure 10.2.[45,46]

For particles of like charge, the dominant repulsive force between particles is electrostatic, which has a dependence of $(\text{distance})^{-2}$. The attractive forces are those comprising the van der Waals interactions (London, Keesom, and DeBye) all of which exhibit a $(\text{distance})^{-6}$ dependence. By convention, forces promoting aggregation are given a negative sign. As can be seen from the resulting net energy curve, particle stability prevails when the distance between particles is such that electrostatic repulsion dominates; aggregation results when particles are close enough for van der Waals forces to dominate. Thus, one can readily envision several practical experimental conditions that will influence aggregation:

1. Particle concentration: As the number of particles in a given volume increase, so does the likelihood of collision through random motion, thus increasing the probability of aggregation.

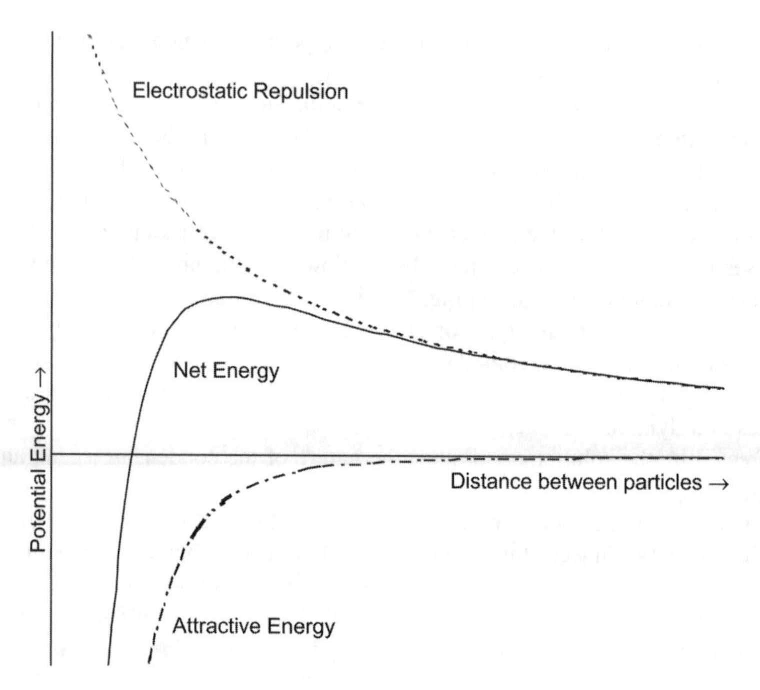

FIGURE 10.2 Representation of the DLVO potential energy between two particles of like charge as a function of the distance between them. Electrostatic repulsion, due to similar charge, is positive, and attraction, due to van der Waals forces, is negative; thus, where net energy is negative, aggregation occurs.

2. Salt concentration: Ions in solution will shield the electrostatic interactions between particles, thus effectively reducing repulsive force and thereby increasing likelihood of aggregation.
3. Agitation, stirring, shear force: Kinetic energy will increase the number of collisions between particles, leading to aggregation.
4. Particle charge: Increasing the electrostatic repulsion between particles improves particle stability.
5. Particle size, shape, mass, and composition: These affect magnitude of the van der Waals attractions.

Clearly, these conditions interact and must be optimized together to achieve a stable nanoparticle solution. But let us focus for the moment on the particle charge, which can be modulated by the condensing agent used.

Toroidal nanoparticles formed from counterion condensation by small, multivalent cations possess a small negative charge, as can be measured by electrophoretic light scattering (ELS).[13] This weakly repulsive force is easily dominated by the collective attractive forces, and thus results in particle aggregation over a broad range of solution conditions. Even in large excess of condensing cation, the nanoparticles cannot achieve positive charge (and, in fact, remain slightly negatively charged) because small, multivalent cations are highly mobile and cannot associate with DNA in excess.[80–82] And although a strong negative charge could serve just as well to facilitate interparticle repulsion, because approximately 90% of DNA phosphates must be charge-neutralized in order for collapse to occur, DNA toroids cannot maintain sufficient negative electrostatic charge facilitate charge-stabilization.[24,27,31]

However, because of their higher charge, cationic polymers possess significantly greater electrostatic affinity and can therefore associate with DNA in excess. Thus, the resulting DNA toroids can acquire a significant positive charge which, depending on the polymer structure, can enable them to exist stably over a broad range of solution conditions (including physiological buffers), which would otherwise result in aggregation for toroids condensed by small, multivalent cations.[16]

But not all cationic polymers are created equal. There appears to be a minimum charge or charge density below which a net positive charge on the condensate cannot be achieved, even when excess cationic agent is used. For example, penta- and octa-lysines, like small, multivalent cations, condense DNA, but the resulting particles do not possess significant charge, as measured by ELS, and therefore, aggregate (personal observations).[16] Early-generation polyamidoamine dendrimers, possessing up to 24 protonable groups, likewise have the ability to condense DNA, but yield nanoparticles with slight charge, which aggregate when directly mixed in physiological buffer conditions (data not published). Later-generation dendrimers, branched cationic polymers (polyethyleneimines), and small linear sugar polymers, with 48 or more protonatable groups are able to associate with DNA in excess, and yield nanoparticles which are reasonably stable in solution over a period of days, under the range of conditions examined.[16] Thus, in solutions of physiological ionic strength, it appears, at least for the limited range of polymers examined, that only compounds with approximately 50 protonatable groups or more appear to possess sufficient electrostatic affinity to associate with DNA in excess and thereby yield positively charged condensates.

DLVO theory alone cannot explain aggregation by high-molecular-weight, linear polymers. For example, linear polylysines (about 180 residues in length) and linear polyethyleneimines can form strongly positive condensates, but these particles still aggregate (personal observations).[16,83] We suspect this may be a specific example of the "bridging" mechanism. Polyelectrolytes have traditionally been used to cause aggregation or flocculation of colloid particle contaminants for easy purification. In the bridging mechanism, polyelectrolytes possessing opposite charge to the contaminating particles are added in excess. They electrostatically attach to the particles at a few points, leaving highly charged, pendant loops extending from the particle into solution. These loops can electrostatically interact and bridge other particles. Thus, the efficiency of aggregation increases with the length of the polyelectrolyte used. And, in fact, the rate and onset of DNA aggregation are observed to correlate with the length of polylysine used.[83] The branched cationic polymers (from the family of PAMAM dendrimers, and branched polyethyleneimines) and short, highly charged polymers may be less susceptible to aggregation because their hydrodynamic radii is smaller and therefore less likely to participate in bridging reactions.[16]

Thus, the structure of the condensing agent can play a significant role in determining the propensity of its DNA condensates toward aggregation. In summary, we have found that cationic compounds possessing on the order of 50 or more charges and a small hydrodynamic radius can be effective in forming DNA nanoparticles which are stable under a wide variety of solution conditions.

Finally, it is important to consider the kinetic component of the toroid formation process. The electrostatic interaction of cationic agents with DNA is very fast. And although the process is reversible, particularly for small, multivalent cations, it is entirely likely that DNA strands can still become entangled and kinetically trapped in a higher thermodynamic state.[82] It is useful to consider that, although the condensation process is spontaneous, viruses still expend considerably energy to facilitate it. For example, the T4 phage packages its DNA by first forming the capsid head, and then expending ATP to reel in the DNA, like winding yarn around a ball.[84] Through this elaborate means, the phage is able to control the condensation process by isolating it in space (by processing it inside its capsid) and time (slowing the process down by tying it to ATP expenditure) and is thus able to prevent aggregation or entanglement of the DNA.

10.6 SUMMARY AND IMPLICATIONS FOR BIONANOTECHNOLOGY

Researchers in nanotechnology have looked to DNA, with its properties of self-recognition, as a solution for accurate, high-throughput, low-cost manipulation and assembly of nanoscale structures. Because of its exquisite capacity for sequence-based molecular recognition, and the sheer vastness of combinatorial sequences it possesses, DNA presents an elegant, versatile solution for directing molecular scale assembly. Moreover, an entire toolbox for synthesizing, excising, splicing, amplifying, or otherwise engineering DNA sequences already exists and can be found in any molecular biology products catalog. Thus, it is easy to see how virtually all of the strategies proposed or developed so far for DNA-directed assembly are built upon sequence-based recognition.

There are, however, many non–base-specific properties of DNA, which might also be exploited to great effect, and which may address some problems that may arise in purely sequence recognition–based schemes. For example, although the number of unique DNA sequences may appear to be virtually limitless, a single DNA molecule of sufficient length to make sequence recognition worthwhile can still be quite large. Its low diffusivity can make the random process of finding its complementary partner prohibitively long. Moreover, sequence-based recognition schemes often suffer from too much sequence diversity and are thus limited by the practical aspects of how to produce and manage the vast library of sequences. These constraints thus generally limit recognition-based schemes to what can be used with high diversity on a very local scale or where a few sequences are used frequently, as in tiling schemes. Applications requiring high-throughput or broader-scale assembly motifs might be better served using non–sequence-based methods.

We also believe that there are many diverse areas in which polyelectrolytes are used, but their unusual properties may not be fully realized. For example, the viscosity of DNA solutions is dramatically reduced upon condensation; this may provide a means for formulating new DNA "inks" for inkjetting or other functions where DNA viscosity is limiting.[85] Electroluminescent polymer inks, for displays or other electronics applications, may also benefit. Counterion condensation may play a role in the assembly of polyelectrolyte multilayers (PEMs), films which are formed by the layer-by-layer deposition of alternating anionic and cationic polyelectrolytes and thought to have broad-ranging applications in photonics, drug delivery, and biosensors.[86,87] We think it might also play a role in describing how diatoms may use silicatein proteins to create the nanoscale silica nodules which form the building blocks for their elaborate skeletal structures.[88,89] Polyelectrolytes are more common than most of us generally think, and there are, no doubt, many opportunities to harness their properties in new ways.

Here, we have presented a brief overview of the polyelectrolyte behavior of DNA with respect to its counterion condensation–induced collapse into toroid nanoparticles. It is hoped this review of some of the forces involved in their formation and solution stabilization may present a recipe for producing these nanoparticles under any of a broad range of conditions that a nanoscience researcher may require. Moreover, as the process is reversible and the DNA remains biologically active, this and other non–sequence-based schemes may very nicely complement sequence-based schemes to provide broadly applicable set of tools for nanoscale assembly. Finally, the concepts outlined here are applicable to any semiflexible polyelectrolyte and thus may present an opportunity for incorporating new, functional polymers in nanoscience.

REFERENCES

1. Newman, J., Tracy, J., and Pecora, R., Dynamic light scattering from monodisperse 2311 base pair circular DNA: ionic strength dependence. *Macromolecules*, 27, 6808, 1994.
2. Schmitz, K.S., Quasi-elastic light scattering studies on T7 DNA in the presence of a sinusoidal electric field. *Chem. Phys.*, 79, 297, 1983.
3. Fulmer, A.W., Benbasat, J.A., and Bloomfield, V.A., Ionic strength effects on macroion diffusion and excess light scattering intensities of short DNA rods. *Biopolymers*, 20, 1147, 1981.
4. Van Winkle, D.H., Davidson, M.W., Chen, W.-X., and Rill, R.L., Cholesteric helical pitch of near persistence length DNA. *Macromolecules*, 23, 4140, 1990.
5. Reich, Z., Wachtel, E.J. and Minsky, A., Liquid-crystalline mesophases of plasmid DNA in bacteria. *Science*, 264, 1460, 1994.
6. Leforestier, A. and Livolant, F., Supramolecular ordering of DNA in the cholesteric liquid crystalline phase: an ultrastructural study. *Biophys. J.*, 65, 56, 1993.
7. Sikorav, J.-L., Pelta, J. and Livolant, F., A liquid crystalline phase in spermidine-condensed DNA. *Biophys. J., 67*, 1387, 1994.

8. Livolant, F., Levelut, A.M., Doucet, J., and Benoit, J.P., The highly concentrated liquid crystalline phase of DNA is columnar hexagonal. *Nature*, 339, 724, 1989.
9. Laemmli, U.K., Characterization of DNA condensates induced by poly(ethylene oxide) and polylysine. *Proc. Natl. Acad. Sci. U.S.A.*, 72, 4288, 1995.
10. Lerman, L.S., The polymer and salt-induced condensation of DNA, in *The Physico-Chemical Properties of Nucleic Acids*, Duchesne, J., Ed., Academic Press, New York, 1973, p. 59.
11. Bloomfield, V.A., Wilson, R.W., and Rau, D.C., Polyelectrolyte effects in DNA condensation by polyamines. *Biophys. Chem.*, 11, 339, 1980.
12. Hud, N.V. and Downing, K.H., Cryoelectron microscopy of λ phage DNA condensates in vitreous ice: the fine structure of DNA toroids, *Proc. Natl. Acad. Sci. U.S.A.*, 98, 14925, 2001.
13. Marx, K.A. and Ruben, G.C., Evidence for hydrated spermidine-calf thymus DNA toruses organized by circumferential DNA wrapping. *Nucleic Acids Res.*, 11, 1839, 1983.
14. Haynes, M., Garrett, R.A., and Gratzer, W.B., Structure of nucleic acid-poly base complexes. *Biochemistry*, 9, 4410, 1970.
15. Olins, D. and Olins, A., Model nucleohistones: the interaction of F1 and F2a1 histones with native T7 DNA. *J. Mol. Biol.*, 57, 437, 1971.
16. Tang, M.X. and Szoka, F.C., Jr., The influence of polymer structure on the interactions of cationic polymers with DNA and morphology of resulting complexes. *Gene Ther.*, 4, 823, 1997.
17. Wagner, E., Cotten, M., Foisner, R., and Birnstiel, M.L., Transferrin-polycation-DNA complexes: the effect of polycations on the structure of the complex and DNA delivery to cells. *Proc. Natl. Acad. Sci. U.S.A.*, 88, 4255, 1991.
18. Boussif, O., Lezoualc'h, F., Zanta, M.A., Mergny, M.D., Scherman, D., Demeneix, B., and Behr, J.-P., A versatile vector for gene and oligonucleotide transfer into cells in culture and *in vivo*: polyethyleneimine. *Proc. Natl. Acad. Sci. U.S.A.*, 92, 7297, 1995.
19. Marx, K. and Ruben, G., A study of phi-X-174 DNA torus and Lambda DNA torus tertiary structure and the implications for DNA self-assembly. *J. Biomol. Struct. Dyn.*, 4, 23, 1986.
20. Lerman, L.S., A transition to a compact form of DNA in polymer solutions. *Proc. Natl. Acad. Sci. U.S.A.*, 68, 1886, 1971.
21. Dorman, B.P. and Maestre, M.F., Experimental differential light-scattering correction to the circular dichroism of bacteriophage T2. *Proc. Natl. Acad. Sci. U.S.A.*, 70, 255, 1973.
22. Maestre, M.F. and Reich, C., Contribution of light scattering to the circular dichroism of deoxyribonucleic acid films, deoxyribonucleic acid-polylysine complexes, and deoxyribonucleic acid particles in ethanolic buffers. *Biochemistry*, 19, 5214, 1980.
23. Chattoraj, D.K., Gosule, L.C., and Schellman, J.A., DNA condensation with polyamines. II. Electron microscope studies. *J. Mol. Biol.*, 121, 327, 1978.
24. Widom, J. and Baldwin, R.L., Monomolecular condensation of lambda-DNA induced by cobalt hexammine. *Biopolymers*, 22, 1595, 1983.
25. Arscott, P.G., Li, A., and Bloomfield, V.A., Condensation of DNA by trivalent cations. 1. Effects of DNA length and topology on the size and shape of condensed particles. *Biopolymers*, 30, 619, 1990.
26. Li, A., Fan, T., and Ding, M., Formation study of toroidal condensation of DNA. *Science in China Series B*, 35, 169, 1992.

27. Plum, G.E., Arscott, P.G., and Bloomfield, V.A., Condensation of DNA by trivalent cations. 2. Effects of cation structure. *Biopolymers*, 30, 631,1990.
28. Fang, Y. and Hoh, J.H., Surface-directed DNA condensation in the absence of soluble multivalent cations, *Nucleic Acids Res.*, 26, 588, 1998.
29. Shen, M.R., Downing, K.H., Balhorn, R., and Hud, N.V., Nucleation of DNA condensation by static loops: formation of DNA toroids with reduced dimensions, *J. Am. Chem. Soc.*, 122, 4833, 2001.
30. Conwell, C.C., Vilfan, I.D., and Hud, N.V., Controlling the size of nanoscale toroidal DNA condensates with static curvature and ionic strength, *Proc. Natl. Acad. Sci., U.S.A.*, 100, 9296, 2003.
31. Wilson, R.W. and Bloomfield, V.A., Counterion-induced condensation of deoxyribonucleic acid. A light scattering study, *Biochemistry*, 18, 2192, 1979.
32. Widom, J. and Baldwin, R.L., Cation-induced toroidal condensation of DNA. Studies with $C(NH_3)_6$, *J. Mol. Biol.*, 144, 431, 1980.
33. Marx, K.A. and Reynolds, T.C., Micrococcal nuclease digestion study of spermidine-condensed DNA, *Int. J. Biol. Macromol.*, 11, 241, 1989.
34. Marquet, R. and Houssier, C., Thermodynamics of cation-induced DNA condensation, *J. Biomol. Struct. Dyn.*, 9, 159, 1991.
35. Ubbink, J. and Odijk, T., Polymer-induced and salt-induced toroids of hexagonal DNA, *Biophys. J.*, 68, 54, 1995.
36. Park, S.Y., Harries, D., and Gelbart, W.M., Topological defects and the optimum size of DNA condensates, *Biophys. J.*, 75, 714, 1998.
37. Grosberg, A.Y. and Zhestkov, A.V., On the toroidal condensed state of closed circular DNA, *J. Biomol. Struct. Dyn.*, 3, 515, 1985.
38. Grosbert, A.Y. and Zhestkov, A.V., On the compact form of linear duplex DNA — globular states of the uniform elastic (persistent) macromolecule, *J. Biomol. Struct. Dyn.*, 3, 859, 1986.
39. Reich, Z., Ghirlando, R., and Minsky, A., Nucleic acids packaging processes: effects of adenine tracts and sequence-dependent curvature. *J. Biomol. Struct. Dyn.*, 9, 1097, 1992.
40. Ma, C., Sun, L., and Bloomfield, V.A., Condensation of plasmids enhanced by Z-DNA conformation of d(CG)n inserts, *Biochemistry*, 34, 3521, 1995.
41. Plum, G.E. and Bloomfield, V.A., Structural and electrostatic effects on binding of trivalent cations to double-stranded and single-stranded poly[d(AT)], *Biopolymers*, 29, 13, 1990.
42. Matulis, D., Rouzina, I., Bloomfield, V.A., Thermodynamics of DNA binding and condensation: Isothermal titration calorimetry and electrostatic mechanism, *J. Mol. Biol.*, 296, 1053, 2000.
43. Vasilevskaya, V.V., Khokhlov, A.R., Kidoaki, S., and Yshiokawa, K., Structure of collapsed persistent macromolecule: toroid vs spherical globule, *Biopolymers*, 41, 51, 1997.
44. Hud, N.V., Allen, M.J., Downing, K.H., Lee, J., and Balhorn, R., Identification of the elemental packing unit of DNA in mammalian sperm cells by atomic force microscopy, *Biochem. Biophys. Res. Commun.*, 193, 1347, 1993.
45. Hiemenz, P.C., *Principles of Colloid and Surface Chemistry,* 2nd ed., Marcel Dekker, New York, 1986.
46. Israelachvili, J.N., *Intermolecular and Surface Forces: With Applications to Colloidal and Biological Systems,* 2nd ed., Academic Press, New York, 1992.
47. Eickbush, T.H. and Moudrianakis, E.N., The compaction of DNA helices into either continuous supercoils or folded-fiber rods and toroids, *Cell*, 13, 295, 1978.

48. Krasnow, M.A. and Cozzarelli, N.R., Catenation of DNA rings by topoisomerases, *J. Biol. Chem.*, 257, 2687, 1982.
49. Baeza, I., Gariglio, P., Rangel, L. M., Chavez, P., Cervantes, L., Arguello, C., Wong, C., and Montanez, C., Electron microscopy and biochemical properties of polyamine-compacted DNA, *Biochemistry*, 26, 6387, 1987.
50. Pingoud, A., Urbanke, C., Alves, J., Ehbrecht, H., Zabeau, M., and Gualerzi, C., Effect of polyamines and basic proteins on cleavage of DNA by restriction endonucleases, *Biochemistry*, 23, 5697, 1984.
51. Ma, C. and Bloomfield, V.A., Condensation of supercoiled DNA induced by $MnCl_2$, *Biophys. J.*, 67, 1678, 1994.
52. Marx, K.A. and Reynolds, T.C., Spermidine-condensed phiX174 DNA cleavage by micrococcal nuclease: Torus cleavage model and evidence for unidirectional circumferential DNA wrapping, *Proc. Natl. Acad. Sci. U.S.A.*, 79, 6484, 1982.
53. Manning, G.S., Limiting law and counterion condensation in polyelectrolyte solutions. IV. The approach to the limit and the extraordinary stability of the charge fraction, *Biophys. Chem.*, 7, 95, 1977.
54. Manning, G.S., Thermodynamic stability theory for DNA doughnut shapes induced by charge neutralization, *Biopolymers*, 19, 37, 1980.
55. Li, A.Z. and Marx, K.A., The isocompetition point for counterion competition binding to DNA: calculated multivalent versus monovalent cation binding equivalence, *Biophys. J.*, 77, 114, 1999.
56. Ma, C. and Bloomfield, V.A., Gel electrophoresis measurement of counterion condensation on DNA, *Biopolymers*, 35, 211, 1995.
57. Manning, G.S., Limiting laws and counterion condensation in polyelectrolyte solutions. 7. Electrophoretic mobility and conductance, *J. Phys. Chem.*, 85, 1506, 1981.
58. Oosawa, F., *Polyelectrolytes*, Marcel Dekker, New York, 1971.
59. Manning, G.S. and Ray, J., Counterion condensation theory revisited, *J. Biomol. Struct. Dyn.*, 16, 461,1998.
60. Imai, N. and Onishi, T., Analytical solution of Poisson-Boltzmann equation for the two-dimensional many-center problem, *J. Chem. Phys.*, 30, 1115, 1959.
61. Manning, G.S., Limiting laws and counterion condensation in polyelectrolyte solutions. I. Colligative properties, *J. Chem. Phys.*, 51, 924, 1969.
62. Ruben, G.C., The iso-competition point, a new concept for characterizing multivalent versus monovalent counterion competition, successfully describes cation binding to DNA, *Biophys J.*, 77, 1, 1999.
63. Bloomfield, V.A., DNA condensation by multivalent cations, *Biopolymers*, 44, 269, 1997.
64. Post, C.B. and Zimm, B.H., Light-scattering study of DNA condensation: competition between collapse and aggregation, *Biopolymers*, 21, 2139, 1982.
65. Post, C.B. and Zimm, B.H., Theory of DNA condensation: collapse versus aggregation. *Biopolymers*, 21, 2123, 1982.
66. Grosberg, A.Yu. and Khokhlov, A.R., *Statistical Physics of Macromolecules,* API Press, New York, 1994.
67. Ivanov, V.A., Stukan, M.R., Vasilevskaya, V.V., Paul, W., and Binder, K., Structures of stiff macromolecules of finite chain length near the coil-globule transition: A Monte Carlo simulation, *Macromol. Theory Simul.*, 9, 488, 2000.
68. Flory, P.J., *Principles of Polymer Chemistry*, Cornell University Press, Ithaca, NY, 1953
69. Flory, P.J., *Statistical Mechanics of Chain Molecules,* Wiley Interscience Publishers, New York, 1969.

70. Sobel, E.S. and Harpst, J.A., Effects of Na+ on the persistence length and excluded volume of T7 bacteriophage DNA, *Biopolymers*, 31, 1559, 1991.
71. Schurr, J.A. and Allison, S.A., Polyelectrolyte contribution to the persistence length of DNA, *Biopolymers*, 20, 251, 1981.
72. Rinehart, F. and Hearst, J., The ionic strength dependence of the coil dimensions of viral DNA in NH_4Ac solutions, *Arch. Biochem. Biophys.*, 152, 723, 1972.
73. Rau, D.C. and Parsegian, V.A., Direct measurement of the intermolecular forces between counterion-condensed DNA double helices, *Biophys. J.*, 61, 246,1992.
74. Bloomfield, V.A., Condensation of DNA my multivalent cations: considerations on mechanism, *Biopolymers*, 31, 1471,1991.
75. Schellman, J.A. and Parthasarathy, N., X-ray diffraction studies on cation-collapsed DNA, *J. Mol. Biol.*, 175, 313, 1984.
76. Manning, G.S., Is the counterion condensation point on polyelectrolytes a trigger of structural transition? *J. Chem. Phys.*, 89, 3772, 1988.
77. Baumann, C.G., Smith, S.B., Bloomfield, V.A., and Bustamente, C., Ionic effects on the elasticity of single DNA molecules, *Proc. Natl. Acad. Sci., U.S.A.*, 94, 6185, 1997.
78. Tang, J.X., Käs, J.A., Shah, J.V., and Janmey, P.A., Counterion-induced actin ring formation, *Eur. Biophys. J.*, 30, 477, 2001.
79. Maurstad, G., Danielson, S., and Stokke, B.T., Analysis of compacted semiflexible polyanions visualized by atomic force microscope: influence of chain stiffness on the morphology of polyelectrolyte complexes, *J. Phys. Chem. B.*, 107, 8172, 2003.
80. Braunlin, W.H., Strick, T.J., and Record, M.T., Jr., Equilibrium dialysis studies of polyamine binding to DNA, *Biopolymers*, 21, 1301, 1982.
81. Latt, S.A. and Sober, H.A., Protein-nucleic acid interactions. II. Oligopeptide-polyribonucleotide binding studies, *Biochemistry*, 6, 3293, 1967.
82. Porschke, D., Dynamics of DNA condensation, *Biochemistry*, 23, 4821, 1984.
83. Wolfert, M.A. and Seymour, L.W., Atomic force microscopy analysis of the influence of the molecular weight of poly(L)lysine on the size of polyelectrolyte complexes formed with DNA, *Gene Ther.*, 3, 269, 1995.
84. Burbea, M. and Prevelige, P.E., Jr., Self-assembly of bacteriophage, in *Self-assembling Complexes for Gene Delivery*, Kabanov, A.V., Felgner, P.L., and Seymour, L.W., Eds., John Wiley & Sons, New York, 1988, p. 51.
85. Mueller, O., Hahnberger, K., Dittmann, M., Yee, H., Dubrow, R., Nagle, R., and Ilsley, D., A microfluidic system for high-speed, reproducible DNA sizing and quantitation, *Electrophoresis*, 21, 128, 2000.
86. Decher, G., Schlenoof, J.B., Eds., *Multilayer Thin Films: Sequential Assembly of Nanocomposite Materials*, Wiley-VCH, New York, 2003.
87. Dubas, S.T. and Schlenoff, J.B., Swelling and smoothing of polyelectrolyte multilayers by salt, *Langmuir*, 17, 7725, 2001.
88. Cha, J.N., Shimizu, K., Zhou, Y., Christiansen, S.C., Chmelka, B.F., Stucky, G.D., and Morse, D.E., Silicatein filaments and subunits from a marine sponge direct the polymerization of silica and silicones *in vitro*, *Proc. Natl. Acad. Sci., U.S.A.*, 96, 361, 1999.
89. Patwardhan, S.V. and Clarson, S.J., Silicification and biosilicification: Part 1. Formation of silica structures utilizing a cationically charged synthetic polymer at neutral pH and under ambient conditions, *Polym. Bull.*, 48, 371, 2002.

11 Micro- and Nanoelectromechanical Systems in Medicine and Surgery

Michael E. Gertner and Thomas M. Krummel

CONTENTS

11.1 INTRODUCTION

Identifying the medical applications of microelectromechanical systems (MEMS) and the nanoscale counterpart, nanoelectromechanical systems (NEMS), is highly dependent on the definition one uses and how hard one looks. While some devices clearly fit the definition, others utilize a MEMS/NEMS component technology but are not necessarily 100% MEMS/NEMS devices. For example, a thin film is produced using a micro- or nanoscale manufacturing technology, but the thin film is used to deliver a pharmaceutical from a more macroscopic device (e.g., a blood vessel stent). With a relaxation of the definitions, it will be apparent that there is an arsenal of technologies in use and even more in the development phase that address every major disease process.

0-8493-1940-4/05/$0.00+$1.50

The intent of this chapter is to introduce the reader to these technologies, focusing on clinical examples and broad concepts in NEMS and MEMS. It is expected that, by emphasizing broad concepts rather than details, the reader will take away a sense of where and in what context new technologies have and can be applied to medicine. Technological details are largely left to an exhaustive list of references at the end of the chapter or are clarified in the context of specific examples.

An attempt is made in what follows to integrate various technological disciplines into a problem-based approach where the technology is the tool rather than the end in itself. Advances are often made at the intersections of traditional disciplines, and we are currently on the verge of many such advances at the intersection of medicine and miniaturized technologies. Physicians, biomedical scientists, and engineers have the potential to significantly impact the future and need to understand a broad range of technologies at a level of detail where they can be combined to solve clinical problems economically and efficiently, much as has been done in the field of genomics. It will be impossible to master each technology; however, a little knowledge and insight will allow for productive collaborations with the proper experts.

This chapter is organized by disease categories and/or procedural categories. The only "basic science" section is the one that pertains to materials science. Materials science is singled out in order to emphasize its importance in NEMS and MEMS and, for that matter, medicine. Terminology and basic concepts are introduced and linked to their practicality in the clinical world.

11.2 MATERIALS SCIENCE

Materials science is the foundation of MEMS and NEMS and, in reality, has been the foundation for most advanced technologies. The electronics industry was transformed by the ability to uniquely manipulate the electronic properties of silicon. The fiber optic industry evolved from some basic findings regarding the optical properties of glass tubes. In the past, advances in biomaterials have been much slower. This is evidenced by the fact that the same materials have been used for the last four decades with modifications in form but not composition of the materials (stainless steel, cobalt chrome, polytetrafluoroethylene, polyester, polypropylene).[1] As understanding of the limitations of these materials and of the material-biological interaction increases, a biomaterials revolution can be expected.

In the next decade, there will be a revolution in materials science due to the introduction of nanotechnology. An unprecedented diversity of materials and material combinations will allow devices unimaginable in the past. There will be change in every field, and with proper planning, there will be dramatic change in almost every field of medicine, including both diagnostic and therapeutic modalities. A glimpse of the future is provided below.

A great example of the impact of biomaterials can have on medicine comes from the field of general surgery where the relatively simple polypropylene (Marlex™) hernia mesh changed herniorrhaphy forever. It was a decade before the mesh was accepted by the surgical community, but once accepted, it quickly became the standard of care, eliminating decades of debate as to which is the best operation. Another great example is that of nitinol (equiatomic nickel–titanium), the metal alloy with the property of shape memory. Its acceptance required two decades because of fears for its biocompatibility. Now that it is accepted, however, it has spawned a new generation of devices based on the shape memory properties of the nitinol alloy.[2] Again however, despite great success in the initial introduction of these materials, improvements have been slow to nonexistent over 30–40 years.

Devices are produced from materials, which are the enabling technologies behind the devices. Typically, there exists an evolution where materials are developed and characterized, and then devices are produced from them. In nanotechnology, we are at the materials characterization stage, and true "nano" devices do not exist at this point in time. However, some remarkable medical applications are evolving from small changes in materials processing based on an understanding of surface interactions at the nano scale; these will be described in later sections.

In MEMS, more is understood about the materials, thanks to three decades of research in the semiconductor industry. Therefore, both material and device applications exist. An important concept in device manufacturing is that of "top down" versus "bottom up." This distinction can be translated to a conceptual, even philosophical, difference between MEMS and NEMS. "Top down" refers to the concept of starting with a raw material and shaping it into a device or material; for example, in a typical MEMS device, silicon is etched, heated, etched again, doped with various materials, etc.; finally, it is cut into a shape or form.[3] As would be expected, the processing steps require tremendous energy and capital; more importantly, as the scale is reduced further to the submicron level, the costs increase further, and the margin for error decreases.

In the nascent field of nanotechnology, the underlying conceptual principle is "self-assembly"; here, the ingredients are placed together in a thermodynamically favorable environment and consequently self-assemble into materials. This is much like a biological system (e.g., a cell membrane). An example of self-assembly from the organic, nonbiological world is the spontaneous, single-molecule array thiol-organic molecules (containing a sulfur-hydrogen functional group) form on gold surfaces.[4]

The classic engineering materials are metals, polymers, and ceramics. Semiconductors are also considered a class of engineering materials in and of themselves,[5] although, as a class of materials, they are distinguished by their technological applications, as opposed to their fundamental structure. As nanotechnology enters into the semiconductor world (or vise versa), organic molecules are now being used as semiconductor elements (i.e., organic-inorganic composites).[6,7]

Composite materials consist of combinations of basic materials (traditionally in layers), whereas alloys consist of homogenous elemental combinations. An alloy is a mixture of elements and a material in its own right with a homogenous structure, whereas composites retain individual structures within one larger structure.[8,9] Composite materials are indeed fascinating, because the unique advantages of two different materials are combined to produce a novel material. For example, sintered orthopedic biomaterials are composed of a first bulk metal, on top of which a layer of micron beads is welded, or "sintered," via heat treatment.[10] The bead layer allows for ingrowth of bone, while the bulk material is responsible for the mechanical stability of the structure. The perfect biocompatible material may in fact be one in which a bulk material is artificial and the surface is seeded with cells (e.g., an endothelialized PTFE vascular graft).[11]

In the case of orthopedic implants, the "sintered" surface results in a higher rate of bony ingrowth than the metal implant alone.[12] This was an almost serendipitous finding as the prevailing dogma until that time was that smoother surfaces were superior for cellular integration. As it turns out, osteoblasts migrate into sintered pores and deposit collagen matrix, whereas this does not occur as efficiently on smooth surfaces. In some cases, with sintered materials, bone cement is not needed for implant fixation, because the apposition induced by the sintered beads can alleviate the need for extra fixation.[10,13,14]

Hydroxyapatite-coated implants are another recent advance in orthopedic biomaterials. Rather than a sintered coating, the ceramic hydroxyapatite[15] coating induces bony ingrowth by mimicking ("biomimicry") the crystalline nature of bone.[16] Despite years of research, however, the hydroxyapatite crystals or grains remain much larger (micron scale) than those found in natural bone (nanometer scale). Research is currently directed toward decreasing the size of the crystal grains for further "mimicry" and more uniform bony apposition. Further attempts at biomimicry involve the attachment of biologically relevant molecules to the structural surfaces above. In the case of orthopedic implants, the RGD peptide (the major cell attachment site in many structural proteins) and bone morphogenic protein have been shown to enhance cell attachment and spreading when they are integrated into the hydroxyapatite surface.[17]

An example of an alloy is nickel titanium. Whereas nickel or titanium alone do not have suitable properties for use in many implants, their combination into an approximately 50:50 alloy (called nitinol) creates a new material with unique "shape memory properties" and the ability to recover its original shape after being strained.[2] For many years, nitinol was considered bio-incompatible due to its nickel content; however, as was learned through advanced surface characterization techniques, titanium is the predominate surface element and forms an oxide at its surface which is biocompatible and corrosion resistant compared to pure nickel.[18] Similarly, stainless steel is "stainless" because chromium complexes with oxygen to form a very stable chromium oxide bond at the surface.[19,20]

The next example is illustrative of the intersection between nanotechnology and conventional materials science: the thickness of the surface oxide layer on both stainless steel and nitinol is on the order of 20 nm (nanometers or 20 billionths of a meter). Although the oxide layer is considered biocompatible, repeated mechanical stress in the chloride-rich *in vivo* environment eventually leads to degradation and device failure.[21] Nickel-titanium appears to be more susceptible to this type of attack than is stainless steel. Attempts to purify the surface oxide using various electrochemical processes has resulted in improved passivation and corrosion properties over the past few years.[22] The reason for the deterioration of the titanium oxide layer in nitinol is two-fold: (1) the brittleness of the titanium oxide surface oxide layers[22] and (2) the issue of "repassivation time constant," which refers to the ability and time for the oxide layer to reform on the surface after it is mechanically abraded. Recent work has attempted to further improve the brittleness and repassivation time constant of the surface oxide layer by incorporating metallic nanoparticles into the surface oxide or adding a thin, (i.e., nanometer thickness) polymer film to prevent electrochemical corrosion.[23–25]

As important as the nature and composition of material is, an equally important concept is to understand how materials are processed and how these steps are factored into obtaining the desired result. These issues become particularly relevant when composites such as endothelialized vascular grafts are considered or stents that elute drug; in this case, certain processing techniques (e.g., high temperature) will destroy the biologic component.

As touched on above, MEMS and NEMS, in their true definition, are manufacturing processes related to either "top down" or "bottom up" manufacturing processes. For example, in a MEMS device, one starts with a substrate (silicon, gold, etc.), and a structure is produced from it. The order of scale in MEMS is .5–10 microns. A way to conceptualize MEMS devices is as a series of films deposited on top of one another. As the layers are successively added, a three-dimensional structure is created. The particular material chosen typically has a desired functional property associated with it. For example, a silicon thin film exhibits piezoelectric activity[26] (i.e., the ability to change shape with applied voltage). A very real problem with MEMS is that the processing steps are not necessarily compatible with biology. Silicon MEMS was taken from the microchip industry where there are three decades of experience. There is in fact much evidence that silicon is not tolerated well systemically[27] and that it will not protect the device from biological corrosion.[28]

Biologic modifications (i.e., composites) of the silicon may allow for biocompatibility with cells as has been shown with islet cells.[29] There is also evidence that careful processing of silicon materials (i.e., to allow for more stable oxides) is critical to improvements.[30] Future MEMS devices may be derived directly from polymers[31,32] or composite devices and metals coated with conducting polymers.[24] Alternatively, polymer structures can be derived from silicon structures by first using the silicon structure as a template. Indeed, polymer replicas of porous silicon materials have been devised for drug delivery applications.[33]

If the biologic constraints are engineered around, future hybrid materials will combine traditional materials with technology to produce a new class of medical devices that transmit data, actively release pharmaceuticals, etc. (for example, a stent that transmits data about its surrounding mechanical forces, or an orthopedic or podiatric implant that transmits stress and strain data so that further device iterations can be based on *in vivo* data).[34]

The manufacturing processes in NEMS are potentially much more compatible with biology. In NEMS, the order of scale is 1–500 nm. As mentioned above, true nanoelectromechanical systems do not exist, and therefore, the term "current NEMS technology" denotes materials which exploit unique properties observed at the 10- to 500-nm size scale. In medicine and biology, the major advantage at this size scale is to enable the materials (or particles) to find places (e.g., body compartments) they otherwise would not be able to find. Much of NEMS today is about nanoparticles, materials with dimensions on the order of 100 nm or less. The fact is, NEMS is a scaled down version of particle research that has been ongoing for decades. The first use of microparticle composites was copier paper, where specks of dye (carbon black—i.e., carbon copy) were attached to a micro bead and then attached to paper.

The micro bead allowed for attachment of the carbon while allowing the paper to retain its flexible structure.[35]

As alluded to above, the materials science philosophy is completely different from MEMS. NEMS materials are produced from "self-assembly" processes in which mixtures of materials are allowed to condense into particles or materials or composites.[36] This type of process is a "bottom up" process in contrast to "top down" MEMS. NEMS processing starts with a nonsolid phase, typically a solution, and by manipulating the environment, materials are created in the solution.

More recently, biologic molecules (e.g., DNA and proteins) have been used to stabilize nanoparticle crystals and create materials with unique properties, opening the door to tremendous diversity in the next generation of nanoparticles and materials formed from the nanoparticles.[37,38] These processes mimic nature's ability to produce such fantastic materials as pearls, coral, calcite microlenses, and collagen.[39] Gold nanoparticles have been incorporated into the active centers of large enzymes such as glucose oxidase allowing for electron conduction an order of magnitude higher than what is seen without the nanoparticle.[40,41]

As an extension of these definitions, the quantum dot is defined. Quantum dots are metals and semiconducting particles on the scale of 10–20 nm. Their name is derived from the fact that they exhibit special "quantum" effects at this size scale. Without delving into the physics, an example of a quantum property is emission of a specific wavelength of light by a particle when stimulated with ordinary white light. The same material would not possess this property as a larger particle. Fluorescent tags currently used in the biological arena have equal intensity; however, the fluorescence fades with time. In the case of quantum dots, the fluorescence is permanent as long as the particle remains a particle. Quantum dots are also much smaller than organic dyes, allowing their attachment to biomolecules without major structural changes.[42,43] The quantum dots can be organized into *in vitro* patterned assay systems which detect very small amounts of tumor markers such as prostate specific antigen.[44]

Finally, it must be said that NEMS and MEMS are not mutually exclusive. There is no doubt that in the future, the two will complement one another. MEMS devices may produces NEMS devices, and NEMS materials will provide the material structure for MEMS devices.[45] This point relates to the earlier one regarding the intersection of different technologies as being paramount to new discoveries.

11.3 DISEASE-SPECIFIC MICROTECHNOLOGIES

11.3.1 DIABETES

An extraordinary amount of effort has gone into the ability to detect glucose. As every physician is aware, tight control of glucose leads to fewer diabetic complications. More recent data even reveal that tight glucose control in the relatively short peri-operative ICU setting leads to lower mortality.[46,47] With 16 million Americans currently affected by diabetes and 20 million expected to be afflicted by 2010, the ability to efficiently detect and deliver insulin has major public health implications and an estimated savings to society in the billions.

The most common detection technique at the present time is an electrochemical method that utilizes the enzyme glucose oxidase, which reacts with glucose to generate hydrogen peroxide. Hydrogen peroxide is easily detected by a direct electrochemical method which generates a current.[48,49] Heller et al.[40] developed a method of "wiring" glucose oxidase to an electrode by incorporating the enzyme into a conductive hydrogel which is in contact with a larger gold electrode. This development allowed for smaller quantities of glucose to be detected and formed the foundation for the company Therasense™. Smaller quantities of blood mean that the patient can sample blood from places other than the fingertip (i.e., less painful spots), improving the quality of life for millions of diabetic patients. Newer technologies are further developing this concept, incorporating glucose-sensing microparticles into conductive films to create electrodes with useful structures.[48] Commercialization of such a technique may one day lead to devices such as a contact lens with glucose detection properties.[50]

An independent aspect of the detection problem is that of obtaining the sample. Physicians and patients have been using the finger stick method for sample acquisition over the past five decades. However, with advanced technology, this will change rapidly.

Iontophoresis (Figure 11.1) is the application of an electrical current across a membrane such as the skin. The current attracts positively charged ions such as sodium and is strong enough to pull sodium through the skin.[51] Experiments with this technique led to the observation that water and uncharged molecules such as glucose and urea were pulled across the membrane with the ions such as sodium and chloride; this effect is known as solvent drag. Interstitial glucose was found to correlate well with serum glucose, and this observation allowed for the development of a novel method to obtain samples of glucose through the skin with minimal invasiveness.[52] Low-frequency ultrasound has further been used to modify the per-

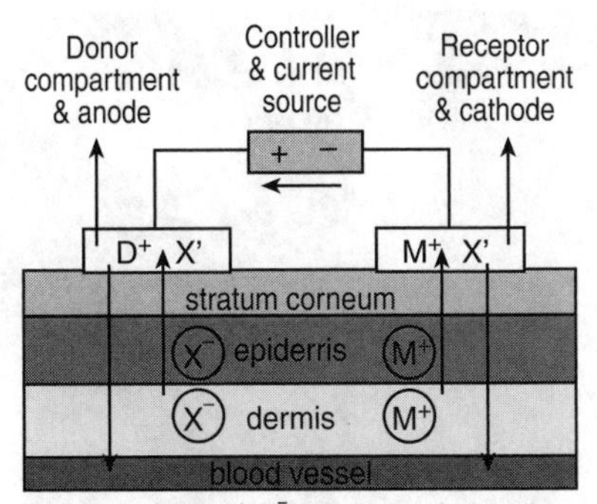

FIGURE 11.1 Schematic of an iontophoresis system. The cathode (negative charge) attracts positively charged ions from the interstitial compartment. To complete the circuit protons from the battery are delivered back to the interstitial compartment. The positively charged ions "drag" uncharged molecules and water with them. (Adapted from Subramony, J., Iontophoresis: application of electrochemical materials and methods for therapeutics and diagnostics. *Interface*, 2003, 12(4), 27. Reproduced by permission of The Electrochemical Society, Inc.)

meability of the stratum corneum so that a larger sample of interstitial fluid can be obtained.[53,54]

A more direct method is that of perforating the stratum corneum with an array of microneedles (Figure 11.2). The stratum corneum is 10–15 microns thick, and it provides most of the resistance to external fluids. Just below this layer is a noninnervated layer filled with cells and interstitial tissue. In theory, and now proven in practice, penetration into this layer decreases the barrier to fluid transfer yet is not painful during injection.[55] The example in Figure 11.2 was produced exclusively with a microfabrication process (top down) borrowed from the electronics industry. The microneedle array was manufactured via a series of etching steps beginning with a silicon wafer. The microneedles can also be manufactured by an electrodeposition (electroplating) process,[56–59] where structures are built from successive deposition of metallic films on a substrate. This is another example of a bottom up process.

Indeed, the Glucowatch™ made by Cygnus corporation is a recently approved device and represents a major advance in diabetes management. The watch utilizes three different MEMS technologies, each of which takes the form of a thin film and each of which performs a different function in the overall device. The device is organized as a series of layers on the back side of a watch, with one layer contacting the skin. Glucose is drawn from the skin using an iontophoretic process. The sensing aspect consists of glucose oxidase incorporated into a conductive hydrogel; the hydrogel retains the activity of the enzyme active and permits a conduction path. Platinum–graphite is used as an electrode, as is a silver electrode for reference (this for the electrochemical detection scheme for the generated hydrogen peroxide). This is the conducting layer, which leads to the electronic portion of the device.[60–62]

More advanced technologies will possess the ability to actively control transport through the pores. For example, a recent published article measured glucose via an

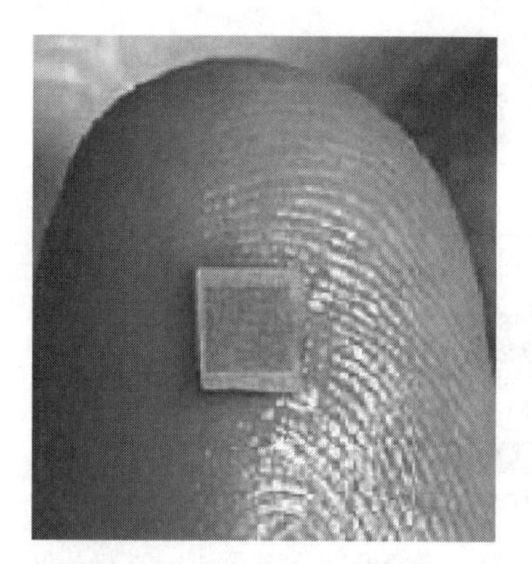

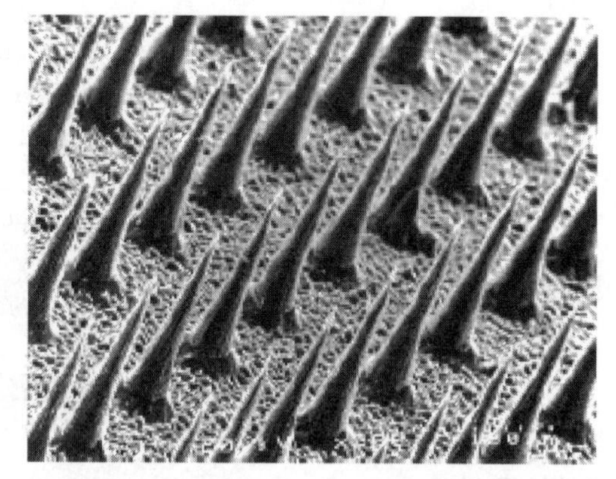

FIGURE 11.2 Microfabricated needles for drug delivery. The microchip shown on the left contains thousands of the small needles shown on the right. Each needle penetrates the stratum corneum to extract or inject fluid. (From Henry, S., et al., Microfabricated microneedles: a novel approach to transdermal drug delivery. *J. Pharm. Sci.*, 1999, 88(9), 948. With permission from Lippincott Williams and Wilkins.)

ultrasound device, which delivered low-frequency ultrasonic energy to the stratum corneum. The stratum corneum underwent a presumed rearrangement, which increased the permeability of the skin severalfold so that interstitial glucose concentration could be determined.[63]

Insulin *delivery* has likewise seen very little change over the past several decades and is currently seeing an extraordinary amount of effort devoted to novel delivery methods. Every possible route of delivery is or has been attempted or is in the product phase (i.e., pulmonary, transdermal, oral, nasal, implanted, etc.); pulmonary delivery is in advanced phase III clinical trials. Most of these technologies involve micro- and/or nanoscale materials and/or devices in one way or another.

In the pulmonary route, micron-sized insulin particles are aerosolized and deposited in the distal alveoli of the lung. The device has two components: the large handheld portion to activate delivery from the canister and the microtechnological methods required to create the aerosols. Two aerosol compositions and associated methods currently utilized in current trials are categorized as either *wet* or *dry*. Whether wet or dry, there are several requirements for successful delivery to the blood stream:

1. To reach the tracheobronchial tree and not be trapped in the proximal nasopharynx, the size requirement for the particles is 1–3 μm.
2. They must flow freely without aggregation.
3. They must have a controllable degradation rate.
4. The particle surface must interface with the pulmonary epithelium in such a way so as to be able to cross the barrier.[64]

In the dry aerosol composition, particles in the form of a flowable polymer powder are formed first, and a dry gas is used to create an aerosol plume.[64] In the wet composition, aerosolized particles are formed through a nozzle upon exit from the device. A liquid composition containing insulin (or any drug, for that matter) is forced through a membrane with microfabricated pores in order to form an aerosol.[65] While these technologies are specific to insulin, they represent a set of novel methods to produce micro- and nanoscale particles which can be translated to other fields, including nonmedical fields. They are further applied to delivery of other types of pharmaceuticals.

The dry particle formulations include a proprietary enhancer molecule for entry across the pulmonary epithelium and a second molecule to prevent particle aggregation.[64] The formation of the particles is achieved by a variety of methods, depending on the starting liquid formulation. In one method, an aqueous solution is pushed through a stream of a gas such as carbon dioxide. The mechanical interaction between the two streams produces a mixture of particles (droplets), which are initially aqueous but then quickly dry as the liquid carbon dioxide evaporates, leaving dry particles containing enhancer, stabilizer, and drug.[66] Alternatively, biodegradeable particles are created in solution and dried by allowing the solvent to evaporate.[67] The art of this technique is finding an enhancer and stabilizer that will migrate to the surface as the particle dries.

In the competing "wet" methodology, an aqueous solution of insulin is forced through a screen with micron-sized perforations.[65,68,69] In this case, the pharmaceutical aerosol is generated within the delivery device as the patient inhales. The screen inside the device contains microfabricated pores created by a laser micromachining process. As a viable technology, this method is more difficult to develop and obtain FDA approval for, because the stored "wet" solution is more prone to infection and the onset of action is slower than the dry powder.

The holy grail for insulin delivery is a closed loop diabetes management system (i.e., one in which the glucose is detected and insulin delivered from the same system), the so called "artificial pancreas." There is no doubt that one day in the near future, a closed loop glucose control system will be realized. What is unclear however, is which technology will achieve this goal; however, microscale technologies are being applied in a variety of ways. Some technologies concentrate on an integrated and completely microfabricated delivery[70] and detection system, whereas others utilize live pancreatic islet cells with a nanoporous membrane to protect the cells from the immune system yet allow for glucose and insulin transport to and from the system (Figure 11.3).[29] In other cases, polymeric delivery systems have been created which increase or decrease their porosity in direct proportion to glucose concentration.[71]

In one example of a microfabricated device, a system of microfabricated pumps and valves are utilized.[72] Silicon MEMS is utilized, in which insulin-containing fluid is pumped by applying energy to a piezoelectric material (silicon dioxide in this case). Alternatively, a fluid-vapor system is used; in this system, the fluid quickly vaporizes through applied heat or electrical current. The gas pressure imparts force on a membrane, which pushes the pharmaceutical-containing solution. The gas quickly becomes a liquid once heat or electrical current is removed (this example is a miniature version of the larger chemotherapy pumps used for continuous hepatic artery infusions). Another method is a miniaturized electrolysis system whereby gas pressure is gener-

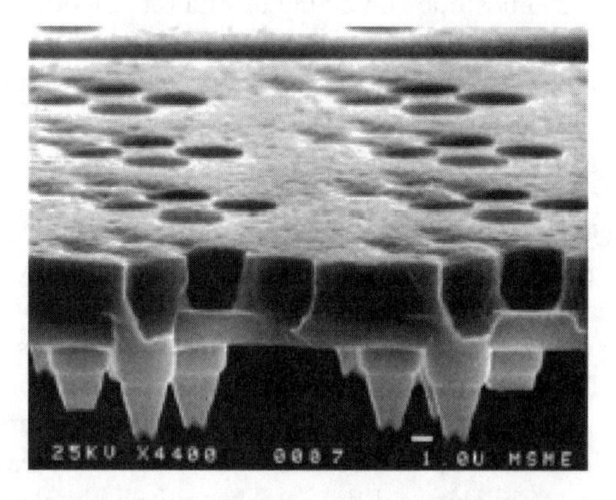

FIGURE 11.3 Capsules fabricated in silicon. Islet cells trapped in these capsules were effectively "immunoisolated" and protected, able to survive for a longer period of time than without the capsules. (Adapted from Desai, T., et al., Nanopore technology for biomedical applications. *Biomed. Microdevices*, 1999, 2(1), 11–40. With permission from Kluwer Academic.)

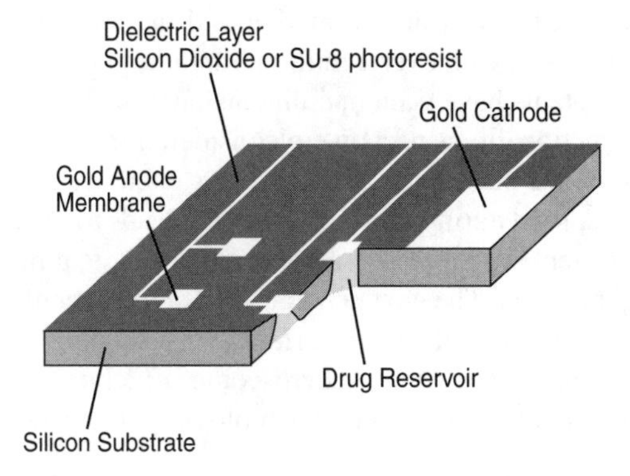

FIGURE 11.4 A depiction of an implantable drug delivery system in which drug is stored in the microfabricated wells. When a current is applied to gold caps on the wells, the gold membrane dissolves and the drug is released. (Adapted from Voskerician, G., et al., Biocompatibility and biofouling of MEMS drug delivery devices. *Biomaterials*, 2003, 24, 1959–1967. Copyright 2003, with permission from Elsevier.)

ated through the splitting of water into its component gases, hydrogen and oxygen.[70] Figure 11.4 depicts another silicon MEMS device currently in clinical trials. A pharmaceutical formulation is stored in micromachined wells covered by a gold membrane. At the specified time, the membrane is electrochemically degraded by application of a current across the well, thereby releasing the pharmaceutical.

Another approach to the closed loop problem is to utilize live islet cells and encapsulate them into a membrane to prevent rejection. The membrane is small enough to allow for transport of insulin and glucose yet prevent immune proteins and cells from entering the structure. Indeed, a micromachined, nanoporous silicon membrane has been devised with pore sizes on the order of 10–30 nm (Figure 11.3).[29,73] In this instance, the pores are ultimately formed by creation of a 20-nm oxidation layer sandwiched in between two silicon wafers; when etched, the space in between the sandwich becomes a 20-nm hole. This technology is a great example of the intersection between the micro, nano, and biological world.

Review of the application of microtechnologies in the field of diabetes is an example of the power of "micro" and "nano" when applied to one specific disease. These techniques are the "tools" and, with adaptation, can be applied to many other problems in medicine. This type of "field crossover" is particularly relevant in the next section.

11.3.2 Cardiovascular Disease

The broad field of cardiovascular medicine stands to benefit tremendously from advances in miniaturized technologies. Technologies related to the cardiovascular system include diagnosis, drug discovery, drug delivery, remote monitoring, minimally invasive therapies, and electrophysiology. Despite the progress in the past few decades with cardiovascular devices, there remain significant areas (heart failure and sudden cardiac death are two of the biggest) for improvement or entirely new devices. Microtechnologies again will be at the forefront of enabling these changes.

The second largest medical device market and the most commonly prescribed medical device is the stent, with over 1 million implants per year in the United States alone. Traditionally, stents have been one-dimensional in their function, in that only their mechanical functionality aspects were considered important. Stents were considered to simply be "a steel cage used to prop open arteries." As it turns out, however, the injury caused during stent implantation leads to restenosis, or a scarring reaction around the stent. There is also evidence that the stent material also contributes to the scarring process.[1] The next generation of stent technology is now reaching the clinic, and it is focused on the stent surface.[74,75] Although the major mechanical function of a stent is related to its macroscopic structure and properties of its composition metals, some of the important biological interactions (i.e., restenosis) occur at the stent surface.[76,77]

The surface is the interface between the bulk material and the environment. In completely inert, homogeneous metals such as gold, the surface is the same as the bulk. Most elements are not homogenous and/or inert. Stainless steel, as mentioned above, has a high concentration of chromium in its bulk, which ultimately migrates to the surface of the material during processing. Typically, the transition from the bulk to the surface occurs over a thickness of tens of nanometers. It is not surprising that this nanolayer is critical to the mechanical properties of the stent and ultimately its longevity.

Drug eluting stents are a second-generation technology and are changing the practice of interventional cardiology. The current generation of drug-eluting stents have a 5- to 10-μm coating made from a polymer (an entirely new surface as far as the blood vessel is concerned) which releases a drug beginning at the time of implantation.[78,79] The polymer coating is very thick on these first-generation stents (approximately 10 μm) and is a permanent coating. The drugs, rapamycin and paclitaxel, diffuse into the surrounding milieu to prevent the fibrotic reaction, which occurs after stent implantation. Early clinical data show that drug-eluting stents have restenosis rates under 10%.[79] It remains to be seen, however, whether longer-term data, where more chronic processes may be involved, will hold up as well. Indeed, many questions remain regarding the ability of the polymer to adhere to the stent struts.

Early success notwithstanding, there will be many improvements along the way before the ideal drug-coated stent is discovered. The ideal stent will likely be engineered to optimize the integrated device, including drug, coating, and stent. In this respect, the three components will act synergistically, rather than independently as they do now. Current polymer coatings have poor adherence, tending to flake upon handling and delaminate from the stent during implantation; it also becomes increasingly difficult to coat stents as the size of the stent struts is decreased. Newer stent designs incorporate MEMS or MEMS-like technologies to customize the stent surface to control specific processes by incorporating drug directly into the substance of the stent. These technologies borrow manufacturing processes from the semiconductor industry to deposit coatings on the order of microns onto the surface of the stents. In one example, an entire stent is manufactured through the process of gold electroforming, where an electroplating process is used to coat a mandrel in a predefined pattern; the mandrel is then dissolved away leaving a solid gold stent.[80]

A similar process with polymers may yield smaller, thinner stents which are entirely composed of a polymeric material.[45]

Next-generation anti-restenosis stents also focus more exclusively on the surface interactions and the individual elements on the surface, rather than providing a drug to overcome the intrinsic deficiencies of the stent.[76] Shown on the left in Figure 11.5 is an electron micrograph of a typical surface from a commercially available stainless steel stent.[25] These imperfections in the surface are a result of the extensive laser-based cutting the stents undergo. These large pits serve as the epicenters for crevice and pitting corrosion wherein metallic elements from the stent substance are released.[81,82] They also serve as weak links that are thought ultimately to lead to cracking seen in larger devices.[21] The figure on the right depicts the same surface after a metallic thin film (less than 1 μm thickness) is deposited via electroless electrodeposition.[25] Note the small grains, approximately 30–50 nm in diameter. The small, uniform grains dramatically decrease the rate of crevice corrosion by spreading the corrosion process in many (almost infinite directions). The added metallic film, which is approximately 1–3 μm, is also porous on a nanometer scale, with a porosity of up to 25%, depending on which ingredients are used.

The further advantage of the porosity is that drug can be loaded into the pores. Similarly, other substances can be introduced into the pores, such as radio-opaque materials (i.e., barium, bismuth, and calcium), pharmaceuticals, and PTFE (for increased lubricity). Because the films can be individually deposited, each at smaller than one micron, a coating can be created with varying beneficial properties in each layer (i.e., radio-opacity, delivery of drug x, delivery of drug y). The film is also a metallic film, so that the film itself carries mechanical properties, in contrast to a polymer film that adds bulk but not mechanical strength. This example is another of "bottom up" engineering principles and the ability to add or subtract properties to and from the surface coatings.[83] Additional advantages include ease of manufacturing and greater flexibility. Future coatings will incorporate novel polymers activated at the surface of the stents by electric fields.[32,84]

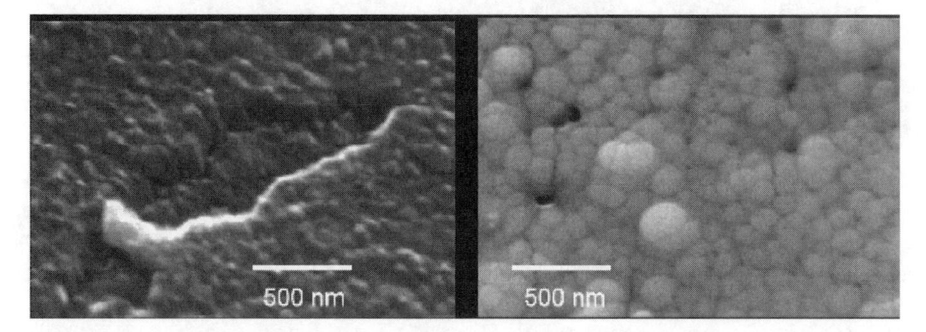

FIGURE 11.5 On the left is a high magnification SEM of a finished metallic surface from a commercial stainless steel stent. The large micron-sized crevices provide a nidus for corrosion, particularly under bending loads. On the right is a prototype surface modification in which metallic grains <100 nanometers cover the defects. Small, relatively uniform grains spread out, or diffuse, surface stresses and deformation stresses so that corrosion effects are delayed by orders of magnitude. (From Nanomedical Technologies, Inc.)

Many investigators, including Dr. Julio Palmaz (the inventor of the modern stent), consider the stent material and surface properties to be paramount to restenosis.[76] These hypotheses are leading to a new generation of stent materials[85] based on controlling the purity of the stent surface. Such developments involve new and sometimes older processing technologies such as plasma vapor deposition (a top down technology). Once MEMS technologies to produce the stents are put into place, the microscopic topography can also be modified to accommodate healing processes such as endothelialization. In this context, the topography of the surface is modified on the micron scale to accommodate certain cell shapes. Cell shape, as related to differentiation and attachment, has been widely studied and is an attractive approach to selecting cell types on a surface.[86,87] The surface thrombogenicity can be further modified by linking the anticoagulant to the surface that in early studies with nitinol has shown an antithrombotic effect.[88]

11.3.3 Remote Sensing

Implantable vascular devices are being taken to the next stage of miniaturization, with MEMS taking the lead in allowing these advances. Newer devices have miniature sensors incorporated into them to provide real-time telemetric feedback to physicians for patient care. Medtronic Corp. will introduce an intravascular pulmonary artery pressure transducer to monitor fluid status in patients with congestive heart failure. The device is chronically implanted and transmits data to a physician over the Internet so that diuretics may be adjusted appropriately.[89] Initial experience with chronic pulmonary artery monitoring showed that patients tolerated the device well but clinical efficacy needs to be established.

Clinical trials are also under way with a pressure transducer to monitor the extraluminal pressure on an aortic stent graft in order to detect endoleaks. In August 2003, the first such device produced by Remon Medical Technologies was implanted in a patient at Mount Sinai Medical Center in New York.[90] Remon is also developing an intracardiac pressure monitor. In the research phase is a MEMS device in which cardiac cells attach to a force transducer. Strain on the device is then measured and relayed to an external monitor. Such a device can be used to optimize the amount of an ionotropic drug a patient needs.[91] Such a device is used in the research setting as a rapid assay for new ionotropic drugs.

This is only the beginning for this type of technology. Future coronary stents with flow sensors will detect restenosis as it occurs. They may have dual uses, such as acting as a Holter monitor or detecting arrhythmias. Potentially, stents will allow for drug release in response to a physiologic variable, as opposed to passive release.[36] In the orthopedic and dental arenas, strain gauges can be used to predict failure of implants or to monitor *in vivo* strains to optimize implant design.[34,92]

On the technology side, the components of these devices are not new (i.e., radiofrequency transmission of information, sensors, etc.). What has allowed for the advances is the miniaturization of energy creation and advances in biocompatibility (i.e., materials science). Along these lines and almost in the realm of science fiction is a fuel cell or battery which derives energy from glucose.[93–95] It takes advantage of the same system as is used for glucose detection. In this case, the electrons generated by the oxidation of glucose are utilized as an energy source.

Nanoparticle technology will advance current diagnostic capabilities. In one example, an anti-myosin antibody is attached to an iron oxide nanoparticle[96] and specific localization to infarcted myocardium is demonstrated by immunohistochemistry. Without label, the iron nanoparticle does not localize to the infarcted myocardium.

11.3.4 Gastrointestinal Tract

Perhaps the only true MEMS device in clinical use at this point is the capsule colonoscopy by Givens™ Technology. Contained in a 1-inch package are two silver oxide batteries, white light emitting diodes to illuminate the camera, and a metal oxide detector array with 256 × 256 bits. The capsule transmits 50,000 images over the 7 hours in which it passes through the GI tract. The images are transmitted externally via a radiofrequency communicator to a receiver belt, which is worn on the patient's waist. A physician then downloads the pictures for review.[97,98] This device is a revolutionary advance and truly deserves to be called a "BioMems" device.

The photosensing metal oxide array is created using a microdeposition process. The metallic stent film shown in Figure 11.3 is in fact a metal oxide layer. In the case of the Givens device, the metallic film collects photons and coverts them to electrical current. The deposition process used in this stent surface is similar to that used in creation of the capsule. The energy source is a standard battery power supply, which lasts 7.5 hours. If the battery were able to last longer, the device would be useful in the colon as well[98,99] as residence time could be prolonged. Looking forward, it is not difficult to envision a device similar to this one, which is inserted into the amniotic sac of a developing fetus, or for that matter, a blood vessel, or the brain. These devices will be powered by fuel cells, which represent the next generation of battery technology enabling such devices to last longer *in vivo*.

A similar type of device (in form but with a novel function) is under development by a group at Ohio State in collaboration with the company IMEDD.[100] Depicted in Figure 11.6 is a prototype particle, a MEMS particle in this case. The basic concept is a particle fabricated from silicon with a chemically modified surface designed to adhere to the small bowel mucosa. The surface contains permeability-enhancing molecules. The goal of this device is to co-localize a pharmaceutical and permeation enhancers to enable the particle to stick to the small bowel mucosa and thereby allow medications to enter through the mucosal membrane that otherwise would be impermeable. Initial animal results are promising that such a device can increase the bioavailability of drugs otherwise unable to pass through the bowel mucosa.[56,101] With further imagination and pulling some tools from the developing arsenal, one can imagine utilizing a method of iontophoresis similar to the glucowatch to push pharmaceuticals through the membrane. Indeed, such a pill is being developed by Elan Pharmaceuticals.[102]

11.3.5 Oncology

Perhaps more than in any other field, micro- and nanoscale technologies will provide the oncology fields with rapid therapeutic advances over current therapies. Historically, oncology involves a multimodality approach for both diagnosis and therapy. Oncology teams include surgeons, radiologists, radiation oncologists, medical oncol-

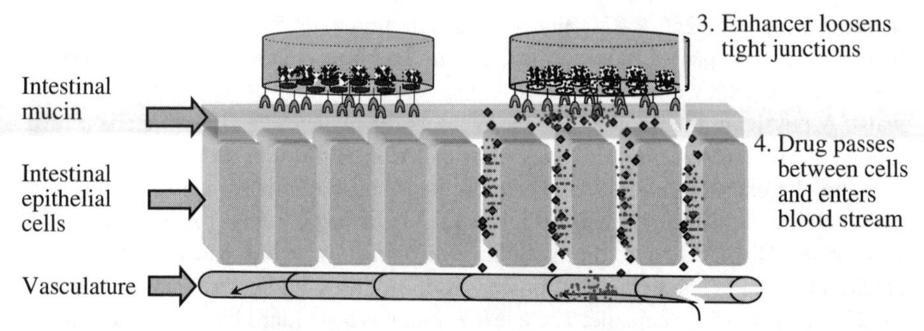

FIGURE 11.6 Depicted is a microfabricated particle whose surface is prepared to accept biologic molecules. A lectin (polysaccharide) is attached to the particle surface to enhance binding to the membrane of an intestinal mucosal cell. The enhancer prepares the tight junctions to accept large protein molecules. (From IMEDD Inc. With permission.)

ogists, pharmacologists, and engineers. Micro- and nanoscale therapies for oncology will likely utilize many modalities—on one particle.

Figure 11.7 depicts a more complex nanoparticle[103] than that described earlier for insulin delivery. This 20- to 30-nm (actually the thickness is variable up to the 100-nm range) particle has a defined structure intended for easy modification via its silicon dioxide shell. Parenthetically, note that silicon dioxide is a major player in MEMS and BioMEMS technology, and this example is illustrative of the intersection between MEMS and NEMS. The silicon dioxide coating can further be functionalized with an organic molecule such as an antibody, a tumor-specific ligand, or a coating to improve biocompatibility or targeting specificity.[104,105]

In the case where the shell is not specifically functionalized for specific uptake, these particles, when injected systemically, specifically localize to tumors because their vasculature has large gaps in its endothelium, permeable to nanoparticles. Normal endothelial vasculature does not contain such gaps, which accounts for the relative tumor specificity. When the core of the material is a magnetic one such as iron, a magnetic field alternating in the radiofrequency range can generate heat in the particle and destroy the tumor cells.[105] A similar system incorporates a pharmaceutical into a hydrogel surrounding the iron particle, and another system even incorporates the pharmaceutical into a hydrogel microparticle.[106] In the case of the

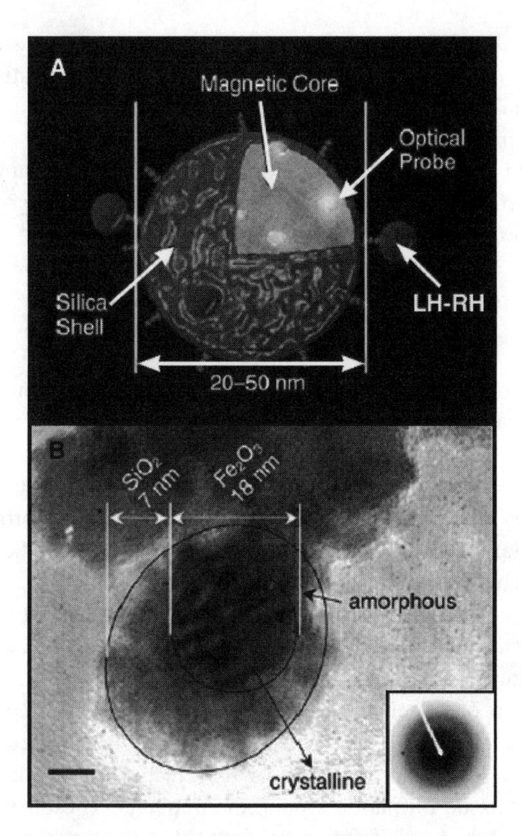

FIGURE 11.7 A schematic of a nanoparticle. An iron oxide core is surrounded by a silicon oxide shell. Ligands attached to the silicon oxide can target the iron oxide to a specific site or potentially a tumor. The iron oxide can be heated in a magnetic field. Alternatively, the iron oxide may carry a toxin, a gene, or pharmaceutical.

iron-hydrogel composite, energy absorbed by the iron particle heats the hydrogel and releases the pharmaceutical locally rather than systemically.[107]

A similar technology can be used for diagnostic purposes and, in some cases, both diagnosis and therapy with the same platform particle. In one diagnostic scenario, the nanoparticles are composed of a gold shell[107,108] and a core material, typically gold-sulfide or silica. The surface of the gold continues to be biocompatible and lend itself to conjugation with bioactive molecules. The size of the inner shell, the outer shell, and the size of the overall particle can be altered to tune the absorptive resonance frequency of the particle. Variation of absorption frequency with particle size is a unique ability possessed by nanometer-sized particles. A near infrared light source (this wavelength chosen because it easily penetrates tissues with little interference) can then be used to detect the tumors.

A recent article published in the *New England Journal of Medicine*[103] detailed one of the first human NEMS trials. Iron oxide nanoparticles were used to help identify metastatic lymph nodes in patients with prostate cancer. These particles were not functionalized but took advantage of a property of metastatic lymph nodes in that they will not take up the iron particles when metastatic carcinoma has infiltrated the nodal tissue. The authors demonstrated increased sensitivity and specificity in identifying lymph nodes, which ultimately contained metastatic cancer. Further work with magnetic nanoparticles functionalized with tumor-specific antibodies will lead to more specific uptake by tumors as well as by small tumors that do not have a blood supply yet.

11.3.6 Applications to Epilepsy, Hydrocephalus, and Neurotrauma

The field of neurosurgery and neurology also has a tremendous amount to gain through the application of sensors, actuators, telemetry, and other micro- and nanoscale technologies.[28] Neural prostheses are being developed in almost every major branch of neuroscience (epilepsy, migraines, Parkinson's disease, paralysis, retinal blindness, memory disorders, Alzheimer's disease, etc.); these prostheses are comprised of a complex of electrodes and circuitry to provide specific anatomic stimulatory patterns. The neural prostheses can be programmed to either enhance or eliminate specific neurologic functions.[28] In one advanced technology in early human trials, a neuroprosthesis is used to both detect seizure signals and deliver an impulse to abrogate a seizure — a pacemaker for the brain.

The scale down of the electrode complex has been possible only with MEMS technology (Figure 11.8). Two problems hindering the further scale down and development of these systems are corrosion of the electrodes and deterioration of the signal due to scar tissue formation around the implants. These issues are similar to those encountered for stents. There is no doubt that nanotechnology and advances in biomimetic surface chemistry will allow for solutions to these problems and even

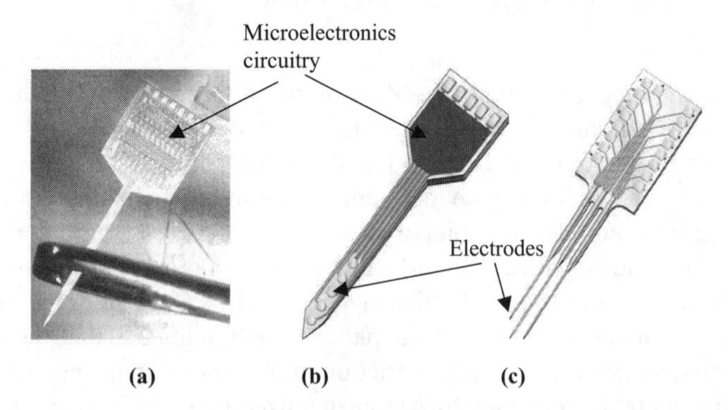

FIGURE 11.8 An electron micrograph (EM) and schematic of an implantable neural electrode for deep cortical stimulation. The EM on the left illustrates the scale of the device as it is passed through a needle. Each electrode stimulates tens of neurons and can also sense seizure activity. (Adapted from Roy, S., et al., Microelectromechanical systems and neurosurgery: a new era in a new millenium. *Neurosurgery*, 2001, 49(4), 779–798. With permission from Lippincott.)

further scale down and integration within the natural system. Indeed, *in vitro* work has shown the ability to select for specific neural cell lineages using MEMS devices coated with biologic ligands.[109]

Neural regeneration touches on the field of tissue engineering and is amenable to a MEMS approach. It is also a great example of the way in which MEMS component technologies can cross over into other advanced technology fields (i.e., tissue engineering). MEMS can be integrated into neural tissue engineering in several ways:

1. The ability to create defined channels through which neurons can grow and/or regenerate
2. The ability to release growth factors or pattern them along the surface on which the neurons are growing
3. The ability to control fluid flow of nutrients to create chemical gradients for neural growth
4. The ability to use the gradients in conductivity to guide the neural elements and/or sense neural signals on the same wafer as 1–3[28]

A biodegradeable, conducting polymer has recently been shown to enhance neurite outgrowth in a model system.[110] The system shown in Figure 11.9 is an evolving clinical application of this research.

Another emerging technology is that of a closed loop shunt for hydrocephalus. Currently, we await a change in neurologic status as an indicator that there is a problem in a typical ventriculo-peritoneal shunt used for hydrocephalus. With an integrated pressure sensor, the intracranial pressure would be quickly detected and a valve adjusted for increased flow. Miniaturized pressure and oxygen sensors are also quite useful for detection of intracranial pressures and oxygen tensions.

Pressure sensing is one area where MEMS has already made a large impact in medicine. Prior to silicon-based transducers, arterial pressure measurements were made using a liquid transduction system in which pressure was transduced through a column of liquid to a pressure sensor.[3] It was then discovered that silicon dioxide possesses piezoelectric properties (that is, voltage generates movements and vice versa).[26] This observation paved the way for the first MEMS device in medicine, the arterial pressure transducer. The fundamental theory is that the silicon dioxide diaphragm is deflected with changes in pressure relative to the atmosphere; the piezoelectric properties lead to a detectable electrical signal, which is processed and calibrated into a pressure reading. For an implantable unit as would be required for a hydrocephalus shunt, the sensor is the same and the signal processing could even be performed externally by a physician. In fact, the processing could take place at a central location (i.e., physician's office) with the raw signal transmitted through an Internet data connection.

11.3.7 Surgical Technology

Surgical technology is a very broad term, encompassing technologies such as ablation devices, scalpels, coagulation devices, laparoscopic technology, robotic micro-

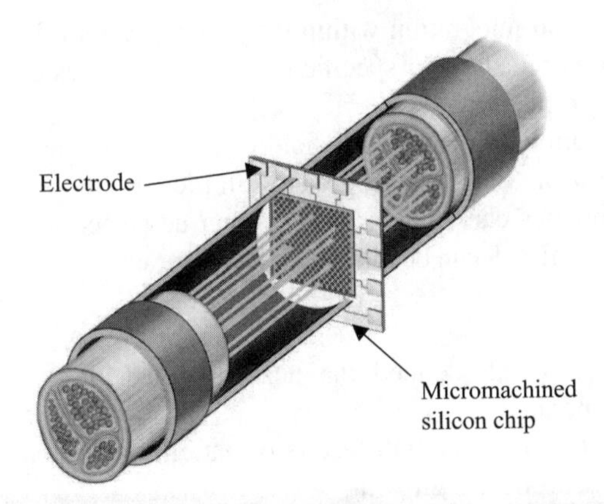

FIGURE 11.9 A neural prosthesis for spinal cord regeneration. The microchip in the center is designed to direct neuron regeneration. (Adapted from Roy, S., et al., Microelectromechanical systems and neurosurgery: a new era in a new millenium. *Neurosurgery*, 2001, 49(4), 779–798. With permission from Lippincott.)

manipulators, and tissue engineering. More recently, engineered biomaterials such as adhesive glues and tissue sealants have come to market and are rightfully placed in this category. These glues include materials such as Tisseal™, which contains a combination of thrombin and fibrinogen, or Coseal™, an artificial version in which two polyethylene glycol molecules polymerize when mixed.

On the horizon are more "engineered" glues and tissue sealants. "Engineered" refers to the enhancement of biomaterials through rational design; tissue engineering refers to whole organs engineered through these techniques. With microfabricated, biomimetic particles becoming mainstream, it is expected that uses for these particles in surgical technology will follow. For example, the increase in surface area afforded by these particles will prove useful in hemostasis. Improved surface area and ligand bonded particles can also lead to enhanced bonding in, for example, a sutureless anastamosis.

Surgical robots incorporate sensors to detect and plan complex movements. MEMS technologies will ultimately provide for force feedback, producing a more realistic simulation of the environment.[3] Hopefully, the point will be reached where the robotic effectors can be placed at the end of a laparoscope or disposable system. Scale down of the robotic effectors will lead to widespread utilization of surgical robotics. Current systems such as the Da Vinci™ system from Intuitive Surgical can only be used in tertiary care centers at the present time.

In the practice of ophthalmology, phacoemulsification is the standard for cataract removal. In this technique, the cataract is actually emulsified by an ultrasonic blade vibrating at 50 MHz. A relatively common complication is puncture of the posterior capsule beneath the lens. Therefore, a force transducer has been placed at the tip of the blade. The transducer is sensitive to changes in the backward force generated by the tissue being cut. If and when the cutting needle advances past the very hard, brittle calcified cataract into the soft posterior capsule, the blade stops.[55] Similarly, in surgical ablation of larger tumors in the abdomen or lungs, such as with an

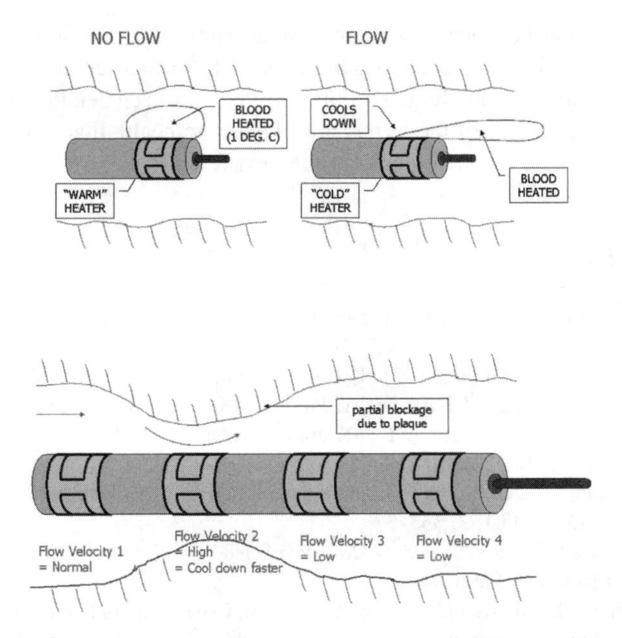

FIGURE 11.10 Flow sensor. *Upper*: A catheter is shown with an array of custom manufactured thin film sensors. On the proximal end of the catheter, blood is heated and on the distal end of the catheter, the temperature of the blood is measured via a MEMS-based thermocouple; the time for the heated blood to reach the distal end of the catheter is inversely proportional to the blood flow. *Lower*: Blockage, wherein the time the heated blood takes to reach the distal end of the catheter is decreased due to the high blood flow created by the blockage. (Courtesy of Verimetra, Inc., Pittsburgh, PA.)

ultrasonic cutting instrument or radio frequency ablation, a force feedback type probe would also be beneficial in locating the edge of the tumors.

MEMS technology has recently made its way to catheter-based systems. Figure 11.10 depicts a catheter with an integrated velocity sensor. The catheter is comprised of a series of electrodeposited or vapor-deposited conducting elements, which are not connected electrically but are connected by the conductivity of the fluid (blood) passing over it. The resistance across the elements is proportional to the speed of the blood above. This catheter is being tested for use in tumor embolization and ablations. It detects the point in time when blood flow stops, the most objective endpoint for these procedures. The catheter is also being utilized in fetal heart operations. The same velocity sensors in this case are also useful for localizing specific intracardiac velocity gradients to map the flows in a congenital cardiac anomaly.

11.4 CONCLUSIONS

A lot of ground has been covered in this chapter. Bulk material properties, nanoparticles, surface chemistry, drug delivery, remote sensing, micromachining, and self-assembly have been described in the context of concrete medical applications.

The common tools between these apparent disparate subjects should be evident. Going forward, micro- and nanotechnology will further create commonalities between these subjects and clinical problems. What has been detailed in this chapter has scratched the surface of the tremendous benefit to health that will be seen from micro- and nanoscale technologies over the next few decades.

REFERENCES

1. Palmaz, J.C., Review of polymeric graft materials for endovascular applications. *J. Vasc. Interv. Radiol.*, 1998, 9(1), 7–13.
2. Rabkin, D.J., E.V. Lang, and D.P. Brophy, Nitinol properties affecting uses in interventional radiology. *J. Vasc. Interv. Radiol.*, 2000, 11(3), 343–350.
3. Bloom, M., A. Salzberg, and T. Krummel, Advanced technology in surgery. *Curr. Probl. Surg.*, 2002, 39(8), 779–781.
4. Whitesides, G., et al., Soft lithography in biology and biochemistry. *Annu. Rev. Biomed. Eng.*, 2001, 3, 335–373.
5. Shackelford, J., *Introduction to Materials Science for Engineers*, 2nd ed, 1988, New York, Macmillan Publishing.
6. Kagan, C., D. Mitzi, and C. Dimitrakopoulos, Organic-inorganic hybrid materials as semiconducting channels in thin-film field-effect transistors, *Science*, 1999, 286(5441), 945–947.
7. Crone, B., et al. Large-scale complementary integrated circuits based on organic transistors. *Nature*, 2000, 403(3), 521–523.
8. Williams, D., *Williams Dictionary of Biomaterials*, Liverpool University Press, 1999.
9. Zelikin, A., et al. Erodible conducting polymers for potential biomedical applications. *Angew. Chem. Int. Ed.*, 2002, 41(1), 141–144.
10. Pilliar, R., Powder metal-made orthopedic implants with porous surface for fixation by tissue ingrowth, *Clin. Orthop.*, 1983, Jun(176), 42–51.
11. Nugent, H. and E. Edelman, Tissue engineering therapy for cardiovascular disease. *Circ. Res.*, 2003, 92, 1068–1078.
12. Spector, M., Historical review of porous-coated implants. *J. Arthroplasty*, 1987, 2(2), 163–177.
13. Pilliar, R., Porous-surfaced metallic implants for orthopedic applications. *J. Biomed. Mat. Res.*, 1987, 21 Suppl. A1, 1–33.
14. Hanker, J. and B. Giammara, Biomaterials and biomedical devices. *Science*, 1988, 242(4880), 885–888.
15. Billotte, W., Ceramic biomaterials, in *The Biomedical Engineering Handbook*, 2nd ed., Bronzino, J., Ed., CRC Press, Boca Raton, 2000.
16. Geesink, R., Osteoconductive coatings for total joint arthroplasty. *Clin. Orthop.*, 2002, 395, 53–65.
17. Shin, H., S. Jo, and A. Mikos, Biomimetic Materials for tissue engineering. *Biomaterials*, 2003, 24, 4353–4364.
18. Sun, E., S. Fine, and W. Nowak, Electrochemical behavior of nitinol alloy in Ringer's solution. *J. Mater. Sci. Mater. Med.*, 2002, 13, 959–964.
19. Shabalovaskaya, S., Surface, corrosion, and biocompatibility aspects of nitinol as an implant material. *Biomed. Mater. Eng.*, 2002, 12, 69–109.

20. Shabalovskaya, S., et al., Surface conditions of nitinol wires, tubing, and as-cast alloys. the effect of chemical etching, aging in boiling water, and heat treatment. *J. Biomed. Mater. Res. B Appl. Biomater.*, 2003, 65B, 193–203.
21. Heintz, C., et al. Corroded nitinol wires in explanted aortic endografts. *J. Endovasc. Ther.*, 2001, 8, 248–253.
22. O'Brien, B., W. Carroll, and M. Kelly, Passivation of nitinol wire for vascular implants — a demonstration of the benefits. *Biomaterials*, 2002, 23, 1739–1748.
23. Villermaux, F., et al., Corrosion resistance improvement of NiTi osteosynthesis staples by plasma polymerized tetrafluoroethylene coating. *Biomed. Mater. Eng.*, 1996, 6(4), 241–254.
24. De Giglio, E., et al., Electropolymerization of pyrrole on titanium substrates for the future development of new biocompatible surfaces. *Biomaterials*, 2001, 22(19), 2609–2616.
25. Gertner, M. and M. Schlesinger, Electrochemistry and medicine devices: friend or foe? *Interface*, 2003, 12(3), 20–25.
26. Madou, M., *Fundamentals of Microfabrication*. 1997, Boca Raton, FL, CRC Press.
27. Voskerician, G., et al., Biocompatibility and biofouling of MEMS drug delivery devices. *Biomaterials*, 2003, 24, 1959–1967.
28. Roy, S., et al., Microelectromechanical systems and neurosurgery: a new era in a new millenium. *Neurosurgery*, 2001, 49(4), 779–798.
29. Desai, T., et al., Nanopore technology for biomedical applications. *Biomed. Microdevices*, 1999, 2(1), 11–40.
30. Kotzar, G., et al., Evaluation of MEMS material of construction of implantable medical devices. *Biomaterials*, 2002, 23, 2737–2750.
31. Lavan, D., P. George, and R. Langer, Simple, three-dimensional microfabrication of electrodeposited structures. *Angew. Chem. Int. Ed.*, 2003, 42(11), 1262–1265.
32. Lahann, J., et al., Reactive polymer coatings: a first step toward surface engineering of microfluidic devices. *Anal. Chem.*, 2003, 75, 2117–2122.
33. Li, Y., et al., Polymer replicas of photonic porous silicon for sensing and drug delivery applications. *Science*, 2003, 299, 2045–2047.
34. Townsend, C., M. Hamel, and S. Arms, Telemetered sensors for implantable and wearable activity and performance monitoring, in 31st Neural Prosthesis Workshop. 2000, Bethesda, MD.
35. Lednicer, D., A capsule history. *American Heritage of Invention and Technology*, 2003, 19(2), 50–55.
36. Lavan, D., T. McQuire, and R. Langer, Small-scale systems for *in vivo* drug delivery. *Nat. Biotechnol.*, 2003, 21(10), 1184–1191.
37. Seeman, N. and A. Belcher, Emulating biology: building nanostructures from the bottom up. *Proc. Natl. Acad. Sci. U.S.A.*, 2002, 99(Suppl. 2), 6451–6455.
38. Mao, C., et al., Viral assembly of oriented quantum dot nanowires. *Proc. Natl. Acad. Sci. U.S.A.*, 2003, 100(12), 6946–6951.
39. Zhang, S., Fabrication of novel biomaterials through molecular self-assembly. *Nat. Biotechnol.*, 2003, 21(10), 1171–1178.
40. Heller, A., Plugging metal connectors into enzymes. *Nat. Biotechnol.*, 2003, 21(6), 631–632.
41. Xiao, Y., et al., "Plugging into enzymes": nanowiring of redox enzymes by a gold nanoparticle. *Science*, 2003, 299, 1887–1881.
42. Klarreich, E., Biologists join the dots. *Nature*, 2001, 413, 450–452.
43. Chan, W., et al., Luminescent quantum dots for multiplexed biologic detection and imaging. *Curr. Opin. Biotechnol.*, 2002, 13, 40–46.

44. Fu, L., et al., Synthesis and patterning of magnetic nanostructures. *Eur. Cell. Mater.*, 2002, 3(Suppl. 2), 156–157.
45. Chou, S., C. Keimel, and J. Gu, Ultrafast and direct imprint of nanostructures in silicon. *Nature*, 2002, 417, 835–837.
46. Hirsch, I.B. and A. Coviello, Intensive insulin therapy in critically ill patients. *N. Engl. J. Med.*, 2002, 346(20), 1586–1588; author reply, 1586–1588.
47. van den Berghe, G., et al., Intensive insulin therapy in the critically ill patients. *N. Engl. J. Med.*, 2001, 345(19), 1359–1367.
48. Burrin, J. and C. Price, Measurement of blood glucose. *Ann. Clin. Biochem.*, 1985, 22, 327–342.
49. Abel, P. and T. Woedtke, Biosensors for *in vivo* glucose measurement: can we cross the experimental stage. *Biosens. Bioelectron.*, 2002, 17, 1059–1070.
50. Alexeev, V., et al., High ionic strength glucose-sensing photonic crystal. *Anal. Chem.*, 2003, 75(10), 2316–2323.
51. Subramony, J., Iontophoresis: application of electrochemical materials and methods for therapeutics and diagnostics. *Interface*, 2003, 12(4), 27.
52. Glikfeld, P., R. Hinz, and R. Guy, Noninvasive sampling of biologic fluids by iontophoresis. *Pharm. Res.*, 1989, 6(11), 988–990.
53. Mitragotri, S., et al., Analysis of ultrasonically extracted interstitial fluid as a predictor of blood glucose levels. *J. Appl. Physiol.*, 2000, 89(3), 961–966.
54. Mitragotri, S., et al., Transdermal extraction of analytes using low-frequency ultrasound. *Pharm. Res.*, 2000, 17(4), 466–470.
55. Polla, D., et al., Microdevices in medicine. *Annu. Rev. Biomed.Eng.*, 2000, 2, 551–576.
56. Tao, S.L. and T.A. Desai, Microfabricated drug delivery systems: from particles to pores. *Adv. Drug Deliv. Rev.*, 2003, 55(3), 315–328.
57. Prausnitz, M.R., Overcoming skin's barrier: the search for effective and user-friendly drug delivery. *Diabetes Technol. Ther.*, 2001, 3(2), 233–236.
58. Kaushik, S., et al., Lack of pain associated with microfabricated microneedles. *Anesth. Analg.*, 2001, 92(2), 502–504.
59. Henry, S., et al., Microfabricated microneedles: a novel approach to transdermal drug delivery. *J. Pharm. Sci.*, 1999, 88(9), 948.
60. Tierney, M.J., et al., Clinical evaluation of the GlucoWatch biographer: a continual, non-invasive glucose monitor for patients with diabetes. *Biosens. Bioelectron.*, 2001, 16(9–12), 621–629.
61. Pitzer, K.R., et al., Detection of hypoglycemia with the GlucoWatch biographer. *Diabetes Care*, 2001, 24(5), 881–885.
62. Tierney, M.J., et al., The GlucoWatch biographer: a frequent automatic and noninvasive glucose monitor. *Ann. Med.*, 2000, 32(9), 632–641.
63. Kost, J., et al., Transdermal monitoring of glucose and other analytes using ultraound. *Nat. Med.*, 2003, 6(3), 343–350.
64. Patton, J., J. Bukar, and S. Nagarajan, Inhaled insulin. *Adv. Drug Deliv. Rev.*, 1999, 35, 235–247.
65. Schuster, J., et al., The AERX aerosol delivery system. *Pharm. Res.*, 1997, 14(3), 354–357.
66. Huang, E., et al., Nanoparticle and microparticle generation with super- or near-critical carbon dioxide, in Intl. Soc. for Applications of Supercritical Fluids. 2003, Versailles, France.
67. Gref, R., et al., Surface-engineered nanoparticles for multiple ligand coupling. *Biomaterials*, 2003, 24, 4529–4537.
68. Insulin inhalation (Aradigm Corporation). NN 1998. Drugs R D, 1999, 2(2), 110–111.

69. Henry, R., et al., Inhaled insulin using the AERx Insulin diabetes management system in healthy and asthmatic subjects. *Diabetes Care*, 2003, 26(3), 764–769.
70. Olthius, S. and P. Bergveld, An integrated micromachined electrochemical pump and dosing system. *J. Biomed. Microdevices*, 1999, 1(2), 121–130.
71. Kost, J. and R. Langer, Responsive polymeric delivery systems. *Adv. Drug Deliv. Rev.*, 2001, 46, 125–148.
72. Liepman, D., A. Pisano, and B. Sage, Microelectromechanical systems technology to deliver insulin. *Diabetes Technol. Ther.*, 1999, 1(4), 469–472.
73. Desai, T., Micro- and nanoscale structures for tissue engineering constructs. *Med. Eng. Phys.*, 2000, 22, 595–606.
74. Carter, A.J., Drug-eluting stents for the prevention of restenosis: standing the test of time. *Catheter Cardiovasc. Interv.*, 2002, 57(1), 69–71.
75. Hiatt, B.L., A.J. Carter, and A.C. Yeung, The drug-eluting stent: is it the Holy Grail? *Rev. Cardiovasc. Med.*, 2001, 2(4), 190–196.
76. Palmaz, J.C., et al., Influence of stent design and material composition on procedure outcome. *J. Vasc. Surg.*, 2002, 36(5), 1031–1039.
77. Hamuro, M., et al., Influence of stent edge angle on endothelialization in an *in vitro* model. *J. Vasc. Interv. Radiol.*, 2001, 12(5), 607–611.
78. Honda, Y. and P. Fitzgerald, Stent thrombosis: an issue revisited in a changing world. *Circulation*, 2003, 108, 2–5.
79. Morice, M., et al., A randomized comparison of a sirolimus-eluting stent with a standard stent for coronary revascularization. *N. Engl. J. Med.*, 2002, 346(23), 1773–1780.
80. Hines, R., Process for Making Electroformed Stents, Patent no. 6019784, 2000, Electroformed Stents, Inc.: USA.
81. Pourbaix, M., Electrochemical corrosion of metallic biomaterials. *Biomaterials*, 1984, 5, 122–134.
82. Kruger, J., Fundamental aspects of the corrosion of metallic implants, in *Corrosion and Degradation of Implant Materials*. 1979, Philadelphia. p. 107–127.
83. Gertner, M. and M. Schlesinger, Drug delivery from electrochemically deposited thin films. *Electrochem. Solid State Lett.*, 2003, 6(4), J4–J6.
84. Schmelner, U., et al., Surface-initiated polymerization on self-assembled monolayer: amplification of patterns on the micrometer and nanometer scale. *Angew. Chem. Int. Ed.*, 2003, 42(5), 559–563.
85. Palmaz, J.C., et al., Endoluminal device exhibiting improved endothelialization and method of manufacture thereof, Patent no. US 2001 0001834A1, 2002, Advanced Bioprosthetic Surfaces.
86. Mata, A., et al., Growth of connective tissue progenitor cells on microtextured polydimethylsiloxane surfaces. *J. Biomed. Mater. Res.*, 2002, 62, 499–506.
87. Voldman, J., Building with cells. *Nat. Mater.*, 2003, 2, 433–434.
88. Kong, X., et al., Effect of biologically active coating on biocompatibility of nitinol devices designed for the closure of intra-atrial communications. *Biomaterials*, 2002, 23(8), 1775–1783.
89. Steinhaus, D., et al., Initial experience with an implantable hemodynamic monitor. *Circulation*, 1996, 93, 745–752.
90. Remon Medical Announces First Implants of New Device for Monitoring Abdominal Aortic Aneurysm Repair. 2003, Remon Medical Website (www.remonmedical.com).
91. Ochoa, E. and J. Vacanti, An Overview of the pathology and approaches to tissue engineering. *Ann. N. Y. Acad. Sci.*, 2002, 979, 10–26.

92. Townsend, C., S. Arms, and M. Hamel, Remotely powered, multichannel, microprocessor based telemetry systems for smart implantable devices and smart structures, 31st Neural Prothesis workshop, Bethesda, MD, Oct. 25–27, 2000.
93. Mano, N., F. Mao, and A. Heller, Characteristics of a miniature compartment-less glucose-O_2 biofuel cell and its operation in a living plant. *J. Am. Chem. Soc.*, 2003, 125, 6588.
94. Chen, T., et al., *In vivo* monitoring with miniature "wired" glucose oxidase electrodes. *Anal. Sci.*, 2001, 17S(i297–i300).
95. Cheng, T., et al., *In vivo* glucose monitoring with miniature "wired" glucose oxidase electrodes. *Anal. Sci.*, 2001, 17S, 297–300.
96. Weissleder, R., et al., Antimyosin-labeled monocrystalline iron oxide allows detection of myocardial infarct: MR antibody imaging. *Radiology*, 1992, 182(2), 381–385.
97. Mylonaki, M., A. Fritscher-Ravens, and P. Swain, Wireless capsule endoscopy: a comparison with push enteroscopy in patients with gastroscopy and colonoscopy negative gastrointestinal bleeding. *Gut*, 2003, 52, 1122–1126.
98. Lewis, B. and N. Goldfarb, Review article: the advent of capsule endoscopy — a not-so-futuristic approach to obscure gastrointestinal bleeding. *Aliment. Pharmacol. Ther.*, 2003, 17(9), 1085–1096.
99. Lewkowcz, S., et al., Device and Method For Examining a Body Lumen, Patent no. W0 03005877A2, 2003, Given Imaging Ltd.: World.
100. Foraker, A.B., et al., Microfabricated porous silicon particles enhance paracellular delivery of insulin across intestinal Caco-2 cell monolayers. *Pharm. Res.*, 2003, 20(1), 110–106.
101. Ahmed, A., C. Bonner, and T. Desai, Bioadhesive microdevices for drug delivery: a feasability study. *Biomed. Microdevices*, 2001, 3(2), 89–96.
102. Leonard, M., et al., Iontophoresis-enhanced absorptive flux of polar molecules across intestinal tissue *in vitro*. *Pharm. Res.*, 2000, 17(4), 2000.
103. Harisinghani, M., et al., Noninvasive detection of clinically occult lymph-node metastases in prostate cancer. *N. Engl. J. Med.*, 2003, 2003, 2491–2499.
104. Whitesides, G., The "right" size in nanobiotechnology. *Nat. Biotechnol.*, 2003, 21(10), 1161–1165.
105. Berry, C. and A. Curtis, Functionalisation of magnetic nanoparticles for applications in biomedicine. *J. Phys. D Appl. Phys.*, 2003, 36, R198–R206.
106. Nsereko, S. and M. Amiji, Localized delivery of paclitaxel in solid tumors from biodegradable chitin microparticle formulations. *Biomaterials*, 2002, 23, 2723–2731.
107. Sershen, S., et al., Temperature-sensitive polymer-nanoshell composites for photothermally modulated drug delivery. *J. Biomed. Mater. Res.*, 2000, 51, 293–298.
108. West, J. and N. Halas, Applications of nanotechnology to biotechnology. *Curr. Opin. Biotechnol.*, 2000, 11, 215–217.
109. Cox, J., et al., Surface passivation of a microfluidic device to glial cell adhesion: a comparison of hydrophobic and hydrophilic SAM coatings. *Biomaterials*, 2002, 23, 929–935.
110. Rivers, T., T. Hudson, and C. Schmidt, Synthesis of a novel, biodegradeable, electrically conducting polymer for biomedical applications. *Adv. Funct. Mater.*, 2003, 12(1).

12 Imaging Molecular and Cellular Processes in the Living Body

Christopher H. Contag

CONTENTS

0-8493-1940-4/05/$0.00+$1.50

12.1 OVERVIEW

Biomedical research has traditionally taken a reductionist approach, where the layers of a specific biological process are removed until a single gene, protein, or mutation is identified as the biochemical basis of the observed effect. High-density screening methods and large-scale sequencing projects have taken this to the next level, where multiple single genes or proteins are analyzed, and approaches to assemble these data into coherent patterns of multiple single elements can point to regulatory networks that control a given process. These approaches have generated a wealth of information that now needs to be analyzed using integrated approaches where the biological context is retained and the gene, protein, or mutation is analyzed in the living body. High-content cell-based assays are one step toward systems integration; however, interrogation of intact living subjects at the level of cells and molecules is the ultimate level of systems biology and integration. The emergence of the nascent field of *in vivo* cellular and molecular imaging marks the beginning of an era of integrated biomedical research.

Assays that can be used *in vivo* are slower and generally more expensive than those performed in cell culture, or in comparison to biochemical assays. However, *in vivo* assays provide relevant information, because the data are obtained in context and from living systems. The intent, for imaging in animal models, is to develop assays that provide more predictive information about a biological processes than can be obtained in cell culture. Until the development of imaging tools, many *in vivo* studies were performed by serially sacrificing animals at predetermined time points and performing labor-intensive assays on predetermined tissues. This type of approach does not permit observations to be made in real time or to assess dynamic or cascading effects in the animal. In addition, the inherent variability in each animal and groups of animals creates statistical variation in the data that may not be useful or may not model the human response well enough to predict the human response. However, conventional models are well known and understood; therefore imaging strategies that can be superimposed on existing animal models are more powerful than those that require significant modifications to a procedure. A diversity of imaging modalities and approaches are needed to provide these capabilities.

A number of imaging approaches have been developed for noninvasive analyses of a diverse set of molecular and cellular events in the living subjects. These include following the levels of gene expression, assessing enzymatic activity, monitoring tumor growth and response to therapy, determining immune cell trafficking patterns, and observing the progression of infectious diseases (Table 12.1).[1–4] Many of these tools are based on imaging modalities that are already used clinically, and several are based on modalities that have been specifically developed for the study of animal models.[5–9] These approaches have been used to visualize the efficacy and safety of drugs within the animal models and as such have accelerated the preclinical phase of drug development. Furthermore, their utility in humans as an outcome measure for optimizing therapeutic intervention has been investigated.[10] Real-time access to cellular and molecular information in intact tissues and organs of animal models enables the researcher to observe mechanisms of action or cascading events within the animal that would not otherwise be detected using conventional methods. These

TABLE 12.1
Modalities, Resolution, and Applications

Modality	Resolution	Application/Advantages	Disadvantages
MRI	10–200 nm	Fine structural imaging with excellent soft tissue contrast	Long scan and analysis times
MRI/MRS (functional)	0.8–2 mm	Metabolic information for small set of metabolites, potential for development of reporter genes	Contrast agent chemistry not developed to the point of molecular sensing
PET	3 mm	Metabolic and physiologic measurements, tools available for reporter gene imaging	Handling radioactive tracers, synthesis of radionuclides on site due to short half lives, long scan times
SPECT	Lower or similar to PET	Metabolic and physiologic measurements, tools available for reporter gene imaging, reagents for some key molecular targets have been developed	Handling radioactive tracers, sensitivity is limited due to need for collimator
CT	100 µm	Structural image and available contrast agents	X-rays required, contrast agents still limiting
Ultrasound	100 µm	Structural, some probes developed for function	Local and not whole body
General optical imaging	1–5 mm	Ability to sense intrinsic changes in a tissue, excellent resolution near surface, clinical utility	Lower resolution and poor sensitivity in deep tissues
Optical imaging of reporter genes	1–5 mm	Inherently good for imaging of reporter genes, gene expression studies, functional studies, excellent in animal models with good sensitivity	Lower resolution, limited clinical utility
Optical imaging in the nir	1–5 mm	Variety of dyes and quenchers for developing molecular sensors, clinical imaging is likely, some dyes approved for clinical use, assays of enzyme activity have been developed and tested.	Lower resolution and sensitivity in deep tissues, limited clinical utility

advances will have significance in drug studies where it is necessary to evaluate the process in models where the biological pathways interact with the potential therapies in a manner that cannot be modeled in culture.

The basis of this field is to develop biological sensors that assess function at the cellular and molecular level, and that can be externally detected through tissues that can severely attenuate the signals. Real-time analyses of biological events in the living body through imaging holds tremendous potential for developing and deploying targeted therapies and evaluating outcome, as the emerging approaches will provide more and significantly higher quality information about the therapeutic targets and physiologic responses. Although the focus of this chapter will be the most recent advances in the emerging field of *in vivo* cellular and molecular imaging, the imaging technologies that have enabled these studies will be introduced as well as selected applications used for demonstration of the accomplishments to date and the challenges for the future.

12.2 ADVANCES IN IMAGING HAVE THEIR FOUNDATION IN CELLULAR AND MOLECULAR BIOLOGY

12.2.1 Reductionist to Intregrative Approaches

Large-scale sequencing efforts and tremendous advances in high-density screens and gene expression studies have identified many key elements in physiology and disease, and have revealed a number of new potential targets for therapy.[11] Evaluation of individual components of a given process and dissecting the regulatory networks that control them requires integrative approaches where the context of living cells and tissues are preserved. This is the driving force behind the development of methods for studying biology in live cells and thick tissues,[12–15] and in the emerging field of molecular imaging where cellular and molecular changes can be studied in the living body. Given the complexity of biological mechanisms most of which are comprised of multiple interrelated genes, proteins and biological pathways, understanding the regulatory networks that control even one of the 30,000 identified genes and their protein products requires significant effort. Reconstructing the pathways and networks that are comprised of multiple single elements will require technological advances and a systems approach to biology. Development of imaging approaches for evaluating biochemical events in living cells, tissues, and whole organisms is an integral part of systems biology, and the need to understand complex biological process in the context of living tissues is what will continue to drive the advances in imaging science.

Integration in the field of cell biology is characterized by a transition from assays in cell lysates and extracts to live cell assays. These assays incorporate reporter genes and vital dyes in attempts to preserve the dynamic milieu in which these events occur and study labeled processes in the context of the living cell. Advances in this area of investigation have led to assays where the activity of reporter molecules can be measured almost continually, often through imaging.[16–27] Such methods have improved our understanding of genetic regulation, protein trafficking, and protein function by enabling the changes to be studied over time in the context of the complex physiology of intact living cells. Important temporal changes and patterns of activity have been revealed in live cells using these methods, However, these studies have typically used single markers, and although the influences on these reporters is multifold, the readout remains singular.

The next transition in this field is toward multiplexed assays and high content cell-based assays. Such approaches include vital staining with detection by multihead confocal microscopes in multiwell cell culture dishes[28–31] and multiparameter intracellular staining followed by flow cytomtery.[32] These cell-based methods are being used to provide the intracellular data necessary for understanding the relationships between two, or more, molecular events within cells. These single and multiplexed assays provide the foundation for the techniques that are being explored for imaging cellular and molecular events in the living body. It is the extension of these techniques to intact living animals, and eventually humans, that will enable us to look into the abyss of the living body and probe gene regulation and protein function in the context of live tissues and organ systems through advances in imaging technologies.

12.2.2 Structure vs. Function

As imaging in cell sciences matured, it transitioned through methods specifically designed for analyzing structure to those that assess function. The early imaging methods were directed at detecting organelles and other subcellular structures, and the more recent advances were aimed at studying cell physiology, cell signaling, and cell–cell interactions. For this purpose, molecular probes have been developed, and are continuing to be developed, that can sense physiological changes through binding to yield concentration changes, conversion from quenched to active forms, or reporter genes as indicators of transcription.[33–38]

Imaging sciences, as they relate to imaging in living subjects, have gone through similar transitions. Initially, tools like X-ray imaging were developed to see dense structures such as bone; then magnetic resonance imaging (MRI) was developed to get soft tissue contrast and ultrasound for imaging of structure. These modalities were designed to reveal structural information and have become indispensable clinical tools. The recent developments in imaging science related to structure are certainly aimed at increasing the resolution, refining image analyses and increasing the speed of data acquisition. However, we are at a transition in this field, from imaging structure and using inherent contrast to get physiologic measurements to the use of molecular probes to obtain functional information relating to cellular and molecular changes in the living body. The advances in whole-body imaging that are aimed at revealing functional information, at the level of cells and molecules, is where the nanosciences interface with imaging sciences and is hence the focus of this chapter.

12.2.3 Correlative Cell Culture Assays

Since the basis of many of the new imaging approaches has its roots in molecular and cell biology, there is a set of ready-made validation assays that can be used to support the *in vivo* imaging data. In fact it is usually essential that correlative cell culture assays be used to develop an imaging strategy prior to *in vivo* use of the imaging approach, and then applied again for post-imaging validation. These assays are typically amenable to high throughput formats, which serves to accelerate development of new imaging reagents. Post-imaging validation can also be performed using biochemical assays on tissue lysates using well-established measures of mRNA and protein levels.[39,40] The times and tissues for analyses can be selected based on the image data, and this image guidance can be used to improve the data set by confining the study to the relevant times and tissues. The new advances in imaging are based on advances in cellular and molecular biology, and the tools of these fields are essential components of the field of molecular imaging.

12.3 ESTABLISHED AND EMERGING TECHNOLOGIES FOR *IN VIVO* MOLECULAR ANALYSES

The emerging field of molecular imaging combines knowledge from genetics, biology, chemistry, and physics to develop integrated imaging strategies that are capable of interrogating cellular and molecular changes in the context of intact biological systems, including the bodies of animals, as models of human biology, and of humans.[7–10] Some of these strategies use instrumentation that was specifically designed for detecting molecular markers *in vivo*, and other approaches have used imaging modalities that were originally designed for imaging structure.[41–48] The nuclear medicine techniques of positron emission tomography (PET) and single photon computed tomography (SPECT) were developed for localizing radiotracers in the body and are, by design, modalities that measure function — originally physiology and metabolism, and more recently gene expression and protein distribution.[49–51] Because mammalian tissue is only relatively opaque, new tools for imaging in the body are being developed that are based on visible and near infrared (nir) light.[7,8,52–55] These optically based modalities can take advantage of some of the same molecular probes that are used in cell biology, such as green fluorescent protein (GFP) and related proteins and luciferase and its related proteins.[52,56] Additional optical probes based on dyes that fluoresce in the nir are being explored, and there are several imaging approaches being developed that specifically target function using these dyes in the living body.[55] A summary of these imaging modalities follows.

12.3.1 Clinical Imaging Modalities

12.3.1.1 CT and Ultrasound

These two modalities were initially developed for imaging structure and changes in structure that are associated with various pathologic conditions, and they have established clinical utility. However, with the development of miniaturized versions of CT and ultrasound instruments, imaging of laboratory rodents has become possible, and there has been a concomitant increased interest in the development of novel contrast agents that reveal physiologic and metabolic changes using these modalities.[43,45,57,58]

There have been several advances made in CT contrast agents especially in the area of gastrointestinal imaging.[44,45,58] A significant advance was demonstrated using intravenous (i.v.) injection of a hepatocyte-selective contrast agent for monitoring murine hepatic tumors.[43] The selective mechanism in this example was receptor-mediated transport into hepatocytes by apoE. This process is selective for normal liver parenchyma relative to malignant tissue, and disruption of the normal liver structure by tumor growth could be measured. The results from this study indicated that assessing tumor burden may be possible using this approach, however standardization is crucial, given the nature of selective uptake. Perhaps the greatest utility of CT is in combination with nuclear medicine approaches to enhance image reconstruction.[59–62] By combining structural information from CT with functional data, images generated by PET and SPECT can be dramatically improved and localization of radiotracers more accurately assessed. Combination of CT with other functional imaging modalities such as nir fluorescence and *in vivo* bioluminescence imaging will likely be developed in the near term and serve to enhance these modalities through providing methods for effective attenuation correction.

Ultrasound contrast agents are typically micron-sized bubbles that are stabilized with a polymer shell.[63,64] These microbubbles have been used to estimate flow rate in the microvasculature[64] and, with modifications, as drug delivery tools.[65,66] Additional approaches for generating ultrasound contrast includes the use of liquid-filled nanoparticles and liposomes. Although ultrasound contrast is typically generated by passive concentration, it is possible to target some of these agents to specific cell surface markers. Nonetheless, the size of these reagents usually restricts them to the vasculature, and they may therefore have their greatest utility in targeting endothelial cell surfaces or markers on circulating cells in the blood. The ability to both image particle distribution and use focused ultrasound for regional destruction of loaded particles offers the advantage of being able to use a single agent for both imaging and for drug delivery. This may represent the greatest strength of ultrasound contrast agents.

12.3.1.2 Magnetic Resonance Imaging

MRI was developed for clinical use and has more recently been adapted for small-animal imaging. Given its history, MRI is a modality where molecular imaging approaches, developed in animal models, may be translated to clinical imaging, and hence has been an area of intense investigation.[41,42,47,67–69] In this modality, the subject to be imaged is placed in a strong static magnetic field, which aligns nuclei of certain elements, such as spin-1/2 hydrogen, preferentially along the direction of the magnetic field. Irradiation of the organs and tissues with proper radio frequency causes transitions between energy states of the nuclei, resulting in net excitation of the nuclear spin. Relaxation of the spins back to the lower energy state generates a signal (the nuclear magnetic resonance signal) that is characteristic of the nuclei and their chemical and physical environment. With suitable magnetic gradient coils and Fourier Transform signal processing, signals generated by nuclear magnetic resonance can be localized to produce an MR image. Although the images are generated using molecular changes, the sensitivity of this modality is generally poor, despite excellent resolution, and changes in cellular physiology are not detectable. Many refinements of instrumentation for MRI have been made. These images of tissue structure can be used for the detection of tumors, and system modifications allow measurements of tissue oxygenation and selected metabolites.[70–74] Recently, several approaches have been developed for MRI that may enable imaging of cell trafficking and gene expression.[46–48,75,76] Many of these studies used cells in culture or model systems that can be manipulated by microinjection; however, whole-body imaging of rodent models and humans remains tantalizing.

12.3.1.3 Nuclear Medicine Techniques of PET and SPECT Imaging

PET and SPECT are two modalities in the area of nuclear medicine that use radioactive emission from radionuclides as tracers to reveal changes in physiology and, more recently, in patterns of gene expression.[77–80] Imaging of emission from the radiotracers is performed using scintillation cameras or other devices, with applications in animal models[81] and in the clinic. Radionuclides such as ^{18}F, ^{15}O, ^{13}N, and ^{124}I that emit positrons are used in PET imaging, which takes advantage of the pairs of coincident 511 keV photons emitted by positron–electron collisions to decrease background and scattered signal. Radionuclides that emit photons are utilized in SPECT, and their emission can be directly imaged using gamma cameras and pinhole collimators to enable computed tomography. The detectors can be ordered in a ring around the subject or rotated around a subject to detect emission from multiple angles to locate the radiotracers in the body. The spatial resolution of PET and SPECT are optimally about 1 mm,[82] and the deep penetration of the high-energy emission from radionuclides enables imaging of large subjects and tomographic reconstruction.

Radioactive emission can not be modulated, and therefore imaging of physiologic changes is dependent on the ability to concentrate the radiotracer in specific cells. This is accomplished using the increased metabolic activity of tumor cells, and other metabolically active tissues, to concentrate radiolabeled glucose or by ectopic expression of cell surface receptors or cytosolic proteins that trap labeled compounds (see below). Activation strategies that can take an inactive compound to its active form can greatly improve signal-to-noise ratios (SNR); this strategy is not available when using nuclear medicine technologies and comprises one of the significant advantages of other imaging modalities over those that use emission from radioactive tracers.

12.3.2 Technologies Best Suited for Imaging in Animal Models

12.3.2.1 Optical Imaging Techniques

The relative opacity of tissue permits the transmission of visible light through the body, and although largely attenuated by both absorption and scattering, light in the visible and near infrared (nir) region of the spectrum have been successfully used for imaging of biological process in living subjects.[7,8,52,55,83–85] The instrumentation for detection of optical signals in the body generally consist of charge couple devices (CCD) cameras or other detector arrays. These devices tend to be less expensive than those of other imaging modalities, and the variety of optical reporters and dyes have made optical modalities more versatile than any one of the other modalities for imaging of animal models. However, the extreme attenuation of light, in the visible and nir regions of the spectrum, by human tissues will be limiting in the translation of optical imaging modalities to the clinic.

Fluorescent signals can be detected *in vivo*,[54–56,83,86–91] and the versatility of dyes and reporters that fluoresce has resulted in a wide range of approaches and uses. Unfortunately, light of the wavelengths that are needed to excite many of these fluors, as well as the wavelengths of emission, are blue and green, which have limited penetration through mammalian tissues due to absorption, largely by hemoglobin. For this reason fluorescent techniques are best used to study cells in culture, smaller transparent organisms or biological processes that occur at superficial tissue sites in mammals. Nonetheless, a number of approaches have been described that improve the detection of fluorescent labels in mammalian models, and indeed the use of fluors which excite and emit in the nir may have clinical utility.[92] Red light penetrates tissues more efficiently than green or blue, and dyes that absorb and fluoresce in the red, such as indocyanin green (ICG[93,94]), have been described and used in animal models for *in vivo* detection and analysis of tumor growth and biology.[54,76,86–88,95–98]

Activatable probes have been generated using these nir dyes by separating a fluor and a quenching dye with a peptide that contains a specific proteolytic cleavage site.[54,55,76,88] Cleavage of such molecules *in vivo* will reveal regions of the body where a the targeted protease is over expressed. Over expression of proteases is associated with a wide variety of disease states, and thus there are numerous applications for these molecular "light switches," including cancer detection, assessing the extent of programmed cell death in response to therapy and in localizing inflammation.[99–102]

Advances in confocal or two-photon detection have had a dramatic impact on cell culture assays and fluorescence detection in thick biological tissues. However, the utility to these advances in animal models and the clinic has been limited by the large cumbersome instrumentation and the extremely small volumes of tissue that can be analyzed at high resolution. Two advances that will bring these approaches into widespread use will be coupled macro- and microscopic imaging devices that provide both a wide field of view and high-resolution imaging, and miniaturization of the microscope to enable implantation and endoscopy.[103–106] Multiphoton systems greatly improve detection of fluorescent signals, as the excitation is accomplished with longer wavelengths yielding deeper tissue penetration.

12.3.2.2 Combination Approaches

Combinations of instruments that provide structural information and functional information represent the most robust imaging approaches. The reason for this is that, as the contrast agents become more and more specific, only the targeted biological change will be apparent in the image, and the structural data will be needed to serve as a reference image such that the biological change, the functional image, can be localized to a specific tissue, organ, or region of the body. In addition, the structural information can be used for attenuation correction and can improve the functional image. Many of the established clinical imaging modalities are now available in combined instrument (e.g., PET-CT and SPECT-CT systems). For development of new imaging approaches similar systems have been developed for imaging of laboratory rodents. An important, and as of yet largely underutilized, aspect of the combined instruments is that, since each modality is based on a specific energy or set of energies, combination instruments can be used to develop multiplexed assays where two different biological functions are imaged using different energies.

Optical imaging modalities are unique among the various imaging tools in that optical detectors are generally sensitive to a wide range of energies in the visible to nir regions of the spectrum. Since each wavelength, or energy, within this range can be used with a specific reagent, it is possible to multiplex optically-based systems. This is the basis for 12-color flow cytometry and multiplexed fluorescent-based cell imaging. The limitation in applying these tools to whole body imaging is the differential penetration of different optical wavelengths by mammalian tissue. Since the body's primary absorbers are hemoglobin and water, and these absorb in the blue and infrared regions of the spectrum, respectively, there is a window between 600 nm and 1300 nm where absorption of the signal is minimal and optical signals are lost due to scatter. Thus, this region of the spectrum is where multiplexed assays, based on compounds that use this energy, can be designed.

The future of the imaging sciences is the development of multimodality instrumentation, and multifunctional contrast reagents that are designed to sense cellular events and display the data in the context of ultrafine structural information. The field is in its infancy, and yet the advances have been significant, and the potential of these tools is well appreciated. There are some key examples that demonstrate the significance of the advances in this field, and some of these are highlighted below.

12.4 REPORTER GENES FOR IMAGING *IN VIVO* GENE EXPRESSION PATTERNS

The sensitivity of most imaging modalities is not sufficient to sense changes in the intrinsic properties of cells and tissues; therefore contrast agents are developed, and in the context of molecular imaging, agents with molecular specificity are designed and evaluated. This is in contrast to blood pool agents or compounds that produce contrast through nonspecific accumulation. One of the molecular targeting approaches, for imaging gene expression patterns, is to tag specific genetic regulatory elements (i.e., promoters) with genes that encode proteins with distinct signatures that can be detected externally. Such approaches have been used in cell biology for decades, and now, with imaging instrumentation specifically designed for small animals, the transition to *in vivo* assays has been made for several of the modalities. In fact, PET reporter genes have been evaluated in clinical trials for optimization of gene transfer to malignant cells.[10,107] Optimal reporter genes for *in vivo* studies should encode well-characterized gene products, proteins that can result in specific generation or accumulation of deeply penetrating emission and a high signal-to-noise ratio. Sensitive detectors that are both accessible and versatile are desirable for *in vivo* gene expression studies. Several such reporters have been described in the literature, and selected examples are described below. If reporter genes are to be used as indicators of tumor burden or for cell trafficking studies, genes that are integrated into the genome and whose detectable signal is inextricably linked to the metabolic activity of the target cell population can circumvent problems of loss of signal due to dilution, as well as the confounding detection of signals that have dissociated from the viable target cell population.

12.4.1 PET Reporter Genes

The best examples of reporter genes used for PET imaging are the surface receptor dopamine type 2 receptor and the tracer 3-(2′-[18F] fluoroethyl) spiperone (FESP), and the cytosolic enzyme thymidine kinase (TK) from herpes simplex virus with labeled nucleoside analogs (e.g., ganciclovir).[51,77,79,80,108] Each of these has its strengths and weaknesses and has been utilized in a variety of applications. The thymidine kinase (TK) from herpes viruses utilizes nucleoside analogs with different specificity than the TK enzymes from mammalian cells. Thus, this difference can be targeted by analogs that have been specifically designed to inhibit the viral enzyme and, if radiolabeled, used as reporters for PET or SPECT imaging. Phosphorylation of the radiolabeled compound by the kinase results in a molecule that can no longer escape the cell, and this trapping provides contrast relative to tissues that are not expressing the viral TK gene. Not only has this reporter been used for animal models, it has also been evaluated in the clinic as a measure of effective gene transfer to a tumor site.[10,107]

12.4.2 SPECT Reporter Genes (NIS)

The Na^+/I^- symporter (NIS) is the plasma membrane glycoprotein that mediates the active uptake of I^- in the thyroid, which is a key step in thyroid hormone biosynthesis. This is the molecular basis of thyroid radioiodine therapy. The NIS protein is also functional in other tissues, such as salivary glands, gastric mucosa, and lactating (but not nonlactating) mammary glands and can mediate I^- uptake in these tissues. The expression of NIS on malignant cells of thyroid origin and the transport of I^- have been used to detect and target these cells for destruction with therapeutic doses of radioiodide. Exogenous delivery of the NIS gene as a reporter gene has been demonstrated in a variety of cell lines, including those derived from human glioma, myeloma, hepatoma, melanoma (A375), human prostatic adenocarcinoma (LNCaP), and cervical carcinoma (SiHa).[109] Transfection of NIS into the human prostatic adenocarcinoma cell line, LNCaP, resulted in significant uptake and retention of radioiodide, indicating that there may be a potential for using a variety of substrates of NIS such as ^{188}Rhenium (^{188}ReO) and ^{211}Astatine (^{211}At).[109] These alternates have different decay properties and a shorter physical half-life than ^{131}I, offering important differences with both diagnostic and therapeutic applications.[109,110]

12.4.3 Fluorescent Proteins

Several reporter genes have been developed that can be used to confer a unique optical signature on cells that express the marker protein. The green fluorescent protein (GFP) from the jellyfish *Aequorea victoria* and its fluorescent homologs from Anthozoa corals have been broadly utilized in cell biology and are being used *in vivo* for gene expression studies. There is significant spectral and chromophore diversity among this family of proteins with about 100 cloned members. These proteins share common structural, biochemical, and photophysical features and have all been derived from marine organisms. Among these reporters, monomeric red and dimeric far-red fluorescent proteins derived from Anthozoa GFP-like proteins hold the greatest potential *in vivo* because of their red-shifted excitation and emission. Although the Anthozoa GFP-like proteins provide this and other advantages over GFP, they are not without their limitations, which can include obligate oligomerization and slow or incomplete fluorescence maturation. Although modifications to all of these proteins have been made to alter excitation and emission light or to change other functions, there remain proteins that are better suited for one application over another.[111] The challenge here is to select reporters of the appropriate wavelength and stability for the application.

12.4.4 Luciferase and Related Proteins

Luciferases are also naturally occurring enzymes that have been found in a wide range of organisms from several different genera. There are three basic biochemistries among the enzymes characterized to date, and these reactions utilize different substrates and conditions.[112] Representative of each of these have been cloned, and their chemistries characterized to a point where they can be routinely used in the laboratory. These include the luciferases from firefly (coleoptera), jellyfish and sea pansies (cnidaria), and bacteria (*Vibrio spp.* and *Photorhabdus luminescens*). Luciferases as a class of enzymes all require energy, oxygen, and a specific substrate (commonly known as luciferin), and may also require cofactors for light production.

The luciferases from fireflies and related insects are a single polypeptide related to the CoA ligase family of proteins[113,114] and use a benzothiazole luciferin substrate (commonly called "luciferin"), ATP, and oxygen to generate light. The gene encoding firefly luciferase (*fluc*) has been cloned, and the coding sequence optimized for expression in mammalian cells. In the presence of luciferin, cells expressing this enzyme emit a yellow-green light with an emission peak at ~560 nm at room temperature and 615 nm at 37°C. This protein offers the advantage of broad-spectrum emission and sufficient red component to the spectrum that enables detection of small numbers of cells such that even minimal residual disease that persists after therapy can be imaged (Figure 12.1). Luciferases from both the sea pansy (*Renilla*

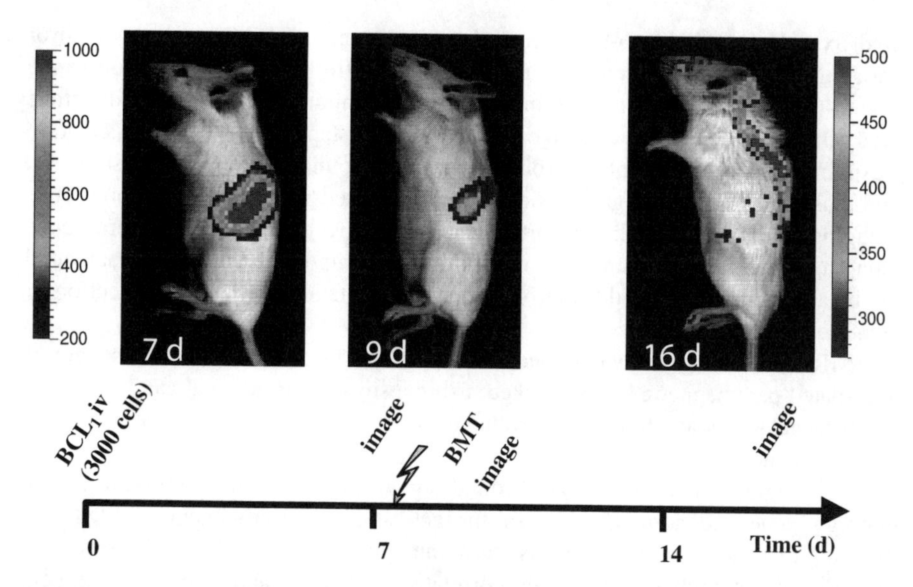

FIGURE 12.1 Imaging of minimal residual disease in a lymphoma model after radiation treatment and bone marrow transplantation. The lymphoma was initiated with 3000 cells (see time line at bottom of figure), and 7 days later the animals were imaged, treated with whole body radiation, and given a syngeneic bone marrow transplant. Two days later (9 days after initiation of tumor), the signal from the bioluminescent lymphoma cells has diminished, but by 16 days after initiation of tumor, the signal from residual cells in the spinal column indicates persistence of minimal residual disease.[134]

reniformis)[115–117] and the jellyfish (*Aequorea aequorea)*[118–120] have been cloned, characterized, and used as *in vivo* reporter genes. Both of these enzymes use coelenterazine as the substrate, but differ in that the hydrozoan *Aequorea* enzyme (aequorin) requires a calcium ion for light production, leading to its use as a calcium sensor. A codon-optimized version of the *Renilla* enzyme has been developed to allow functional expression of the protein in mammalian cells, and this enzyme is routinely used as a monitor of gene expression in cell lines. The advantages of having enzymes with different spectra of light emission and different chemistries for luciferases is that both reporters can be used in the same cell line to monitor the expression of two different genes. Since the *Renilla* enzyme does not require cofactors provided by the host cell, this protein may have the additional utility of serving as a sensor on the cell membrane.

12.4.5 Lac Z and Activation of Gd-Containing Compounds

Beta galactosidase (beta-gal) has been widely used as a transgene reporter enzyme in cell biology, and several attempts have been made to extend the use of this enzyme to *in vivo* imaging studies.[87,121] Several different fluorescent substrates are available for its detection, and these have been used for flow cytometry, and extension of this powerful reporter into *in vivo* studies would offer a significant tool for biological research. Conjugates of the beta-galactoside substrates and 7-

hydroxy-9H-(1,3-dichloro-9,9-dimethylacridin-2-one) (DDAO), are both chromogenic and red fluorescent.[87] Activation events offer a powerful solution to the signal-to-noise problems in imaging, and these compounds offered the advantage of a 50-nm red shift upon cleavage. This activation step enabled specific detection despite the presence of intact probe in the background tissues. In this study 9L gliomas expressing beta galactosidase were detected using *in vivo* fluorescence imaging. This offers another reporter gene for *in vivo* application.[87] Fluorescence imaging has limitations, and use of gadolidium compounds that can be activated by beta galactosidase would offer significant advantages by enabling detection of gene expression by MRI.[122–124]

MRI contrast agents, where access of water to the first coordination sphere of a chelated paramagnetic ion is blocked, using a sugar molecule that can be cleaved by beta galactosidase have been evaluated *in vivo*.[123] The presence of the sugar residue silences the MRI signature, and thus enzymatic activation provides a switch that can significantly reduce background signal. Heterologous expression of the reporter gene and administration of the activatable contrast agent enabled MR imaging of gene expression. In this study the approach was validated by revealing that regions of higher MR signals correlated with cells and tissues where beta galactosidase was expressed.

12.4.6 Multifunctional Reporter Genes

Multimodality imaging can strengthen a study by providing both structural and function information, in addition, reporter genes can be linked through genetic means to generate reagents that can be detected by two or more modalities. This provides greater versatility than single reporters. Gene fusions consisting of luciferase and green fluorescent protein (GFP)[125] or related proteins can link *in vivo* measurements using luciferase to *ex vivo* assays such as flow cytometry and fluorescence microscopy. This can greatly increase the utility of the reporter gene construct and offers validation measures that are not possible using a single-function reporter.[39,49,126] Several genetic reporters have been developed that use three reporter genes with signatures for fluorescence, bioluminescence, and nuclear imaging techniques.[98,127]

As a source of labeled cells of multiple types, transgenic mice have been developed that express the dual-function reporter genes, thus serving as universal donors for transplantation and trafficking studies.[128] Stem cells from this transgenic donor have been transplanted and analyzed (Figure 12.2). Since the reporter genes are integrated into the genome, the signals are not lost over time, and the labeled cells can be followed throughout the life of the animal. Using transgenic animals that express the multifunctional reporter genes in all tissues will be a useful tool for developing new therapeutic strategies, since this method enables tracking any cell population of interest over time and links *in vivo* and *ex vivo* assays. Combining this reporter animal with knockout and transgenic animals will enable analysis of the role that specific genes play in cell migration and mammalian development.

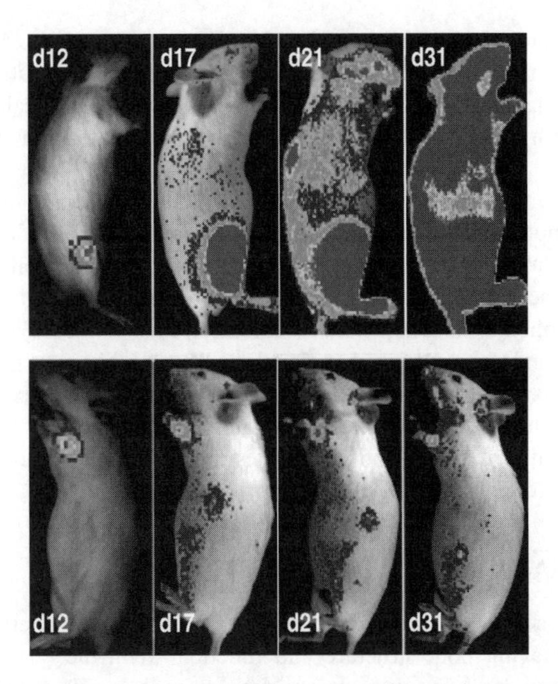

FIGURE 12.2 Pushing the limits of imaging in a stem cell transplant model. Single hematopoietic stem cells from a labeled transgenic donor mouse were transferred to irradiated unlabeled recipient mice, and the patterns of engraftment imaged using *in vivo* bioluminescence imaging.[128] Despite identical cells, the patterns of engraftment and reconstitution were very different in the two mice shown. In one animal the labeled cell participated fully in reconstitution (top), and in another animal there was limited participation. Although single stem cells were not visible, the foci arising from these cells were visible at 12 days post transfer. Advancing the technologies to detect fewer and fewer cells will enable analysis of these very early time points.

12.5 DYES AND REAGENTS

12.5.1 Quantum Dots

Fluorescent proteins, and even dyes, are limited in the numbers of available colors; this is especially true in the region of the spectrum that is well-suited for *in vivo* imaging. One alternative to this shortfall is to use semiconductor quantum dots.[129] Quantum dots of different sizes and surface coatings can fluoresce at different colors, over a very broad region of the spectrum, and the emission is very narrow, lending this reagent to multiplexed assays, especially for cells in culture. The quantum dots tested *in vivo* include amphiphilic poly(acrylic acid), short-chain (750 Da) methoxy-PEG and long-chain (3400 Da) carboxy-PEG quantum dots, which had fairly short circulating half lives (less than 12 minutes), and long-chain (5000 Da) methoxy-PEG quantum dots which had long circulating half lives of about 70 minutes.[130] In the *in vivo* study, localization of labeled quantum dots was assessed by fluorescence imaging of living animals. This was followed by analyses at necropsy using frozen tissue sections for optical microscopy, and by electron microscopy.[130] The versatility of this approach from live animals to nanometer resolution suggested that this may be an ideal means of linking *in vivo* and *ex vivo* assays. The localization of these labels varied with surface coatings and, although not evaluated, may also be variable with size.

Molecules with structures that may impart a detectable signature on cells that can also be amplified due to their composition provide a basis on which future imaging agents may be built.[131] The power of DNA amplification methods, polymerase chain reaction (PCR), and isothermal approaches for *in vitro* assays is being realized and applied to tissues,[132] and its extension to *in vivo* assays may not be too far off. For example, the power of molecular self-assembly may incorporated into imaging reagents as a means of spontaneously forming complex structures with detectable signatures in the body based on simple components.[133] Such approaches would have significant utility, with modalities that are high resolution but generally poor sensitivity (e.g., MRI).

12.6 SUMMARY AND FUTURE OUTLOOK

Imaging sciences have advanced to the point where we have a variety of instruments available for imaging both structure and function. With the convergence of cell biology, chemistry, and imaging we are developing a range of tools for probing biology *in vivo* and as a basis for imaging molecular changes in the human body. Medical imaging techniques such as PET, SPECT, and MRI show great promise in the detection of a wide range of functions and as modalities that are well established in the clinic, and now the availability of instruments for small animals will accelerate the development of new tools for these modalities. *In vivo* bioluminescence imaging has potential, due to the ability of luciferases to emit light without external excitation and their utility as genetic tags, to reveal biological changes that were not previously accessible for investigation. Advances in fluorescent dyes and other reagents and the ability for tomographic reconstruction of images obtained using optical signatures *in vivo* will greatly enhance our animal models. The exploitation of these techniques in the study of biology will greatly accelerate and refine our studies of biology and improve our analyses of new therapeutic approaches through refinement of animal models. With the advent of many small animal imaging systems, it is possible to more rapidly test new imaging approaches and refine these applications such that, in translation to the clinic, we will have a greater understanding of the properties of imaging agents and the utility for making outcome measures in the clinic.

REFERENCES

1. Luker, G.D., Pica, C.M., Song, J., Luker, K.E., and Piwnica-Worms, D., Imaging 26S proteasome activity and inhibition in living mice. *Nat. Med.* 9, 969–973, 2003.
2. Luker, G.D. et al., Molecular imaging of protein-protein interactions: controlled expression of p53 and large T-antigen fusion proteins *in vivo*. *Cancer Res.* 63, 1780–1788, 2003.

3. Luker, G.D., Sharma, V., and Piwnica-Worms, D., Visualizing protein-protein interactions in living animals. *Methods* 29, 110–122, 2003.
4. Luker, G.D. et al., Noninvasive imaging of protein-protein interactions in living animals. *Proc. Natl. Acad. Sci. U.S.A.* 99, 6961–6966, 2002.
5. Germano, G. et al., Use of the abdominal aorta for arterial input function determination in hepatic and renal PET studies. *J. Nucl. Med.* 33, 613–620, 1992.
6. Choi, Y. et al., Parametric images of myocardial metabolic rate of glucose generated from dynamic cardiac PET and 2-[18F]fluoro-2-deoxy-d-glucose studies. *J. Nucl. Med.* 32, 733–738, 1991.
7. Contag, C.H. et al., Photonic detection of bacterial pathogens in living hosts. *Mol. Microbiol.* 18, 593–603, 1995.
8. Contag, C.H. et al., Visualizing gene expression in living mammals using a bioluminescent reporter. *Photochem. Photobiol.* 66, 523–531, 1997.
9. Tjuvajev, J.G. et al., Imaging the expression of transfected genes *in vivo*. *Cancer Res.* 55, 6126–6132, 1995.
10. Jacobs, A.H. et al. Imaging in gene therapy of patients with glioma. *J. Neurooncol.* 65, 291–305, 2003.
11. Nielsen, T.O. et al., Molecular characterisation of soft tissue tumours: a gene expression study. *Lancet* 359, 1301–1307, 2002.
12. Krogsgaard, M., Huppa, J.B., Purbhoo, M.A., and Davis, M.M., Linking molecular and cellular events in T-cell activation and synapse formation. *Semin. Immunol.* 15, 307–315, 2003.
13. Davis, M.M. et al., Dynamics of cell surface molecules during T cell recognition. *Annu. Rev. Biochem.* 72, 717–742, 2003.
14. Huppa, J.B., Gleimer, M., Sumen, C., and Davis, M.M., Continuous T cell receptor signaling required for synapse maintenance and full effector potential. *Nat. Immunol.* 4, 749–755, 2003.
15. Stephens, D.J. and Allan, V.J., Light microscopy techniques for live cell imaging. *Science* 300, 82–86, 2003.
16. Hooper, C.E., Ansorge, R.E., and Rushbrooke, J.G. Low-light imaging technology in the life sciences. *J. Biolumin. Chemilumin.* 9, 113–22, 1994.
17. Hooper, C.E., Ansorge, R.E., Browne, H.M., and Tomkins, P., CCD imaging of luciferase gene expression in single mammalian cells. *J. Biolumin. Chemilumin.* 5, 123–130, 1990.
18. Rutter, G.A., White, M.R., and Tavare, J.M., Involvement of MAP kinase in insulin signalling revealed by non-invasive imaging of luciferase gene expression in single living cells. *Curr. Biol.* 5, 890–899, 1995.
19. White, M.R. et al., Real-time analysis of the transcriptional regulation of HIV and hCMV promoters in single mammalian cells. *J. Cell Sci.* 108 (Pt 2), 441–455, 1995.
20. White, M.R., Wood, C.D., and Millar, A.J., Real-time imaging of transcription in living cells and tissues. *Biochem. Soc. Trans.* 24, 411S, 1996.
21. Stanewsky, R., Jamison, C.F., Plautz, J.D., Kay, S.A., and Hall, J.C., Multiple circadian-regulated elements contribute to cycling period gene expression in Drosophila. *Embo J.* 16, 5006–5018, 1997.
22. Plautz, J.D., Kaneko, M., Hall, J.C., and Kay, S.A., Independent photoreceptive circadian clocks throughout Drosophila. *Science* 278, 1632–1635, 1997.
23. Plautz, J.D. et al., Quantitative analysis of Drosophila period gene transcription in living animals. *J. Biol. Rhythms* 12, 204–217, 1997.
24. Brandes, C. et al., Novel features of drosophila period transcription revealed by real-time luciferase reporting. *Neuron* 16, 687–692, 1996.

25. Plautz, J.D. et al., Green fluorescent protein and its derivatives as versatile markers for gene expression in living Drosophila melanogaster, plant and mammalian cells. *Gene* 173, 83–87, 1996.
26. Sala-Newby, G.B., Taylor, K.M., Badminton, M.N., Rembold, C.M., and Campbell, A.K., Imaging bioluminescent indicators shows Ca^{2+} and ATP permeability thresholds in live cells attacked by complement. *Immunology* 93, 601–609, 1998.
27. Sala-Newby, G.B. et al., Targeted bioluminescent indicators in living cells. *Methods Enzymol.* 305, 479–498, 2000.
28. Giuliano, K.A., Haskins, J.R., and Taylor, D.L., Advances in high content screening for drug discovery. *Assay. Drug Dev. Technol.* 1, 565–577, 2003.
29. Abraham, V.C., Taylor, D.L., and Haskins, J.R., High content screening applied to large-scale cell biology. *Trends Biotechnol.* 22, 15–22, 2004.
30. Olson, K.R. and Olmsted, J.B., Analysis of microtubule organization and dynamics in living cells using green fluorescent protein-microtubule-associated protein 4 chimeras. *Methods Enzymol.* 302, 103–120, 1999.
31. Giuliano, K.A., High-content profiling of drug-drug interactions: cellular targets involved in the modulation of microtubule drug action by the antifungal ketoconazole. *J. Biomol. Screen.* 8, 125–135, 2003.
32. Krutzik, P.O., Irish, J.M., Nolan, G.P., and Perez, O.D., Analysis of protein phosphorylation and cellular signaling events by flow cytometry: techniques and clinical applications. *Clin. Immunol.* 110, 206–221, 2004.
33. Nitin, N., Santangelo, P.J., Kim, G., Nie, S., and Bao, G., Peptide-linked molecular beacons for efficient delivery and rapid mRNA detection in living cells. *Nucleic Acids Res.* 32, e58, 2004.
34. Santangelo, P.J., Nix, B., Tsourkas, A., and Bao, G., Dual FRET molecular beacons for mRNA detection in living cells. *Nucleic Acids Res.* 32, e57, 2004.
35. Dooley, C.M. et al., Imaging dynamic redox changes in mammalian cells with green fluorescent protein indicators. *J. Biol. Chem.* 279, 22284–22293, 2004.
36. Tsien, R.Y., Bacskai, B.J., and Adams, S.R., FRET for studying intracellular signalling. *Trends Cell Biol.,* 3, 242–245, 1993.
37. Babendure, J.R., Adams, S.R., and Tsien, R.Y., Aptamers switch on fluorescence of triphenylmethane dyes. *J. Am. Chem. Soc.* 125, 14716–14717, 2003.
38. Hasegawa, S., Jackson, W.C., Tsien, R.Y., and Rao, J., Imaging Tetrahymena ribozyme splicing activity in single live mammalian cells. *Proc. Natl. Acad. Sci. U.S.A.* 100, 14892–14896, 2003.
39. Edinger, M. et al., Revealing lymphoma growth and the efficacy of immune cell therapies using *in vivo* bioluminescence imaging. *Blood* 101, 640–648, 2003.
40. Lipshutz, G.S. et al., In utero delivery of adeno-associated viral vectors: intraperitoneal gene transfer produces long-term expression. *Mol. Ther.* 3, 284–292, 2001.
41. Modo, M. et al., Mapping transplanted stem cell migration after a stroke: a serial, *in vivo* magnetic resonance imaging study. *Neuroimage* 21, 311–317, 2004.
42. Allen, M.J., MacRenaris, K.W., Venkatasubramanian, P.N., and Meade, T.J., Cellular delivery of MRI contrast agents. *Chem. Biol.* 11, 301–307, 2004.
43. Weber, S.M. et al., Imaging of murine liver tumor using microCT with a hepatocyte-selective contrast agent: accuracy is dependent on adequate contrast enhancement. *J. Surg. Res.* 119, 41–45, 2004.
44. Doerr-Stevens, J.K. et al., Imaging efficacy of a hepatocyte-selective polyiodinated triglyceride (DHOG-LE) for contrast-enhanced CT. *Acad. Radiol.* 9 Suppl. 1, S200–S204, 2002.

45. Bakan, D.A., Lee, F.T., Jr., Weichert, J.P., Longino, M.A., and Counsell, R.E., Hepatobiliary imaging using a novel hepatocyte-selective CT contrast agent. *Acad. Radiol.* 9 Suppl. 1, S194–S199, 2002.
46. Tsourkas, A., Hofstetter, O., Hofstetter, H., Weissleder, R., and Josephson, L., Magnetic relaxation switch immunosensors detect enantiomeric impurities. *Angew. Chem. Int. Ed. Engl.* 43, 2395–2399, 2004.
47. Perez, J.M., Josephson, L., and Weissleder, R., Use of magnetic nanoparticles as nanosensors to probe for molecular interactions. *Chembiochem.* 5, 261–264, 2004.
48. Schellenberger, E.A., Reynolds, F., Weissleder, R., and Josephson, L., Surface-functionalized nanoparticle library yields probes for apoptotic cells. *Chembiochem.* 5, 275–279, 2004.
49. Mandl, S. et al. Multi-modality imaging identifies key times for annexin V imaging as an early predictor of therapeutic outcome. *Mol. Imaging* 3(1), 1–8, 2004.
50. Blankenberg, F.G. et al., Tumor imaging using a standardized radiolabeled adapter protein docked to vascular endothelial growth factor. *J. Nucl. Med.* 45(8), 1373–1380, 2004.
51. Tjuvajev, J.G. et al., Noninvasive imaging of herpes virus thymidine kinase gene transfer and expression: a potential method for monitoring clinical gene therapy. *Cancer Res.* 56, 4087–4095, 1996.
52. Contag, P.R., Olomu, I.N., Stevenson, D.K., and Contag, C.H., Bioluminescent indicators in living mammals. *Nat. Med.* 4, 245–247, 1998.
53. Weissleder, R. and Ntziachristos, V., Shedding light onto live molecular targets. *Nat. Med.* 9, 123–128, 2003.
54. Ntziachristos, V., Tung, C.H., Bremer, C., and Weissleder, R., Fluorescence molecular tomography resolves protease activity *in vivo*. *Nat. Med.* 8, 757–760, 2002.
55. Weissleder, R., Tung, C.H., Mahmood, U., and Bogdanov, A., Jr., *In vivo* imaging of tumors with protease-activated near-infrared fluorescent probes. *Nat. Biotechnol.* 17, 375–378, 1999.
56. Chishima, T. et al., Visualization of the metastatic process by green fluorescent protein expression. *Anticancer Res.* 17, 2377–2384, 1997.
57. Patel, D.N., Bloch, S.H., Dayton, P.A., and Ferrara, K.W., Acoustic signatures of submicron contrast agents. *IEEE Trans. Ultrason. Ferroelectr. Freq. Control* 51, 293–301, 2004.
58. Bakan, D.A. et al., Imaging efficacy of a hepatocyte-selective polyiodinated triglyceride for contrast-enhanced computed tomography. *Am. J. Ther.* 8, 359–365, 2001.
59. Kapoor, V., McCook, B.M., and Torok, F.S., An introduction to PET-CT imaging. *Radiographics* 24, 523–543, 2004.
60. Stahl, A. et al., PET/CT molecular imaging in abdominal oncology. *Abdom. Imaging*, 29(3), 388–397, 2004.
61. Pelosi, E. et al., Value of integrated PET/CT for lesion localisation in cancer patients: a comparative study. *Eur. J. Nucl. Med. Mol. Imaging* 31(7), 932, 2004.
62. Zangheri, B. et al., PET/CT and breast cancer. *Eur. J. Nucl. Med. Mol. Imaging.*, 2004.
63. Stride, E. and Saffari, N., Microbubble ultrasound contrast agents: a review. *Proc. Inst. Mech. Eng. [H]* 217, 429–447, 2003.
64. Dayton, P.A. and Ferrara, K.W., Targeted imaging using ultrasound. *J. Magn. Reson. Imaging* 16, 362–377, 2002.
65. Mitragotri, S., Edwards, D.A., Blankschtein, D., and Langer, R.A., mechanistic study of ultrasonically-enhanced transdermal drug delivery. *J. Pharm. Sci.* 84, 697–706, 1995.

66. Chomas, J.E., Dayton, P., Allen, J., Morgan, K., and Ferrara, K.W., Mechanisms of contrast agent destruction. *IEEE Trans. Ultrason. Ferroelectr. Freq. Control* 48, 232–248, 2001.
67. Morawski, A.M. et al., Targeted nanoparticles for quantitative imaging of sparse molecular epitopes with MRI. *Magn. Reson. Med.* 51, 480–486, 2004.
68. Pirko, I. et al., *In vivo* magnetic resonance imaging of immune cells in the central nervous system with superparamagnetic antibodies. *FASEB J.* 18, 179–182, 2004.
69. Yan, F. et al., Synthesis and characterization of silica-embedded iron oxide nanoparticles for magnetic resonance imaging. *J. Nanosci. Nanotechnol.* 4, 72–76, 2004.
70. Pathak, A.P. et al., Molecular and functional imaging of cancer: advances in MRI and MRS. *Methods Enzymol.* 386, 1–58, 2004.
71. Taylor, S.F. et al., A functional neuroimaging study of motivation and executive function. *Neuroimage* 21, 1045–1054, 2004.
72. Kim, D.S. et al., Spatial relationship between neuronal activity and BOLD functional MRI. *Neuroimage* 21, 876–885, 2004.
73. Duyn, J., Specificity of high-resolution BOLD and CBF fMRI at 7 T. *Magn. Reson. Med.* 51, 644–645; author reply 646–647, 2004.
74. Matthews, P.M. and Jezzard, P., Functional magnetic resonance imaging. *J. Neurol. Neurosurg. Psychiatry* 75, 6–12, 2004.
75. Kim, D.E., Schellingerhout, D., Ishii, K., Shah, K., and Weissleder, R., Imaging of stem cell recruitment to ischemic infarcts in a murine model. *Stroke*, 35(4), 952–957, 2004.
76. Shah, K. et al., *In vivo* imaging of HIV protease activity in amplicon vector-transduced gliomas. *Cancer Res.* 64, 273–278, 2004.
77. Blasberg, R., PET imaging of gene expression. *Eur. J. Cancer* 38, 2137–2146, 2002.
78. Ray, P. et al., Monitoring gene therapy with reporter gene imaging. *Semin. Nucl. Med.* 31, 312–320, 2001.
79. Sun, X. et al., Quantitative imaging of gene induction in living animals. *Gene Ther.* 8, 1572–1579, 2001.
80. Gambhir, S.S. et al., Imaging transgene expression with radionuclide imaging technologies. *Neoplasia* 2, 118–138, 2000.
81. Green, L.A. et al., Noninvasive methods for quantitating blood time-activity curves from mouse PET images obtained with fluorine-18-fluorodeoxyglucose. *J. Nucl. Med.* 39, 729–734, 1998.
82. Townsend, D.W., Carney, J.P., Yap, J.T., and Hall, N.C., PET/CT today and tomorrow. *J. Nucl. Med.* 45 Suppl. 1, 4S–14S, 2004.
83. Chen, Y. et al., Metabolism-enhanced tumor localization by fluorescence imaging: *in vivo* animal studies. *Opt. Lett.* 28, 2070–2072, 2003.
84. Tamura, M., Hazeki, O., Nioka, S., and Chance, B., *In vivo* study of tissue oxygen metabolism using optical and nuclear magnetic resonance spectroscopies. *Annu. Rev. Physiol.* 51, 813–834, 1989.
85. Schellenberger, E.A., Weissleder, R., and Josephson, L., Optimal modification of annexin v with fluorescent dyes. *Chembiochem.* 5, 271–274, 2004.
86. Doubrovin, M. et al., Development of a new reporter gene system — dsRed/xanthine phosphoribosyltransferase-xanthine for molecular imaging of processes behind the intact blood-brain barrier. *Mol. Imaging* 2, 93–112, 2003.
87. Tung, C.H. et al., *In vivo* imaging of beta-galactosidase activity using far red fluorescent switch. *Cancer Res.* 64, 1579–1583, 2004.
88. Kircher, M.F., Weissleder, R., and Josephson, L., A dual fluorochrome probe for imaging proteases. *Bioconjug. Chem.* 15, 242–248, 2004.

89. Naumov, G.N. et al., Cellular expression of green fluorescent protein, coupled with high-resolution *in vivo* videomicroscopy, to monitor steps in tumor metastasis. *J. Cell Sci.* 112 (Pt. 12), 1835–1842, 1999.
90. Hoffman, R.M., Visualization of GFP-expressing tumors and metastasis *in vivo. Biotechniques* 30, 1016–1022, 1024–1026, 2001.
91. Hoffman, R., Green fluorescent protein imaging of tumour growth, metastasis, and angiogenesis in mouse models. *Lancet Oncol.* 3, 546–556, 2002.
92. Hsu, E.R. et al., A far-red fluorescent contrast agent to image epidermal growth factor receptor expression. *Photochem. Photobiol.* 79, 272–279, 2004.
93. Sakatani, K. et al., Noninvasive optical imaging of the subarachnoid space and cerebrospinal fluid pathways based on near-infrared fluorescence. *J. Neurosurg.* 87, 738–745, 1997.
94. Reynolds, J.S. et al., Imaging of spontaneous canine mammary tumors using fluorescent contrast agents. *Photochem. Photobiol.* 70, 87–94, 1999.
95. Zheng, G. et al., Tricarbocyanine cholesteryl laurates labeled LDL: new near infrared fluorescent probes (NIRFs) for monitoring tumors and gene therapy of familial hypercholesterolemia. *Bioorg. Med. Chem. Lett* .12, 1485–1488, 2002.
96. Becker, A. et al., Receptor-targeted optical imaging of tumors with near-infrared fluorescent ligands. *Nat. Biotechnol.* 19, 327–331, 2001.
97. Becker, A. et al., Macromolecular contrast agents for optical imaging of tumors: comparison of indotricarbocyanine-labeled human serum albumin and transferrin. *Photochem. Photobiol.* 72, 234–241, 2000.
98. Ponomarev, V. et al., A novel triple-modality reporter gene for whole-body fluorescent, bioluminescent, and nuclear noninvasive imaging. *Eur. J. Nucl. Med. Mol. Imaging* 31, 740–751, 2004.
99. McIntyre, J.O. et al., Development of a novel fluorogenic proteolytic beacon for *in vivo* detection and imaging of tumour-associated matrix metalloproteinase-7 activity. *Biochem. J* 377, 617–628, 2004.
100. Funovics, M., Weissleder, R., and Tung, C.H., Protease sensors for bioimaging. *Anal. Bioanal. Chem.* 377, 956–963, 2003.
101. Jaffer, F.A., Tung, C.H., Gerszten, R.E., and Weissleder, R., *In vivo* imaging of thrombin activity in experimental thrombi with thrombin-sensitive near-infrared molecular probe. *Arterioscler. Thromb. Vasc. Biol.* 22, 1929–1935, 2002.
102. Thatte, H.S. et al., Acidosis-induced apoptosis in human and porcine heart. *Ann. Thorac. Surg.* 77, 1376–1383, 2004.
103. Wang, T.D., Mandella, M.J., Contag, C.H., and Kino, G.S. Dual-axis confocal microscope for high-resolution *in vivo* imaging. *Opt. Lett.* 28, 414–416, 2003.
104. Sokolov, K. et al., Endoscopic microscopy. *Dis. Markers* 18, 269–291, 2002.
105. Sokolov, K. et al., Optical systems for *in vivo* molecular imaging of cancer. *Technol. Cancer Res. Treat.* 2, 491–504, 2003.
106. Drezek, R.A. et al., Optical imaging of the cervix. *Cancer* 98, 2015–2027, 2003.
107. Jacobs, A. et al., Positron emission tomography-based imaging of transgene expression mediated by replication-conditional, oncolytic herpes simplex virus type 1 mutant vectors *in vivo*. *Cancer Res.* 61, 2983–2995, 2001.
108. Gambhir, S.S., Barrio, J.R., Herschman, H.R., and Phelps, M.E., Assays for noninvasive imaging of reporter gene expression. *Nucl. Med. Biol* .26, 481–490, 1999.
109. Shen, D.H. et al., Effects of dose, intervention time, and radionuclide on sodium iodide symporter, NIS-targeted radionuclide therapy. *Gene Ther.* 11, 161–169, 2004.
110. Dadachova, E. and Carrasco, N., The Na/I symporter (NIS): imaging and therapeutic applications. *Semin. Nucl. Med.* 34, 23–31, 2004.

111. Liu, Z.M. et al., Upregulation of heme oxygenase-1 and p21 confers resistance to apoptosis in human gastric cancer cells. *Oncogene* 23, 503–513, 2004.
112. Hastings, J.W., Chemistries and colors of bioluminescent reactions: a review. *Gene* 173, 5–11, 1996.
113. Franks, N.P., Jenkins, A., Conti, E., Lieb, W.R., and Brick, P., Structural basis for the inhibition of firefly luciferase by a general anesthetic. *Biophys. J.* 75, 2205–2211, 1998.
114. Conti, E., Franks, N.P., and Brick, P., Crystal structure of firefly luciferase throws light on a superfamily of adenylate-forming enzymes. *Structure* 4, 287–298, 1996.
115. Karkhanis, Y.D. and Cormier, M.J., Isolation and properties of *Renilla reniformis* luciferase, a low molecular weight energy conversion enzyme. *Biochemistry* 10, 317–326, 1971.
116. Matthews, J.C., Hori, K., and Cormier, M.J., Purification and properties of *Renilla reniformis* luciferase. *Biochemistry* 16, 85–91, 1977.
117. Srikantha, T. et al., The sea pansy *Renilla reniformis* luciferase serves as a sensitive bioluminescent reporter for differential gene expression in *Candida albicans. J. Bacteriol.* 178, 121–129, 1996.
118. Wang, Y., Wang, G., O'Kane, D.J., and Szalay, A.A., A study of protein-protein interactions in living cells using luminescence resonance energy transfer (LRET) from *Renilla luciferase* to Aequorea GFP. *Mol. Gen. Genet.* 264, 578–587, 2001.
119. Shimomura, O. and Johnson, F.H., Chemical nature of bioluminescence systems in coelenterates. *Proc. Natl. Acad. Sci. U.S.A.* 72, 1546–1549, 1975.
120. Greer, L.F., 3rd, and Szalay, A.A., Imaging of light emission from the expression of luciferases in living cells and organisms: a review. *Luminescence* 17, 43–74, 2002.
121. Alam, J. and Cook, J.L. Reporter genes: application to the study of mammalian gene transcription. *Anal. Biochem.* 188, 245–254, 1990.
122. Modo, M. et al., Tracking transplanted stem cell migration using bifunctional, contrast agent-enhanced, magnetic resonance imaging. *Neuroimage* 17, 803–811, 2002.
123. Louie, A.Y. et al., *In vivo* visualization of gene expression using magnetic resonance imaging. *Nat. Biotechnol.* 18, 321–325, 2000.
124. Jacobs, R.E., Ahrens, E.T., Meade, T.J., and Fraser, S.E., Looking deeper into vertebrate development. *Trends Cell Biol.* 9, 73–76, 1999.
125. Day, R.N., Kawecki, M., and Berry, D., Dual-function reporter protein for analysis of gene expression in living cells. *Biotechniques* 25, 848–850, 852–84, 856, 1998.
126. Edinger, M., Hoffmann, P., Contag, C.H., and Negrin, R.S., Evaluation of effector cell fate and function by *in vivo* bioluminescence imaging. *Methods* 31, 172–179, 2003.
127. Ray, P., De, A., Min, J.J., Tsien, R.Y., and Gambhir, S.S., Imaging tri-fusion multimodality reporter gene expression in living subjects. *Cancer Res.* 64, 1323–1330, 2004.
128. Cao, Y.A. et al., Shifting foci of hematopoiesis during reconstitution from single stem cells. *Proc. Natl. Acad. Sci. U.S.A.* 101, 221–226, 2004.
129. Mattheakis, L.C. et al., Optical coding of mammalian cells using semiconductor quantum dots. *Anal. Biochem.* 327, 200–208, 2004.
130. Swenson, D.L. et al., Generation of Marburg virus-like particles by co-expression of glycoprotein and matrix protein. *FEMS Immunol. Med. Microbiol.* 40, 27–31, 2004.
131. Liu, D., Park, S.H., Reif, J.H., and LaBean, T.H., DNA nanotubes self-assembled from triple-crossover tiles as templates for conductive nanowires. *Proc. Natl. Acad. Sci. U.S.A.* 101, 717–722, 2004.

132. Embretson, J. et al., Analysis of human immunodeficiency virus-infected tissues by amplification and in situ hybridization reveals latent and permissive infections at single-cell resolution. *Proc. Natl. Acad. Sci. U.S.A.* 90, 357–361, 1993.
133. Shih, W.M., Quispe, J.D., and Joyce, G.F., A 1.7-kilobase single-stranded DNA that folds into a nanoscale octahedron. *Nature* 427, 618–621, 2004.
134. Edinger, M. et al., Advancing animal models of neoplasia through *in vivo* bioluminescence imaging. *Eur. J. Cancer* 38, 2128, 2002.

[illegible]

13 Tissue Engineering and Artificial Cells

R. Lane Smith

CONTENTS

13.1 INTRODUCTION

Individual specialized cells are the fundamental units of function for any organ or tissue and, as a collective body, represent the organism, whether yeast or human. Fabrication of artificial cells with functional properties and system-wide compatibility equal to that of any particular differentiated cell would represents a major goal for tissue engineering.[1]

The working concept of artificial cells was pioneered by T. M. Chang, who proposed that an artificial cell-based approach could be used to circumvent limitations inherent in hemodialysis.[2] This early concept for fabrication and use of artificial

0-8493-1940-4/05/$0.00+$1.50

cells was partially realized in various experimental designs that incorporated agents for detoxification of blood within artificial membranes.[3–5]

At a time when cell donor availability is a paramount concern, protocols for reproducibly creating artificial cells as unit structures to provide autonomous function when placed *in vivo* would significantly impact clinical therapeutics. To be fully successful, the design criteria for artificial cells must incorporate methods to reproduce organized enzyme systems that accomplish specialized tasks while establishing a self-contained structure.

The capacity to augment metabolic function using artificial cells would enable repair of diseased tissues and correction of genetic abnormalities without the restrictions inherent in tissue availability and immune compatibility. The convenience of having manufactured cells that satisfy design specifications to fulfill enzymatic, structural or neurological function would bring the field of tissue engineering closer to off-the-shelf products for tissue and organ replacement.[6]

For purposes of this review, an artificial cell will be considered as a single autonomous construct bounded by a membrane structure and capable of carrying out a specialized function. In that capacity, any type of microvesicles enclosed by a boundary-limiting membrane that encapsulates proteins, nucleic acids, or other reactive materials and provides metabolic activity would meet these criteria.[7–9] Similarly, a stable liposome containing a viral construct for delivery of genetic information to a secondary cell through targeted fusion also qualifies.[10–12] However, a biological cell simply encapsulated within a secondary surface coating will not be considered as an artificial cell here, since the inner working components are not subject to exogenous design and fabrication.[13–15]

The potential advantages of artificial cells extend across most of the major issues that confront tissue engineering. A major obstacle for the repair of damaged tissues or organs is sufficient donor material to serve as a source of cells.[16] The current estimate for liver transplantation is that approximately 30,000 individuals will die due to insufficient donor material. Other tissue diseases where transplantation is a mainstay for chronic treatment, such as kidney failure, are also experiencing donor shortages. At present, approaches to generate stable cell lines, to use transformed cells, or to apply allogeneic or xenogeneic cells have not satisfied the demand or shown evidence of long-term function.[17–21]

The ability to produce artificial cells for incorporation into a liver assist device to carry out the major hepatic functions of detoxification and albumin production would greatly alleviate donor shortage. In addition, without concern regarding loss of cell viability in the liver assist device, supply could follow demand, since scale-up production of the artificial hepatocytes could provide off-the-shelf availability. The appropriate design of the artificial cells might also alleviate major problems inherent in thrombolytic reactions and could potentially eliminate the need for plasmapheresis approaches. Finally, in the case of liver function, the scale-up of the device size to reproduce normal organ capacity would not be limited by availability of donor cells, and *in vivo* implantation could eliminate the necessity for bulky exogenous perfusion devices.

13.2 ARTIFICIAL CELLS AS UNIT STRUCTURES

Establishing design criteria for artificial cells demands a consideration of the properties of the naturally occurring biological templates, the cells of archea, prokaryotes, and eukaryotes.[22] Biological cells represent the smallest unit that exists for unassisted replication and constitute standalone miniature bioreactors that derive energy from the environment, convert it to products, and disperse waste products. For an artificial cell to function successfully, internal cytoplasmic enzyme systems must be present within organized membrane-based mechanisms to facilitate uptake of metabolites and provide pathways for export of degradation products. The normal mechanisms supporting these transport processes rely on membrane cycling events including endocytosis to achieve uptake and exocytosis for secretion. The details of these processes are complex and will be discussed below.

Biological cells establish multiple differential gradients for molecules that are maintained in microenvironments inside and outside the plasma membrane.[23,24] The establishment of gradients generates ionic compartmentalization, electrical potentials, and osmotic pressures.[25,26] In normal cells, the differential concentrations between intra- and extracellular compartments established for ionic molecules, such as potassium, sodium, and calcium, require expenditure of energy and specialized enzymatic mechanisms to maintain gradients.[27,28] In prokaryotic and eukaryotic cells, the ATPase pump transports potassium ions inside and sodium ions outside.[29] Nerve cells and muscle cells have developed specialized uses for the membrane potential by being able to undergo transient, rapid changes in membrane depolarization.[30,31] The potential differences in gradients permit electrical signals to be transmitted across distances so that nervous tissue can receive and transmit messages and muscle can undergo changes in the contractile apparatus.

13.2.1 Plasma Membranes

The primary element of a functional cell is the boundary membrane that establishes an inside and outside environment.[32,33] The most primitive cells achieve this function with limited molecular components that serve to establish an ionic gradient between inside and outside boundaries.[34,35] The evolution of cells has been accompanied by the emergence of a universal lipid bilayer that serves as the boundary structure separating interior and exterior environments. A similarity exists for eukaryotic and prokaryotic cellular membrane with respect to structural organization and chemical components. The evolution of cellular membranes has been accompanied by inclusion of membrane-spanning proteins that cross the lipid bilayer and form the basis for ion transport, chemical reception, and enzymatic reactivity. The current representation of biological membranes is referred to as the fluid mosaic model. In this model, membrane proteins are recognized as mobile components that are integral elements to the structure and function of the cell boundary.[36,37] Modifications of the membrane proteins with various carbohydrate substitutions have given rise to different types of binding reactions with plasma membrane components. The addition of carbohydrate groups to the proteins provides mechanisms that ensure recognition and adhesion between cells of like and unlike varieties.[38,39]

The cell plasma membrane is composed of phospholipids that are oriented to form a bilayer that is hydrophobic within the interior regions and hydrophilic along the outside and inside layers.[40,41] Most membranes contain some percentage of cholesterol, which contributes to the stability and fluidity of the lipid bilayer. Plasma membranes have a trilaminar appearance when examined by electron microscopy, due to the organization of the lipid bilayers with the ionic components facing outward and the hydrophobic components localizing in the interior region.[42] The interior hydrophobic region of the membrane serves a major barrier function to hydrated and polar molecules. However, nonpolar molecules may cross the lipid bilayer, depending on solubility within the lipid components. The dimensions of all plasma membranes are in the order of 7 to 10 nm as visualized by electron microscopy. The most abundant components of the plasma membrane are the lipids, which can be fractionated into three prominent types, glycerophospholipids, sphingolipids, and cholesterol. The glycerophospholipids have a glycerol molecule as the backbone structure to which phosphoric acid and two long-chain fatty acids are esterified at the alpha and two remaining carbons, respectively.[43] Esterification of hydroxyl-containing compounds to the phosphate molecule can include choline, ethanolamine, serine, glycerol, and inositol to generate a family of glycerophospholipids. The long-chain fatty acid chains attached to carbon 1 typically consist of saturated fatty acid, whereas the carbon 2 usually has unsaturated fatty acid chains attached.

Variation in content of the fatty acid chains imparts unique physicochemical properties to the plasma membranes of different cell types.[44] The existence of outer polar groups linked to the phosphate group at the alpha carbon coupled with the nonpolar hydrophobic hydrocarbon chains of the fatty acid chains creates an amphipathic molecule. The assembly of the glycerophospholipids into the bilayer conformation results from the amphipathic properties of the lipid moieties, with the chemical composition determining the physiological state of the membranes. Mixtures of glycerophospholipids and sphingolipids interact in aqueous systems to form spheres, termed micelles, due to the amphipathic nature of the lipids. Micelles of specific diameters and properties can be formed with appropriate lipid concentrations as the hydrophobic tails organize to exclude water and the polar heads orient to accommodate the aqueous environment.

The self-assembling nature of the phospholipids and the sphingolipids makes it possible to produce artificial cells with selective properties for transport and stability. The specific lipid concentration required to form spherical structures, the critical micelle concentration, varies depending on whether a single lipid or a mixture of lipids is applied. Spherical lipid structures formed by spontaneous aggregation and reorientation possess excellent physical stability due to strong hydrophobic interactions between the hydrocarbon chains. The lipid molecules may undergo lateral diffusion in the plane of the membrane, but migration from one monolayer to another does not readily occur. The monolayer stability is due to thermodynamic constraints on movement of the polar head group across the hydrophobic boundary.[45] However, mechanisms exist that can catalyze a lipid-based flip-flop of the polar groups for transporter function in some biologic membranes.

Recent evidence shows that lipid segregation occurs in the plane of the membrane to form microdomains or "rafts."[46,47] The role of membrane rafts appears to be critical for specific processes of signal transduction, membrane transport and ligand trafficking.[48] The rafts have a local high concentration of cholesterol, and the role of cholesterol sequestration in the formation and function of lipid rafts remains an area of intensive study.

13.2.2 Membrane Trafficking

Transport of larger substances across membranes occurs in conjunction with rearrangement in the lipid bilayer so that vesicle formation occurs.[49,50] Endocytosis is a process of inclusion of polar molecules by the lipid bilayer to internalize the material. In some instances, endocytosis provides a means by which materials cross the cell boundary within internalized vesicles that can fuse with other membranes at different locations inside the cells. Exocytosis is the process for export of a material from the interior to the outside of the cell. With exocytosis, polar materials, such as integral membrane proteins, may be inserted into the membrane at the time of fusion or exported from the cell. In biological systems, transfer of membrane through endocytotic and exocytotic processes involves mechanisms for membrane conservation to ensure that the surface area and volume of the cells remain stable.

Endocytosis results in the internalization of materials present at the surfaces of cells and occurs through a process of vesicle formation.[51,52] The process involves invagination of the selected membrane region and cleavage of the vesicle releasing it to the interior. In some cases, binding of materials involves specific receptors, receptor-mediated endocytosis. In other cases, fluid is enclosed for internalization, a process termed pinocytosis. In both processes, the internalized vesicle may then fuse to other vesicles, Golgi membrane, or lysosomes to support the needs of the cell, either metabolically or defensively. Invagination of the membrane requires a coating of the interior membrane proteins, including clathrin, clathrin assembly proteins (AP-1, AP-2), and small GTP binding proteins in the Rab family.[53–55]

For artificial cells to exhibit selectivity of uptake, membrane design needs to include provision for protein-mediated vesicle formation. In cells, separation of the formed vesicle is dependent on the action of dynamin, a protein that forms rings surrounding the neck of the invagination to contract and severe the membranous connection.[56] Active transport of phosphatidylserine and phosphatidylethanolamine from the outer to the inner boundaries of the plasma membrane is catalyzed by the enzyme, aminophospholipid translocase.[57] The active transport of the lipid moieties may drive membrane folding under appropriate conditions without the participation of the protein components of the clathrin-associated mechanism. This would fit with the observation that vesicles can form in yeast not able to express clathrin.[58]

Exocytosis involves fusion with vesicles containing materials for export that are in the size range of 50 nm to 1 μm with the plasma membrane.[59] The size differential between the diameter of the small fusion vesicles and the diameter of the cell make the fusion process dependent on multiple molecular processes. Physical parameters that influence fusion pore formation include membrane composition, pH, ion concentration, temperature, and vesicle localization. Model systems provide evidence that protein networks are required to create a fusion locus in the plasma membrane.[60–62] Use of liposomal vesicles and osmotic pressure to drive the fusion reaction has shown that channel proteins are involved but may not be completely necessary. Surface-immobilized, unilamellar liposomes, and nanotube networks have been applied to show that expansion of the fusion pore to the final stage of exocytosis can occur in the absence of protein.[63–65]

With artificial cells, absence of protein receptors would be expected to limit selectively of uptake of materials. However, for certain applications where exclusion of large materials is needed, selectivity for membrane transport could be provided by lipid solubility. In addition to fusion of membranes for import/export, membrane repair may also be a critically important process in the event of a rupture due to mechanical insult. In biological systems, the "resealing" process that occurs in response to an influx of calcium into the cells causes vesicles present in the cytoplasm to fuse with each other and the adjacent plasma membrane.[66]

13.2.3 Prototypic Cell Progenitors

The earliest progenitor cells were able to respond to environmental changes through mechanosensitive channels.[67–69] Phylogenetic analysis shows that a common ancestral molecule resembling the bacterial MscL channel protein was established in the lipid bilayer of the earliest life forms. The ubiquity of lipid channel gating mechanisms extends from the Archaea through prokaryotes and to eukaryotic organisms with evidence of common structural motifs.[70] The probability is high that the common ancestral cell was a thermophilic microorganism, based on the presence of many thermophilic organisms at the root of the universal phylogenetic tree.[71] This observation has lead to the hypothesis that the structure of the lipids of the common ancestral cell included the tetraether type glycerophospholipids with C40 isoprenoid chains. The C40 isoprenoid chains can form covalent bonds, and the tetraether lipids may undergo cyclization. Protein insertion would have followed later after formation of lipid boundaries to enclose the cell. Determination of the membrane compositions of primitive organisms may provide insights for the development of artificial cells having simple enzymes systems that can work together to provide multiple function.

The functional specialization of modern prokaryotic and eukaryotic cells also provide models for the creation of artificial cells that can carry out specialized secretory or detoxification functions. The myriad of specialized functions carried out by living cells is achieved through the expression of genetic information to generate the protein phenotypes that support metabolism. The protein complement of the cells is recognized as the primary target of evolutionary change.[72] The cellular proteome provides the biochemical properties integral to the cellular specialization and, as such, is constantly impacted by natural selection. Evolutionary pressures on proteins set a premium on the balance between changes in amino acid sequence that impact energy utilization versus changes that impact the rate and accuracy of synthesis.[72] Improved understanding of the fundamental properties of protein conformation will enhance the design of artificial cells.

13.3 ARTIFICIAL CELL PROTOTYPES AND DESIGN CONSIDERATIONS

Initial efforts to create an artificial cell centered on providing a means of removing soluble toxins in the body. Early experiments revealed that numerous microvesicles (semipermeable microcapsules) provide a way to capitalize on the efficacy associated with the large gain in surface area.[73] A volume of 10 ml of 20-μm diameter microcapsules was calculated to have a total surface area of 2.5 m^2. A larger particle of 100 μm would have a comparable surface area in 33 ml. What was considered more important was the fact that the membrane thickness would be 0.02 μm, 400 times thinner that standard hemodialysis membrane, that would allow permeant metabolites to cross the membrane 1,250 times faster than in a standard 1 m square hemodialysis machine.

The hypothesis advanced as a result of these studies was that encapsulation of reactive elements, such as enzymes, ion exchange resin, activated charcoal, or other materials, would provide a means for development of artificial organs using the concept of artificial cells.[74] The use of artificial membranes composed of fabric for encapsulation of charcoal was rapidly followed by the use of various molecular coatings of the charcoal to establish artificial cellular trapping systems for metabolite purification. This approach encountered difficulties due to the changes in particle properties in the *in vivo* setting as serum proteins bound to the microvesicles.

Despite difficulties in the approach, a number of studies were conducted using different routes of administration of the artificial cells. These routes included direct implantation by intramuscular, subcutaneous, or intraperitoneal routes.[75] As an alternative route, the microvesicles were adminstered directly into the blood stream. When injected intravenously, the artificial cells larger than 2 μm in diameter were filtered by the pulmonary capillaries, while smaller microvesicles were cleared by the reticuloendothelial networks of the liver and spleen. Other routes include oral application for activity of the cells to be carried out during transit through the gastrointestinal tract and direct topical application of microencapsulated enzymes to provide function without generation of an inflammatory response that might arise from the presence of free enzyme.

13.3.1 Intracellular Membranes and Enzyme Systems

A significant requirement for functional artificial cells will entail the formation of membrane-linked enzyme systems to enable transport and directionality necessary for separation of substrate and product.[76,77] Within the microsomal fraction of cells, two major membrane systems provide the compartmentalization for internally controlled synthetic process. The membrane system that constitutes the endoplasmic reticulum provides the surfaces and fluid filled spaces for the support of protein synthesis.[78] The smooth endoplasmic reticulum is a network of tiny interconnected tubules, whereas the rough endoplasmic reticulum is characterized by bound ribosomes. The smooth endoplasmic reticulum carries the enzymes that produce the lipid for membrane formation, and the rough endoplasmic reticulum is associated with protein synthesis and release of the proteins into the fluid-filled spaces that connect to the Golgi.

The Golgi is recognized to consist of three compartments, each having an internal pattern of movement of molecules and a characteristic set of enzyme systems involved in processing of proteins, carbohydrate, and lipids for export.[79] The extent of the Golgi complex will vary with cell type, but generally consists of three regions, cis, medial, and trans, each of which carries out uniquely different posttranslational modifications to proteins. The movement of products through the Golgi is dependent on budding and fusing of vesicles or progressive movement of an individual vesicle along the cisternae. The finished products are targeted to secretory vesicles that permit concentration of secretory proteins or to transport vesicles that fuse to the plasma membrane to release products. The localization of the vesicles for export of products occurs in association with docking proteins, the v-SNAREs that bind t-SNARE, that provide specificity of binding.[80]

A breakthrough essential for the recreation of membrane systems is to establish technology for production of purified enzyme that can then be inserted into membrane systems with concomitant function. This task has recently been achieved for the acyl-CoA binding protein.[81] The acyl-CoA binding protein stimulates utilization of long-chain fatty acyl-CoA through multiple enzyme systems within the microsomal fraction. The native mouse acyl-CoA binding protein was expressed recombinantly as a fusion protein that, when cleaved of the fusion tag, exhibited full activity for binding cis-parinaroyl-CoA, stimulating oleoyl-CoA utilization by microsomal glycerol-3-phosphate acyltransferase, and protecting oleoyl-CoA from microsomal acyl-CoA hydrolase. Exposure of the recombinant enzyme to anionic and neutral unilamellar phospholipid rich vesicles, either small or large, revealed that interaction occurred preferentially with the anionic membrane. The activity of the enzyme was also significantly increased, with the small vesicles showing that the highly curved membranes facilitated transfer of the ligands bound to the acyl-CoA binding protein. This study shows that functional reconstitution of the membrane-ligand binding protein can be achieved with the proper geometry and charge of the target vesicles. A similar approach using an isolated antiporter of the plant Arabinosis confirms that sodium/potassium transport function can be transferred to a liposomal construct.[82]

13.3.2 Channel Proteins

Membrane selectivity and environmental responsiveness rests with the specialized integral membrane proteins that extend across the lipid bilayer and carry out specialized tasks, primarily through conformation change.[83–85] The types of protein responses include ion transport, ion dependent gating, growth factor recognition and binding, extracellular matrix macromolecule recognition and binding, phosphorylation, ion scavenging, sugar transport, and pore formation. Proteins that penetrate the lipid bilayer, such as the transferrin family members, transferrin, ovotransferrin, and lactoferrin, alter ion transfer as they undergo conformational change.[86] Bacterial membranes have porins that are trimers of beta-barrels that form ion channels that are gated by an extracellular loop that folds back onto the channel cavity. Analysis of spontaneous mutants and site-directed mutants showed that gating of the channel could be modified by changes in protein structure and by the ionic interactions between charged residues within the constriction zone of the porin.[87]

13.3.3 Ion Transport

Membrane channels and pores that control the flow of ions in and out of cells are composed of large variety of proteins, peptides, and other chemical components.[88,89] Transport of ions into and out of artificial cells can be controlled in part by the lipid composition of the membrane and by the inclusion of specific protein components at the time of production. Purification of endogenous cell membrane component has permitted transfer of function into synthetic vesicles. Reconstitution of human erythrocyte membrane protein into soybean phospholipid membrane vesicles provided efficient uptake of sodium ion in exchange for hydrogen ion.[90] The transfer of ions only occurred in the presence of native protein but not denatured proteins or vesicles alone. Incorporation of the acetylcholine receptor permitted reestablishment of sodium and potassium transport that occurred at physiological rates and was sensitive to agonist.

Creating artificial ion channels for incorporation into membrane-delimited structures that effectively mimic biological function has proven a difficult task.[91–93] The difficulty lies in part with the incomplete understanding of the precise geometry of the proteins that form the pore. With some ion channels, a molecular model that encompasses a barrel-stave configuration is proposed.[94] Other biological properties of ion channel function suggest that orientation of the lipid head group is a critical element of pore stability. In support of a protein-dominant role in function, one strategy for design of functional membrane channels applied the concept of self-assembly of cylindrical beta-sheet peptide rings to penetrate the membrane. Stacks of these cylindrical peptide rings incorporated into liposomes displayed good channel-mediated ion transport behavior that was equal to or better than naturally occurring counterparts.[95]

Support for a lipid prominent component for pore formation comes from model systems showing that nonesterified fatty acids participate in transfer of cations.[96,97] The efficacy of ion transport depends on membrane concentration, degree of unsaturation, and chain length of the selected fatty acid. Internal pH changes in the presence of a potassium ion gradient increased with unsaturation but decreased with increased chain length of the fatty acid. The use of different composition of lipids having defined amounts of saturated and unsaturated fatty acids may significantly improve the functional selectively of the transport function needed to bring in metabolites and/or exclude other molecules that could be inhibitory to the artificial cell specialization.

13.4 ARTIFICIAL CELLS AS LIPOSOMES

Liposomes are vesicular structures that are composed of one or more phospholipid bilayers.[98] The creation of the bilayer may occur through mechanical agitation in an aqueous environment such as sonication. Formation of vesicles by mechanical induction follows dissolution of the lipid in a solvent system that is subsequently removed under an inert atmosphere to prevent oxidation. The mixtures of the lipid components will depend on their relative solubility in the solvent of choice.

Liposomes are attractive constructs for the encapsulation of bioactive components and for the delivery of drugs, therapeutic proteins, and diagnostic agents.[99,100] Packaging of materials within liposomes affords improved protection of drugs from metabolic degradation and clearance so that pharmacological effects can be extended.[101] With certain antibiotics and neoplastic agents, the use of liposomes for delivery reduces cytotoxicity. The difficulty with liposomal-based drug delivery is the rapid clearance of the vesicles from the blood.[102] Clearance of liposomes is influenced by size, charge, membrane rigidity, and polar group constituency.[103] In addition, binding of specific serum proteins, such as apolipoprotein E, to the liposomes serves as a critical determinant in the interaction with hepatocytes that destroy the vesicles.

The use of liposomes as a nonviral gene delivery system is considered advantageous, since the membranes can be developed to provide a range in the type of embedded agent to be delivered.[104–106] An extension of this strategy is that the liposome could also be targeted to a particular type of cell. This approach is of critical importance in the implementation of suicide gene therapy, where the expression of a toxic gene optimally only occurs within a cancer cell. Direct incorporation of bacterial and other microorganisms, viruses of various types, and naked DNA in liposomes does not provide the specificity needed to achieve precise targeting. However, the use of viral vectors may ensure the insertion and incorporation of the DNA into the cellular transcriptional machinery of specific cells. The major concern with suicide gene therapy, in addition to target specificity, is the safety for administration in humans and efficiency of transduction. Transduction efficiencies vary among a number of systems.

A major advance in the production of gene delivery systems takes advantage of a cationic liposome/micelle-based structure.[107–109] The liposome component includes a single synthetic cationic amphile (the cytofectin, cyto for cell and fectin for transfection) that may be combined with a neutral lipid.[110,111] Unilamellar vesicles are formed using reverse-phase evaporation and dehydration-rehydration from the cytofectin and neutral lipid combination. Micelles formed by dispersion of the cytofectin in water or aqueous organic solvents provide an alternative delivery system. The unilamellar vesicles/micelles are combined with nucleic acids to form cationic liposome–nucleic acid complex mixtures with nanometric properties. The fusion of the cationic liposome–nucleic acid complex occurs by endocytosis with subsequent release of the nucleic acid within the cellular cytoplasm.

The development of cationic liposomes has also been applied as a delivery vehicle for intracellular deposition of superparamagnetic iron oxide particles into stem cells.[112] Delivery of the iron oxide into rabbit skeletal myoblasts through transfection liposomes resulted in cells that provide contrast-inducing images using MRI, but without the level of toxicity observed using endocytosis as the process for particle uptake. The use of cationic liposomes increased labeling efficiency approximately 100-fold and appeared to provide a valuable method for labeling and tracking stem cell migration by MRI.

13.4.1 Nuclear Function

Creation of a membrane within a membrane to localize nucleic acid in a liposomal artificial cell has not been achieved. However, some approaches have used viral-like constructs incorporated in lipids to achieve targeting and transfer of nucleic acid from one cell to another.[113,114] Combinations of cationic lipids, 1,2-dioleoyl-3-trimethylammonium propane (DOTAP) and dimethyldioctadecylammonium bromide (DDAB), with or without equimolar amounts of cholesterol (CHOL) or 1,2-dioleoylphosphatidylethanolamine (DOPE) did serve as efficient delivery systems for transgene expression. Improved formulations of cationic liposome–nucleic acid complexes will likely provide critically important reagents for specificity in nucleic acid transfer, cellular fusion, and transgene expression.

13.4.2 Enzymatic Activity

Fabrication of artificial cells containing a complement of microsomal-like enzyme systems has had limited success. The major difficulty lies with establishing membrane polarity and substrate availability.[115] However, liposomes have been developed that carry antisense oligonucleotides and enzymes for the dissolution of vesicular compartments to increase intracellular delivery and the therapeutic efficacy of the nucleic acid.[116] This delivery strategy takes advantage of the inclusion of the enzyme listeriolysin, an endosomolytic hemolysin from *Listeria monocytogenes*, to catalyze the release of the oligonucleotide inhibitor from the endocytic compartments. In the presence of hemolysin-containing liposomes, suppression of activation-induced expression of the target gene was increased when compared to delivery of inhibitory nucleotide without vesicle release.

13.4.3 Immune Recognition

The acceptance of artificial cells *in vivo* will be determined by the immunological response of the host organism. This is a process that involves recognition of foreign matter (antigens) and destruction or neutralization of the foreign or "nonself" substance. The defensive position of the host is catalyzed by one of two distinct processes inherent in an immune response. The innate immune response represents a nonspecific clearance mechanism, whereas the adaptive immune response resides within a dynamic and specific reactivity to material previously encountered. The control of the latter process is critical to the long efficacy of administration of artificial cells on multiple occasions.

Harnessing the power of the membrane recognition process provides a two-edged approach for achieving specificity of cell delivery. This approach is exemplified by gene targeting in the retina of the eye, using a liposomal delivery system applied through noninvasive intravenous administration.[117] The goal was to transfect the cells of the retina with an exogenous gene under conditions permitting passage through the blood–brain barrier and the plasma membrane of the retina cells. The effective procedure used nonviral expression plasmids encapsulated in pegylated immunoliposomes. The plasmid DNA was encapsulated in the interior of 75- to 100-nm liposomes for protection from endogenous nucleases. The liposomes were protected from rapid clearance by addition of several thousand strands of 2000-Da polyethylene glycol (PEG). Cell binding specificity was achieved by the addition of receptor-specific monoclonal antibody to the exterior ends of the PEG chains. The monoclonal antibody was targeted to the transferrin receptor, which is expressed at the blood–brain barrier and in plasma membrane of the retina cells. Coupling the delivery system with the use of cell-specific gene promoter provided a means for achieving widespread expression of an exogenous gene to treat genetic disease.

13.5 ARTIFICIAL CELLS AND NANOSTRUCTURED MEMBRANE SYSTEMS

Extensions of the lipid bilayer model of a plasma membrane to nanostructured systems may be achieved by the use of self-assembling monomers that organize into supramolecular networks.[118–120] Fabrication of supramolecular systems with properties mimicking the lipid bilayer would provide selectively permeable membrane, ionic, protonic, and electronic conductors and chemical catalysis. In self-assembling structures, substituent materials will provide the functionality of the lipid moieties and the integral membrane proteins. Examples of first-generation devices that achieve an element of this organization are the networks formed by methacrylate monomers that form a continuous phase. The methacrylate network designs are based on a paraffin barrier material that contains ion-active channels constructed from protected oligooxyethylene units.[121]

13.6 RED BLOOD CELL SUBSTITUTES

Early efforts to create a red blood cell substitute proved to be overambitious.[122–124] The first constructs were based on encapsulation of hemoglobin together with the red blood cell enzymes, catalase and superoxide dismutase, and other proteins. This has been followed by different strategies, one being cell-less, using cross-linking polymeric hemoglobin or other polymers for direct oxygen exchange, and a second using newer approaches for synthetic red blood cells.[125,126]

These strategies evolved from three experimental approaches for creation of artificial red blood cells. The first approach used synthetic membranes and cross-linked protein membranes to replace the natural red blood cell membrane to encapsulate hemoglobin. The oxygen dissociation was comparable to authentic red blood cells, and coating of the surface with an organic liquid preserved the integrity of the membrane. Since the membrane had no blood group antigens, the artificial cells did not form aggregates. However, the cells were rapidly removed from the circulation following intravenous infusion.

A second approach uses 200-nm diameter lipid membrane vesicles to encapsulate the hemoglobin. The use of lipid containing polyethylene glycol (PEG) to form the microvesicle membrane has extended the circulation time but has resulted in formation of methhemoglobin. In some cases, hemoglobin has been cross-linked to yield a membrane with discerning permeability between dissolved hemoglobin and small molecules.

A third approach employs biodegradable polymers, such as polylactide, to form the boundary for encapsulation of hemoglobin. The advantage of polymeric formulation is that varying the composition of the monomeric units can provide a means to establish defined rates of degradation. The polymeric materials can be used to provide nanoscale vesicles ranging in size from 80 to 200 nm. The formulation of the polymer has also contributed to the ability to include multi-enzyme systems in an effort to prevent the formation of methhemoglobin and to scavenge oxygen radicals.

Expansion of efforts to create functional artificial cell blood cells will likely rely on new materials and improved techniques for encapsulation of hemoglobin to establish a longer circulatory half-life.[127]

13.7 SAFETY

Currently, a broad spectrum of liposomes has been used for treatment of patients, primarily for delivery of antibiotics or chemotherapeutics. In general, the results have been good. Success with liposomal delivery has been extended to vector delivery for gene therapy.[128,129] The safety and effectiveness of a hybrid liposome vector composed of hemagglutinating virus of Japan–artificial viral envelope liposomes for human therapeutic gene transfer was tested in a series of experiments carried out in cynomolgus monkeys.[130] The liposomal-based constructs were delivered by repetitive intramuscular injections in 2-ml suspension or a single intravenous injection of a 10-ml suspension. The animals were observed up to 29 days. The results showed that the liposomal preparation was well tolerated and demonstrated the safety, feasibility, and therapeutic potential of the use of the liposomal transfection vehicle.

13.8 CONCLUSIONS AND FUTURE DIRECTIONS

Development of membrane-based artificial cells from early liposomes that only incorporated receptors and transport proteins to the development of cell-targeting vesicles is continuing at a rapid pace.[131–134] Recent studies show that liposomal membrane structures can be selectively constructed of lipids that exhibit highly selective transporter function. Advances in understanding plasma membrane ultrastructure and cellular signaling mechanisms anticipate the development of artificial cells that will further expand an already diverse array of specialized function.[135,136] Current designs incorporate synthetic helical and cyclic polypeptides to permit insertion of ion transport channels into the artificial cell membrane.

The anticipation for the future is that selectivity and specificity of membrane transport will improve by coupling peptide design criteria with methods for modification of lipid components. These design parameters have already led to artificial cells based on hollow polyelectrolyte microcapsules coated with different compositions of phosphatidylcholine/phosphatic acid and ion channel forming peptides.[137] The short-term future of artificial cells will most likely remain for the constructs to serve as delivery vehicles.[138–140] The long-term expectation is that artificial cells will be fabricated as true self-contained functionally equivalent biological units that support organized tissue repair and regeneration.

ACKNOWLEDGMENTS

This work was supported by a Department of Veterans Affairs Research Career Award, a VA Medical Merit Review Grant, NIH R01 AR45788, and the Stanford Orthopaedic Research Fund.

REFERENCES

1. Langer, R. and Vacanti, J.P., Tissue engineering, *Science,* 260, 920, 1993.
2. Chang, T.M.S., Hemoglobin corpuscles, Report of research project for B.Sc. Honours in Physiology, McGill University, Montreal, 1957.
3. Chang, T.M.S., Semipermeable microcapsules, *Science,* 146, 524, 1964.
4. Cousineau, J. and Chang, T.M.S., Formation of amino acid from urea and ammonia by sequential enzyme reaction using a microencapsulated multi-enzyme system, *Biochem. Biophys. Res. Commun.*, 79, 24, 1977.
5. Campbell, J. and Chang, T.M.S., The recycling of NAD+ (free and immobilized) within semipermeable aqueous microcapsules containing a multi-enzyme system, *Biochem. Biophys. Res. Commun., 69, 562, 1976.*
6. Langer, R. and Vacanti, J.P., Special report: the promise of tissue engineering, *Sci. Am.,* 280, 38, 1999.
7. Chang, T.M.S., Biodegradable semipermeable microcapsules containing enzymes, hormones, vaccines, and other biologicals, *J. Bioeng.,* 1, 25, 1976.
8. Chang, T.M., Artificial cells: 35 years, *Artif. Organs,* 16, 8, 1992.
9. Klein, H.G., Red blood cell substitutes, *N. Engl. J. Med.,* 342, 1666, 2000.
10. Pohorille, A. and Deamer, D., Artificial cells: prospects for biotechnology, *Trends Biotechnol.,* 20, 123, 2002.
11. Barron, L.G., Uyechi, L.S., and Szoka, F.C., Jr., Cationic lipids are essential for gene delivery mediated by intravenous administration of lipoplexes, *Gene Ther.,* 6, 1179, 1999.
12. Zhang, Y. et al., Receptor-mediated delivery of an antisense gene to human brain cancer cells, *J. Gene Med.,* 4, 183, 2002.
13. Chang, T.M.S. and Prakash, S., Procedure for microencapsulation of enzymes, cells and genetically engineered microorganisms, *Mol. Biotechnol.,* 17, 249, 2001.
14. Rokstad, A.M. et al., Evaluation of different types of alginate microcapsules as bioreactors for producing endostatin, *Cell Transplant.,* 12, 351, 2003.
15. Visted, T. et al., Prospects for delivery of recombinant angiostatin by cell-encapsulation therapy, *Hum. Gene Ther.,* 14, 1429, 2003.
16. Parenteau, N.L. and Hardin-Young, J., The use of cells in reparative medicine, *Ann. N. Y. Acad. Sci.,* 961, 27, 2002.
17. Hoekstra, R. and Chamuleau, R.A., Recent developments on human cell lines for the bioartificial liver, *Int. J. Artif. Organs.,* 25, 182, 2002.
18. Mclaughlin, B.E. et al., Overview of extracorporeal liver support systems and clinical results, *Ann. N. Y. Acad. Sci.,* 875, 310, 1999.
19. Sauer, I.M. et al., Primary human liver cells as source for modular extracorporeal liver support — a preliminary report, *Int. J. Artif. Organs.,* 25, 1001, 2002.

20. Duvivier-Kali, V.F. et al., Complete protection of islets against allorejection and autoimmunity by a simple barium-alginate membrane, *Diabetes,* 50, 1698, 2001.
21. Chen Z. et al., Bioartificial liver inoculated with porcine hepatocyte spheroids for treatment of canine acute liver failure model, *Artif. Organs,* 27, 613, 2003.
22. Pohorille, A. and Wilson, M.A., Molecular dynamics studies of simple membrane-water interfaces: structure and functions in the beginnings of cellular life, *Orig. Life Evol. Biosph.,* 25, 21, 1995.
23. Marsh, D., Peptide models for membrane channels, *N. Eng. J. Med.,* 328, 1244, 1993.
24. Lasic, D.D. et al., Transmembrane gradient driven phase transitions within vesicles: lessons for drug delivery, *Biochim. Biophys. Acta.,* 1239, 145, 1995.
25. Osterfield, M., Kirschner, M.W., and Flanagan, J.G., Graded positional information: interpretation for both fate and guidance, *Cell,* 113, 425, 2003.
26. Catterall, W.A., Structure and function of voltage-gated ion channels, *Annu. Rev. Biochem.,* 64, 493, 1995.
27. Mackinnon, R., New insights into the structure and function of potassium channels, *Curr. Opin. Neurobiol.,* 1, 14, 1991.
28. Neher, E., Nobel lecture. Ion channels for communication between and within cells, *Science,* 256, 498, 1992.
29. Kaplan, J.H., Biochemistry of Na,K-ATPase, *Annu. Rev. Biochem.,* 71, 511, 2002.
30. Rutecki, P.A., Neuronal excitability: voltage-dependent currents and synaptic transmission, *J. Clin. Neurophysiol.,* 9, 195, 1992.
31. Scriven, D.R. et al., The molecular architecture of calcium microdomains in rat cardiomyocytes, *Ann. N. Y. Acad. Sci.,* 976, 488, 2002.
32. Simons, K. and Wandinger-Ness, A., Polarized sorting in epithelia, *Cell,* 62, 207, 1990.
33. Mostov, K. et al., Plasma membrane protein sorting in polarized epithelial cells, *J. Cell Biol.,* 116, 577, 1992.
34. Heginbotham, L., Kolmakova-Parensky, L., and Miller, C., Functional reconstitution of a prokaryotic K+ channel, *J. Gen. Physiol.,* 111, 741, 1998.
35. Cavalier-Smith, T., Obcells as proto-organisms: membrane heredity, lithophosphorylation, and the origins of the genetic code, the first cells, and photosynthesis, *J. Mol. Evol.,* 53, 555, 2001.
36. Devaux, P.F., Static and dynamic lipid asymmetry in cell membranes, *Biochemistry* 30, 1163, 1991.
37. Singer, S.J. and Nicolson, G.L., The fluid mosaic model of the structure of cell membranes, *Science,* 175, 720, 1972.
38. Engund, P.T., The structure and biosynthesis of glycosyl phosphatidylinositol protein anchors, *Ann. Rev. Biochem.,* 62, 121, 1993.
39. Saez, J.C. et al., Plasma membrane channels formed by connexins: their regulation and functions, *Physiol. Rev.,* 83, 1359, 2003.
40. Glaser, M., Lipid domains in biological membranes, *Curr. Biol.,* 3, 475, 1993.
41. Marsh, D., Lipid interactions with transmembrane proteins, *Cell Mol. Life Sci.,* 60, 1575, 2003.
42. Lee, A., Membrane structure, *Curr. Biol.,* 11, R811, 2001.
43. Spector, A. A. and Yorek, M. A., Membrane lipid composition and cellular function, *J. Lipid Res.,* 26, 1015, 1985.
44. Brown, D., Structure and function of membrane rafts, *Int. J. Med. Microbiol.,* 291, 433, 2002.
45. Daleke, D.L., Regulation of transbilayer plasma membrane phospholipid asymmetry, *J. Lipid Res.,* 44, 233, 2003.

46. Hekman, M. et al., Associations of B- and C-Raf with cholesterol, phosphatidylserine, and lipid second messengers: preferential binding of Raf to artificial lipid rafts, *J. Biol. Chem.,* 277, 24090, 2002.
47. Edidin, M., The state of lipid rafts: from model membranes to cells, *Annu. Rev. Biophys. Biomol. Struct.,* 32, 257, 2003.
48. Van Voorst, F. and De Kruijff, B., Role of lipids in the translocation of proteins across membranes, *Biochem. J.,* 347, 601, 2000.
49. Bonifacino, J.S. and Traub, L.M., Signals for sorting of transmembrane proteins to endosomes and lysosomes, *Annu. Rev. Biochem.,* 72, 395, 2003.
50. Jahn, R. and Sudhof, T.C., Membrane fusion and exocytosis, *Annu. Rev. Biochem.,* 68, 863, 1999.
51. Devaux, P.F., Is lipid translocation involved during endo- and exocytosis? *Biochimie,* 82, 497, 2000.
52. Evan, P.R. and Owen, D.J., Endocytosis and vesicle trafficking. *Curr. Opin. Struct. Biol.,* 12, 814, 2002.
53. Lefkir, Y. et al., Involvement of the AP-1 adaptor complex in early steps of phagocytosis and macropinocytosis, *Mol. Biol. Cell.,* 15, 861, 2004.
54. Beck, K.A. et al., Clathrin assembly protein AP-3 induces aggregation of membrane vesicles: a possible role for AP-2 in endosome formation, *J. Cell Biol.,* 119, 787, 1992.
55. Haucke, V. and Krauss, M., Tyrosine-based endocytic motifs stimulate oligomerization of AP-2 adaptor complexes, *Eur. J. Cell Biol.,* 81, 647, 2002.
56. Hill, E. et al., The role of dynamin and its binding partners in coated pit invagination and scission, *J. Cell Biol.,* 152, 309, 2001.
57. Bevers, E. M. et al., Lipid translocation across the plasma membrane of mammalian cells. *Biochim. Biophys. Acta.,* 1439, 317, 1999.
58. Pomorski, T. et al., Drs2p-related P-type ATPases Dnf1p and Dnf2p are required for phospholipid translocation across the yeast plasma membrane and serve a role in endocytosis, *Mol. Biol. Cell.,* 14, 1240, 2003.
59. Steyer, J.A., Horstmann, H., and Almers, W., Transprort, docking and exocytosis of single secretory granules in live chromaffin cells, *Nature,* 388, 474, 1997.
60. Subczynski, W.K., and Kusumi, A., Dynamics of raft molecules in the cell and artificial membranes: approaches by pulse EPR spin labeling and single molecule optical microscopy, *Biochim. Biophys. Acta,* 1610, 231, 2003.
61. Banerjee, R.K. and Datta, A.G., Proteoliposome as the model for the study of membrane-bound enzymes and transport proteins, *Mol. Cell Biochem.,* 50, 3, 1983.
62. Calakos, N. and Scheller, R.H., Synaptic vesicle biogenesis, docking, and fusion: a molecular description, *Physiol. Rev.,* 76, 1, 1996.
63. Zimmerberg, J., Vogel, S.S., and Chernomordik, L., Mechanisms of membrane fusion, *Annu. Rev. Biophys. Biomol. Struct.,* 22, 433, 1993.
64. Woodbury, D.J., Building a bilayer model of the neuromuscular synapse, *Cell Biochem. Biophys.*, 30, 303, 1999.
65. Can, A.-S. et al., Artificial cells: Unique insights into exocytosis using liposomes and lipid nanotubes, *Proc. Natl. Acad. Sci. U.S.A.,* 100, 400, 2003.
66. McNeil, P.L. and Terasaki, M., Coping with the inevitable: how cells repair a torn surface membrane, *Nat. Cell Biol.,* 3, E124, 2001.
67. Perozo, E. et al., Open channel structure of MscL and the gating mechanism of mechanosensitive channels. *Nature,* 418, 942, 2003.
68. Stop, P., Bass, R., and Rees, D. C., Prokaryotic mechanosensitive channels, *Adv. Protein Chem.,* 63, 177, 2003.

69. Kloda, A. and Martinac, B., Mechanosensitive channels in archaea, *Cell Biochem. Biophys.,* 34, 349, 2001.
70. Kloda, A., Martinac, B., Mechanosensitive channels of bacteria and archaea share a common ancestral origin, *Eur. Biophys. J.,* 31, 14, 2002.
71. Itoh, Y.H., Sugai, A., Uda, I., and Itoh, T., The evolution of lipids, *Adv. Space Res.,* 28, 719, 2001.
72. Akashi, H., Metabolic economics and microbial proteome evolution, *Bioinformatics,* 19, II15, 2003.
73. Chang, T.M., Artificial cells for artificial kidney, artificial liver and detoxification, in *Artificial Kidney, Artificial Liver, and Artificial Cells*., T. M. Chang, Ed., Plenum Press, New York, pp. 57–77.
74. Chang, T.M., Artificial cells in medicine and biotechnology, *Appl. Biochem. Biotechnol.,* 10, 5, 1984.
75. Chang, T.M., Artificial cells as bioreactive biomaterials, *J. Biomater. Appl.,* 3, 116, 1988.
76. Malide, D. et al., The export of major histocompatibility complex I molecules from the endoplasmic reticulum of rat brown adipose cells is acutely stimulated by insulin, *Mol. Biol. Cell.,* 12, 101, 2001.
77. Eisfeld, K. et al., Endocytotic uptake and retrograde transport of a virally encoded killer toxin in yeast. *Mol. Microbiol.,* 37, 926, 2000.
78. Soltys, B.J., Falah, M., and Gupta, R.S., Identification of endoplasmic reticulum in the primitive eukaryote Giardia lamblia using cryoelectron microscopy and antibody to Bip, *J. Cell Sci,.,* 109, 1909, 1996.
79. Lee, T.H and Linstedt, A.D., Osmotically induced cell volume changes alter anterograde and retrograde transport, Golgi structure, and COPI dissociation, *Mol. Biol. Cell.,* 10, 1445, 1999.
80. Kierszenbaum, A.L., Fusion of membranes during the acrosome reaction: a tale of two SNAREs, *Mol Reprod. Dev.,* 57, 309, 2000.
81. Chao, H. et al., Membrane charge and curvature determine interaction with acyl-CoA binding protein (ACBP) and fatty acyl-CoA targeting, *Biochemistry,* 41, 10540, 2002.
82. Venema, K. et al., The arabidopsis Na+/H+ exchanger AtNHX1 catalyzes low affinity Na+ and K+ transport in reconstituted lipsomes, *J. Biol. Chem.*, 277, 2413, 2002.
83. Anner. B.M., Moosmayer, M., and Imesch, E., Na,K-ATPase characterized in artificial membranes. 1. Predominant conformations and ion-fluxes associated with active and inhibited states, *Mol. Membr. Biol.,* 11, 237, 1994.
84. Anner, B.M. and Moosmayer, M., Na,K-ATPase characterized in artificial membranes. 2. Successive measurement of ATP-driven Rb-accumulation, ouabain-blocked Rb-flux and palytoxin-induced Rb-efflux, *Mol. Membr. Biol.,* 11, 247, 1994.
85. Cowan, S.W., Bacterial porins: lessons from three high-resolution structures, *Curr. Biol.,* 3, 501, 1993.
86. Aguilera, O., Quiros, L.M., and Fierro, J.F., Transferrins selectively cause ion efflux through bacterial and artificial membranes, *FEBS Lett.,* 548, 5, 2003.
87. Liu, N. et al., Effects of pore mutations and permeant ion concentration on the spontaneous gating activity of OmpC porin, *Protein Eng.,* 13, 491, 2000.
88. Clapham, D.E., TRP channels as cellular sensors, *Nature,* 426, 517, 2003.
89. Chen, F.Y., Lee, M.T., and Huang, H.W., Evidence for membrane thinning as the mechanism for peptide-induced pore formation, *Biophys. J.,* 84, 3751, 2003.
90. Weinman, E.J. and Chamras, H., Reconstitution of human red blood cell Na/H and Na/Na exchange transport, *Am. J. Med. Sci.,* 312, 47, 1996.

91. Anner, B.M., Reconstitution of the Na+, K+-transport system in artificial membranes, *Acta Physiol. Scand. Suppl.,* 481, 15, 1980.
92. Riquelme, G. et al., A chloride channel from human placenta reconstituted into giant liposomes, *Am. J. Obstet. Gynecol.*, 173, 733, 1995.
93. Gokel, G.W., Artificial cation-conducting channels: design, synthesis, and characterization, *Cell Biochem. Biophys.,* 35, 211, 2001.
94. Yang, L. et al., Barrel-stave model or toroidal model? A case study of melittin pores, *Biophys. J.,* 81, 1475, 2001.
95. Ghadiri, M.R., Granja, J.R., and Buehler, L.K., Artificial transmembrane ion channels from self-assembling peptide nanotubes, *Nature,* 369, 301, 1994.
96. Sharpe, M.A., Cooper, C.E., and Wrigglesworth, J.M., Transport of K+ and other cations across phospholipid membranes by nonesterified fatty acids, *J. Membr. Biol.,* 141, 21, 1994.
97. Zeng, Y. et al., Nonesterified fatty acids induce transmembrane monovalent cation flux: host-guest interactions as determinants of fatty acid-induced ion transport, *Biochemistry,* 37, 9497, 1998.
98. Yamashita, Y. et al., A new method for the preparation of giant liposomes in high salt concentrations and growth of protein microcrystals in them, *Biochim. Biophys. Acta.,* 1561, 129, 2002.
99. Lian, T. and Ho, R.J., Trends and developments in liposome drug delivery systems, *J. Pharm. Sci.,* 90, 667, 2001.
100. Wassef, N.M., Alving, C.R., and Richards, R.L., Liposomes as carriers for vaccines, *Immunomethods,* 4, 217, 1994.
101. Hussein, M.A. et al., A Phase II trial of pegylated liposomal doxorubicin, vincristine, and reduced-dose dexamethasone combination therapy in newly diagnosed multiple myeloma patients, *Cancer,* 95, 2160, 2002.
102. Scherphof, G.L. and Kamps, J.A., The role of hepatocytes in the clearance of liposomes from the blood circulation, *Prog. Lipid Res.,* 40, 149, 2001.
103. Allen, T.M. et al., Uptake of liposomes by cultured mouse bone marrow macrophages: influence of liposome composition and size, *Biochim. Biophys. Acta.,* 1061, 56, 1991.
104. Ledley, F.D., Non-viral gene therapy, *Curr. Opin. Biotechnol.,* 5, 626, 1994.
105. Smyth Templeton, N., Liposomal delivery of nucleic acids *in vivo*, *DNA Cell Biol.,* 21, 857, 2002.
106. Johanning, F.W. et al., A Sindbis virus mRNA polynucleotide vector achieves prolonged and high level heterologous gene expression *in vivo*, *Nucleic Acids Res.*, 23, 1495, 1995.
107. Kao, M.C. et al., *In vitro* gene transfer in mammalian cells via a new cationic liposome formulation, *Oncol. Rep.,* 5, 625, 1998.
108. Wheeler, J.J. et al., Stabilized plasmid-lipid particles: construction and characterization. *Gene Ther.,* 6, 271, 1999.
109. Kikuchi, H. et al., Gene delivery using liposome technology, *J. Control Release,* 62, 269, 1999.
110. Lee, C.H. et al., Synergistic effect of polyethylenimine and cationic liposomes in nucleic acid delivery to human cancer cells, *Biochim. Biophys. Acta.,* 1611, 55, 2003.
111. Miller, A.D., Nonviral liposomes, *Methods Mol. Biol.,* 90, 107, 2004.
112. Van den Bos, E.J. et al., Improved efficacy of stem cell labeling for magnetic resonance imaging studies by the use of cationic liposomes, *Cell Tranplant.,* 12, 743, 2003.
113. Anwer, K. et al., Cationic lipid-based delivery system for systemic cancer gene therapy, *Cancer Gene Ther.*, 7, 1156, 2000.

114. Wiethoff, C.M. et al., Compositional effects of cationic lipid/DNA delivery systems on transgene expression in cell culture, *J. Pharm. Sci.*, 93, 108, 2004.
115. Fischer, A., Franco, A., and Oberholzer, T., Giant vesicles as microreactors for enzymatic mRNA synthesis, *Chembiochem.*, 3, 409, 2002.
116. Mathew, E. et al., Cytosolic delivery of antisense oligonucleotides by listeriolysin O-containing liposomes, *Gene Ther.,* 10, 1105, 2003.
117. Zhu, C., Zhang, Y., and Pardridge, W.M., Widespread expression of an exogenous gene in the eye after intravenous administration. *Invest. Opthalmol. Vis. Sci.*, 43, 3075, 2002.
118. Takei, K. et al., Generation of coated intermediates of clathrin-mediated endocytosis on protein-free liposomes, *Cell*, 94, 131, 1998.
119. Hahya, E. et al., Reconstitution of membrane proteins into giant unilamellar vesicles via peptide-induced fusion, *Biophys. J.,* 81, 1464, 2001.
120. Kim, K.H., Ahn, T., and Yun, C.H., Membrane properties induced by anionic phospholipids and phosphatidylethanolamine are critical for the membrane binding and catalytic activity of human cytochrome P450 3A4, *Biochemistry,* 42, 15377, 2003.
121. Percec, V. and Bera, T.K. Cell membrane as a model for the design of ion-active nanostructured supramolecular systems, *Biomacromolecules,* 3, 167, 2002.
122. Mobed, M. and Chang, T.M., Preparation and surface characterization of carboxymethylchitin-incorporated submicron bilayer-lipid membrane artificial cells (liposomes) encapsulating hemoglobin, *Biomater. Artif. Cells Immobil. Biotechnol.,* 19, 731, 1991.
123. Chang, T.M., Modified hemoglobin-based blood substitutes: crosslinked, recombinant and encapsulated hemoglobin, *Vox Sang.*, 74, Suppl. 2, 233, 1998.
124. Ogata, Y., el al., Development of neo red cells (NRC) with the enzymatic reduction system of methemoglobin, *Artif. Cells Blood Substit. Immobil. Biotechnol.,* 25, 417, 1997.
125. Chang, T.M., Red blood cell substitutes, *Baillieres Best Pract. Res. Clin. Haematol.,* 13, 651, 2000.
126. Chang, T.M., Powanda, D., and Yu, W.P., Analysis of polyethylene-glycol-polylactide nano-dimension artificial red blood cells in maintaining systemic hemoglobin levels and prevention of methemoglobin formation, *Artif. Cells Blood Substit. Immobil. Biotechnol.,* 31, 231, 2003.
127. Chang, T.M., Future generations of red blood cell substitutes, *J. Intern. Med.,* 253, 527, 2003.
128. Bekersky, I. et al., Pharmacokinetics, excretion, and mass balance of 14C after administration of 14C-cholesterol-labeled AmBisome to healthy volunteers, *J. Clin. Pharmacol.*, 41, 963, 2001.
129. Walsh, T.J. et al., Safety, tolerance, and pharmacokinetics of high-dose liposomal amphotericin B (AmBisome) in patients infected with Aspergillus species and other filamentous fungi: maximum tolerated dose study, *Antimicrob. Agents Chemother.,* 45, 3487, 2001.
130. Tsuboniwa, N. et al., Safety evaluation of hemagglutinating virus of Japan — artificial viral envelope liposomes in nonhuman primates, *Hum. Gene Ther.,* 12, 469, 2001.
131. Wu, W.C., Moore, H.P., and Raftery, M.A., Quantitation of cation transport by reconstituted membrane vesicles containing purified acetylcholine receptor, *Proc. Natl. Acad. Sci. U.S.A.,* 78, 775, 1981.
132. Tranum-Jensen, J. et al., Membrane topology of insulin receptors reconstituted into lipid vesicles, *J. Membr. Biol.,* 140, 215, 1994.

133. Sekler, I. et al., Sulfate transport mediated by the mammalian anion exchangers in reconstituted proteoliposomes, *J. Biol. Chem.,* 270, 11251, 1995.
134. Cragg, P.J., Artificial transmembrane channels for sodium and potassium, *Sci. Prog.* 85, 219, 2002.
135. Rachev, E. et al., Efficacy and safety of phospholipid liposomes in the treatment of neuropsychological disorders associated with the menopause: a double-blind, randomised, placebo-controlled study, *Curr. Med. Res. Opin.,* 17, 105, 2001.
136. Rivera, E. et al., Phase II study of pegylated liposomal doxorubicin in combination with gemcitabine in patients with metastatic breast cancer, *J. Clin. Oncol.,* 21, 3249, 2003.
137. Tiourina, O.P. et al., Artificial cell based on lipid hollow polyelectrolyte microcapsules: channel reconstruction and membrane potential measurement, *Membr. Biol.,* 190, 9, 2002.
138. Hu, Q., Bally, M.B., and Madden, T.D., Subcellular trafficking of antisense oligonucleotides and down-regulation of bcl-2 gene expression in human melanoma cells using a fusogenic liposome delivery system, *Nucleic Acids Res.,* 30, 3632, 2002.
139. Wichmann, C. et al., Liposomes for microcompartmentation of enzymes and their influence on catalytic activity, *Biochem. Biophys. Res. Commun.,* 310, 1104, 2003.
140. Pedroso de Lima, M.C. et al., Cationic liposomes for gene delivery: from biophysics to biological applications, *Curr. Med. Chem.*, 10, 1221, 2003.

14 Artificial Organs and Stem Cell Biology

R. Lane Smith

CONTENTS

0-8493-1940-4/05/$0.00+$1.50

14.1 INTRODUCTION

Advances in cell biology and immunology firmly establish that populations of cells exist in humans that can be recruited to proliferate and differentiate for repair of damaged or diseased tissues.[1–5] The extent to which such immature progenitor cells, collectively referred to as stem cells, can restore tissue damage remains an area of intense investigation.[6–8] Different classes of stem cells, whether embryonic, adult, or engineered, exhibit a wide range of commitment levels and proliferative capacity.[9,10] Levels of stem cell commitment and expansion potential depend in part on local factors present in the microenvironment of sites of origin, such as bone marrow or olfactory bulb.[11,12]

The presence of progenitor cells in organisms that contribute to renewal of differentiated tissues has been understood for many years.[13–15] Throughout the life of the organism, repair of skin, remodeling of bone, renewal of blood, and regenerative healing of organs, such as the liver, represents normal physiology. The existence of less-restricted progenitor cells in higher organisms in sufficient numbers for isolation and culture was less well established. Verification that stem cells exist at multiple sites immediately generated hope that cell-based approaches could be used to treat damaged organs or create artificial organs.[16,17] Variability in competency of the different types of stem cells implied that exogenous stimulation could selectively drive the cells down specific differentiation pathways. The concept of targeted differentiation was further advanced with the discovery of progenitor cells having seemingly unlimited potential for differentiation. Having cells with unrestricted capacity raises the possibility for totally directed differentiation that could extend to humans a capacity for regenerative healing only experienced by certain amphibians.

This chapter will examine the current state of knowledge regarding the major classes of stem cells, how they are defined, and what possible mechanisms serve to limit their levels of developmental plasticity. The paper will also examine how scaffold design and nanofabrication methods may be applied to enhance the goal of tissue repair and regeneration with or without artificial organs. The review will describe how facets of stem biology may best fit with tissue engineering approaches to manufacture organized cellular systems that replicate certain organs. The hypothesis is advanced here that artificial extracellular matrices with nanoscale features that model cell-specific ligands will provide an informative "smart" framework for directed cell organization. "Smart" extracellular matrices may avoid inefficiencies inherent with *in vivo* methods relying solely on recruitment and expansion of progenitor cells into compromised tissues.

14.1.1 Definitions

Stem cells are defined by their capacity for self-renewal.[18] Replication of stem cells to yield daughter cells involves a molecular decision that determines the subsequent fate of the progeny.[19] One type of molecular decision establishes that one of the daughter cells will assume a state of commitment distinct from that of the parent progenitor cell. A second type of decision leaves both daughter cells with equal competency for self-renewal. Variations associated with stem cells of differing origins determine the total number of replications that the original progenitor cell will undergo and the different types of cells that will arise from the progeny.[20] A stem cell giving rise to a single cell type is considered unipotent, whereas a stem cell giving rise to a wide spectrum of differentiated progeny is considered to be either pluripotent or totipotent. A totipotent stem cell can give rise to all cells of the organism, whereas a pluripotent cell supports development of most but not all cells of the organism.[21] More restricted cell lineages arise from multipotent stem cells, such as the mesenchymal stem cells that reside in bone marrow stroma and give rise to a limited set of cell types, including adipocytes, chondrocytes, osteoblasts, and muscle.

Recent studies document that stem cells can be isolated from a number of different sources.[22–25] The earliest recognized stem cell was considered the fertilized egg, and ablation experiments showed that in regulative embryos, blastocysts isolated up to the eight-cell stage could serve as totipotent stem cells to support development of the organism.[26] Sources of pluripotent cells include cells derived from some cancers (EC), embryonic cells derived from preimplantation embryos (ES), and primordial germ cells (EG).[27,28] Pluripotent cells exhibit select cell surface antigens, have a normal and stable karyotype, and can grow as colonies in semisoft support medium.[29] Multipotent cells arise from various tissues within an embryo or adult and require isolation and separation from companion cell types.

14.1.2 Need

The need for stem cells has expanded due to the impetus to develop tissue-engineering approaches for replacement of tissues and organs. Increased demand for organ transplants and the recurrent shortage of appropriate donors serves as a driving force for artificial tissue reconstruction.[30,31] Current levels of donor organs for kidney, liver, and heart disease is such that an estimated 30,000 to 40,000 individuals die before transplant surgery. Other surgical approaches to relieve disability are also limited by available graft tissue, including bone for large tumor resection, cartilage, tendon, ligament, and meniscus. The increasing prevalence of adult onset diseases such as diabetes further expands the need for organ donors and, importantly, suitable donor cells to support fabrication of pancreatic islets for implantation.

14.1.3 Availability

Isolation techniques have been reported for a number of different types of human multipotent stem cells.[32–35] As a result, availability of this type of stem cell is only limited by the procurement of donor site tissue. The situation is more complicated with respect to human embryonic stem cells, since a government moratorium has restricted university research on this type of cells. However, industrial affiliates have access to the limited number of human embryonic cell lines that were established prior to regulatory prohibitions. An area of research that seeks to circumvent ethical issues regarding embryonic stem cells is the use of reagents to remove restrictions on DNA so that an adult stem cell might be reverse-engineered to a pluripotent or totipotent state.[36] Restrictions on the use of embryonic stem cells will continue with respect to issues of cloning of organisms. However, generation of adult-derived cells capable of extensive replication and plasticity of differentiation will greatly expand strategies for tissue repair and may be expected to progress rapidly with information gained from the human genome.

14.2 STEM CELL COMPETENCE

14.2.1 PLURIPOTENT

The concept of pluripotency, as it relates to any one isolated stem cell candidate, is verified by development of cell types found in one of the three embryonic germ layers.[37] The importance of the embryonic germ layers comes from the need to recreate tissues having specific tissue attributes that originate early during development. In many cases, the expression of specific cell membrane proteins under the control of developmentally regulated master genes sets the competency of the cells.[38] The cell membrane proteins include ion channel proteins, membrane pumps such as the intestinal cell glucose transporter with its companion sodium/potassium ATPase, and cell adhesion molecules such as the cadherins. Other membrane proteins integral to determination include the integrins, where one of the two subunits will establish the basis for cell recognition and activate intracellular signaling pathways that set function.[39–41]

The presence or absence of cell adhesion proteins influences the stability of the tissue organization and determines whether epithelial–mesenchymal or mesenchymal–epithelial transformations occur within populations of daughter cells.[42] The occurrence of these shifts in cellular type is exemplified by the differentiation of a murine embryonic stem cell line that expresses markers for each of the three germ layers.[43] The induction of cells that represent the individual germ layers occurs with spontaneous embryoid body formation induced by withdrawal of the growth factor, leukemia inhibitory factor. Pluripotent stem cell markers, Oct-3/4, SSEA-1, and EMA-1, persist for 7 days, while an endoderm marker requires up to 6 days before appearing. Mesodermal cells are also present based on expression of beta-tubulin and sarcomeric myosin in the cultures.

Approaches to understand the molecular basis for control of stem cell pluripotency have utilized a variety of methods to study cell-state.[44–46] Levels of gene expression have been categorized using cDNA microarray technology and differential mRNA expression analysis. Microarray analysis has generated evidence that multiple genes (up to 895) are elevated in human embryonic stem cells and human embryonal carcinoma cells when compared to somatic cell lines.[47] The question of whether a limited set or even a single gene could be a master switch for competency remains unanswered. If regenerative states are to be subject to genetic methods, the determination of a master gene pool will be essential for use of genetic methods to alter differentiation potential of somatic cells.

14.2.2 Totipotent

Evidence for a cell that can restore the full sequence of events for development of a complete organism with functional germ line cells is limited. In general, embryonic stem cells in culture are considered pluripotent, rather than totipotent, since germ line cells are not induced during differentiation. A recent study shows that mouse embryonic stem cells can develop into oogonia that enter meiosis, recruit adjacent cells to form follicle-like structures, and support blastocyst development.[48]

Somatic cell nuclear transfer is being used to establish cell lines that exhibit totipotency similar to that of embryonic stem cells. However, this is currently restricted to mouse cells.[49] Continued efforts in this area may make the technology applicable to human cells. A characteristic set of transcription factors are associated with the totipotency of the stem cells, but the mechanisms by which these proteins act to set competency remains unknown.[50] Analysis of mouse gene expression in gonad primordial cells has shown that a prominent transcript, DPPA3, is also associated with human germ cell tumors and may be associated with pluripotentiality.[51] The discovery of other similar transcription factors could lead to formulation of genetic strategies to alter multipotent adult cells and endow them with pluripotency.

14.2.3 Immortalized

Immortalization of stem cells represents a working strategy for circumventing difficulties encountered either in obtaining sufficient amounts of autologous or allogeneic human cells or impediments in the use of xenogeneic stem cells.[52–54] The availability of a reversible immortalization system, such as the Cre-Lox P site for specific recombination reactions, has yielded a safe human hepatocyte cell line capable of differentiation and expandability.[55] These cells form the basis of a hybrid bioartificial liver that relies on incorporation of living liver cells into a hollow fiber and nonwoven fabric substrate. Other approaches have included viral transformation of neural progenitor cells isolated from murine neonatal brain to generate immortalized cells that are multipotent and capable of differentiating into all neural cell types.[56] Implantation of immortalized neural stem cells into the injured spinal cord has also shown promise for providing some level of nervous system repair.[57]

14.3 EMBRYONIC STEM CELLS

14.3.1 Germ Layers

During embryogenesis, early cell division generates a blastocyst with cells that exhibit varying levels of potency. With continued cell proliferation, morphogenetic processes that involve cell sorting, movement, folding, and invagination result in the formation of three germ layers that serve as the progenitors of the major tissue and organs of the adult. Factors derived from one or more of the germ layers are involved in the emergence of organ precursor cells and provide signals for morphogenetic processes such as delamination or epithelial–mesenchymal transformation.[58] Human and mouse embryonic stem cells are able to generate all three germ layer derivatives (endodermal, mesodermal, and ectodermal), while maintaining the ability for indefinite self-renewal.[59]

The capacity for self-renewal involves activation of the canonical Wnt pathway.[60] As the embryonic stem cells undergo differentiation, endogenous activation of Wnt signaling is downregulated. Involvement of Wnt activation was demonstrated in human and mouse embryonic stem cells by pharmacological inhibition of glycogen synthase-3, an enzyme active in intracellular signaling. As embryonic stem cells undergo differentiation, numerous agents or processes acting at various levels of signaling control expression of specific proteins. Transforming growth factor-beta, induction of expression of MyoD and Myf5, activation of the transcription factor Tcf4, activation of mitogen-activated protein kinase cascade, and neurogenin3 represent some of the factors and processes.[61–65] These factors or processes, alone or combined, may be critical determinants for setting competency or determining a differentiation pathway in stem cells. Manipulation of transcription factors to direct or set stem cell differentiation or competency will require either secondary methods of direct delivery or epigenetic approaches for induction of their synthesis and activation.

14.3.2 Neural Crest

The development of mesenchymal tissues such as the heart, skeletal muscle, bones, and cartilage originates from the mesoderm. However, components of the cardiac outflow tract are formed by neural crest.[66] The neural crest cells are ectodermal derivatives that generate a broad spectrum of tissues that include the peripheral nervous system (ganglia), endocrine cells, melanocytes, and connective tissue, bone, and cartilage of the face and ventral neck. As development proceeds, the neural crest cells shift from a pluripotent state to a more restricted state in response to growth factors such as neurotrophin-3.[67] Of interest, the changes in developmental potential of the neural crest cells is not restricted to the neural structures but also affects the cardiac components. A question still remains whether neural crest cells persist in adult tissues as a result of migration into various regions along the anterior/posterior axis of the organism during early embryogenesis.

14.4 ADULT STEM CELLS

Adult stem cells in mammalians are associated with organs that function on a day-to-day basis through high levels of cellular replacement.[68–70] These organs include tissues where turnover rates of the cells can be measured in days, as in the intestinal lining, or in terms of several years, as for bone. The bone marrow contributes progenitors for final maturation into erythrocytes that ultimately are removed from the circulation within 3 months. Cells of the epidermis have a turnover time in the range of 2 months, and replacement persists throughout life. To generate tissue-specific structures, adult stem cells give rise to several types of differentiated progeny. The cellular offspring may then undergo morphogenetic movements and rearrangements to constitute organ systems. Recent studies suggest that all organs may contain stem cells that may or may not be activated during normal homeostasis. Of the populations of adult stem cells, three major classes, mesenchymal, hematopoietic, and neural, have been recognized to be present in sufficient quantities to enable study.[71,72]

14.4.1 Mesenchymal

Isolation of mesenchymal stem cells from adult organisms has focused primarily on tissues that exhibit normal regenerative activity, such as healing of bone after fracture.[73] Currently, a number of different types of differentiated tissues serve as a source of adult mesenchymal stem cells, including adipose tissue, muscle, and bone marrow. Adipose-derived stromal cells from the mouse inguinal fat pads are capable of expressing bone-specific proteins and cartilage-specific macromolecules when maintained in an osteogenic or chondrogenic culture medium.[74] In the bone marrow, the stromal cells serve as the primary source of a pluripotent cell that can give rise to cartilage, tendon and ligament, bone, muscle, and fat cells. Muscle-derived stem cells require longitudinal culture to establish a cell line that exhibits pluripotent properties.[75] The muscle-derived cell lines originate from satellite cells situated adjacent to the basal lamina that surrounds the multinucleated muscle cells. The recruitment of these cells may be solely dependent on tissue damage incurred with repetitive muscle contraction.

14.4.2 Hematopoietic

Characterization of hematopoietic stem cells that give rise to the blood system firmly established that renewal of an adult tissue follows a progressive pattern of cell differentiation.[76,77] Studies from infusion of bone marrow cells into animals irradiated to abolish the blood system showed that reconstitution of the blood system could save the animals.[78] Analysis of spleens of irradiated animals revealed that colonies of cells in that organ could be traced to a single cell.[79] The observation that some of the spleen colonies contained a mixed population of cells meant that the progenitor cell was at least multipotent. Subsequent experiments established that subsets of hematopoietic stem cells are present, which represent multipotent populations, and a few cells exist that are pluripotent.[80–82] Although these classes of stem cells were predicted by animal reconstitution experiments, it took new technologies, such as fluorescent cell sorting and preparation of population-specific monoclonal antibodies, to establish their existence.[83,84] Establishing the identity of the various types of hematopoietic stem cells ranks as one of the major scientific achievements in biological research in the last decade.

14.4.3 Neural

Neural stem cells were hypothesized to exist because subsets of neurons in the olfactory bulb and hippocamus were continuously produced.[85] Multipotent neuronal stem cells were then demonstrated to emerge in culture in response to addition of growth factors under conditions where substrate attachment did not occur.[86,87] The cells that emerged were capable of forming neurons, astrocytes, and oligodendrocytes. Neural stem cells are now recognized to be localized within the subventricular zone of the brain.[88,89] The existence of a neural stem cell population in the brain raises the possibility for regenerative healing following traumatic insults and degenerative diseases that impair cognitive and motor performance. Targets for cell-based therapeutic approaches include Parkinson's disease, Huntington's disease, and multiple sclerosis.[90–92] Traumatic events such as spinal cord or peripheral nerve injury have been the target of efforts to establish neuronal regeneration or proliferation of the accessory cells, such as the glial cells, that might guide nerve fiber migration.[93,94] Although application of growth factors and other neurotrophic substances provides a means for recruitment of endogenous stem cells, the challenge is to be able to maintain a sustained source of these factors until networks of neurons are reestablished.

14.5 MESENCHYMAL STEM CELLS

Adult-derived mesenchymal stem cells have been commercially prepared from bone marrow stromal cells.[95] Evidence that other sites such as adipose tissue are also repositories of stem cells has expanded investigations for cell sources.[96] The periosteal layer surrounding cortical bone has been used as a source of osteoprogenitor cells for connective tissue repair.[97] Transfection of these cells with either bone morphogenetic protein-7 or sonic hedgehog genes enhanced the repair of full-thickness cartilage defects.[98] Treatment of human bone marrow–derived mesenchymal stem cells with bone morphogenetic-2 and transforming growth factor-beta3 was sufficient to induce cartilage-specific gene expression. Conversion of mesenchymal stem cells to a cartilage phenotype was not accompanied by evidence of bone-specific proteins, such as osteocalcin, or adipocyte-specific fatty acid binding protein, which would represent adipogenesis.[99]

14.5.1 CARTILAGE

The major thrust for cartilage repair has centered on the use of mesenchymal stem cells derived primarily from bone marrow stromal cells.[100,101] Pluripotent embryonic stem cells have been programmed to differentiate into chondrocytes by coculture with progenitor cells from the limb buds of a developing embryo.[102] The embryonic stem cells exhibited a downregulation of the pluripotential transcription factor, Oct-4, as the cells became committed to a chondrogenic phenotype as evidenced by expression of Sox-9, type II collagen, and proteoglycans.

14.5.2 BONE

The most prominent of healing responses is that associated with bone repair following fracture. The healing of bone can proceed by direct bone formation from migration of adjacent osteoblasts or can follow a pattern of endochondral ossification. In endochondral ossification, the healing response follows the embryonic process, in that a cartilage anlage forms at the site of fracture that is replaced by incursion of bone cells. The formation of the cartilage follows the fibrin clot formation that results from bleeding at the site. Successful mineralization and remodeling at the fracture site depend on the mechanical environment.[103,104] Excessive motion will lead to a fibrous non-union due to the differentiation of the progenitor cells to a fibroblastic phenotype. Efforts to enhance bone healing have included development of fibrin scaffolds for delivery of isolated mesenchymal stem cells.[105]

Bone marrow stromal cells serve as progenitor cells that can give rise to many mesenchymal lineages, including chondroblasts, adipocytes, and osteoblasts. Administration of transforming growth factor species can shift the lineage-specific behavior of the cells by acting on transcription factor expression to inhibit adipocyte conversion and promote osteoblast differentiation.[106] Bone progenitor cells respond to addition of osteoblast stimulation factor-1 (osf-1) in a biomimetic scaffold in the presence of osteoinductive factors derived from a bone cell line (Saos-2) with increased adhesion, migration, expansion, and differentiation.[107] Mesenchymal stem cells also produce more mineralized matrix in a porous scaffold that releases ascorbate-2 phosphate and dexamethasone.[108]

14.5.3 Tendon

Repair of connective tissue following traumatic injury remains a major focus in orthopedics. The use of mesenchymal stem cells is of significant interest as a way to speed healing and shorten return to normal function. Collagen gels seeded with varying numbers of rabbit bone marrow mesenchymal cells implanted into central defects of patellar tendons improved the biomechanical properties of tendon repair tissues.[109] Of interest, the mesenchymal stem cell–bearing grafts induced bone at the regenerating repair site in 28% of the cases, suggesting that control of differentiated function is an issue in the use of multipotent cells. The use of mesenchymal cells for tissue repair may depend on the extent to which transcription factors can be expressed that are instrumental for acquisition of differentiated function during stem cell maturation. With respect to tendon, scleraxis is a transcription factor whose expression becomes restricted to tendon. The pattern of scleraxis synthesis follows a temporal and spatial association that coincides with tendon formation during embryonic development.[110]

14.6 Hematopoietic Stem Cells

14.6.1 Identification

Recognition of a population of cells present in bone marrow that renew the blood system under normal conditions for the life of the organism defined the concept of stem cells.[111,112] The pioneering investigations that showed colony-dependent proliferation and the role of growth factors in this process propelled cell biology into a new era of cell biology.[113] Experimental culture conditions and development of cell surface markers established that discrete maturational steps underlie cell differentiation.[114,115]

The fundamental mechanisms that determine the self-renewing properties of the hematopoietic stem cells are the center of numerous investigations. A number of studies have examined factors that determine the longevity of the cells, including telomere length, cell cycle control, and apoptotic proteins. Other equally important questions concern the genes involved in the control of the progenitor state of the daughter cells.

14.6.2 Isolation

One major observation that has implications for application of stem cells for tissue engineering is the fact that stem cells survive in the presence of a minimal number of accompanying cells.[116] In spite of the fact that a single cell may restore the blood system, efforts to isolate that cell have only been accomplished by the isolation of a population of approximately one thousand cells that may represent members of the hematopoietic stem cell niche within the bone marrow.[117]

14.6.3 Bone Marrow

The results of developing antibodies to cell surface molecules has permitted the delineation of stages of hematopoietic maturation that has led to the recognition that a single cell can restore the blood–forming capacity and survival to lethally irradiated animals.[118] It is also recognized that the primary stem cell is localized to a minimal number of accompanying cells that represent a critical microenvironment or niche.[119] The antibody-recognized surface markers are c-kit+, SCA-1+, and linlo/-Thy1lo.[120] These hematoprogenitor cells will localize to the appropriate stromal microenvironment, where the cell proliferation and maturation process is initiated to restore all the cells of the blood system. The processes that underlie prolonged cell proliferation required to provide an animal with sufficient complement of red and white blood cells continue to be studied.[121,122] The availability of a number of mouse models have contributed to the ability to distinguish roles of individual classes of cells.[123,124]

14.7 NEURAL STEM CELLS

14.7.1 Endogenous Cells

A number of approaches have been tried to reestablish function through repopulation of specific regions of the brain through stem cell delivery or recruitment. Endogenous radial glial cells serve both as a precursor cell and as scaffolds for neuron migration.[125] In a study designed to investigate whether radial glial cells can be generated from embryonic forebrain neural stem cells, the addition of epidermal growth factor was sufficient to support the generation and differentiation of morphologically, antigenically, and functionally defined radial glial cells.[126] When applied *in vivo*, epidermal growth factor promoted adult forebrain ependymal cells to dedifferentiate and exhibit a radial morphology, suggesting that neural stem cells can give rise to radial glial cells so that neural migration can be recapitulated in the adult central nervous system.

One approach for neuroregeneration has focused on establishing a microenvironment for neural stem cells that is designed to direct their development down a specific pathway. Stem cells have their fate determined by soluble factors that are accessible within a defined niche and also by contact-mediated pathways.[127,128] In the latter case, the cells interact with the extracellular matrix proteins through cell surface receptors that trigger intracellular signaling pathways and activate transcription factors that drive differentiation. To reproduce the neural microenvironment, artificial proteins have been developed to act as inducers of neuronal specificity.[129] Acquisition of an artificial protein phenotype follows development of an artificial gene designed to contain the characteristic features of the artificial protein needed for cell recognition. Work completed to date using this approach has centered on the Notch signaling pathway, which is important for gliogenesis and control of neurite outgrowth.[130] The target proteins include the active domains of hJagged1 and hDelta1, both Notch receptor ligands, placed into an elastin backbone.[131] The goal is to formulate these proteins into an extracellular matrix scaffold that will reproduce the microenvironment for neural cell development and differentiation with the artificial proteins acting as ligands for the Notch receptor.

14.7.2 Blood–Brain Barrier

The brain is protected in part by the blood–brain barrier so that transit of cells that might serve as progenitors will be restricted. Some cells such as T cells of the immune system are able to cross an intact blood–brain barrier,[132] and in a IL-1 knock-out SCID mouse, increased numbers of T cells improved neuroregenerative processors for an axotomized facial motor nucleus.[133] The absence of IL-2 did not appear to influence neuronal regeneration and may have decreased microglial phagocytic clusters. These data suggest that local release of growth factors and cytokines by cells that cross the blood–brain barrier may impact stem cell recruitment and function. In some instances of injury, transport systems for cytokines such as tumor necrosis factor-alpha are upregulated.[134] Other factors do not cross the blood–brain barrier, including glial cell line-derived neurotrophic factor, which promotes recovery of injured nigrostriatal dopamine system and improves motor function in models of Parkinson's disease.[135] Administration of this neurotrophic factor requires intracerebral administration until new methods of factor transport across the blood–brain barrier are developed. Nanogels permit oligonucleotides to the blood–brain barrier by using specific targeting molecules, insulin or transferrin, in a formulation based on a nanoscale network of cross-linked poly(ethylene) glycol and polyethylenimine.[136] Extension of these techniques to bioactive peptides or to genetic expression systems may enable the endogenous differentiated cells of the brain to generate the protein factors needed to induce expansion of neuronal stem cells.

14.8 STEM CELL NICHE AND TISSUE REPAIR

The importance of stromal microenvironments as deterministic elements in stem cell behavior emerged from efforts to repopulate tissues with differentiated cells. It became obvious that stem cell survival requires a specific microenvironment or niche. Stem cells interact with the extracellular matrix through structural components that bind plasma membrane receptors or as a source of growth factors. Within the defined microenvironment for skin stem cells, division rarely occurs, but once the cells leave the matrix their properties change so that proliferation then follows. While localized within the matrix, the gene expression of the stem cells involves surface receptors and proteins that preserve their location, survival, and quiescence.[137] The localization of hematopoietic stem cells within a bone marrow microenvironment proved to be dependent on a transmembrane isoform of stem cell factor (tm-SCF). In the absence of this factor, adhesion of transplanted stem cells within the hemopoietic niche was significantly impaired.[138] Osteoblasts contribute to the hematopoietic niche by contributing to Notch activation. Other stem cell niches include the limbal basal epithelium that contain corneal epithelial stem cells, microenvironments for neural cells, and the basal lamina surrounding muscle cells.[139]

Recent evidence indicates that the ventricular zone of the developing brain persists as a specialized subventricular zone (SVZ) that supports neurogenesis in adult life. The region is characterized by a subpopulation of astrocytes, which in adult mammals appear to serve as neuronal precursors.[140,141] The astrocytes are derived from early postnatal radial glia that are surrounded by ependymal cells that migrate into the region but continue to extend as single cilium similar to neuroepithelial cells and radial cells. This zone appears essential for stem cell function, and in response to epidermal growth factor (EGF), a subset of SVZ cells grow to form neurospheres that are multipotent and self-renewing.

The role of the local cellular microenvironment on the recruitment of stem cells is linked to tissue damage. Pathophysiology is associated with the release of cytokines or other soluble factors that signal the stem cells to migrate to sites of injury and take up residence for tissue repair.[142,143] In muscle, local damage induced by stretch combined with electrical stimulation activated satellite cells, the muscle stem cell, through the release of a splice variant of insulin-like growth factor-I, MGF. The continued differentiation of the cells was maintained by later expression of IGF-Iea, a second splice variant of IGF-I.

Transplantation experiments demonstrate that, as stem cells such as those derived from bone marrow arrive at a disparate location, fusion with other cells may occur. Bone marrow–derived stem cells fuse *in vivo* with hepatocytes in liver, Purkinje neurons in brain, and cardiac muscle in heart.[144] The fusion of the two cell types results in an apparent "transdifferentiation," since the resulting cells demonstrate properties of the host tissue. There is evidence that a limited number of bone marrow stem cells may contribute to tissue formation in the absence of fusion. The frequency is low as shown for brain (0.1%) and for one specific muscle, the panniculus carnosus (5%).[145,146] However, determining the full extent to which such a process occurs and developing methods to increase the frequency of occurrence remain major areas of investigation in cell biology and fields focused on tissue repair.

A number of tissue-specific uses of stem cells have been tried for repair and regeneration of diseased tissues. Although the approaches exhibit limited levels of success, the efforts represent clinically relevant areas where methods for nanofabrication of tissue substrates may vastly increase efficiency and outcome of cell transplantation.

14.8.1 Liver Regeneration

Conventional approaches for liver repair have focused on preparation of hepatocytes that can be applied to substrates to create a liver assist device.[147,148] These devices are designed as filtration systems where the use of cells is use to improve detoxification rather than replacement of critically important hepatocyte derived proteins, such as albumin and glucagon. The other major role of the liver is to contribute to the regulation of glucose in the blood stem through glucogenesis and glycogenesis. Glucogenesis culminates in the formation of fatty acids that are released into the blood stem and stored in the adipose tissue. Glycogen is stored in the liver for rapid access. The major goal of tissue engineering for liver disease would be to create an organized tissue in which the hepatocyte would be able to monitor and control the intermediary metabolism. However, the limitation to date has been due to the fragile nature of the mature hepatocytes and the inability to prevent their coalescence into spheroid bundles that undergo necrosis due to limitations in nutrient supply.[149]

Recently, candidate stem cells have been identified in the human liver using markers such as OV6, CD34, c-kit, and NCAM holding some promise that cell selection may be extended to these cells.[150] It is of interest that the hepatic oval cells that support some forms of liver regeneration express markers associated with hematopoietic stem cells.[151] In addition, bone marrow cells can be isolated and stimulated to produce subsets of epithelial cells that represent oval cells, hepatocytes, and duct epithelium. However, the bone marrow cells appear to contribute to liver regeneration through the elaboration of hematopoietic cells, including lymphocytes, neutrophils, macrophages, and platelets, that supply factors accelerating healing.

14.8.2 Muscle Repair

Repair of muscle function may be amenable to the use of isolated stem cells. In the case of cardiac injury, there is hope that repopulation of damaged tissue by injection of stem cells will culminate in organization of contractile tissue under control of neuromuscular junctions.[152] Incorporation of muscle-derived stem cells into small intestinal submucosa as a scaffold resulted in establishment of contractile activities in 100% of the constructs after 8 weeks that were blocked by removal of calcium.[153] The contractile response was also diminished by addition of succinylcholine, which acts on nicotinic receptors, suggesting that the constructs could function as replacement tissue for a deficient sphincter or to augment bladder contraction.

14.8.3 Pancreatic Beta Cells

Islet transplantation has provided a successful clinical avenue for treatment of patients with type I diabetes. The challenge is to expand the approach by development of a source of cells to expand the number of patients that can be treated. Insulin-producing cells have been prepared from adult pancreatic tissues. Evidence exists that mature pancreatic duct epithelial cells that have regressed or lost the mature phenotype may serve as a source of beta-cells.[154] The challenge is to develop fundamental information that will permit generation of new beta cells from stem cells.

14.9 SUBSTRATE-DEPENDENT DIFFERENTIATION

The development of scaffolds for tissue engineering applications has generated an array of novel formulations. The rationale for scaffold development is that an appropriate representation of the native extracellular matrix will provide a structural and functional framework that will contribute to the properties of the tissue in question.[155] The role of the scaffold has been studied, examining the behavior of osteoprogenitor cells placed in materials ranging from gels of extracellular matrix proteins such as type I collagen to more rigid skeleton of a marine sponge. Skeletons of marine sponges were examined because the native tissue organization satisfied important criteria applicable to the design and formulation of multiple tissues.[156] The marine skeleton is attractive because of its potential for hydration, open but interconnected channels, collagenous composition, and diverse fiber architecture. Human osteoprogenitor cells adhere to this natural substrate and surround themselves with an extracellular matrix containing the bone-specific proteins, alkaline phosphatase, and type I collagen. Spongin is the major protein component of the marine skeleton and has some analogous sequences with collagens. The marine skeleton absorbed growth factors, as evidenced by addition of BMP-2. This observation suggests that growth factor delivery may be used in conjunction with the natural product scaffold to enhance specific types of tissue formation.

14.9.1 NATURAL PRODUCTS

Type I collagen scaffolds have been used extensively for many different types of bone stem cells as well as other cell types.[157–159] Type I collagen represents a primary use of natural materials that can be isolated, purified, and inserted as a tissue replacement. The use of collagen as scaffold is a choice of material that follows the easy availability of the material, since bovine skin preparations have a long history in human applications. The history of type I collagen as an injectable used for skin reconstruction is one specific example where governmental regulation has been satisfied. Type I collagen is a fibrillar collagen that is characterized by an alpha helical structure of individual peptides that bind together to create a fibril consisting of three polypeptide chains, two having identical amino acid composition (alpha1(I) chains) and a third with a different composition (alpha2(I) chain). The abundance of type I collagen in the body and the relative conservation across species contributes to its immunological tolerance. In addition, a number of chemical steps have been included to remove minor contaminants, such as other collagens and carbohydrate residues, to further improve immune acceptance.

14.9.2 POLYMERICS

Polymeric materials for use as cell substrates have been under investigation for years. The polymers fall into two major classes that can be used to create scaffolds that have a permanent quality for long-term implantation, whereas other polymers are designed to be degradable.[160] Both types of material, whether permanent or degradable, have unique issues that impact their biocompatibility. Permanent materials must not act as reservoirs for release of toxic components and should not serve as a nidis for bacterial attachment or thrombogenic events and should not shed particulates due to excessive corrosion or electrolysis.

Degradable polymers that exhibit stability and biocompatibility include polylactic acid, polyglycolic acid, polyhydroxybutyrate (PHB), polyhydroxyvalerate (PHV), polycaprolactone, polyanhydrides, poly(ortho esters), and poly(amino acids). Of these materials, polyglycolic and polylactic acid are the most investigated polymers. Mixtures of these components to create a variety of copolymers have been instrumental in the development of resorbable sutures and other fixation devices. The major difficulty with polyglycolic and polylactic acid formulations is due to the release of acidic components that contribute to tissue inflammation and, in some instances, necrosis.[161,162] The copolymer formulations of polyglycolic acid and polylactic acid impart properties to the material that are not predicted from the behavior of the pure polymeric material, such as the rate of bioerosion.

14.9.3 Bioglasses and Ceramics

Permanent materials for cell scaffolds include ceramics, glasses, and glass-ceramics. The tissue response to the permanent materials depends in large part on the surface energy and the degree of porosity of the material.[163] The microstructures of the materials depend on the type of thermal processing steps required for production. Alumina and calcium phosphate–based materials are typically formed from fine-grained particulate solids that are mixed with water and pressed into a mold. Increased temperature removes water, and if a porous structure is being produced heating removes a sacrificial component during the finishing step. The liquid-phase sintering process, vitrification, can be adjusted to produce varying degrees of crystallinity. Modern techniques of powder formation permit high-density and high-purity ceramics that exhibit corrosion resistance and good biocompatibility.[164] Adjustment in the types of additives combined during the sintering process has generated a diverse array of bioactive glasses and glass-ceramics for application as preformed structures.

14.10 NANOTECHNOLOGY AND SCAFFOLD FABRICATION

Current efforts for construction of polymeric scaffold are focused on methods to produce high porosities (up to 98%) with defined connectivity using modified leaching techniques coupled with freeze-drying techniques.[165] Other procedures for scaffold development, particularly for bone, are focused on creation of a composite material that includes a mineral component linked to organic constituents, such as gelatin and chitosan.[166] The ideal scaffold would be one that represents a biomimetic surface that triggers cell recognition as well as adhesive reactions.[167,168] In some instances, the formulation may require unique sintering conditions to establish a defined geometry.[169]

14.10.1 Fixed Surface Adhesion

Cells interact with surfaces through integral membrane receptors. The major classes include the integrins, the collagen receptors and the cell surface proteoglycans. Integrins constitute a family of dimeric proteins, an alpha and beta subunit, which together interact with extracellular matrix cell adhesion proteins and with intermediate and microfilaments of the intracellular architecture.[170,171] A porous polyvinyl resin was used to create a three-dimensional scaffold to establish a culture system for expansion of bone marrow cells by maintaining the cells in the presence of erythropoietin, interleukin-3, stem cell factor, and interleukin-6. The combination of scaffold with growth factors significantly outperformed comparable cultures maintained in culture dishes.[172]

Culture of mesenchymal stem cells on poly-D-lysine, poly-L-lysine, collagen, laminin, fibronectin, and Matrigel surfaces was carried out to determine which surface coating would improve neuronal differentiation.[173] Of the materials examined, Matrigel at a coating density of 50 μg per square centimeter enhanced differentiation and improved cell expansion. The fact that Matigel is a composite material containing laminin and collagen together suggests that the membrane protein and substrate interaction may directly influence the differentiation of the mesenchymal stem cells in culture. The ability to couple surface modification with growth factor presentation may enable controlled expansion of stem cells under conditions where alternative cell types are not generated.

14.10.2 Biodegradable Surfaces

Biodegradable polymer scaffolds have been formatted in a three-dimensional framework as a means for promoting human embryonic stem cell growth and differentiation.[174] Addition of growth factors to the stem cells placed in either a poly(lactic-co-glycolic acid) or poly(L-lactic acid) polymer scaffold induced neural tissues, cartilage, and liver, depending on the specific inducer. Implantation of the scaffolds into severe combined immunodeficient mice resulted in continuation of human protein expression and appeared to incite neovascularization.

14.10.3 Microfluidics

Nutrition of cells within tissues *in vivo* is under precise control, primarily by fluctuations that occur in the capillary beds that supply blood to the cells. Development of tissue-engineered constructs results in an immediate problem that centers on delivery of nutrients and export of waste. In the absence of blood vessels, nutrient supply can only occur by simple or, with agitation, facilitated diffusion. Although the kinetics of fluid transport by passive diffusion is a highly developed science, the principles for solute transport are primarily applied to chemical reactions that take place in chemostats, not to organized tissues. Techniques and devices to improve nutrient and growth factor delivery to cells embedded in multidimensional scaffold represents a major area for research and development. One approach that is being explored involves development of automated perfusion systems to serve as bioreactors to provide oxygen and nutrient delivery. A number of parameters require optimization in order to deliver nutrients at appropriate pressures, flow rates, type of flow (pulsatile/steady), viscosity, and ionic concentrations of the perfusate solutions.[175]

The technology applied in methods for nanofabrication of cellular substrates will be of major importance in permitting development of relatively large bioartificial tissues and organ systems. Electrospinning has already been applied to develop a novel nanofibrous structure for cell culture.[176] Ultrafine fibers (500 to 800 nm) of poly(D,L-lactide-co-glycolide) were formed in a high-voltage electrostatic field to yield a structure to mimic the extracellular matrix of natural tissues. The electrospun structure supported cell attachment and proliferation, making the formulation procedure attractive for tissue engineering of tissues. Molecular nanotechnology works to create biomimetic membranes by reproducing two-dimensional crystalline bacterial S-layers composed of identical protein or glycoprotein subunits.[177] The resulting isoporous protein lattices provide substrates for molecular sieving, stabilizing liposomes, and patterning and immobilization of functional molecules.

14.11 CONCLUSIONS AND FUTURE DIRECTIONS

The impending intersection between studies of stem cell biology and smart scaffold production using methodologies emerging from nanotechnology will make possible tissue engineering of artificial organs, not feasible only a few years past. Advances in the technology of nanofabrication will permit physical features to be designed into cellular substrates with morphological features at molecular dimensions.[178,179] A new generation of cellular scaffolds will be available that are not restricted to surfaces composed of simple coatings of polymeric cell adhesion molecules, such as polylysine, fibronectin, or collagen. Rather, the surface topology of the manufactured substrates will be designed to recreate the thousands of constituents of extracellular matrix macromolecules. Deposition of surface features that have a charge and density that exactly duplicate the extracellular or pericellular microenvironment of a stem cell may be used to induce differentiation.[180] Precise nanofabrication of external surfaces will allow extracellular signals that trigger intracellular regulation of gene expression to be used for recapitulation of embryogenesis in a tissue engineering application. The future of organ and tissue repair and regeneration can expected to hinge on atomic-level features that create a readable instruction set by which stem cells of all classes, embryonic, adult, or immortalized can be resources for construction of damaged or diseased tissues. Recent studies show that precise regimens of mechanical loading are integral elements in the regulation of differentiated cell gene expression.[181–185] Isolated autologous stem cells placed in appropriate three-dimensional surface-specific molecular scaffolds, exposed to the proper mechanical environment, and nourished with adequate nutrients and growth factors will generate "artificial organs" and will permit individuals to serve as their own tissue and organ donor.

ACKNOWLEDGMENTS

This work was supported by VA Research Career Award, Medical Merit Review Grant, NIH 5R01 AR45788 and the Stanford Orthopaedic Research Fund.

REFERENCES

1. Gage, R.H., Mammalian neural stem cells, *Science,* 287, 1433, 2000.
2. Bunting, K.D. and Hawley, R.G., Integrative molecular and developmental biology of adult stem cells, *Biol. Cell*, 95, 563, 2003.
3. Sharpless, N.E. and DePinto, R.A., Telomeres, stem cells, senescence and cancer, *J. Clin. Invest.*, 113,160, 2004.
4. Sylvester, K.G. and Longaker, M.T., Stem cells: review and update, *Arch. Surg.,* 139, 93, 2004.
5. Young, H.E., Existence of reserve quiescent stem cells in adults, from amphibians to humans, *Curr. Top. Microbiol. Immunol.*, 280, 71, 2004.
6. Shen, G. et al., Tissue engineering of blood vessels with endothelial cells differentiated from mouse embryonic stem cells, *Cell Res.,* 13, 335, 2003.
7. Zwaginga, J.J. and Doevendans, P., Stem cell-derived angiogenic/vasculogenic cells: possible therapies for tissue repair and tissue engineering, *Clin. Exp. Pharmacol. Physiol.,* 30, 900, 2003.
8. Zimmermann, W.H., Melnychenko, I., and Eschenhagen, T., Engineered heart tissue for regeneration of diseased hearts, *Biomaterials*, 25, 1639, 2004.
9. Tanade, V.M. et al., Human stem-progenitor cells from neonatal cord blood have greater hematopoietic expansion capacity than those from mobilized human blood, *Exp. Hematol.* 30, 816, 2002.
10. Li, A. et al., Extensive tissue-regenerative capacity of neonatal human keratinocyte stem cells and their progeny, *J. Clin. Invest.* 113, 390, 2004.
11. Kyba, M. and Daley, G.O., Hematopoiesis from embryonic stem cells: lessons from and for ontogeny. *Exp. Hematol.*, 31, 1015, 2003.
12. Alison, M.R. et al., Recipes for adult stem cell plasticity: fusion cuisine or readymade? *J. Clin. Pathol.* 57, 113, 2004.
13. Zepeda, M.L, Chinoy, M.R., and Wilson, J.M., Characterization of stem cells in human airway capable of reconstituting a fully differentiated bronchial epithelium. *Somat. Cell Mol. Genet.,* 21, 61, 1995.
14. Turksen, K. and Troy, T.C., Epidermal cell lineage, *Biochem. Cell Biol.*, 76, 889, 1998.
15. Ghazizadeh, S. and Taichman, L.B., Multiple classes of stem cells in cutaneous epithelium: a lineage analysis of adult mouse skin, *EMBO J.*, 20, 1215, 2001.
16. Loeffler, M. and Roeder, I., Tissue stem cells: definition, plasticity, heterogeneity, self-organization and models — a conceptual approach, *Cells Tissues Organs*, 171, 8, 2002.
17. Hammerman, M.R., Tissue engineering the kidney, *Kidney Int.*, 63, 1195, 2003.
18. Lansdorp, P.M., Developmental changes in the function of hematopoietic stem cells, *Exp. Hematol.,* 23, 187, 1995.
19. Okabe, M. et al., Intrinsic and extrinsic determinants regulating cell fate decision in developing nervous system, *Dev. Neurosci.,* 19, 9, 1997.
20. Perron, M. and Harris, W.A., Determination of vertebrate retinal progenitor cell fate by the Notch pathway and basis helix-loop-helix transcription factors, *Cell Mol. Life Sci.*, 57, 215, 2000.
21. Ivanoa, N.B. et al., A stem cell molecular signature, *Science*, 298, 601, 2002.
22. Nathan, S. et al., Cell-based therapy in the repair of osteochondral defects: a novel use for adipose tissue, *Tissue Eng.,* 9, 733, 2003.
23. Gojo, S. and Umezawa, A., Plasticity of mesenchymal stem cells-regenerative medicine for diseased hearts, *Hum. Cell*, 16, 23, 2003.

24. Hu, Y. et al., Isolation and identification of mesenchymal stem cells from human fetal pancreas, *J. Lab Clin. Med.* 141, 342, 2003.
25. Passier, R. and Mummery, C., Origin and use of embryonic and adult stem cells in differentiation and tissue repair, *Cardiovas. Res.* 58, 324, 2003.
26. Li, M. et al., Isolation and culture of embryonic stem cells from porcine blastocysts, *Mol. Reprod. Dev.* 130, 4235, 2003.
27. Bacigalupo, A., Frassoni, F., and Van Lint, M.T., Bone marrow or peripheral blood as a source of stem cells for allogeneic transplantation, *Haematologica,* 87, 4, 2002.
28. Albano, R.M., Groome, N., and Smith, J.C., Activins are expressed in preimplantation mouse embryos and in ES and EC cells and are regulated on their differentiation. *Development,* 117, 711, 1993.
29. Carpenter, M.K., Rosler, E., and Rao, M.S., Characterization and differentiation of human embryonic stem cells, *Cloning Stem Cells,* 5, 79, 2003.
30. Seaman, D.S., Adult living donor liver transplantation: current status, *J. Clin. Gastroenterol.,* 33, 97, 2001.
31. Davis, G.L. et al., Projecting future complications of chronic hepatitis C in the United States, *Liver Transpl.,* 9, 331, 2003.
32. Prelle, K., Zink, N., and Wolfe, E., Pluripotent stem cells-model of embryonic development, tool for gene targeting, and basis of cell therapy, *Anat. Histol. Embryol.,* 31, 169, 2002.
33. Saito, S. et al., Isolation of embryonic stem-like cells from equine blastocyts and their differentiation *in vitro*, *FEBS Lett.,* 531, 389, 2002.
34. Lin, H. et al., Multilineage potential of homozygous stem cells derived from metaphase II oocytes, *Stem Cells,* 21, 152, 2003.
35. Gimble, J. and Guilak, F., Adipose-derived adult stem cells: isolation, characterization, and differentiation potential, *Cytotherapy,* 5, 362, 2003.
36. Bunting, K.D. and Hawley, R.G., Integrative molecular and developmental biology of adult stem cells, *Biol. Cell,* 95, 563, 2003.
37. Turnpenny, L. et al., Derivation of human embryonic germ cells: an alternative source of pluripotent stem cells, *Stem Cells,* 21, 598, 2003.
38. Niwa, H., Miyazaki, J., and Smith, A.G., Quantitative expression of Oct-3/4 defines differentiation, dedifferentiation or self-renewal of ES cells, *Nat. Genet.,* 24, 372, 2000.
39. Hynes, R.O., Integrins: bidirectional, allosteric signaling machines, *Cell,* 110, 673, 2002.
40. Giancotti, F.G. and Ruoslahti, E., Integrin signaling, *Science*, 285, 1028, 1999.
41. Danen, E.H.J. and Sonnenberg, A., Integrins in regulation of tissue development and function, *J. Pathol.*, 200, 471, 2003.
42. Boyer, B., Valles, A.M., and Edme, N., Induction and regulation of epithelial-mesenchymal transitions, *Biochem. Pharmacol.,* 60, 1091, 2000.
43. Toumadje, A. et al., Pluripotent differentiation *in vitro* of murine ES-D3 embryonic stem cells, *In Vitro Cell Dev. Biol. Anim.,* 39, 449, 2003.
44. Seshi, B., Kumar, S., and King, D., Multilineage gene expression in human bone marrow stromal cells as evidenced by single-cell microarray analysis, *Blood Cells Mol. Dis.,* 31, 268, 2003.
45. Luo, Y. et al., Designing, testing and validating a focused stem cell microarray for characterization of neural stem cells and progenitor cells, *Stem Cells*, 21, 575, 2003.
46. Ota, J. et al., Proteomic analysis of hematopoietic stem cell-like fractions in leukemic disorders, *Oncogene*, 22, 5619, 2003.

47. Sperger, J.M. et al., Gene expression patterns in human embryonic stem cells and human pluripotent germ cell tumors, *Proc. Natl. Acad. Sci. U.S.A.*, 100, 13350, 2003.
48. Hubner, K. et al., Derivation of oocytes from mouse embryonic stem cells, *Science*, 300, 1251, 2003.
49. Wilmut, I. and Paterson, L., Somatic cell nuclear transfer, *Oncol. Res.*, 13, 303, 2003.
50. Pan, G.J. et al., Stem cell pluripotency and transcription factor Oct4, *Cell Res.*, 12, 321, 2002.
51. Bowles, J. et al., Dppa3 is a marker of pluripotency and has a human homologue that is expressed in germ cell tumours, *Cytogenet. Genome Res.*, 101, 261, 2003.
52. Meng, X.L. et al., Brain transplantation of genetically engineered human neural stem cells globally corrects brain lesions in the mucopolysaccharidosis type VII mouse, *J. Neurosci. Res.*, 74, 266, 2003.
53. Mihara, K. et al., Development and functional characterization of human bone marrow mesenchymal cells immortalized by enforced expression of telomerase, *Br. J. Haematol.*, 120, 846, 2003.
54. Fukuda, K., Development of regenerative cardiomyocytes from mesenchymal stem cells for cardiovascular tissue engineering, *Artif. Organs*, 25, 187, 2001.
55. Kobayashi, N. et al., Hybrid bioartificial liver: establishing a reversibly immortalized human hepatocyte line and developing a bioartificial liver for practical use. *J. Artif. Organs*, 6, 236, 2003.
56. Suetake, K. et al., Expression of gangliosides in an immortalized neural progenitor/stem cell line, *J. Neurosci. Res.*, 74, 769, 2003.
57. Mitsui, T. et al., Immortalized neural stem cells transplanted into the injured spinal cord promote recovery of voiding function in the rat. *J. Urol.*, 170, 1426, 2003.
58. Sugi, Y. and Markwald, R.R., Endodermal growth factors promote endocardial precursor cell formation from precardiac mesoderm, *Dev. Biol.*, 263, 35, 2003.
59. Sato, N. et al., Maintenance of pluripotency in human and mouse embryonic stem cells through activation of Wnt signaling by a pharmacological GSK-3-specific inhibitor, *Nat. Med.*, 10, 55, 2004.
60. Reya, T. et al., A role for Wnt signalling in self-renewal of haematopoietic stem cells, *Nature*, 423, 409, 2003.
61. Roelen, B.A. and Dijke, P., Controlling mesenchymal stem cell differentiation by TGFBeta family members, *J. Orthop. Sci.*, 8, 740, 2003.
62. Kablar, B. et al., Myf5 and MyoD activation define independent myogenic compartments during embyronic development, *Dev. Biol.*, 258, 307, 2003.
63. Kardon, G., Harfe, B.D., and Tabin, C.J., A Tcf4-positive mesodermal population provides a prepattern for vertebrate limb muscle patterning, *Dev. Cell*, 5, 937, 2003.
64. Yao, Y. et al., Extracellular signal-regulated kinase 2 is necessary for mesoderm differentiation, *Proc. Natl. Acad. Sci. U.S.A.*, 100, 12759, 2003.
65. Vetere, A. et al., Neurogenin3 triggers beta-cell differentiation of retinoic acid-derived endoderm cells, *Biochem. J.*, 371, 831, 2003.
66. Burgess, A.W., Growth control mechanisms in normal and transformed intestinal cells, *Philos. Trans. R. Soc. Biol. Sci.*, 353, 831, 1998.
67. Boot, M.J. et al., Spatiotemporally separated cardiac neural crest subpopulations that target the outflow tract septum and pharyngeal arch arteries, *Anat. Rec.*, 275A, 1009, 2003.
68. Seiber-Blum, M., Cardiac neural crest cells, *Anat. Rec.*, 276A, 34, 2004.
69. Fuchs, E. and Raghavan, S., Getting under the skin of epidermal morphogenesis, *Nat. Rev. Genet.*, 3, 199, 2002.

70. Mackenzie, I.C., Retroviral transduction of murine epidermal stem cells demonstrates clonal units of epidermal structure, *J. Invest. Dermatol.,* 109, 377, 1997.
71. Okamoto, R. and Watanabe, M., Molecular and clinical basis for the regeneration of human gastro-intestinal epithelia, *J. Gastroenterol.*, 39, 1, 2004.
72. Fibbe, W.E. and Noort, W.A., Mesenchymal stem cells and hematopoietic stem cell transplantation, *Ann. N.Y. Acad. Sci.*, 996, 235, 2003.
73. Jakel, R.J., Schneider, B.L., and Svendsen, C.N., Using human neural stem cells to model neurological disease, *Nat. Rev. Genet.*, 5, 136, 2004.
74. Noel, D., Djouad, F., and Jorgense, C., Regenerative medicine through mesencymal stem cells for bone and cartilage repair, *Curr. Opin. Investig. Drugs*, 3, 1000, 2002.
75. Ogawa, R. et al., Osteogenic and chondrogenic differentiation by adipose-derived stem cells harvested from GFP transgenic mice, *Biochem. Biophys. Res. Commun.*, 313, 871, 2004.
76. Bailey, P., Holowacz, T., and Lassar, A.B., The origin of skeletal muscle stem cells in the embryo and the adult, *Curr. Opin. Cell Biol.*, 13, 679, 2001.
77. Ford, C.E., et al., Cytological identification of radiation chimaeras, *Nature*, 177, 452, 1956.
78. Till J. and McCulloch E., A direct measurement of the radiation sensitivity of normal mouse bone marrow cells. *Radiat. Res.*, 14, 213, 1961.
79. Becker, A., McCulloch, E., and Till, J., Cytological demonstration of the clonal nature of spleen colonies derived from transplanted mouse marrow cells, *Nature*, 197, 452, 1963.
80. Wu, A.M. et al., A cytological study of the capacity for differentiation of normal hemopoietic colony-forming cells, *J. Cell Physiol.*, 69, 177, 1967.
81. Visser, J.W. M. et al., Isolation of murine pluripotent hemopoietic stem cells. *J. Exp. Med.,* 59, 1576, 1984.
82. Goodell M.A. et al., Isolation and functional properties of murine hematopoietic stem cells that are replicating *in vivo. J. Exp. Med.,* 183, 1797, 1996.
83. Osawa, M. et al., Long-term lymphohematopoietic reconstitution by a single CD34-low/negative hematopoietic stem cell. *Science,* 273, 242, 1996.
84. Uchida, N., Weissman, I.L. Searching for hematopoietic stem cells: evidence that Thy-1.1lo Lin-Sca-1+ cell are the only stem cells in C57BL/Ka-Thy1.1 bone marrow. *J. Exp. Med.,* 175, 175, 1992.
85. Morrison, S.J., Weissman, I.L. The long-term repopulation subset of hematopoietic stem cells is deterministic and isolatable by phenotype. *Immunity,* 1, 661, 1994.
86. Doetsch, F., Garcia-Verdugo, J.M., and Alvarez-Buylla, A., Cellular composition and three-dimensional organization of the subventricular germinal zone in the adult mammalian brain, *J. Neurosci.,* 17, 5046, 1997.
87. Palmer, T.D., Takahashi, J., and Gage, F.H., The adult rat hippocampus contains primodial neural stem cells, *Mol. Cell Neurosci.,* 8, 389, 1997.
88. Lim, D.A. and Alvarez-Buylla, A., Interaction between astrocytes and adult subventricular zone precursors stimulates neurogenesis, *Proc. Natl. Acad. Sci. U.S.A.,* 96, 7526, 1999.
89. Garcia-Verdugo, J.M. et al., Architecture and cell types of the adult subventricular zone: in search of the stem cells, *J. Neurobiol.,* 36, 234, 1998.
90. Alvarez-Buylla, A., Garcia-Verdugo, J.M., and Tramontin, A.D., A unified hypothesis on the lineage of neural stem cells, *Nat. Rev. Neurosci.*, 2, 287, 2001.
91. Apuzzo, M.L.J. et al., Cellular and molecular neurosurgery: fetal grafting to treat Parkinson's disease, *Neurosurgery,* 49, 575, 2001.

92. Rosser, A.E. et al., Staging and preparation of human fetal striatal tissue for neural transplantation in Huntington's disease, *Cell Transplant.*, 12, 679, 2003.
93. Muraro, P.A. et al., Hematopoietic stem cell transplantation for multiple sclerosis: current status and future challenges, *Curr. Opin. Neurol.*, 16, 299, 2003.
94. Morest, D.K. and Silver, J., Precursors of neurons, neuroglia, and ependymal cells in the CNS: what are they? Where are they from? How do they get where they are going? *Glia*, 43, 6, 2003.
95. Murakami, T. et al., Transplanted neuronal progenitor cells in a peripheral nerve gap promote nerve repair, *Brain Res.*, 974, 17, 2003.
96. Murphy, J.M. et al., Stem cell therapy in a caprine model of osteoarthritis, *Arthritis Rheum.* 48, 3464, 2003.
97. Huang, J.I. et al., Chondrogenic potential of multipotential cells from human adipose tissue, *Plast. Reconstr. Surg.*, 113, 585, 2004.
98. O'Driscoll, S.W. and Fitzsimmons, J.S., The role of periosteum in cartilage repair, *Clin. Orthop.*, 391 Suppl., S190, 2001.
99. Grande, D.A. et al., Stem cells as platforms for delivery of genes to enhance cartilage repair, *J. Bone Joint Surg.*, 85-A, Suppl. 2, 111, 2003.
100. Schmitt, B. et al., BMP2 initiates chondrogenic lineage development of adult human mesenchymal stem cells in high-density culture, *Differentiation*, 71, 567, 2003.
101. Fuchs, J.R. et al., Fetal tracheal augmentation with cartilage engineered from bone marrow-derived mesenchymal progenitor cells, *J. Pediatr. Surg.*, 38, 984, 2003.
102. Barry, F.P., Mesenchymal stem cell therapy in joint disease, *Novartis Found. Symp.*, 249, 86, 2003.
103. Sui, Y. et al., Limb bud progenitor cells induce differentiation of pluripotent embryonic stem cells into chondrogenic lineage, *Differentiation*, 71, 578, 2003.
104. Mammone, J.F. and Hudson, S.M., Micromechanics of bone strength and fracture, *J. Biomech.*, 26, 439, 1993.
105. Liu, Z. et al., Molecular signaling in bone fracture healing and distraction osteogenesis, *Histol. Histopathol.*, 14, 587, 1999.
106. Bensaid,W. et al., A biodegradable fibrin scaffold for mesenchymal stem cell transplantation, *Biomaterials*, 24, 2497, 2003.
107. Ahdjoudj, S., Fromique, O., and Marie, P.J., Plasticity and regulation of human bone marrow stromal osteoprogenitor cells: potential implication in the treatment of age-related bone loss, *Histol. Histopathol.*, 19, 151, 2004.
108. Yang, X.B. et al., Novel osteoinductive biomimetic scaffolds stimulate human osteoprogenitor activity-implications for skeletal repair, *Connect. Tissue Res.*, 44, Suppl. 1, 312, 2003.
109. Kim, H., Kim, H.W., and Suh, H., Sustained release of ascorbate-2-phosphate and dexamethasone from porous PLGA scaffolds for bone tissue engineering using mesenchymal stem cells, *Biomaterials*, 24, 4671, 2003.
110. Awad, H.A. et al., Repair of patellar tendon injuries using a cell-collagen composite, *J. Orthop. Res.*, 21, 420, 2003.
111. Asou, Y. et al., Coordinated expression of scleraxis and Sox9 genes during embryonic development of tendons and cartilage, *J. Orthop. Res.*, 20, 827, 2002.
112. Morrison, S.J., Uchida, N., and Weissman, I.L., The biology of hematopoietic stem cells, *Annu. Rev. Cell Biol.*, 11, 35, 1995.
113. Uchida, N. et al., Rapid and sustained hematopoietic recovery in lethally irradiated mice transplanted with purified Thy-1.1lo Lin-Sca-1+ hematopoietic stem cells, *Blood*, 83, 3758, 1994.

114. Moore, M.A., Cytokine and chemokine networks influencing stem cell proliferation, differentiation, and marrow homing, *J. Cell Biochem., Suppl.*, 38, 29, 2002.
115. Randall, T.D. et al., Expression of murine CD38 defines a population of long-term reconstituting hematopoietic stem cells, *Blood*, 87, 4057, 1996.
116. Uchida, N., Jerabek, L., and Weissman, I.L., Searching for hematopoietic stem cells. II. The heterogeneity of Thy-1.1(lo)Lin(-/lo)Sca1+ mouse hematopoietic stem cells separated by counterflow centrifugal elutriation, *Exp. Hematol.*, 25, 649, 1996.
117. Tamayo, E. et al., A quantitative assay that evaluates the capacity of human stromal cells to support granulomonopoiesis *in situ*, *Stem Cells*, 12, 304, 1994.
118. Pazianos, G., Uqoezwa, M., and Reya, T., The elements of stem cell self-renewal: a genetic perspective, *BioTechniques*, 35, 1240, 2003.
119. Domen, J. and Weissman, I.L., Self-renewal, differentiation or death: regulation and manipulation of hematopoietic stem cell fate, *Mol. Med. Today,* 5, 201, 1999.
120. Weissman, I.L., The road ended up at stem cells, *Immunol. Rev.*, 185, 159, 2002.
121. Christensen, J.L., and Weissman, I.L., Flk-2 is a marker in hematopoietic stem cell differentiation: a simple method to isolate long-term stem cells, *Proc. Natl. Acad. Sci. U.S.A.*, 98, 14541, 2001.
122. Wright, D.E. et al., Physiological migration of hematopoietic stem and progenitor cells, *Science*, 294, 1933, 2001.
123. Wright, D.E. et al., Hematopoietic stem cells are uniquely selective in their migratory response to chemokines, *J. Exp. Med.*, 195, 1145, 2002.
124. Larochelle, A. et al., Identification of primitive human hematopoietic cells capable of repopulating NOD/SCID mouse bone marrow: implications for gene therapy, *Nat. Med.*, 2, 1329, 1996.
125. Dao, M.A. and Nolta, J.A., Immunodeficient mice as models of human hematopoietic stem cell engraftment, *Curr. Opin. Immunol.,* 11, 532, 1999.
126. Clarke D.L. et al., Generalized potential of adult neural stem cells, *Science*, 288, 1660, 2000.
127. Gregg, C.T., Chojnacki, A.K., and Weiss, S., Radial glial cells as neuronal precursors: the next generation?, *J. Neurosci. Res.*, 69, 708, 2002.
128. Li, W., Cogswell, C.A., and LoTurco, J.J., Neuronal differentiation of precursors in the neocortical ventricular zone is triggered by BMP, *J. Neurosci.*, 18, 8853, 1998.
129. Morrison, S.J., Shah, N.M., and Anderson, D.J., Regulatory mechanisms in stem cell biology, *Cell*, 88, 287, 1997.
130. Furukawa, T. et al., rax, Hes1 and notch1 promote the formation of Muller glia by postnatal retinal progenitor cells, *Neuron*, 26, 383, 2000.
131. Gray, G.E. et al., Human ligands of the Notch receptor, *Am. J. Pathol.*, 154, 785, 1999.
132. Liu, C. et al., Engineering of the extracellular matrix: working toward neural stem cell programming and neurorestoration-concept and progress report, *Neurosurgery,* 42, 1154, 2003.
133. Laschinger, M. and Engelhardt, B., Interaction of alpha4-integrin with VCAM-1 is involved in adhesion of encephalitogenic T cell blasts to brain endothelium but not in their transendothelial migration *in vitro*, *J. Neuroimmunol.*, 102, 32, 2000.
134. Petitto, J.M. et al., IL-2 gene knockout affects T lymphocyte trafficking and the microglial response to regenerating facial motor neurons, *J. Neuroimmunol.*, 134, 95, 2003.
135. Pan, W. and Kastin, A.J., Upregulation of the transport system for TNF-alpha at the blood brain barrier, *Arch. Physio. Biochem.*, 109, 350, 2001.
136. Gash, D.M., Zhang, Z., and Gerhardt, G., Neuroprotective and neurorestorative properties of GDNF, *Ann. Neurol.*, 44, S121, 1998.

137. Vinogradov, S.V., Batrakova, E.V., and Kabanov, A.V., Nanogels for oligonucleotide delivery to the brain, *Bioconjug. Chem.*, 15, 50, 2004.
138. Tumbar, T. et al., Defining the epithelial stem cell niche in skin, *Science*, 303, 359, 2004.
139. Driessen, R.L., Johnston, H.M., and Nilsson, S.K., Membrane-bound stem cell factor is a key regulator in the initial lodgment of stem cells within the endosteal marrow region, *Exp. Hematol.*, 31, 1284, 2003.
140. Espana, E.M. et al., Stromal niche controls the plasticity of limba and corneal epithelial differentiation in a rabbit model of recombined tissue, *Invest. Opthalmol.*, 44, 5130, 2003.
141. Wichterle, H. et al., Young neurons from medial ganglionic eminence disperse in adult and embryonic brain, *Nat. Neurosci.*, 2, 461, 1999.
142. Seri, B. et al., Astrocytes give rise to new neurons in the adult mammalian hippocampus, *J. Neurosci.*, 21, 7153, 2001.
143. Hattori, K., Hessig, B., and Rafii, S., The regulation of hematopoietic stem cell and progenitor mobilization by chemokine, SDF-1, *Leuk. Lymphoma*, 44, 575, 2003.
144. Alvarez-Dolado, M. et al., Fusion of bone-marrow-derived cells with Purkinje neurons, cardiomyocytes, and hepatocytes, *Nature*, 425, 968, 2003.
145. Weimann, J.M. et al., Stable reprogrammed heterokaryons form spontaneously in Purkinje neurons after bone marrow transplant, *Nat. Cell Biol.*, 5, 959, 2003.
146. Brazelton, T.R., Nystrom, M., and Blau, H.M., Significant differences among skeletal muscles in the incorporation of bone marrow-deficient cells, *Dev. Biol.*, 262, 64, 2003.
147. Mizumoto, H. and Funatsu, K., Liver regeneration using a hybrid artificial liver support system, *Artif. Organs*, 28, 53, 2004.
148. Alison, M.R., Characterization of the differentiation capacity of rat-derived hepatic stem cells, *Semin. Liver Dis.*, 23, 325, 2003.
149. Hasbe, Y. et al., Plasminogen activator/plasmin system regulates formation of the hepatocyte spheroids, *Biochem. Biophys. Res. Commun.*, 308, 852, 2003.
150. Strain, A.J. et al., Human liver-derived stem cells, *Semin. Liver Dis.*, 23, 373, 2003.
151. Gompe, M., The role of bone marrow stem cells in liver regeneration, *Semin. Liver Dis.*, 23, 363, 2003.
152. Herreros, J. et al., Autologous intramyocardial injection of cultured skeletal muscle-derived stem cells in patients with non-acute myocardial infarction, *Eur. Heart J.*, 24, 2012, 2003.
153. Lu, S.H. et al., Muscle-derived stem cells seeded into acellular scaffolds develop calcium-dependent contractile activity that is modulated by nicotinic receptors, *Urology*, 61, 1285, 2003.
154. Bonner-Weir, S., Stem cells in diabetes: what has been achieved, *Horm. Res.*, 60, Suppl., 10, 2003.
155. Bottaro, D. P., Liebmann-Vinson, A., and Heidaran, M.A., Molecular signaling in bioengineered microenvironments, *Ann. N.Y. Acad. Sci.*, 961, 143, 2002.
156. Green, D. et al., Natural marine sponge fiber skeleton: a biomimetic scaffold for human osteoprogenitor cell attachment, growth and differentiation, *Tissue Eng.*, 9, 1159, 2003.
157. Xiao, Y. et al., Tissue engineering for bone regeneration using differentiated alveolar bone cells in collagen scaffolds, *Tissue Eng.*, 9, 1167, 2003.
158. Radhika, M., Babu, M., and Sehgal, P.K., Cellular proliferation on desamidated collagen matrices, *Comp. Biochem. Physiol. Toxicol. Endocrinol.*, 124, 131, 1999.

159. Brown, R.A. et al., Enhanced fibroblast contraction of 3D collagen lattices and integrin expression by TGf-beta1 and beta3: mechanoregulatory growth factors? *Exp. Cell Res.,* 274, 310, 2002.
160. Levenberg, S. et al., Differentiation of human embryonic stem cells on three-dimensional polymer scaffolds, *Proc. Natl. Acad. Sci. U.S.A.,* 100, 12741, 2003.
161. Pariente, J.L., Kim, B.S., and Atala, A., *In vitro* biocompatibility assessment of naturally derived and synthetic biomaterials using normal human urothelial cells, *J. Biomed. Mater. Res.*, 55, 149, 2001.
162. Iwasaki, Y. et al., Reduction of surface-induced inflammatory reaction on PLGA/MPC polymer blend, *Biomaterials,* 23, 3897, 2002.
163. Qiu, Q.Q., Ducheyne, P., and Ayyaswamy, P.S., New bioactive, degradable composite microspheres as tissue engineering substrates, *J. Biomed. Mater. Res.,* 52, 66, 2000.
164. Zyman, Z.Z. et al., Preparation and properties of inhomogenous hydroxyapatite ceramics, *J. Biomed. Mater. Res.,* 46, 135, 1999.
165. Hou, Q., Grijpma, D.W., and Feijen, J., Preparation of interconnected highly porous polymeric structures by a replication and freeze-drying process, *J. Biomed. Mater. Res.*, 67B, 732, 2003.
166. Yin, Y. et al., Preparation and characterization of macroporous chitosan-gelatin/beta-tricalcium phosphate composite scaffolds for bone tissue engineering, *J. Biomed. Mater. Res.*, 67A, 844, 2003.
167. Tessmar, J., Mikos, A., and Gopferich, A., The use of poly(ethylene glycol)-block-poly(lactic acid) derived copolymers for the rapid creation of biomimetic surfaces, *Biomaterials*, 24, 4475, 2003.
168. Hacker, M. et al., Towards biomimetic scaffolds: anhydrous scaffold fabrication from biodegradable amine-reactive diblock copolymers, *Biomaterials*, 24, 4459, 2003.
169. Takezawa, T., A strategy for the development of tissue engineering scaffolds that regulate cell behavior, *Biomaterials*, 24, 2195, 2003.
170. Williams, M.J. et al., The inner world of cell adhesion: integrin cytoplasmic domains, *Trends Cell Biol.*, 4, 109, 1994.
171. Wehrle-Haller, B. and Imhof, B.A., Integrin-dependent pathologies, *J. Pathol.*, 200, 481, 2003.
172. Tun, T. et al., Effect of growth factors on *ex vivo* bone marrow cell expansion using three-dimensional matrix support, *Artif. Organs*, 26, 333, 2002.
173. Qian, L. and Saltzman, W.M., Improving the expansion and neuronal differentiation of mesenchymal stem cells through culture surface modification, *Biomaterials,* 25, 1331, 2004.
174. Takahashi, Y. and Tabata, Y., Homogenous seeding of mesenchymal stem cells into a nonwoven fabric for tissue engineering, *Tissue Eng.,* 9, 931, 2003.
175. Sistino, J.J., Bioreactors for tissue engineering — a new role for perfusionists? *J. Extra. Corpor. Technol.,* 35, 200, 2003.
176. Li, W.J. et al., Electrospun nanofibrous structure: a novel scaffold for tissue engineering, *J. Biomed. Mater. Res.,* 60, 613, 2002.
177. Sleytr, U.B., Pum, D., and Sara, M., Advances in S-layer nanotechnology and biomimetics, *Adv. Biophys.,* 34, 71, 1997.
178. Yamato, M. et al., Nanofabrication for micropatterned cell arrays by combining electron beam-irradiated polymer grafting and localized laser ablation, *J. Biomed. Mater. Res.,* 67A, 1065, 2003.
179. Vogel, V. and Baneyx, G., The tissue engineering puzzle: a molecular perspective. *Annu. Rev. Biomed. Eng.,* 5, 441, 2003.

180. Webster, T.J. et al., Osteoblast response to hydroxyapatite doped with divalent and trivalent cations, *Biomaterials,* 25, 2111, 2004.
181. Smith, R.L. et al., *In vitro* stimulation of articular chondrocyte mRNA and extracellular matrix synthesis by hydrostatic pressure. *J. Orthop. Res.,* 14, 53, 1996.
182. Smith, R.L. et al., Time-dependent effects of intermittent hydrostatic pressure on articular chondrocyte type II collagen and aggrecan mRNA expression. *J. Rehab. Res. Dev.,* 37, 153, 2000.
183. Ikenoue T. et al., Mechanoregulation of human articular chondrocyte aggrecan and type II collagen expression by intermittent hydrostatic pressure *in vitro. J. Orthop. Res.,* 21, 110, 2003.
184. Lee M.S. et al., Intermittent hydrostatic pressure inhibits shear stress-induced nitric oxide release in human osteoarthritic chondrocytes *in vitro. J. Rheumatol.,* 30, 326, 2003.

15 Microbial Biofilms

Alfred M. Spormann, Kai Thormann, Renee Saville, Soni Shukla, and Plamena Entcheva

CONTENTS

15.1 INTRODUCTION

Microbial biofilms are very broadly defined as a consortium of interacting microorganisms that are associated to an environmental interface, typically a solid–liquid interface, by a self-produced polymeric matrix (Figure 15.1). At almost any aqueous interface, biofilms can be found to have diverse impacts on the surrounding environments. The first recognition of microbial biofilms was made by environmental (sanitary) engineers in the late 1800s.[1] Used as "trickling filters" for biological treatment (Figure 15.2), sewage was applied to coke, cinder, crushed stone, gravel, crushed slag, burned clay, and broken brick, and it was noted that "After more than a year's continuous use, a casual observer on taking the crushed stone in his hand would note no change in it. A microscopic examination, however, would reveal a thin film on the surface of the particles. It is in this film that the bacteria live that cause the purification."[2]

0-8493-1940-4/05/$0.00+$1.50

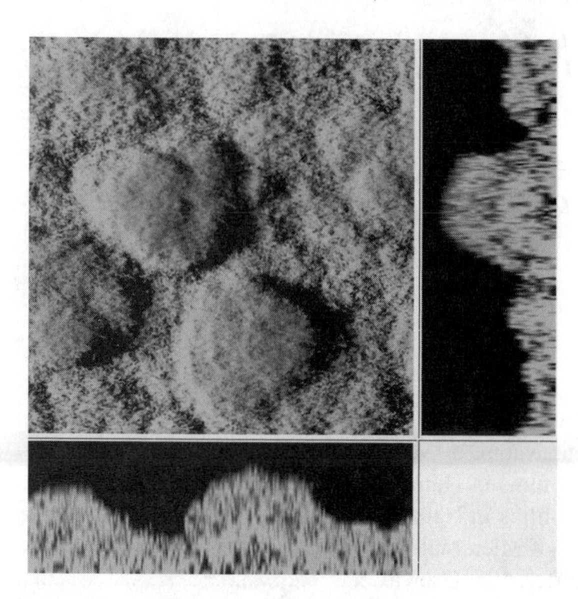

FIGURE 15.1 Biofilm of *Shewanella oneidensis* MR1. Confocal laser scanning microscopic image of a 48-h-old biofilm growing on a lactate minimal medium in a hydrodynamic flow chamber. The image shows the pronounced three-dimensional structure of *gfp*-expressing *S. oneidensis* cells. The central image is a shadow projection of the entire biofilm, reconstructed from CLSM images, and the images on the bottom and the right side are x-z and y-z saggital sections through the biofilm at a selected position, respectively.

15.2 UBIQUITOUS NATURE OF MICROBIAL BIOFILMS

15.2.1 Biofilms in Human Environment

Biofilms are present in numerous locations in the human body, causing both beneficial and adverse effects on human health.[3] Biofilms (e.g., on the skin, the vaginal or intestinal mucosa, and the oral cavity) harbor microbial consortia that often coevolved over the entire life span with each human individual to form stable commensal interactions and provide protection against invasion by allochthonous, potentially pathogenic microbes. The nutrient-rich environment of the human body and the numerous internal and external surfaces in the form of epithelial tissue provide a selective environment for microbial biofilms to develop. Particularly in compartments with a continuous flow of nutrient-rich fluids, microbes can often associate with epithelial surfaces. Such compartments include the cardiovascular system (heart, arteries, and veins), the oral cavity, and the urinary tract. The attenuation of biofilm growth in healthy individuals is probably due to an innate immune response in those compartments, which is sufficient to suppress significant biofilm development. Conversely, nontissue surfaces, for example stainless steel or plastic surfaces, when introduced into the human body in the form of vascular and urinary catheters (Figure 15.3), artificial joints, and prosthetic heart valves and stents, or orthopedic implants, are prone to colonization by microbes, which develop biofilms to a size and organismal complexity that prevents the body of compromised indi-

FIGURE 15.2 Trickling filter for use of wastewater treatment. The pebble stones are covered with microscopically visible microbial biofilms that mediate the degradation of organic and inorganic compounds. (Courtesy of Perry L. McCarty, Stanford University.)

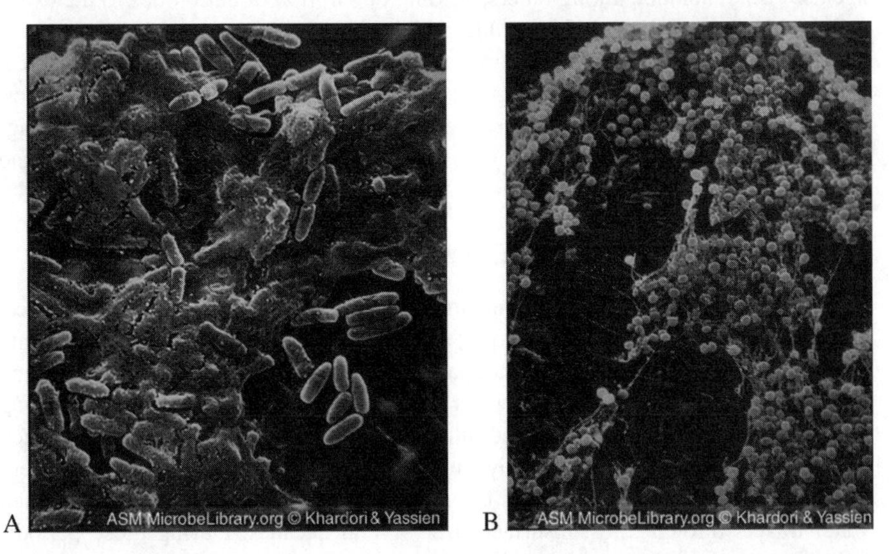

FIGURE 15.3 Microbial biofilms that developed on indwelling vascular catheters. (A) shows scattered and grouped bacteria in microcolonies covered with extracellular polysaccharides adhering to an indwelling vascular catheter. The rod-shaped bacteria are *P. aeruginosa*. (B) shows the spheroid shape of *S. epidermidis* cells, which is the most common cause of bloodstream infections in patients with indwelling vascular catheters. (Courtesy of Mahmoud Yassien and Nancy Khardon.)

viduals from completely eliminating these biofilm populations. As a result, these residual communities can serve as a reservoir for pathogenic microorganisms and can cause chronic infections even in non-immunocompromised patients. In fact, it is estimated that up to 65% of human bacterial infections involve biofilms. In a sense, even a simple infection at a localized region in the human body, where pathogenic microorganisms undergo rapid multiplication leading to high cell density, can be considered as a "biofilm."

Microbial biofilms developing on dental surfaces in the oral cavity are the major cause of dental diseases, including acid-induced dissolution of tooth enamel (caries) as well as periodontal diseases.[4] Oral biofilms form on dental apatite surfaces and/or in periodontal pockets at the tooth–subgingival interface. One of the principal etiological agents of dental caries is *Streptococcus mutans*. However, the microbial community is heterogeneous and includes numerous other microbes, and varies between teeth as well as between solid and soft surfaces.[5,6] Several other human diseases have direct links to biofilms, including kidney stones,[7,8] bacterial endocarditis,[9] and cystic fibrosis.[10,11]

A major significance of medical biofilms is that the biofilm microbes are about 100–1,000 times more resistant to antibiotics and antimicrobial agents than planktonic cells. Therefore, pathogenic microbes, which form or invade biofilms, are extremely difficult to eradicate by such treatment. Consequently, biofilms on medical devices and implants can function as a source of chronic infections. A variety of explanations has been provided to explain the increased resistance to antimicrobial agents. These include, among others, binding or sorption of antibiotics to the extracellular biofilm matrix and, thus, reduced bioavailability of these agents, the expression of specific antibiotic export pumps in the bacteria, and an altered physiological and metabolic state rendering cells (or subpopulations) insensitive to antibiotics. More recently, molecular and genetic evidence in *Pseudomonas aeruginosa* suggested the production of cyclic polyglucans, which appear to specifically bind antibiotics.

15.2.2 Biofilms in Natural Environments

Biofilms developing in nonmedical environments have been also of great importance, especially in the water industry. After the discovery of their usefulness, trickling filters were used heavily in the earlier part of the 1900s for treatment of organic and inorganic waste in municipal wastewater. Today, large bodies of treated water exiting sewage treatment plants are injected into the subsurface in infiltration basins. Such groundwater recharge is particularly important in the freshwater deprived western states of the United States. Driven by gravity and hydraulic gradients, the treated water travels through the subsurface over several decades. During those periods, indigenous soil microbes, immobilized in biofilms of soil and sediment particles, remove the trace nutrients and contaminants, eventually allowing the recovery of clean groundwater. As availability of clean and safe water is becoming more and more important, increased research focus is directed to understanding the interacting metabolic processes occurring in such biofilms. Similar to medical biofilms, some biofilms also have adverse effects in natural and engineered systems. Biofilms that develop in cooling towers, heat exchangers, and the water-exposed surfaces of ships cause serious corrosion damage and increased operating costs due to reduced heat transfer and increased drag, respectively.

15.3 PROPERTIES OF MICROBIAL BIOFILMS

Most insights into the biology of microbial biofilms have been obtained from studies of a few microbial model systems, including *Pseudomonas aeruginosa*, *P. putida*, and *P. fluorescence*, *Vibrio cholerae*, *Staphylococcus aureus*, *S. epidermidis*, *Streptococcus mutants*, and *Escherichia coli*. Biofilms are not planktonic cells. In planktonic systems, such as bioreactors, fermentation systems, open water bodies like rivers and oceans, or even in the laboratory test tube, individual microorganisms are well mixed and exposed to the same physical and chemical constituents of the bulk aqueous phase. Consequently, most members of the microbial population are in a similar metabolic and physiological state and exhibit a similar pattern and level of gene expression. However, the structured environment in a biofilm gives rise to rapidly developing multiscale heterogeneity. Microbial cells are present at high density and in close (i.e., μm) proximity to each other. Biofilm microbes are exposed to steep, self-generated gradients of nutrients including molecular oxygen. Other gradients such as of metabolic end products, toxins, and other compounds also exist. The formation and maintenance of such gradients is the balance of the diffusion rate of chemical compounds from the bulk liquid into the densely packed microbial environment, as well as of the metabolic activities of the microbial community. In addition, extracellular polymeric substances (EPS), which are produced by the biofilm microbes, provide not only a scaffold for positioning and immobilizing cells but also sorption sites for organic and inorganic matter, including metals. Therefore, the metabolic and physiological state of individual cells in the community is a function of a cell's position within the community and can vary on a micrometer scale. Consequently, the densely packed microbial cells in these steep, fluctuating gradients are subjected to substantial environmental stress.

All of the biofilm systems discussed above have several characteristics in common, and these seem to determine the functionality of a biofilm; the only differences may be in the effects that these biofilms cause. The biofilm commonalities include: (1) differential growth, (2) functional heterogeneity, (3) metabolic interactions of community members, (4) signaling interactions, and (5) phase variation and phenotypic adaptation. These characteristics have been the focus of biofilm research and are briefly reviewed in the following sections.

15.3.1 DIFFERENTIAL GROWTH

As for growth of planktonic cells, each microbial cell in a biofilm is powered by a specific metabolic process (e.g., glucose respiration, amino acid fermentation). The catabolic substrate(s) provide the metabolic energy and, thus, the metabolic "driver" for microorganisms to grow and perform their physiological function(s). Consequently, each biofilm environment is more or less selective for a metabolic community of microorganisms that it will accommodate. In general, the extent of growth of biofilm microbes is proportional to the flux of nutrients from the bulk aqueous phase into the biofilm. However, the volumetric rate of substrate utilization varies temporally and spatially in an active biofilm. When biofilms grow from a dispersed layer of single cells to thick layers, the volumetric substrate utilization rate increases over time at a given region in the biofilm. In addition, growth in a biofilm is spatially uneven, and microbes in areas with locally higher biomass will consume substrates at a higher rate than those of a more sparsely populated area. These differential substrate utilization rates will induce changes in local substrate concentrations, which, in turn, will result in altered rates. This simple amplification loop, therefore, is a main factor in the generation of the steep, spatially and temporally varying gradients.

Fusion of the genes encoding the green fluorescence protein, *gfp*, to the growth rate regulated *rrnB*P1 promoter have been used to experimentally assess the heterogeneity of growth activity in hydrodynamic biofilm.[12] In conjunction with unstable GFP protein, it was demonstrated for a benzylalcohol-grown biofilm of *P. putida* that dramatically different growth activities can be observed, which change dynamically with the age of the biofilm (Figure 15.4). In addition, the biofilm architecture depends on the catabolic substrate and is metabolically controlled (Figure 15.5).[13]

15.3.2 Functional Heterogeneity

As a result of the differential growth activity and consequent uneven substrate utilization kinetics, gradients are created and some nutrients may become locally growth limiting. The altered environmental conditions, in turn, induce differential gene expression in the biofilm cells. If the limiting substrate is molecular oxygen (e.g., for aerobic microorganisms), then biofilms will rapidly develop anaerobic "pockets" that can "spread" through the deeper sections of a biofilm. Recent studies on *P. aeruginosa* and *E. coli* biofilms surveying the global gene expression status via DNA microarray analysis revealed that between 1 and 2% of the genes of a genome are differentially expressed in a biofilm.[14,15] The differentially regulated genes include genes of the general stress response and those related to the growth state. These data indicate that growth conditions different from those in the planktonic state exist in the biofilm population, and that at least some fraction of the population is exposed to stress. Notably, also a significant fraction of genes (about 50–100 genes) appears to be specifically expressed under the biofilm conditions tested. Therefore, more genes of a genome are expressed in a biofilm environment than in a planktonic population and, consequently, provide a basis for functional heterogeneity of the overall biofilm.

15.3.3 Metabolic Interactions

Another example of the complexity of a biofilm can be found in metabolic interactions between the community members. Until recently biofilms have been studied mostly in monospecies cultures. However, natural biofilms often consist of communities of functionally and often phylogenetically different microbes. A prototype of metabolic, specifically of competitive and commensal, interactions in biofilms was

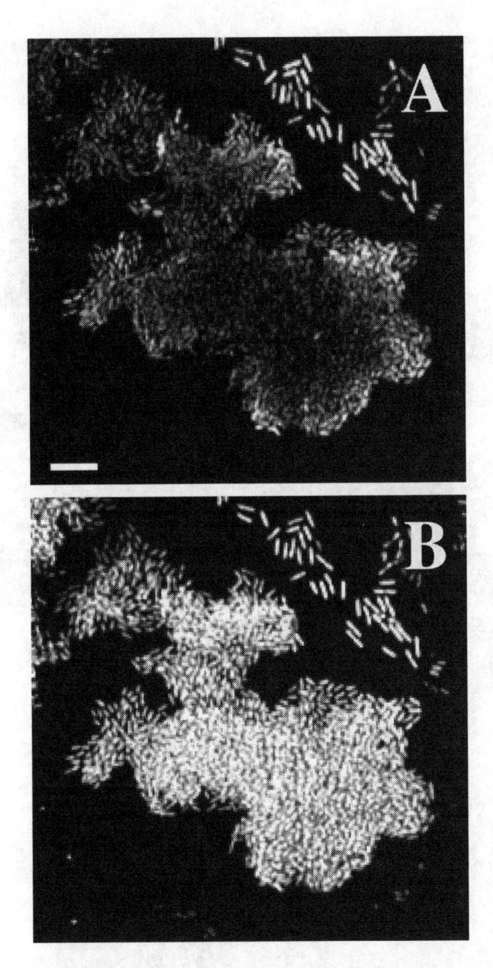

FIGURE 15.4 Differential growth activity in a biofilm. Monolayers of *P. putida* R1 biofilms developing after 20 h of growth in a benzyl alcohol-containing medium in a flow chamber system. (A) *P. putida* R1 cells carry the unstable *gfp*(AAV) under the control of the growth rate-regulated promoter *rrnB*P1. Bright cells indicate cells with high growth activity, whereas dim cells (center of the layer) reveal slow-growing cells. Note that only cells at the perimeter and isolated cells appear to have the highest growth rate. (B) Shown for reference are all *P. putida* cells, stained by FISH with PP986. (From Sternberg, C. et al., Distribution of bacterial growth activity in flow-chamber biofilms. *Appl. Environ. Microbiol.*, 1999, 65(9), 4108–4117. With permission.)

studied in a two-member consortium consisting of an *Acinetobacter* sp. strain C6 and a *Pseudomonas putida* strain[16–18] growing aerobically in the presence of benzyl alcohol as sole electron donor (Figure 15.6; see color insert following page 204). Both microorganisms utilize benzyl alcohol, but *Acinetobacter* mineralizes benzyl alcohol via benzoate as transient intermediate, which can leak into the extracellular environment. *P. putida sp.*, on the other hand, is well adapted for benzoate utilization. Metabolic crossfeeding between the two species via benzoate was demonstrated by

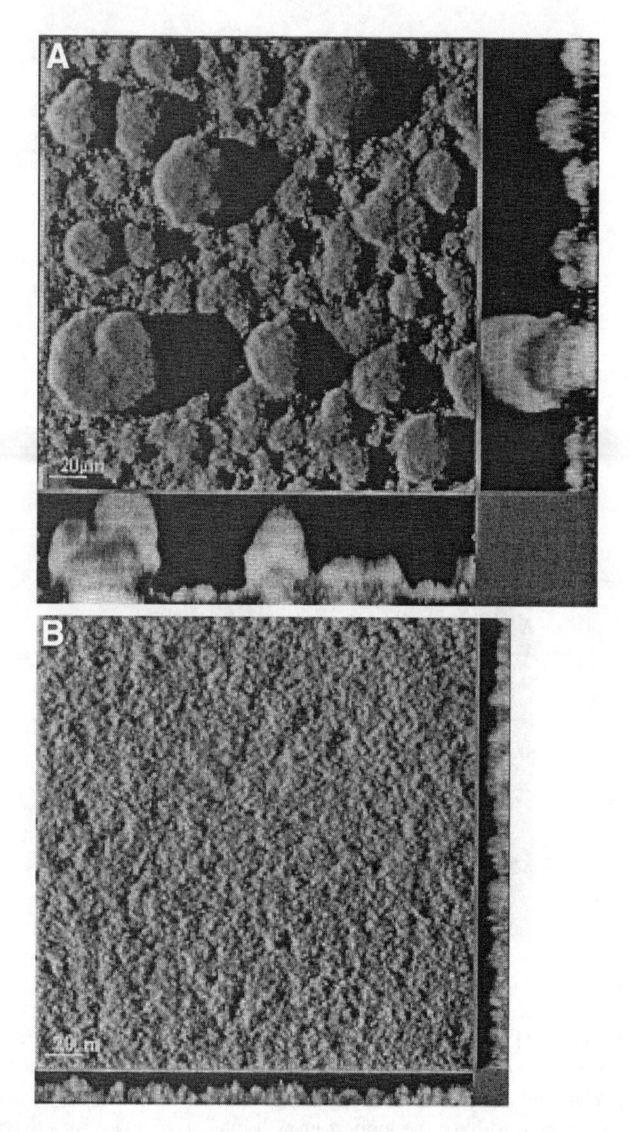

FIGURE 15.5 Metabolic control of biofilm architecture. CLSM images of 5-day-old *P. aeruginosa* biofilms grown in a glucose (A) or citrate (B) containing medium. Glucose-grown biofilms have a pronounced three-dimensional architecture with mushrooms and indents, whereas the biofilm of citrate-grown cells is flat and monotonous. (From Klausen, M. et al., Biofilm formation by *Pseudomonas aeruginosa* wild type, flagella and type IV pili mutants. *Mol. Microbiol.*, 2003, 48(6), 1511. With permission.)

reporter gene fusion experiments with benzylalcohol- and benzoate-inducible promoters (P*u* and P*m*, respectively), using unstable GFP, and revealed specific, dynamically changing architecture of the biofilm community (Figure 15.6). In the unstructured environment of a chemostat, both strains competed with each other for the growth substrate benzyl alcohol, resulting in the dominance of *Acinetobacter* and

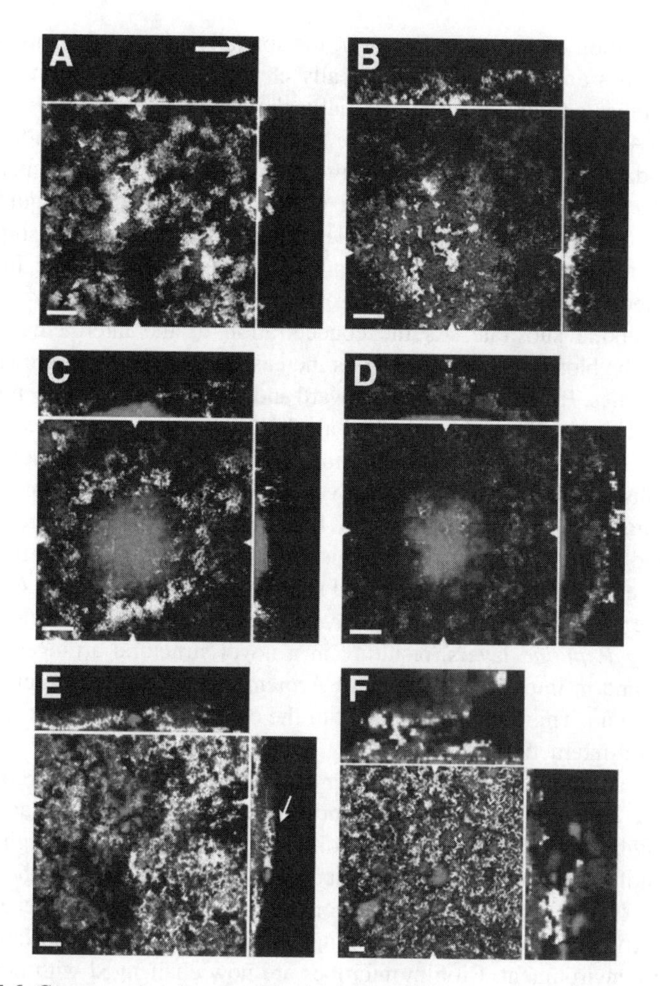

FIGURE 15.6 Commensal and competitive interactions in a two-member biofilm consortium. (See color insert following page 204.) SCLM micrographs show the dynamically changing structural relationships between *Acinetobacter* strain C6 and *P. putida* R1 cells with differential growth activities in a benzyl alcohol containing biofilm. *P. putida* R1 and *Acinetobacter* strain C6 were hybridized with PP986 labeled with CY5 (blue) and ACN449 labeled with CY3 (red), respectively. The actively growing *P. putida* R1 cells can be detected as cells emitting green fluorescence due to the *rrnB*P1-*gfp*^{AGA} fusion inserted in the chromosome of *P. putida* R1. These cells appear as cyan due to the combination of green (GFP) and blue (hybridization). Panels A and B are representative of the biofilm structures observed at day 1 and 2. Panels C and D are examples of the large *Acinetobacter* strain C6 microcolony that developed after 3 days (C) and which later was overgrown by *P. putida* R1 (D). Panels E and F are examples of *Acinetobacter* strain C6 microcolonies (arrow) that were established in the upper part of *P. putida* R1 cell clusters after 2 to 3 days (E), which resulted in production of large macrostructures of associated *P. putida* R1 and *Acinetobacter* strain C6 cells (F). Shown to the right and above the x-y plots are vertical sections through the biofilm collected at the positions indicated by the white triangles. The arrow indicates the direction of flow. Bars, 20 μm. (From Christensen et al., Metabolic commensalism and competition in a two-species microbial consortium *Appl. Environ. Microbiol.*, 2002, 68(5), 2495–2502. With permission.)

the outcompetition of *P. putida*. However, the spatially structured environment of a biofilm provides a platform for dynamically changing, commensal interactions in the two-member consortium. While initially the growth activity of *P. putida* was higher near *Acinetobacter* colonies, these colonies became overgrown by *P. putida* after a few days. Within a short period after, the formation of novel microcolonies of *Acinetobacter* on top of the engulfing *P. putida* and at the bulk liquid–exposed side of the biofilm was observed. The developing structural relationship between these two microorganisms can be rationalized in the following way: In a freshly seeded, mixed-species biofilm, both microbes grow initially with benzyl alcohol as the sole catabolic substrate. As the concentration of the intermediate benzoate increases in the biofilm system, due to its increasing production by growing *Acinetobacter* colonies, *P. putida* cells move toward and engulf the *Acinetobacter* colonies, thereby reducing the transport of benzyl alcohol and oxygen to *Acinetobacter* and, thus, reducing its growth and metabolism. As a consequence, commensal growth and metabolism of *P. putida* on benzoate vanishes temporarily in lieu of competitive benzyl alcohol metabolism, which reduces benzyl alcohol mass transport to the lower layers of the *Acinetobacter* colonies. This diminishes the rate of benzoate release and, thus, the availability of the preferred substrate for *P. putida*. As a consequence of the severe substrate limitation of *Acinetobacter*, microcolonies develop on top of the engulfing *P. putida* layers, resulting in a novel structural arrangement of this consortium and in improved exposure of *Acinetobacter* colonies to benzyl alcohol in the bulk liquid. This local environment of the consortium has now switched again to commensal interactions.

The discussion above illustrates several important aspects of microbial biofilm physiology: (1) the physical-chemical and organismal environments are heterogeneous throughout a biofilm, (2) expression of genes in a biofilm is highly variable within a biofilm and depends on the local environment, and (3) the metabolic interactions of a multispecies biofilm consortium are dynamically changing. All these properties lead to a rapidly evolving structural and functional heterogeneity in a biofilm environment. Biofilm microbes are now challenged with how to adapt best and occupy newly developing ecological niches in a biofilm that form as a function of position and time. It is beginning to emerge that microbes are able to readily adapt phenotypically to such new niches by a variety of mechanisms other than by control of gene expression (see below).

15.3.4 Signaling Interactions

Cell–cell signaling has been recognized as an important mechanism for control of gene expression in bacteria when they are present at high cell density, such as in biofilms. For cooperative microbial behavior, such as feeding on insoluble nutrients, sporulation, swarming, and also pathogenicity, it seems particularly advantageous for microbes to synchronize the behavior of individual cells with members of the same species in a community through gene expression that takes into account the presence and physiological state of all members of the species in a particular environment. Originally discovered as "autoinduction" of light emission in bioluminescent bacteria and subsequently termed "quorum sensing,"[19] low-molecular-weight autoinducer, or signaling, molecules (AI) are produced by individual cells at low rates and excreted into the extracellular environment. As the rate of AI accumulation is proportional to the number of microbes in a similar physiological state, the extracellular concentration of the AI reflects the density of all microbial species members in a spatially constrained environment. At a typically nanomolar threshold concentration, transcription of specific sets of genes (regulons) is initiated in each cell. The regulons are involved in diverse functions (see below), where their coordinated expression in the entire population seems opportune, resulting in cooperative behavior of the entire species.

Quorum sensing in Gram-negative microorganisms has been shown to involve acyl-homoserine lactones (AHL, or AI-1) as signals and a *luxR*-type transcriptional regulator as the response element.[20,21] Signal specificity is conferred by the acyl side chain of the AHL and a cognate LuxR receptor protein. AHL-based cell–cell signaling has been shown to be involved in a diverse set of physiological responses (for review, see Miller and Bassler[21]) including bioluminescence, protease and exoenzyme production, plasmid transfer, antibiotic, cyanide, and pigment production, capsular polysaccharide production, chromosome replication, biofilm formation, rhamnolipid synthesis, and swarmer cell differentiation.

An interspecies signaling system was identified first in *Vibrio harveyi*, but has subsequently been demonstrated or suspected to operate in more than 50 species, including Gram-positive microbes. This signaling molecule (AI-2) has no known structural diversity and is a borofuranone. This important characteristic, together with the abundant presence of AI-2 in diverse microbial species, suggests that the AI-2 cell–cell signaling system is an inter- rather than an intra-species communication system. The AI-2 signal relay is carried out by two-component signal transduction proteins.[22]

In addition to the interspecies AI-2 signaling, Gram-positive bacteria have evolved a species-specific cell–cell signaling system. The system is based on small, 5–17 amino acid–containing peptides as signaling molecules and on a two-component signal-transduction protein pair for signal reception. The peptide-dependent quorum sensing regulons include genes required for competence in *S. pneumoniae, S. gordonii,* and *S. mutans*,[23] sporulation in *Bacillus subtilis*,[24] antibiotic biosynthesis in *Lactobacillus lactis*, and induction of virulence factors in *Staphylococcus aureus*.[25]

Cell–cell signaling has been found to be important for biofilm formation. In *P. aeruginosa*, biofilm architecture, as well as expression of virulence factors, was found to be controlled by AHL signaling,[26] and AHL was demonstrated to be present in sputum of cystis fibrosis patients, where *P. aeruginosa* is believed to exist as biofilm.[27] Also in *S. mutans*, cell–cell signaling affects biofilm formation as well as competence for DNA uptake.[28] Interestingly, none of the mutations impairing cell–cell signaling completely abolished biofilm formation but rather altered biofilm architecture. AI-2 signaling is required for formation of a mixed species biofilm of *S. gordonii* and *Porphyromonas gingivalis.*[29]

Regardless of whether cell–cell signaling is required directly or indirectly for biofilm formation, the mere fact that microbes are confined to high density in biofilm environments is sufficient to facilitate cell–cell signaling and to induce quorum sensing–controlled genes in the biofilm. This, then, enables the biofilm community to express a physiological (quorum sensing–dependent) profile that is different from that of planktonic cells. That quorum sensing *per se* is not essential for biofilm formation is exemplified in *V. cholerae*. Although at least three quorum sensing systems (including AI-1 and AI-2) are involved in biofilm maturation by controlling EPS production, biofilm formation proceeds only in the absence of cell–cell signaling.[30]

15.3.5 Phase Variation and Phenotypic Adaptation

The above-described physiological traits of biofilm cells are due to differential control of gene expression in response to a dynamically changing biofilm environment. Each biofilm cell can, in principle, respond to a particular environmental condition with the same pattern of gene expression (number of controlled genes, expression level) as all other cells. Because of the specific architecture of the regulatory circuitry in a microorganism, this environmental control of gene expression typically confers an improved adaptation to each microbe in the immediate surrounding biofilm environment. However, in addition to this phenotypic adaptation at the level of the entire population, microorganisms can engage in adaptive strategies that lead to diversification of only a subpopulation of cells.[31] The best-studied mechanism to generate diversification is known as phase variation.[32] Phase variation is a reversible, high-frequency ($>10^{-5}$ per generation) switching between two phenotypic states, the ON or OFF state. Historically, phase variation was discovered in and has been subsequently studied predominantly using pathogenic microorganisms. The immune response of a host to bacterial invasion induces a rapidly changing, hostile environment to which microbes have to respond. As phase variation provides a fast response to evade the host's immune response, the targets of phase variation include typically cellular surface structures, such as flagella, pili, surface antigens, and lipooligosaccharides. In contrast to phenotypic adaptation through gradual control of gene expression, phase switching enables a microbial subpopulation to "stochastically choose" preprogrammed physiological routes. The phase switching is an *all or none* response, where cells are physiologically either in one (e.g., the "ON") or the other (e.g., "OFF") state. Both states have distinguishable, mutually exclusive physiological consequences (e.g., either one or another surface antigen is expressed, or pili are in either one or another glycosylation state).[33]

The molecular basis for phase variation includes genetic alterations, such as DNA inversion, homologous recombination,[34] transposon and IS insertions,[35–37] slipped strand repair,[38] and modification of DNA such as DNA methylation.[39–42] Interestingly and importantly for the biofilm context, the switching frequency can be controlled by environmental factors, such as temperature or media composition.[32,43–45]

Phase variation is also an important adaptive mechanism for biofilm microbes to respond to the challenge of life on a surface. Microorganisms, such as *V. cholerae*, *Xanthomonas oryzae*, *Staphylococcus epidermidis*, *Pseudomonas atlantica*, *Vibrio parahaemolyticus*, as well as *P. aeruginosa* and others, can phase switch and generate subpopulations which apparently carry an improved persistence and/or fitness in the biofilm environment.[31] Often, the switch is accompanied by an increased biosynthesis of extracellular polysaccharides, pili, adhesions, or other factors, which confer an improved cellular adhesion. Such variants, for example in *V. cholerae* or *P. aeruginosa*, can easily be detected by an altered colony morphology (Figure 15.7).[46] Interestingly, there seems to be a link between rough small-colony variants, biofilm formation, and the resistance to antibiotics. For *P. aeruginosa*, it was recently demonstrated that rough small-colony variants can be generated in liquid cultures by phase variation in the absence of antibiotics. These variants, when tested for biofilm formation, showed an enhanced biofilm phenotype and increased resistance to several antibiotics.[45] Notably, the switching frequency is modulated by environmental cues in the same way as is the resistance to an antibiotic: the higher the switching frequency to rough small-colony variant, the more those variants are resistant to kanamycin. The switching in *P. aeruginosa* is controlled by a two-component response regulator, PvrR, which primarily induces the reversion from the rough small-colony variant phenotype to wild type. A similar regulation of switching frequency by a regulator was also previously observed in *P. tolaasii*.[47]

The discussion above illustrates that phase switching is one important mechanism found in many phylogenetically diverse bacteria to adapt rapidly to a highly selective biofilm environment. It further emphasizes that the heterogeneously structured biofilm environment is highly selective for certain variants, and probably only a small fraction of the arising variants have been recognized because of their easily detectable variant colony morphology.[31]

In addition to acquiring a preprogrammed phenotype by phase switching, it is becoming apparent that microbes can also induce diversification by generating random mutants at either a constant high frequency (mutator phenotype) or by only transiently elevating the mutation frequency (induced mutations).[31] Microbial mutator strains that carry mutations in the *mutD* or *mutSLH* genes have mutation frequencies that are about two orders of magnitude above wild type. This defect is due to a lack in the proofreading function of DNA polymerase or a defective mismatch repair system, respectively. As the resulting mutations are random and typically irreversible, a larger degree of diversity is generated in the population by this mechanism, compared to phase variation, and facilitates the arising of subpopula-

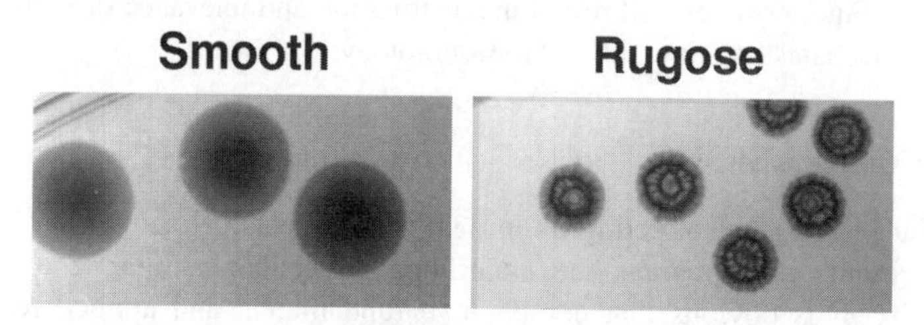

FIGURE 15.7 Colonies of smooth and rugose variants of *V. cholerae* O1 El Tor. Phase variation is a mechanism in *V. cholerae* to shift subpopulations between both phenotypes. The rugose phenotype is associated with carrying an overproduction of EPS and enhanced biofilm formation, whereas smooth variants form less dense and more antibiotic sensitive biofilms. (From Yildiz, F.H. and G.K. Schoolnik, *Vibrio cholerae* O1 El Tor: Identification of a gene cluster required for the rugose colony type, exopolysaccharide production, chlorine resistance, and biofilm formation. *Proc. Natl. Acad. Sci. U.S.A.*, 1999, 30(96), 4028–4033. With permission.)

tions with better fitness in local biofilm environment. In planktonic environments (i.e., liquid growth) mutator strains become less fit over time, due to the high cost of accumulating deleterious mutations. The constantly changing biofilm environment, such as in the lung of cystic fibrosis patients, however, seems to instead select for mutator biofilm strains. Notably, 36% of all CF patients carry *P. aeruginosa* mutator strains, and 43% of all *P. aeruginosa* isolates from CF patients have a mutator phenotype.[48] Apparently, the ecological cost of a mutator phenotype is being outweighted by the enormous selective pressure in the ever-changing environment of the lung of a CF patient.

Because of the mode of action of the *mut* genes, cell growth is required for mutations to be generated by the mutator strain mechanism. In contrast, stress-inducible mechanisms generate mutations only transiently for as long as the stress is active.[49,50] In fact, there is an inverse correlation between stationary phase inducible mutator activity and the mutator phenotype: a strain is either a "good" inducible mutator or a "good" constitutive mutator. The inducible (or adaptive, or stress-inducible) mutation mechanism does not require replication, but depends on RecA as a DNA damage and repair protein, which also activates the SOS response system, as well as DNA pol II, pol I, or pol IV (error-prone DNA polymerases DinB). Notably, in aging colonies but not in starving liquid cultures, inducible mutations resulted from induction of the SOS system and were found to be dependent on cAMP synthesis. This mechanism is described as ROSE (resting organisms in a structured environment). Aging colonies can be approximated as a form of "biofilm," and cellular cAMP concentration increases under starvation, which is required for control of operons regulated by catabolite repression. These observations demonstrate that starvation *per se* is not sufficient to transiently induce mutations, but rather that the structure of the environment holds essential cues that trigger generation of mutations. Although it is currently unknown if a ROSE-type mechanism occurs in a biofilm environment, such transient mutability may be advantageous for biofilm populations; when the molecular mode of mutation generation is uncoupled from the resulting mutant phenotype, generation of diversity can occur without cells accruing the high costs of a mutator phenotype. Future studies will reveal the distribution and relevance of such inducible mutation mechanisms in microbial biofilm biology.

15.4 CONCLUSION

Microbial biofilms are affecting us in nearly all aspects of our lives. We are now only beginning to appreciate the underlying molecular, cellular, and ecological complexity. It is obvious that comparative fundamental and applied research on microbial biofilms using mono- or mixed species, natural or laboratory biofilms, and molecular, genomic, and proteomic approaches, will be important to uncover the biological mechanism at work. Moreover, these insights are certain to provide new opportunities to positively or negatively control microbial biofilms.

ACKNOWLEDGMENT

We thank Soeren Molin for many helpful discussions.

REFERENCES

1. Hazen, A., The Lawrence experiments on the purification of sewage in 1890 and 1891. *Engineering News-Record*, 1892, 28, 559–560.
2. Library, I.C.S.R., *Oxidation in Porous Material and at High Rates, in Water Supply, Sewerage, Purification of Water, Sewage Purification and Disposal, Irrigation*, International Textbook Company, Scranton, Pennsylvania, 1908, p. 125.
3. Costerton, J.W. et al., Microbial biofilms. *Annu. Rev. Microbiol.*, 1995, 49, 711–745.
4. Kolenbrander, P.E., Oral microbial communities: biofilms, interactions, and genetic systems. *Annu. Rev. Microbiol.*, 2000, 54, 413–437.
5. Kolenbrander, P.E. et al., Spatial organization of oral bacteria in biofilms. *Methods Enzymol.*, 1999, 310, 322–332.
6. Palmer, R.J., Jr. et al., Mutualism versus independence: strategies of mixed-species oral biofilms *in vitro* using saliva as the sole nutrient source. *Infect. Immun.*, 2001, 69(9), 5794–5804.
7. Nickel, J.C. et al., Ultrastructural microbiology of infected urinary stone. *Urology*, 1986, 28(6), 512–515.
8. Nickel, J.C. et al., Bacterial colonization of intestinal urinary conduit diversion: a morphologic and bacteriologic experimental study. *Can. J. Surg.*, 1987, 30(4), 273–277.
9. Freedman, L.R., The pathogenesis of infective endocarditis. *J. Antimicrob. Chemother.*, 1987, 20 Suppl. A, 1–6.
10. Koch, C. and N. Hoiby, Pathogenesis of cystic fibrosis. *Lancet*, 1993, 341(8852), 1065–1069,
11. Govan, J.R. and V. Deretic, Microbial pathogenesis in cystic fibrosis: mucoid *Pseudomonas aeruginosa* and *Burkholderia cepacia*. *Microbiol. Rev.*, 1996, 60(3), 539–574.
12. Sternberg, C. et al., Distribution of bacterial growth activity in flow-chamber biofilms. *Appl. Environ. Microbiol.*, 1999, 65(9), 4108–4117.
13. Klausen, M. et al., Involvement of bacterial migration in the development of complex multicellular structures in *Pseudomonas aeruginosa* biofilms. *Mol. Microbiol.*, 2003, 50(1), 61–68.
14. Whiteley, M. et al., Gene expression in *Pseudomonas aeruginosa* biofilms. *Nature*, 2001, 413(6858), 860–864.
15. Schembri, M.A., K. Kjaergaard, and P. Klemm, Global gene expression in *Escherichia coli* biofilms. *Mol. Microbiol.*, 2003, 48(1), 253–267.
16. Moller, S. et al., Activity and three-dimensional distribution of toluene-degrading *Pseudomonas putida* in a multispecies biofilm assessed by quantitative *in situ* hybridization and scanning confocal laser microscopy. *Appl. Environ. Microbiol.*, 1996, 62(12), 4632–4640.
17. Moller, S. et al., *In situ* gene expression in mixed-culture biofilms: evidence of metabolic interactions between community members. *Appl. Environ. Microbiol.*, 1998, 64(2), 721–732.
18. Christensen, B.B. et al., Metabolic commensalism and competition in a two-species microbial consortium. *Appl. Environ. Microbiol.*, 2002, 68(5), 2495–2502.
19. Fuqua, W.C., S.C. Winans, and E.P. Greenberg, Quorum sensing in bacteria: the LuxR-LuxI family of cell density-responsive transcriptional regulators. *J. Bacteriol.*, 1994, 176(2), 269–275.
20. Fuqua, C. and E.P. Greenberg, Listening in on bacteria: acyl-homoserine lactone signalling. *Nat. Rev. Mol. Cell Biol.*, 2002, 3(9), 685–695.

21. Miller, M.B. and B.L. Bassler, Quorum sensing in bacteria. *Annu. Rev. Microbiol.*, 2001, 55, 165–199.
22. Federle, M.J. and B.L. Bassler, Interspecies communication in bacteria. *J. Clin. Invest.*, 2003, 112(9), 1291–1299.
23. Cvitkovitch, D.G., Y.H. Li, and R.P. Ellen, Quorum sensing and biofilm formation in Streptococcal infections. *J. Clin. Invest.*, 2003, 112(11), 1626–1632.
24. Lazazzera, B.A., Quorum sensing and starvation: signals for entry into stationary phase. *Curr. Opin. Microbiol.*, 2000, 3(2), 177–182.
25. Kleerebezem, M. and L.E. Quadri, Peptide pheromone-dependent regulation of antimicrobial peptide production in Gram-positive bacteria: a case of multicellular behavior. *Peptides*, 2001, 22(10), 1579–1596.
26. Davies, D.G. et al., The involvement of cell-to-cell signals in the development of a bacterial biofilm. *Science*, 1998, 280(5361), 295–298.
27. Singh, P.K. et al., Quorum-sensing signals indicate that cystic fibrosis lungs are infected with bacterial biofilms. *Nature*, 2000, 407(6805), 762–764.
28. Li, Y.H. et al., A quorum-sensing signaling system essential for genetic competence in *Streptococcus mutans* is involved in biofilm formation. *J. Bacteriol.*, 2002, 184(10), 2699–2708.
29. McNab, R. et al., LuxS-based signaling in *Streptococcus gordonii*: autoinducer 2 controls carbohydrate metabolism and biofilm formation with *Porphyromonas gingivalis*. *J. Bacteriol.*, 2003, 185(1), 274–284.
30. Hammer, B.K. and B.L. Bassler, Quorum sensing controls biofilm formation in *Vibrio cholerae*. *Mol. Microbiol.*, 2003, 50(1), 101–104.
31. Kirkelund Hansen, S. and S. Molin, Temporal segregation: succession in biofilms, in *Microbial Evolution*, R.V. Miller and M. Day, Eds. 2004, ASM Press, Washington, DC.
32. Henderson, I.R., P. Owen, and J.P. Nataro, Molecular switches—the ON and OFF of bacterial phase variation. *Mol. Microbiol.*, 1999, 33(5), 919–932.
33. Power, P.M. et al., Genetic characterization of pilin glycosylation and phase variation in *Neisseria meningitidis*. *Mol. Microbiol.*, 2003, 49(3), 833–847.
34. Sinha, H., A. Pain, and K. Johnstone, Analysis of the role of recA in phenotypic switching of *Pseudomonas tolaasii*. *J. Bacteriol.*, 2000, 182(22), 6532–6535.
35. Rajeshwari, R. and R.V. Sonti, Stationary-phase variation due to transposition of novel insertion elements in *Xanthomonas oryzae pv. oryzae*. *J. Bacteriol.*, 2000, 182(17), 4797–4802.
36. Ziebuhr, W. et al., A novel mechanism of phase variation of virulence in *Staphylococcus epidermidis*: evidence for control of the polysaccharide intercellular adhesin synthesis by alternating insertion and excision of the insertion sequence element IS256. *Mol. Microbiol.*, 1999, 32(2), 345–356.
37. Perkins-Balding, D., G. Duval-Valentin, and A.C. Glasgow, Excision of IS492 requires flanking target sequences and results in circle formation in *Pseudoalteromonas atlantica*. *J. Bacteriol.*, 1999, 181(16), 4937–4948.
38. Torres-Cruz, J. and M.W. van der Woude, Slipped-strand mispairing can function as a phase variation mechanism in *Escherichia coli*. *J. Bacteriol.*, 2003, 185(23), 6990–6994.
39. Wallecha, A. et al., Dam- and OxyR-dependent phase variation of agn43: essential elements and evidence for a new role of DNA methylation. *J. Bacteriol.*, 2002, 184(12), 3338–3347.
40. van der Woude, M., B. Braaten, and D. Low, Epigenetic phase variation of the pap operon in *Escherichia coli*. *Trends Microbiol.*, 1996, 4(1), 5–9.

41. Braaten, B.A. et al., Methylation patterns in pap regulatory DNA control pyelonephritis-associated pili phase variation in *E. coli*. *Cell*, 1994, 76(3), 577–588.
42. Saunders, N.J., E.R. Moxon, and M.B. Gravenor, Mutation rates: estimating phase variation rates when fitness differences are present and their impact on population structure. *Microbiology*, 2003, 149(Pt 2), 485–495.
43. Gally, D.L. et al., Environmental regulation of the fim switch controlling type 1 fimbrial phase variation in *Escherichia coli* K-12: effects of temperature and media. *J. Bacteriol.*, 1993, 175(19), 6186–6193.
44. Hallet, B., Playing Dr Jekyll and Mr Hyde: combined mechanisms of phase variation in bacteria. *Curr. Opin. Microbiol.*, 2001, 4(5), 570–581.
45. Drenkard, E. and F.M. Ausubel, *Pseudomonas* biofilm formation and antibiotic resistance are linked to phenotypic variation. *Nature*, 2002, 416(6882), 740–743.
46. Yildiz, F.H. and G.K. Schoolnik, *Vibrio cholerae* O1 El Tor: Identification of a gene cluster required for the rugose colony type, exopolysaccharide production, chlorine resistance, and biofilm formation. *Proc. Natl. Acad. Sci. U.S.A.*, 1999, 30(96), 4028–4033.
47. Grewal, S.I., B. Han, and K. Johnstone, Identification and characterization of a locus which regulates multiple functions in *Pseudomonas tolaasii*, the cause of brown blotch disease of *Agaricus bisporus*. *J. Bacteriol.*, 1995, 177(16), 4658–4668.
48. Oliver, A. et al., High frequency of hypermutable *Pseudomonas aeruginosa* in cystic fibrosis lung infection. *Science*, 2000, 288(5469), 1251–1254.
49. Rosenberg, S.M., Evolving responsively: adaptive mutation. *Nat. Rev. Genet.*, 2001, 2(7), 504–515.
50. Bjedov, I. et al., Stress-induced mutagenesis in bacteria. *Science*, 2003, 300(5624), 1404–1409.

16 Nanobiology in Cardiology and Cardiac Surgery

Theo Kofidis and Robert C. Robbins

CONTENTS

0-8493-1940-4/05/$0.00+$1.50

Cardiovascular morbidity and mortality account for a major portion of the socioeconomic deficit in industrialized countries with increasing trends in developing countries.[1] The diversity and complexity of heart disease and the cost and duration aspects of treatment trigger extensive research activity to develop novel therapeutic concepts for the future. Therefore, atherosclerosis, ischemic heart disease, and heart failure constitute vast fields for the integration of new technologies. Although nanobiology and nanotechnology[2] have not been integrated in the clinical practice in cardiology or cardiac surgery to date, they hold great promise for advanced future applications that will reshape cardiovascular medicine. The main domain for nanobiology in cardiology is expected to be interventional drug delivery and the promotion of angiogenesis for the salvage of ischemic tissue. Cardiac surgery will primarily benefit from this advancing technology through applications in tissue engineering and tissue- or cell transfer. However, there is a plethora of pathological entities in contemporary clinical practice that might constitute an immediate field of action for novel nanotechnological applications. These should be subdivided into diagnostic applications and therapeutic applications. Novel diagnostic modalities, based upon nanotechnological accomplishments, will help identify emerging disease processes at a very early stage, accurately capture the pathological process in the three-dimensional space within the affected structure, and help guide specific therapeutic applications. The spectrum of cardiovascular pathologies, which may derive enormous therapeutical benefit from the new discoveries in the sector of nanobiology and nanotechnology, includes ischemia, reperfusion-associated conditions, organ undersupply with nutrients and oxygen, acute cardiovascular trauma and wound healing, and atherosclerosis and its hemodynamic consequences. Furthermore, prevention of the progression of a pathological condition will be lifted to a new level of efficacy, by guided antiproliferative drug delivery and miniaturized “detectives” such as nanorobots and nanomachines.

The potential applications for the daily cardiovascular surgery practice are vast and might redefine the way both acute and chronical conditions are being approached. These possibilities include the therapy of acute bleeding disorders in the perioperative phase, treatment of perforations and ruptures in the great vessels, better control of the healing process following trauma or major cardiac surgery, local hemostatic control without counteracting indispensable systemic anticoagulation, reduction of brain ischemia time during major reconstructive procedures on the ascending aorta and aortic arch, and smoother separation of the patient from cardiopulmonary bypass. With the advent of tissue engineering and the current progress in the field of stem cell transfer, a whole new chapter in the treatment of end-stage myocardial disease is being written. The ability to repair ischemic or dilated heart muscle by targeted cell and tissue transfer appears to be a milestone in cardiovascular treatment. Nanobiology and nanotechnology will be an indispensable part in this process, facilitating a more targeted administration and engraftment of cells and *in vitro* production of bioartificial tissues with the highest possible fidelity.

For the better understanding of the potential that nanobiology and nanotechnology harbor for the specific applications above, we will survey the most recent developments which could be of significance for our clinical practice, as they are introduced by the leading experts of this advancing scientific field.

PART I: DIAGNOSTIC APPLICATIONS OF NANOBIOLOGY AND NANOTECHNOLOGY

16.1 MOLECULAR IMAGING OF ANGIOGENESIS

Molecular imaging of angiogenic vasculature represents an attractive opportunity for noninvasive early detection in conditions such as atherosclerosis and neoplastic disease. Novel molecular imaging techniques using ligand-directed paramagnetic nanoparticles emerge as promising tools to detect, characterize, and quantify neoangiogenesis by targeting $\alpha_v\beta_3$-integrins,[3,4] well-recognized biomarkers of angiogenesis. In 2003, Patrick M. Winter et al. reported the first *in vivo* use of a magnetic resonance (MR) molecular imaging nanoparticles to sensitively detect and spatially characterize neovascularity induced by implantation of the rabbit Vx-2 tumor using a common clinical field strength (1.5T).[3] Later on, Winter et al. formulated paramagnetic nanoparticles for *in vivo* MRI imaging of sparse molecular epitopes of angiogenesis associated with atherosclerosis, thereby targeting $\alpha_v\beta_3$-integrins to permit noninvasive molecular imaging of plaque-associated angiogenesis in hyperlipidemic rabbits.[4] In addition to improving and facilitating early diagnosis, these target-specific nanoparticles might give rise to a variety of noninvasive monitoring opportunities for site-directed drug delivery and progression of the disease during treatment.

16.2 CELLULAR IMAGING

Fluorescent microscopy facilitates functional and molecular studies in living cells, but the standard use of organic labels in this technique encounters several limitations for imaging long-term cellular events. Fluorescent semiconductor nanocrystals, otherwise known as quantum dots (QDs) are inorganic fluorophores made up of substances such as cadmium selenide (CdSe) and present significant advantages over organic fluorophores by combining the advantages of high photobleaching threshold, good photostability, and readily tunable spectral properties.[5,6] These novel optical properties facilitate the long-term tracking of QD-labeled cells and molecules and the capacity to tag and detect simultaneously several different populations of cells and molecules (multicolor imaging).[5-9] Additionally, due to their inorganic nature, QDs have proven to be retained in living cells without causing any cytotoxic effects.[7] Consequently this approach is likely to find widespread application *in vivo*. Recently, some biological applications for these nanocrystal/protein conjugates have been reported and include imaging of small vasculature[10] and tracking QD-labeled cancer cells in living animals.[11]

16.3 ARTIFICIAL MOLECULAR RECEPTORS

There is a new methodology called molecular imprinting that could have various applications in medical science, particularly for the construction of biosensors and chemosensors. Molecular imprinting is used for creating artificial receptors that are replicas of antibodies and cell receptors, and is based on the concept of creating substrate-specific recognition sites in polymeric matrices by means of template polymerization. In the Departments of Bioengineering and Chemical Engineering at the University of Washington in Seattle, Shi and Ratner[12] generated polymer surfaces using a radiofrequency-plasma glow-discharge process to imprint a polysaccharide-like film with nanometer-sized pits in the shape of several proteins of great biological relevance such as albumin, fibrinogen and immunoglobulins. Others[13,14] have used the molecularly imprinted polymers approach to explore clinical applications for artificial receptors such as controlled drug release, drug monitoring devices, and *in vivo* monitoring of cell-based therapies.

16.4 FLUID ACCELERATION SENSORS

With nanotechnolgy and the development of highly sensitive but small sensing systems biological fluid accelerations may also be inferred from measurements of spatial pressure gradients and provide a summary for the ~19,000 km of human arteriovenous vasculature (mostly capillaries), as Robert Freitas Jr. reports in his book:[15] "Micron-scale medical nanorobots traveling across blood vessel surfaces could use pressure sensors capable of 10^{-6} atm resolution to infer local blood flow accelerations to an accuracy of ±0.01 g from sequential measurements taken 1 mm apart." What sounds like science fiction is actually a reality thanks to the efforts of scientists such as those at the University of Pittsburgh in the John A. Swanson Micro and Nanotechnology Laboratory. They are committed to the fabrication of nanoscale micro-electro-mechanical systems (MEMS) for mechanical and biomedical applications such as pressure and acceleration sensors and devices that dispense drugs and perform analysis on tissue and body fluids.[16] Moreover, Hsiai et al.,[17] at USC in Los Angeles, fabricated MEMS similar to endothelial cells in order to link real-time shear stress with monocyte/EC interactions in an oscillatory flow environment, imitating the turbulent separation point at arterial bifurcations, hence providing high spatiotemporal resolution to link biomechanical forces on the microscale with large-scale physiology.

16.5 THERAPEUTIC APPLICATIONS

16.5.1 Targeted Antiproliferative Drug Delivery/ Prevention of Restenosis after Percutaneous Revascularization

A serious complication of coronary angioplasty is restenosis which entails proliferation and migration of vascular smooth muscle cells (VSMCs) from the media to the intima, increased synthesis of extracellular matrix, and remodeling.[18] This complication can be ameliorated by local deposition and prolonged release of appropriate antiproliferative agents. In some studies carried out by Lanza et al.[19,20] it has been demonstrated that tissue factor-targeted nanoparticles specifically bind with high avidity to smooth muscle cell membranes *in vivo*, and can penetrate and bind stretch-activated vascular smooth muscles in the media after balloon injury. This concept of VSMC-targeted nanoparticles has been studied by the mentioned author as a drug-delivery platform for the prevention of restenosis after angioplasty; he targeted nanoparticles to VSMC surface epitopes, thereby increasing nanoparticle-based antiproliferative effectiveness, particularly for the drugs paclitaxel and doxorubicin.[21] These targeted paramagnetic nanoparticles had the additional potential for high-resolution MR molecular imaging of vascular injury in concordance with local dosimetry of antiproliferative therapy.

16.5.2 Smart Drugs

In addition to revolutionary diagnostic applications, nanotechnology will also impact treatment, since therapeutic agents can also be carried and administered by target-specific nanoparticles. Medical nanomaterials may also include "smart drugs" that become medically active only under specific conditions, in contrast to the conventional ones that display limited control of drug delivery rate and precision in reaching the target area. An additional benefit of these submicron systems is that they present a higher intracellular uptake than microsized particles.[22] Recently, different groups have explored this new treatment modality, and have found a wide spectrum of potential applications, such as drugs that unleash their antibiotic effect only in the presence of an infection;[23] encapsulation of sexual hormones in degradable polymer microparticles that are injectable;[24] release of drug contained in the microreservoirs by externally applied electromagnetic waves focused on the target volume;[25] creation of a nanodevice composed of oligonucleotide DNA covalently attached to titanium dioxide nanoparticles with the ability to target, bind, and cleave DNA;[26] and nanoporous microsystems to avoid rejection in the absence of chronic immunosuppression for islet cell replacement for the treatment of Type I diabetes.[27]

16.5.3 Nanorobotics

In the nanomedical era, emerging electro-mechanical devices will become an important therapeutic tool. These advanced nanomachines or "nanobots" will have the capacity to perform computations; sense and respond to environmental stimuli; move; communicate and cooperate; perform molecular assembly; self-repair; and replicate.[28]

Current research focuses on exploring flotation mechanisms to develop a biomimetic swimming robot, inspired by the motility mechanism of microorganisms. Behkam and collaborators[29] proposed a new type of propulsion inspired by the motility mechanism of bacteria with flagellation, such as *Escherichia coli, Salmonella typhimurium,* and *Serratia marcescens*. The authors foresee this robot having the capability to swim to inaccessible areas of the human body and perform complicated, user-directed tasks such as diagnosis of diseases at early stages and targeted drug delivery.

16.6 DNA-BASED NANODEVICES

The organizational capabilities of structural DNA nanotechnology are just beginning to be explored, as is the idea of using DNA to engineer nanoscale objects. This field has been pioneered by Nadrian Seeman at New York University[30] who has used motifs based on branched DNA molecules that are linked together by sticky ends to produce three-dimensional objects, periodic arrays, and nanomechanical devices. Currently, the creation of DNA-based nanodevices has been proposed. These DNA-nanobots will be able to intervene at the cellular level, performing *in vivo* nuclear "cytosurgery". In a simple cytosurgical procedure, a nanorobot would extract existing DNA from a diseased cell and replace it. The replaced chromosomes are suitably demethylated, thus expressing only the appropriate exons that are expressed in the cell type, to which the nanodevice has been targeted.[15]

16.7 ANGIOGENESIS ASSIST DEVICES

The process of angiogenesis encompasses the growth and regression of capillary blood vessels, and the success of therapeutic angiogenesis relies on the preexistence of intact microvessels, containing viable endothelial cells, in tissues or in proximity to the sites where angiogenesis is expected to be induced.[31] Obviously, in many instances in which the local microcirculation is compromised, this condition is not met. In severe cases of tissue ischemia and/or necrosis, where no viable microvessels exist to promote sprouting of new capillaries, the use of a nanodevice capable to deliver angiogenic proteins would be ideal. The pro-angiogenesis device, or angiochip, would facilitate controlled delivery of angiogenic factors useful for the development of both the exogenous and the endogenous microvasculature, mechanical robustness required for tissue implantation and retrieval using surgical means, and the capability of being incorporated into electronic biosensors for *in situ* diagnostics and for the individualized administration of angiogenic therapy.[15]

16.8 THE "RESPIROCYTE"

Medical nanomachines will be among the most extraordinary *in vivo* applications of this novel technology. The respirocyte, or artificial red cell, proposed by Robert Freitas,[15,32] is a spherical, diamondoid, 1000-atm pressure nanomachine made of 78 billion atoms and powered by endogenous serum glucose that performs the same function as red blood cells. Consequently, this nanoengineered device will have a wide range of medical applications,[33] including blood substitution; partial treatment for anemia, perinatal/neonatal and lung disorders; enhancement of cardiovascular procedures, thoracic oncology therapies, and diagnostics; prevention of asphyxia; artificial breathing; and gas delivery in anesthesia. According to Freitas's proposal each respirocyte would be able to detect the concentration of gases in the blood by using sensors on their surface and thereby mimicking the function of the natural, hemoglobin-filled red blood cells. Via molecular sorting rotors, gas molecules would enter the tanks inside the respirocyte. Up to 37% of the surface of each artificial red cell is covered with 29,160 of these molecular sorting rotors[34] which would spin and be able to pick up and drop off oxygen and carbon dioxide molecules. This nanorobot is far more efficient than a red blood cell, as each respirocyte would be able to hold 200 times more gas per unit volume than a natural red blood cell. Thus a few milliliters of respirocyte suspension injected into the human bloodstream would have the power to replace the gas-carrying capacity of the patient's entire 5.4 liters of blood.

16.9 THE "CLOTTOCYTE"

Another example of a widely utilizable nanorobot is the clottocyte or artificial mechanical platelets. The structure and primary functions of the platelet are well known.[35,36] This nanomachine described by Freitas[37] is conceived as a serum oxy-glucose-powered spherical nanorobot of about 2 μm in diameter containing a fiber mesh that is compactly folded onboard (Figure 16.1). Upon command from its control computer, the device promptly unfolds its mesh packet in the immediate vicinity of an injured blood vessel. Natural hemostasis may take from 2–5 min up to 9–10 min.[38-39] In contrast, the artificial mechanical platelet appears to permit clotting in about 1 sec — 1000 times faster — and seems to be about 10,000 times more effective than an equal volume of natural platelets.[37]

Possible applications would be clottocyte infusion for the improvement of coagulation in cases of intra- or postoperative bleeding. An additional application could be wound healing disturbances for the enhancement of clot formation and wound closure. If disseminated intravascular coagulation conditions arise following major cardiac surgery,[40] nanorobots might respond by absorbing and metabolizing the excess thrombin, or by releasing thrombin inhibitors such as antithrombin III, hirudin, argatroban or lepirudin, or anticoagulants that reduce thrombin generation such as danaparoid to interrupt the cascade. For example, a ~0.02% concentration of properly activated nanorobots could replace the entire depleted natural bloodstream content of antithrombin III.[37] In addition, clottocytes administration could be very instrumental in cases of heparin-induced thrombocytopenia after cardiac surgery.[41]

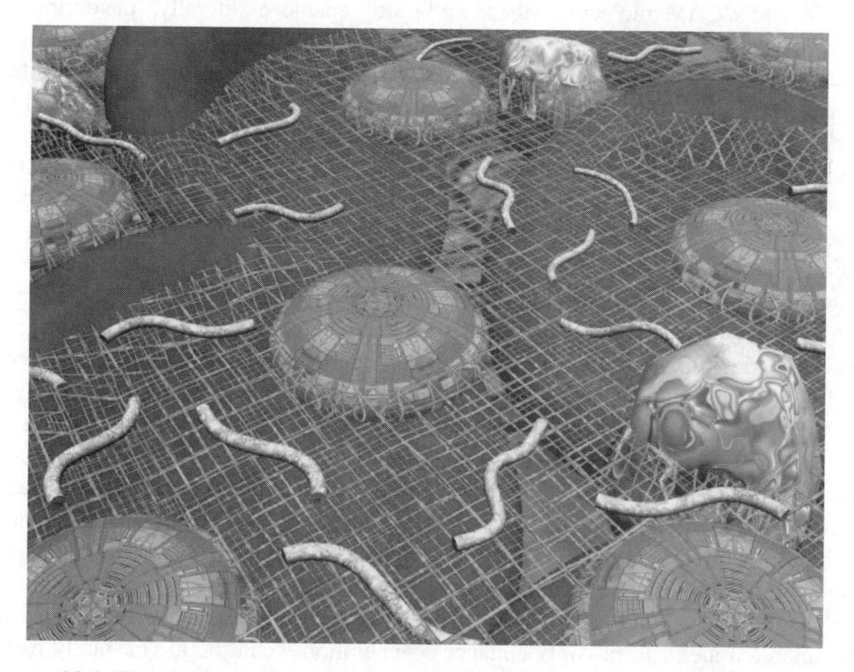

FIGURE 16.1 The clottocyte as it deploys its fibrin meshwork and facilitates wound closure and hemostatsis (permission granted by the designer).

PART II: APPLICATIONS OF NANOBIOLOGY/NANOTECHNOLOGY IN CARDIOLOGICAL AND CARDIOSURGICAL PRACTICE

16.10 APPLICATIONS IN THE THERAPY OF MYOCARDIAL ISCHEMIA

Heart muscle is a hypermetabolic tissue with high energetic demands and limited tolerance to blood undersupply.[42] If blood flow through the coronary arteries falls below 20% of resting levels, the supplied myocardium is in danger of becoming necrotic. In a state of reduced blood flow or incapacity of the erythrocytes to give up enough oxygen to the tissue, cardiac contractility diminishes rapidly (within few heart beats). If enough oxygen is provided to the tissue for its basic metabolic demands, this condition can result in a chronic hypocontractile state known as hibernation.[43] At the microstructural level calcium exchange through the cellular membrane and therefore the intracellular concentration of this ion is insufficient for an adequate contraction.[44] A targeted application of nanodevices at the site of hibernation might provide crucial nutrients and Ca to the tissue at risk. This way the hibernating myocardium can be protected from further damage, since its energy stores cannot be depleted so easily.

If the ischemic insult progresses or is of longer duration, a series of fulminant events occurs, such as free radical formation, xanthin formation, cytokine liberation, and chemoattraction of macrophages and other inflammatory cells. Mediated by ICAM and VCAM molecules, these cells are capable of literally "devouring" a cardiomyocyte within minutes. The final stages of cell damage involve loss of the nucleus, lysis of membranes, and cell death. In consideration of the very limited regenerative capacity of heart muscle, such events are irreversible and may lead to life-threatening complications: extensive myocardial infarction, aneurysm formation, development of a ventricular septal defect, valve malfunction, severe arrhythmias, and finally heart failure. Many of these conditions require cost-ineffective and interventional/surgical therapeutic measures, whereas prevention and limitation of the disease processes could have helped circumvent further progress.

Accurate nanotechnological imaging techniques could help localize the disease process very precisely and guide the delivery of nanoparticles,[3-9] such as angiogenesis assist devices. Both nanocapsules and angiochips[15] could penetrate into the area of lesion within the ischemic heart muscle and deliver their load (growth factors: VEGF, FGF, HGF, etc., as well as antioxidants: ascorbic acid, β-carotene, tocoferol etc.) in a very targeted fashion. This could promote angiogenesis and increase blood flow to the ischemic area, thereby supporting regeneration of the affected heart muscle. Any further therapeutic measures, such as stem cell transfer, or implantation of bioartificial tissue would be facilitated by improved engraftment and metabolic function. Nano-biosensors would be added to monitor the angiogenic treatment.

Finally, the optimal way to limit the disease process within the boundaries of the infarcted area and possibly enhance regeneration would be to constantly oversupply the hibernating tissue and the infarct border zones with oxygen. Respirocytes,[15,32] with a more than 220-fold oxygen-tissue affinity compared to red blood cells, would infiltrate the area at risk and home locally for prolonged periods of time. This would maintain high local levels of oxygen, prevent expansion of the infarction and maintain normal contractility of the affected heart muscle. Other nanoparticles could serve as "recycling bins", attracting tissue waste, free radicals and products of the purine metabolism, thereby significantly reducing the overall toxic effect to the already damaged heart muscle.

16.11 EXTRACORPOREAL CIRCULATION / RESTORATION OF CORONARY FLOW / REPERFUSION, FREE RADICALS, TOXIC OXYGEN SPECIES AND PROMISING NANOTECHNOLOGICAL APPLICATIONS

The conduct of extracorporeal circulation for the accomplishment of complex cardiac surgery involves nonphysiological and potentially harmful phenomena, such as non-pulsatile perfusion of tissues, hypothermia, anticoagulation, activation of host inflammatory cascades and organ damage. Most of these conditions are reversible following termination of cardiopulmonary bypass, may however result in prolonged organ function impairment, intractable bleeding or overconsumption of proteins of the coagulation cascade. Separation of the patient from cardiopulmonary bypass constitutes the last phase of cardiac surgery and requires attention to all of the parameters above and harmonious collaboration between the surgeon, the perfusionist and the anesthesiologist. This particular phase of cardiac surgery could benefit significantly if appropriate nanoparticles could be administered systemically to counteract cytokine liberation by leucocytes and macrophages, support function of platelets and sustain oxygenation of the recovering and reperfused myocardium. This could be done locally, through administration of nanoparticles via the cardioplegia line, prior to its removal, or via direct injection into the bypass grafts, which would result in their distribution to the area at risk.

The majority of interventional and surgical approaches for the treatment of ischemic heart disease entails recanalization of occluded coronary arteries or bypass grafting procedures. These procedures result in acute restoration of blood flow and exposure of the tissue distal to the former occlusion of the coronary artery to reperfusion. With reperfusion, increased generation of oxygen radicals occurs.[45] Oxygen radicals are highly reactive species and tend to propagate in chain reactions. Although radicals are continuously produced by normal metabolic processes, when they exceed the capacity of the normal quenching and scavenging systems they are extremely toxic and affect all cellular processes. Reperfusion injury may also occur through swelling of the cells within the vessel wall and from platelet adherence to the luminal surface of the vessel. Additionally, constant loss of nucleotide bases may impair the ATP supply to the cells. Furthermore, lytic enzymes are being liberated, causing disruption of cellular membranes and cell death. Nanoparticles loaded with scavengers could ameliorate these fulminant effects and protect the reperfused myocardium from extensive damage.[45-47] This can be accomplished locally, via the coronary circulation, or systemically through an intravenous infusion. It would also be possible through intraarterial administration via the angiography catheter following balloon angioplasty or stent placement procedures. Smart drugs could home at the site of stent implantation and prevent in-stent stenosis.

Another form of cardiopulmonary support is the extracorporeal membrane oxygenator (ECMO)[48] and cardiac assist devices. The former resembles the cardiopulmonary bypass but can be maintained for a period of days in the ICU, exposes the organism to a severe stress, and may cause severe hemorrhage and depletion of coagulation factors and disseminated intravascular coagulation known as DIC. Assist devices are implantable machines that take over the heart function by retrieving blood from the apex of the left ventricle and by pumping it back into the aorta. Even though their surfaces are more biocompatible than in the case of the ECMO, they might also lead to severe hemolysis, thrombus formation, and device failure.[49] Nanorobots could be built into the system to maintain the smoothness and biocompatibility of the artificial surfaces. The respirocyte could support the ECMO function by offering a mode of oxygen overload which would facilitate reduction of artificial flow for the supply of the tissues, thereby reducing contact with foreign surfaces, hemolysis, and depletion of coagulation factors. Clottocytes would be administered to replace lost platelets and maintain adequate coagulation.

16.12 NANOTECHNOLOGICAL APPLICATIONS IN TRAUMA / BLEEDING / WOUND HEALING IN CARDIAC SURGERY

The most frequent and often severe complications in cardiological interventions and cardiac surgery is excessive bleeding.[50] It can occur during diagnostic angiography at the site of percutaneous catheter insertion, requiring groin pressure for prolonged periods of time. It can also be acutely necessary following intracoronary procedures in fragile vessels. These cardiological patients are anticoagulated strongly and some form of sealant may become indispensable. Local application of clottocytes,[37] which would infiltrate the area of interest and deploy their network, would significantly shorten the process of hemostasis. This effect is even more desirable intraoperatively, whenever inacceptable blood losses occur. Application of clottocytes at the site of vascular anastomoses could revolutionize hemostatic treatment, maintain good visibility of the site of hemorrhage (as opposed to the local application of solid and visibility-obscurring sealants), and shorten the time on cardiopulmonary bypass. Such an effect may have significant economical advantages as well, beyond the immediate benefit for the patient and his recovery. The field of cardiology and cardiac surgery literally constitutes the main field of focus for the clottocyte, due to the plethora of potential applications in the system of high hemodynamic pressure of the body. In case of valvular surgery and subsequent ICU stay, any invasive procedure in the ICU or catheter laboratory that requires adequate clotting capacity of the blood could now be accomplished without jeopardizing the indispensable anticoagulation that guarantees good valve prosthesis function. Such applications could benefit from the administration of clottocytes as well.

Moreover, bleeding control using clottocytes would reduce the need for blood, fresh frozen plasma, refrigerated clotting factors, and thrombocyte transfusion, and therefore alleviate possible side effects such as anaphylaxis and viral infection. Due to the long-term functionality of clottocytes and their superior efficacy compared with platelets,[37] their utilization may prove even more cost effective in patients with severe bleeding complications and prolonged ICU stay.

Wound healing may constitute a further field of focus for nanobiology and nanotechnology. For the various phases of the healing process of a wound, adequate capillarization is of paramount significance as is appropriate oxygenation of the wounded or incised tissue. The healing process can be accelerated by the local administration of clottocytes (in the early stages of would healing) followed by targeted growth factor release into the area of lesion by nanoparticles or nanospheres, generally, angiogenesis assist devices. Wound healing processes could be improved both in soft tissue (subcutaneous tissue and skin), as well as bone (sternum and chest wall). An improved and accelerated wound healing would be very critical for the early ambulation of the patient and significantly reduce the risk of thrombosis, embolism, and infection.

16.13 NANOTECHNOLOGY AND AORTIC SURGERY

The central nervous system has a high metabolic activity and low energy stores. In the state of reduced circulation, such as during cardiopulmonary bypass or aortic arch surgery with selective cannulation of the carotid and subclavian arteries, the brain needs to be protected.[51] With the introduction of regional cerebral perfusion,[52] mostly antegrade through catheters inserted into the cranial arteries and blocked, many surgeons tend to apply only mild systemic hypothermia. The latter reduces the metabolic rate of the central nervous system and extends the period of maximally tolerated ischemia. The respirocyte with its superb affinity to oxygen could serve as a strong and sustained oxygen supply to the brain, much more effectively than normal blood. This way, the maximally tolerated ischemia (circulatory standstill) time could be significantly reduced. This measure could provide additional time for aortic arch manipulations or reduce the hypothermia requirements for these types of operations.

In cardiovascular surgery practice, an aortic aneurysm or aortic dissection often progresses to affect the aortic arch and branches which supply the brain. In such patients, a vascular prosthesis needs to be placed and sutured within the lumen of the rest of the aorta, via and end-to-end anastomosis. For some considerable period of time, a hypothermic arrest is necessary. Some of the aortic arch replacement procedures involve implantation of a prosthesis in a so-called "elephant trunk" fashion, i.e., invagination of the inverted prosthesis into the distal arch or the descending aorta. The space remainder between the prosthesis and the original aortic wall is then expected to thrombose. Equally critical is the hemostatic effect of the "graft inclusion" technique following replacement of the ascending or abdominal aorta. Clottocytes could be applied into the space between which is intended to thrombose and seal it, by deploying large amounts of fibrin meshwork. This action is rapid and highly efficient (fractions of a second), results in saving of precious time on extracorporeal circulation, and shortens rewarming time. Rewarming constitutes a critical period as well, during which brain cells can be further damaged. A scavenger nanoparticle could help wash out accumulated metabolites, buffer free radicals, and provide substrates for the regeneration of high energy molecules before the resumption of cerebral electrical activity.

16.14 TISSUE ENGINEERING OF BIOARTIFICIAL HEART MUSCLE/STEM CELL TRANSFER FOR MYOCARDIAL RESTORATION AND NANOBIOLOGY

Michael Shefton of Toronto announced in 1998 that humans would be able to produce a fully functional heart in the laboratory in 10 years. Today, 6 years later, tissue engineers consider this prognosis somewhat premature. The heart is a complex, helical structure with unique morphological characteristics at various sites of its body. It is asymmetric, anisotropic, and highly angiotropic. Hence it can be viewed as an organ with more than three dimensions, particularly when the pumping activitiy and rhythmogenicity are considered. These functional-morphological aspects of the heart pose significant problems on the way to achieve a robust and sustained myocardial restoration by means of stem cell and bioartificial tissue transfer for the treatment of ischemic or dilative heart disease.

With the advent of the new science of tissue engineering and the rich body of experience obtained through stem cell transfer, a new era in the treatment of heart disease emerges. The new techniques, some of which will be summarized here, face important limitations. Nanobiology and nanotechnology are expected to help resolve some of those. For a better understanding of the innumerable possibilities that nanotechnology may offer for the field of myocardial restoration, it is important to obtain an overview of what has been achieved so far.

A plethora of stem cell types[53,54] has already been utilized for myocardial restoration, both via intraarterial infusion and direct intramyocardial injection. The use of myoblasts, for instance, is limited by the arrhythmogenicity of these cells following injection of the cells into the heart, which makes the implantation of an automated defibrillator in the same operative session necessary. These cells are terminally differentiated and do not turn into cardiomyocytes in the area of injury of the host, which would naturally form intercellular junctions for the promotion of the electrical signal through the heart. Bone marrow–derived stem cells, on the other hand, lack plasticity, i.e., assume only bone marrow-specific phenotypes following injection into the area of lesion of the heart. Embryonic stem cells (ESC) constitute an attractive alternative option, due to their unlimited capacity to regenerate and give rise to a committed progeny of cells, also cardiomyocytes. However, their clinical use will require pre-implantation treatment of their tumorigenic potential. Furthermore, there is evidence of immunogenicity of ESC,[55,56] which might require immunossupressive therapy. The leading limitation of all cell and tissue transplantation procedures is donor cell death early after the procedure, due to the ischemic and therefore hostile environment within the injured myocardium. The respirocyte constitutes a very hopeful measure in this context. Respirocytes could be injected along with the stem cells or transplanted concurrently with the *in vitro* generated bioartificial tissue and enhance donor cell viability in the targeted area. A secondary benefit of the application of respirocytes would be the maintenance of the thickness of the left ventricular wall of the heart, a factor which is critical for the wall tension and the contractility of the left ventricle. Nanoparticles could counteract the action of free radicals and the production of "danger signals" such as products of the purine metabolism, thus inhibiting the non-ischemic expansion of the infarct.

A random injection of stem cells into the injured myocardium as well as the efficacy of a three-dimensional bioartificial heart muscle cannot be optimal without appropriate vascularization. In fact, any tissue-engineered structure is destined to have limited dimensions without a supplying vascular network. Robbins, Kofidis et al. at Stanford University have developed both solid and liquid tissues for intramyocardial transplantation[57] (Figures 16.2A and 16.2B) and found out that cell and tissue viability is significantly improved following stimulation of paracrine pathways (addition of growth factors to the cell suspension and bioartificial tissues) or maturation of the tissue substitutes within angiogenic bioreactors. The major problem in three-dimensional generation of all bioartificial tissues for the surgical therapies of the future is the lack of a chaotic but still homogenously distributed vascular network. Smart nanorobots can be programmed to generate this pattern in solid as well as collagenous matrices, so that it can be exposed to blood perfusion. It has been shown that angiogenic islets within chick allantoic membranes as well as gels (droplets of angiogenic factors) promote targeted sprouting of capillaries.[58] Kofidis et al. have shown that inoculated cells prefer to form colonies in the vicinity of the oxygen and nutrient source, within a novel bioreactor.[59] Respirocytes and angiogenic nanoparticles (angiochips) could boost this effect. The targeted and remote controlled angiogenic action within three-dimensional structures constitutes a milestone for the cardiological/surgical therapy of the heart in the new millennium. Finally, cardiac rhythm could be generated by nanopacers, particles that act in unison and communicate with each other from various remote parts of the bioartificial tissue or the reconstructed heart muscle. Nanodiagnostics will facilitate monitoring of engraftment of bioarticial tissues, which are based upon living cells and complex matrices, both in the construction phase and following transplantation. Nanotechnology is therefore expected to play a critical role in the progress of this field, toward a totally bioartificial, organic substitution of the human heart.

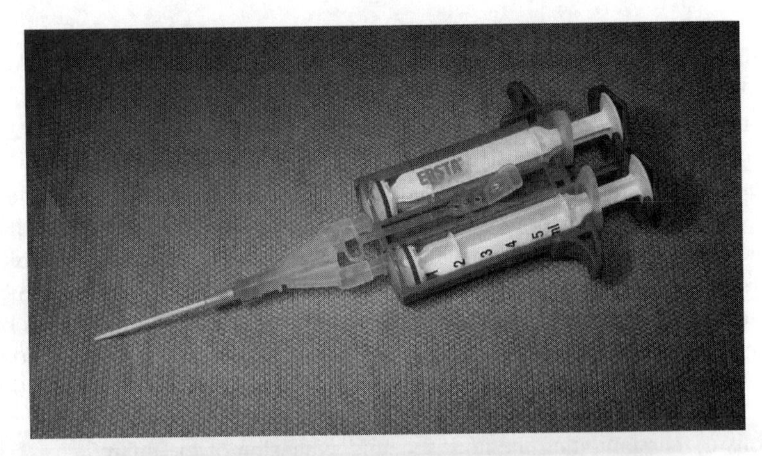

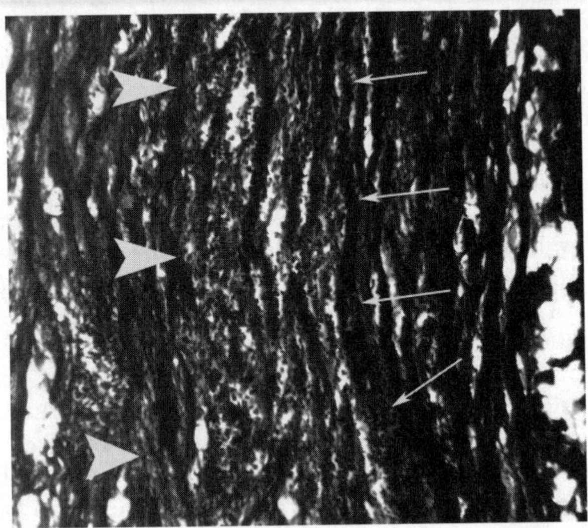

FIGURE 16.2 (A) Injectable bioartificial myocardial tissue in the two-line applicator. (B) The cell-collagen mixture consolidates within injured myocardium and supports myocardial restoration. The respirocyte might significantly enhance engraftment and survival of the transplanted tissue, while clottocytes might facilitate large scale replacement of the free left ventricular wall by this novel restorative method, without significant blood losses (permission granted by the author and inquired at the JTCVS). (From Kofidis, T. el al., *J. Thorac. Cardiovasc. Surg*. 2004; 128. With permission.)

REFERENCES

1. 2004 National Heart, Lung and Blood Institute Morbidity and Mortality Chartbook. From: http://www.nhlbi.nih.gov/resources/docs/cht-book.htm.
2. Drexler, KE. Nanotechnology: from Feynman to funding. *Bull. Sci. Technol. Soc.* 2004; 24(1): 21–27.

3. Winter PM, Caruthers SD, Kassner A, et al. Molecular imaging of angiogenesis in nascent Vx-2 rabbit tumors using a novel alpha(nu)beta3-targeted nanoparticle and 1.5 tesla magnetic resonance imaging. *Cancer Res.* 2003; 63(18):5838–5843.
4. Winter PM, Morawski AM, Caruthers SD, et al. Molecular imaging of angiogenesis in early-stage atherosclerosis with {alpha}v{beta}3-integrin-targeted nanoparticles. *Circulation* 2003; 108:2270–2274.
5. Jaiswal JK, Simon SM. Potentials and pitfalls of fluorescent quantum dots for biological imaging. *Trends Cell Biol.* 2004; 14(9):497–504.
6. Mattoussi H, et al. Self-assembly of CdSe-ZnS quantum dot bioconjugates using an engineered recombinant protein. *J. Am. Chem. Soc.* 2000; 122:12142–12150.
7. Jaiswal K, et al. Long-term multiple color imaging of live cells using quantum dot bioconjugates. *Nat. Biotechnol.* 2003; 21:47–51.
8. Gao X, Nie S. Molecular profiling of single cells and tissue specimens with quantum dots. *Trends Biotechnol.* 2003; 21(9):371–373.
9. Bruchez, M, Moronne, M, Gin, P, Weiss, S, Alivisatos, AP. Semiconductor nanocrystals as fluorescent biological labels. *Science* 1998; 281:2013–2016.
10. Larson DR, Zipfel WR, Williams RM, et al. Water-soluble quantum dots for multiphoton fluorescence imaging *in vivo*, *Science* 2003; 300:1434–1436.
11. Wu XY et al., Immunofluorescent labeling of cancer marker Her2 and other cellular targets with semiconductor QDs. *Nat. Biotechnol.* 2003; 21:41–46.
12. H. Shi, B.D. Ratner, Template recognition of protein-imprinted polymer surfaces, *J. Biomed. Mater. Res.* 2000; 49:1–11
13. Simonova M, Shtanko O, Sergeyev N, Weissleder R, Bogdanov A Jr. Engineering of technetium-99m-binding artificial receptors for imaging gene expression. *J. Gene Med.* 2003; 5(12):1056–1066.
14. Allender CJ, Richardson C, Woodhouse B, et al. Pharmaceutical applications for molecularly imprinted polymers. *Int. J. Pharm.* 2000; 195:39–43.
15. Freitas RA. *Nanomedicine, Volume I: Basic Capabilities*, Landes Bioscience, Georgetown, TX, 1999. See at: http://www.nanomedicine.com
16. Micro Electro Mechanical Systems Laboratory. University of Pittsburgh. See at: http://www.engr.pitt.edu/site/scpi/tech/mems.html
17. Hsiai TK, Cho SK, Wong PK, et al. Micro Sensors: Linking real-time oscillatory shear stress with vascular inflammatory responses. *Ann. Biomed. Eng.* 2004; 32:189–201.
18. Clowes AW, Clowes MM, Fingerle J, et al. Kinetics of cellular proliferation after arterial injury: V. Role of acute distension in the induction of smooth muscle proliferation. *Lab Invest.* 1989; 60:360–364.
19. Lanza GM, Abendschein DR, Hall CH, et al. In vivo molecular imaging of stretch-induced tissue factor in carotid arteries with ligand-targeted nanoparticles. *J. Am. Soc. Echo.* 2000; 13:608–614.
20. Lanza G, Abendschein D, Hall C, et al. Molecular imaging of stretch-induced tissue factor expression in carotid arteries with intravascular ultrasound. *Invest. Radiol.* 2000; 35: 227–234.
21. Lanza GM, Yu X, Winter PM, Abendschein DR, et al. Targeted antiproliferative drug delivery to vascular smooth muscle cells with a magnetic resonance imaging nanoparticle contrast agent: implications for rational therapy of restenosis. *Circulation* 2002; 106(22):2842–2847.
22. Orive G, Hernandez RM, Rodriguez Gascon A, Dominguez-Gil A, Pedraz JL. Drug delivery in biotechnology: present and future. *Curr. Opin. Biotechnol.* 2003; 14: 659–664.

23. Suzuki M, Tanihara M, Nishimura Y, Suzuki K, et al. A new drug delivery system with controlled release of antibiotic only in the presence of infection. *J. Biomed. Mater. Res.* 1998; 42:112–116.
24. Okada, H. One- and three-month release injectable microspheres of the LH-RH superagonist leuprorelin acetate. *Adv. Drug Deliv. Rev.* 1997; 28: 43–70.
25. Pizzi M, De Martiis O, Grasso V. Fabrication of self assembled micro reservoirs for controlled drug release. *Biomed. Microdevices* 2004; 6(2):155–158.
26. Paunesku T, Rajh T, Wiederrecht G, et al. Biology of TiO_2-oligonucleotide nanocomposites. *Nat. Mater.* 2003; 2:343–346.
27. Desai TA, West T, Cohen M, Boiarski T, Rampersaud A. Nanoporous microsystems for islet cell replacement. *Adv. Drug Deliv. Rev.* 2004; 56(11):1661–1673.
28. Nanobots. See at: http://www.medibotics.com/
29. Behkam B, Sitti ME. Coli inspired propulsion for swimming microrobots. Proceedings of 2004 ASME International Mechanical Engineering Conference and Exposition, November 2004, Anaheim, CA (to appear).
30. Seeman NC. DNA engineering and its application to nanotechnology. *Trends Biotechnol.* 1999; 17:437–443.
31. Folkman J. Fundamental concepts of the angiogenic process. *Curr. Mol. Med.* 2003; 3(7):643-51. Review.
32. Freitas RA. Exploratory design in medical nanotechnology: a mechanical artificial red cell. *Artificial Cells, Blood Substitutes, and Immobil. Biotech.* 1998; 26:411-430. See also: http://www.foresight.org/Nanomedicine/Respirocytes.html.
33. Frietsch T, Lenz C, Waschke KF. Artificial oxygen carriers. *Eur. J. Anaesthesiol.* 1998; 15(5):571–584.
34. Drexler KE. *Nanosystems: Molecular Machinery, Manufacturing, and Computation.* John Wiley & Sons, New York, 1992, page 374.
35. Ito E, Suzuki K, Yamato M, et al. Active platelet movements on hydrophobic/hydrophilic microdomain-structured surfaces. *J. Biomed. Mater. Res.* 1998; 42:148–155.R.P.
36. Awadhiya RP, Vegad JL, Kolte GN. Demonstration of the phagocytic activity of chicken thrombocytes using colloidal carbon. *Res. Vet. Sci.* 1980; 29:120–122.
37. Freitas RA. Clottocytes: Artificial Mechanical Platelets. See at: http://www.imm.org/Reports/Rep018.html.
38. HealthGate Medical Tests. Bleeding Time, 27 July 1999, see at: http://www3.healthgate.com/mdx-books/tests/test35.asp.
39. Hertzendorf LR, Stehling L, Kurec AS, et al. Comparison of bleeding times performed on the arm and the leg. *Am. J. Clin. Pathol.* 1987; 87:393–396.
40. Norman KE. Alternative treatments for disseminated intravascular coagulation. *Drug News Perspect.* 2004; 17(4):243–250. Review.
41. Greinacher A. The use of direct thrombin inhibitors in cardiovascular surgery in patients with heparin-induced thrombocytopenia. *Semin. Thromb. Hemost.* 2004; 30(3):315–327.
42. Tamm, C, Benzi R, Papageorgiou I, et al. Substrate competition in postischemic myocardium. Effect of substrate availability during reperfusion on metabolic and contractile recovery in isolated rat hearts. *Circ. Res.* 1994; 75:1103-1112.
43. Rahimtoola SH. The hibernating myocardium. *Am. Heart J.* 1989; 117:211–221.
44. Terrand J, Papageorgiou I, Rosenblatt-Velin N, et al. Calcium-mediated activation of pyruvate dehydrogenase in severely injured postischemic myocardium. *Am. J. Physiol. Heart Circ. Physiol.* 2001; 281(2): H722–730.

45. Kamler M, Wendt D, Pizanis N, et al. Deleterious effects of oxygen during extracorporeal circulation for the microcirculation in vivo. *Eur. J. Cardiothorac. Surg*. 2004; 26: 571–579
46. Barratt-Boyes BG, Harris EA, Kenyon AM, et al. Coronary perfusion and myocardial metabolism during open-heart surgery in man. *J. Thorac. Cardiovasc. Surg*. 1976; 72: 133–141.
47. Valen G, Owall A, Takeshima S, et al. Metabolic changes induced by ischemia and cardioplegia: a study employing cardiac microdialysis in pigs. *Eur. J. Cardiothorac. Surg*. 2004; 25(1): 69–75.
48. Chatzis AC, Giannopoulos NM, Tsoutsinos AJ, et al. Extracorporeal membrane oxygenation circulatory support after cardiac surgery. *Transplant Proc*. 2004; 36(6): 1763–1765.
49. Imasaka K, Masuda K, Masuda M, et al. Mechanical cardiac support system for patients with postcardiotomy cardiogenic shock: analysis of risk factors for survival. *Jpn. J. Thorac. Cardiovasc. Surg*. 2004; 52(4):163–168.
50. Shore-Lesserson L. Monitoring anticoagulation and hemostasis in cardiac surgery. *Anesthesiol. Clin. N. Am*. 2003; 21(3):511–526. Review.
51. Mezrow CK, Sadeghi AM, Gandsas A, et al. Cerebral blood flow and metabolism in hypothermic circulatory arrest. *Ann. Thorac. Surg*. 1992; 54:609–616.
52. Kilpack VD, Stayer SA, McKenzie ED. Limiting circulatory arrest using regional low flow perfusion. *J. Extra Corpor. Technol*. 2004; 36(2):133–138.
53. Min JY, Yang Y, Converso KL, et al. Transplantation of embryonic stem cells improves cardiac function in postinfarcted rats. *J. Appl. Physiol*. 2002; 92, 288–296.
54. Bonaros N, Yang S, Ott H, Kocher A. Cell therapy for ischemic heart disease. *Panminerva Med*. 2004; 46(1):13–23.
55. Drukker M, Benvenisty N. The immunogenicity of human embryonic stem-derived cells. *Trends Biotechnol*. 2004; 22(3):136–141.
56. Drukker M. Immunogenicity of human embryonic stem cells: can we achieve tolerance? *Springer Semin. Immunopathol*. 2004 Jul 29, Epub ahead of print
57. Kofidis T, deBruin JL, Tanaka M, et al. They are not stealthy in the heart: embryonic stem cells trigger cell infiltration, humoral and T-lymphocyte-based host immune response. *JTCVS*, 2004; 128.
58. Ribatti D, Nico B, Morbidelli L, et al. Cell-mediated delivery of fibroblast growth factor-2 and vascular endothelial growth factor onto the chick chorioallantoic membrane: endothelial fenestration and angiogenesis. *J. Vasc. Res*. 2001; 38(4):389–397.
59. Kofidis T, Lenz A, Boublik J, et al. Pulsatile perfusion and cardiomyocyte viability in a solid three-dimensional matrix. *Biomaterials*. 2003; 24(27):5009–5014.

17 Translating Nanotechnology to Vascular Disease

Michael D. Kuo, Jacob M. Waugh, Chris J. Elkins, and David S. Wang

CONTENTS

17.1 INTRODUCTION

Vascular disease is the leading cause of morbidity and mortality worldwide.[1] Great strides in treatment have been realized in recent years, particularly in the realm of minimally invasive catheter-based solutions. Many of these advances were the result of successful translation of seemingly disparate fields and disciplines into practical and clinically driven tools. Indeed, today's catheter-based therapies, such as balloon angioplasty and stenting, are largely the result of such an approach applied by

0-8493-1940-4/05/$0.00+$1.50

interventional radiologists over 30 years ago, in which clinical imaging tools and mechanical engineering solutions were translated into the creation of minimally invasive endovascular therapies. Subsequently, an entirely new field of treatment was developed. However, today's predominantly mechanical solutions are beginning to reach their limits. As we are now understanding disease processes at the molecular level, nanoscale diagnosis and treatment modalities are required to clinically apply such findings. A new generation of translational tools is needed for the treatment of vascular disease.

In this chapter, we discuss how different scientific disciplines will be brought together to translate recent advances in nanotechnology to drive the next revolution in the management of vascular disease. We first discuss the fundamental role that genomics and proteomics will play in providing insights that will ultimately serve as the foundation for the next generation of nanobiology tools in vascular medicine. We then survey novel technologies that will be employed in developing nanoscale biosensors that can be used for large-scale detection of pathologic alterations at the molecular level *in vivo*. Lastly, we discuss several promising micro- and nanoscale vascular disease treatment innovations: therapeutic transport systems, nanoscale tools to manipulate angiogenesis, *in vivo* vascular engineering, and bioactive endovascular stents.

17.2 GENOMICS AND PROTEOMICS FOR VASCULAR DISEASE

17.2.1 Genomics

Ultimately, if we wish to truly impact disease on a molecular level, we need to first understand the disease process on a molecular level. The completion of the Human Genome Project[2,3] has provided us with a detailed map of our genetic makeup. This has allowed us to see in detail the genetic instructions that specify all of the molecular components and their governing in both normal and perturbed states. Indeed, as a result of the sequencing of the human genome, we now essentially have an instruction book to understand the hardware and software of the entire human organism. With this information, we are able to fully interrogate and study the function and interaction of the entire complement of genes in the genome, launching the field of genomics. Although we are still in the infancy of this medical renaissance, it is clear that the impact of genomic medicine will be tremendous by giving us a molecular window into human biology in both normal physiology and development as well as in disease. The information obtained from this initial phase of discovery science will have far-reaching translational implications for the diagnosis and management of human disease. DNA microarrays are a principal genomic tool that will allow us to observe and understand the physiology and pathophysiology of the entire human genome at a molecular level and which will provide insights that will ultimately serve as the foundation for the next generation of nanobiology tools for the diagnosis and treatment of vascular disease.

17.2.2 DNA Microarrays

DNA microarrays are highly ordered microscopic arrays of DNA of known sequences immobilized to a solid surface. Each array is extremely dense with current iterations able to accommodate all of the genes of the human genome on the space of a microscope slide. While there are many different types of DNA microarrays, each array fundamentally consists of thousands of cDNA clones or oligonucleotides printed in a precisely known location on the physical substrate, with each DNA "spot" in the nanoliter range and 50–150 μm in diameter. The DNA printing process incorporates components of nanofabrication technology in order to achieve such spatial resolution such as "spotting" using pins and robotics technology,[4] "printing" using ink-jet printer technology[5] or photolithography.[6,7] The general concept behind the microarray is the ability to perform massively parallel DNA hybridization studies by taking advantage of the known sequence and location of DNA probe on the array allowing us to simultaneously visualize every gene on the array at once.

In order to quantitatively and comparatively read out the gene expression of a biological sample, the RNA must first be extracted. The RNA transcripts of the sample are then amplified if necessary, reverse transcribed and labeled (typically with a fluorescent dye), and then hybridized to the microarray. Through Watson-Crick base-pair interactions, the experimental samples will hybridize to their immobilized DNA complement on the microarray. Quantitative assessment of gene expression in the sample can then be measured as a function of the signal intensity of each "captured" fluorescent probe to the microarray. In this fashion, systematic and quantitative observations can be made of thousands of genes at once in a given cell type or tissue under a variety of experimental conditions (Figure 17.1; see color insert following page 204).

Each cell is largely defined by a specific set of genes governed by a series of specific regulatory programs that, in composite, define the cell's phenotype and function. Further, regulation of cellular genetic programs can be modulated by the downstream effects of interactions of the cell with its microenvironment, whether in response to mechanical, pharmacologic, infectious, or other types of stressors or stimuli, which typically are manifested as changes in gene expression. Therefore, the global gene expression pattern of a given cell can serve as a reflection of the cell's internal state *and* microenvironment at a given moment. Therefore, counter to previous "forward genetics" approaches which attempted to move from an observed phenotype or function to the relevant genes that caused that phenotype, DNA microarrays uniquely allow us to engage in a "reverse genetics" approach that evaluates large-scale gene expression changes in the context of an entire system and can thereby uniquely define a cell process or pathology.

With the information obtained from this technology, a wide range of applications has been suggested with many of these demonstrated in earlier studies. DNA microarrays have been used in such diverse areas as DNA sequencing,[8] genetic mapping of single nucleotide polymorphisms (SNPs) to assess disease susceptibility,[9] disease profiling and classification,[10,11] drug discovery[12,13] and response,[14] and pathway elucidation.[15]

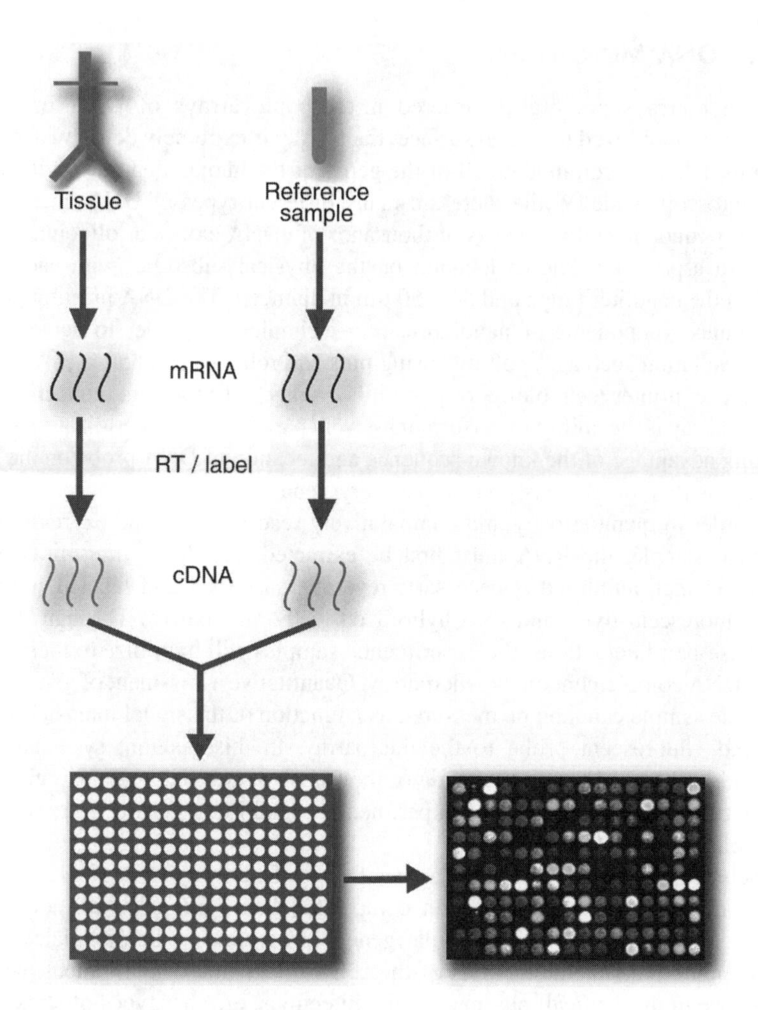

FIGURE 17.1 Vascular gene expression analysis using a DNA microarray. (See color insert following page 204.) Total mRNA is first extracted from the tissue of interest (aorta in this example) and then reverse transcribed (RT) in the presence of red fluorescently labeled nucleotide precursors to cDNA. A reference sample is prepared similarly with a green fluorescent tag. The two fluorescently labeled cDNA populations are then mixed and competitively hybridized with a DNA microarray, where each vascular gene is represented as a distinct spot on the microarray. The fluorescently labeled cDNA sequences representing an expressed transcript hybridize with their corresponding gene sequence target on the microarray. Relative abundance of transcripts is then determined as the ratio of "red" to "green" fluorescence measured for each gene on the array. In this manner, massively parallel quantitative evaluation of gene expression of a tissue of interest can be performed.

17.2.3 Gene Expression Profiling in Vascular Disease

Virtually every disease process is accompanied by alterations in gene expression. Certain disease entities, such as cancer, are driven by genomic instability reflected as fundamental alterations in gene expression programs. In other disease processes, the alterations in gene expression patterns may be a reflection of both innate alterations of the diseased cell type as well as the response of surrounding healthy cells to local and systemic perturbations induced by the disease process. DNA microarrays therefore provide a powerful platform technology to understand human disease because they allow us to visualize and quantify subtle or overt changes in genetic programs in the context of an entire biological system. This molecular phenotyping of underlying molecular heterogeneity has allowed us to understand why certain disease entities, which appear histologically similar, can behave very differently clinically with respect to outcome and treatment response.[10]

Genomic analysis using DNA microarrays is already beginning to have an impact on our understanding of complex vascular disease processes. Microarray analysis of human atherosclerotic lesions in coronary and carotid artery explants has revealed differential regulation of inflammatory and apoptotic genes whose roles were previously unrecognized.[16] Similarly, genomic analysis of endothelial and vascular smooth muscle cells (SMCs) subjected to a variety of experimental conditions such as shear stress and cigarette toxins has afforded us new insights into the roles of reduction-oxidation and shear stress pathways in atherogenesis.[17,18] DNA microarray analysis has also been performed on human in-stent restenosis specimens. This has demonstrated relative upregulation of cyclooxygenase-1, heat shock protein B, as well as FK506-binding protein (the receptor for Rapamycin), all of which may serve as future drug targets for the prevention of restenosis or are current anti-restenosis targets (Rapamycin) after percutaneous transluminal angioplasty.[19] In addition, experimental studies of aneurysm development in rats have also shown that oxidative stress and inflammatory pathways as well as certain matrix/stromal elements play a role in the pathogenesis of elastase-induced abdominal aortic aneurysms.[20] Finally, SNPs, which can be markers of disease susceptibility in complex polygenic diseases (such as atherosclerosis, myocardial infarction, and hypertension), are being identified across the entire human genome using a variety of methods including DNA microarrays. Together, these insights will not only allow us to better understand the molecular mechanisms of these disease processes as well as disease susceptibility, but may also uncover new potential therapeutic targets which may be incorporated in the next generation of nanotherapies.

17.2.4 Translating Genomics to Therapy

Genomics will have a profound impact upon the practice of medicine beyond diagnosis. The human genome project and genomics are rapidly identifying the roughly 35,000 genes, which comprise the composite catalogue of possible therapeutic targets, and their role in disease. Identifying the function of these genes and their interactions in normal and perturbed states will ultimately allow us to impact disease at a molecular level through improved rational design of new classes of micro- and nanoscale molecular therapeutics and bioactive devices. Indeed, we are already beginning to see the impact of DNA microarrays on drug development at all stages of the development process.

Array technologies are powerful tools in drug discovery, because they allow rapid characterization of gene expression of thousands of genes simultaneously for a given condition. Currently, one common approach used in drug discovery is high throughput screening of chemical libraries against individual targets. However, a significant bottleneck in this process has been the identification of a suitable number of targets against which to screen compounds. Microarrays hold potential to identify many candidate genes and thus targets, involved in a given disease process at once. In addition to using more conventional techniques, pharmacogenomic analysis can also be used for target validation by generating response profiles which examine how a particular expression pattern may be altered in response to a particular compound.[13] Microarrays have also been used in similar fashion to identify the mechanism of action of existing drugs or other bioactive compounds[21] as well as to optimize and prioritize candidate compounds.

Another area in drug development where genomics is beginning to have a tremendous impact on is pharmacogenetics (how genetic variations affect individual drug response). SNPs are the most common genetic variant in the human genome and are important markers of disease susceptibility and drug response. Using high throughput sequencing and genotyping, genome-wide SNP[22] and haplotype maps[23] are being generated. By profiling these variations in individuals' DNA using microarrays, we hope to better predict responses to therapeutic intervention and thereby improve individual drug efficacy and safety profiles. These advances will ultimately allow us to understand how and why interindividual drug response variability exists and to then optimize and personalize individual therapy.

17.2.5 Beyond Genomics — Proteomics, the Next Frontier

For any given species, the number of biomolecules and their interactions and assembly into complexes, processes, and pathways, while large, is in theory finite. Genomics aims to capture much of this complexity at the level of the genome by uncovering how the hardwired genetic coding — the operating system for an organism — can drive a given phenotype in both health and disease through systemic profiling of mRNA levels. However, while tremendously powerful, there are clearly limitations to this approach, as we now know that the corresponding protein complement is roughly 6–7 times greater than that of the genome, due to alternate splicing and post-translational modifications. Clearly then, additional complementary approaches are necessary for us to have a comprehensive, integrated understanding of biological systems and methods with which to impact them. Because proteins are involved in virtually every cellular process and regulatory mechanism, are modified either as a cause of or result of every disease process, and thus, ultimately dictate phenotype, understanding the protein complement is an essential component to this strategy.

Proteomics is the systematic study of the protein complement of the genome. The proteome consists of all proteins in a cell at a given time including not only those directly translated from mRNA, but also those modified through alternate splicing and post-translational modification. The goal of proteomics is not only the systematic identification of the protein complement, but also the sequencing, quantification, state of modification, interaction with other proteins, activity, subcellular distribution, and structure of every protein (Figure 17.2). Proteomics, thereore, will provide a tremendously powerful synergistic complement to genomics in terms of

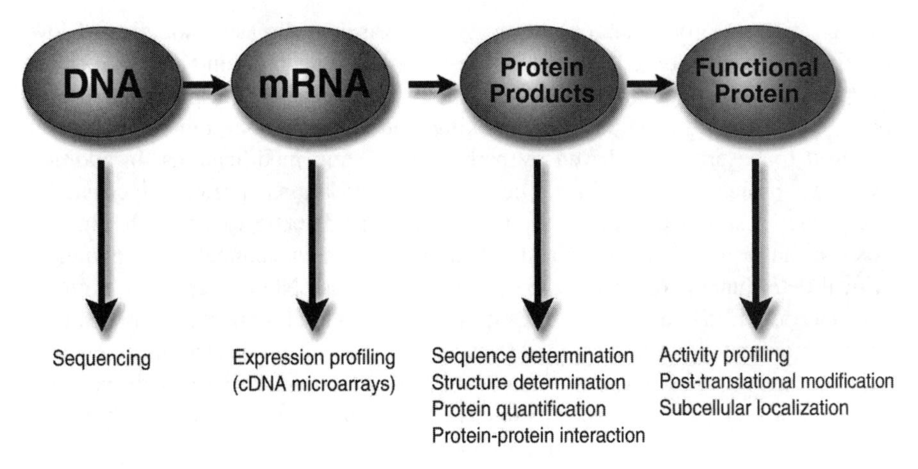

FIGURE 17.2 Biological flow of information from gene to functional protein with associated technologies.

our understanding of the structure, function, and control of complex biological systems.

A number of advances in technology have allowed us to tackle this problem with greater ease. Mass spectrometry is one such advance that has had a significant impact on the field of proteomics.[24,25] In tandem with precision gel and chromatographic protein separation methods, mass spectrometry has allowed us to rapidly identify, characterize, and sequence individual proteins in complex mixtures.[26] Using labeling methods such as isotope-coded affinity tagging[27] and activity-based reagents,[28] we are now also able to quantify protein abundance as well as detect and identify functionally active protein classes. Further, extensions of lessons learned from DNA microarrays are being applied to proteomics with the development of protein, antibody,[29] and peptide microarrays[30] for proteomic assessment, though these technologies are still in their infancy. Finally, methods of detailing protein–protein interaction maps are being developed and include, but are not limited to, technologies such as the yeast two-hybrid system[31,32] and mass spectrometry.[33,34]

17.2.6 Translating Proteomics to Vascular Disease

Given that there is no strict linear relationship between gene and protein complement and that the vast majority of current therapeutics are targeted against proteins, understanding proteins and how they are modified in vascular disease is of paramount importance. In general, proteins are not only more abundant than their corresponding gene complements, but also are much more diverse structurally and functionally. In addition, proteins undergo numerous post-translational modifications including phosphorylation, methylation, complex glycosylation, acetylation, ubiquination, oxidation, and proteolysis. In general, it is believed that acute vascular insults are manifested as post-translational modifications, while chronic vascular conditions are predominantly characterized as changes in protein abundance owing to differential gene expression and isoform switching.

A variety of protein alterations are seen in vascular disease. Oxidation of low-density lipoprotein at the level of oxidative modifications of amino acid side chains of apoprotein B is well known to contribute to atherogenesis. Further, atherothrombotic conditions associated with smoking and hyperhomocysteinemia are also believed to be primarily driven by pathologic protein modifications. In addition, extensive protein modifications, such as post-translational protein glycosylation events play a significant role in the pathogenesis of diabetic vasculopathy. Indeed, post-translational modification and alteration in protein abundance of numerous critical determinants of vascular homeostasis, such as eNOS, heat shock proteins, E-cadherin, NF-κB, and AP-1, in response to acute and chronic perturbations, have been demonstrated in a wide range of vascular diseases. Thus, proteomic assessment will play an essential role in our ability to evaluate complex vascular disease processes, to assess their relationships to environmental factors, and to identify targets to impact with next generation micro- and nanoscale molecular therapeutics.

17.3 VASCULAR BIOSENSORS

As scientists strive to understand biology in greater detail on the cellular and molecular level, more sensitive tools are required. Because virtually every disease process results from or causes alterations in the vascular system, it is ideally situated for detection of cellular and molecular changes. This, however, requires the miniaturization of current biosensor technologies without a concomitant loss of sensitivity or function. With the recent gains made in nanotechnology, such sensors are beginning to take shape. In the following text, we survey recent advances in nanoscale biosensor technology that will soon allow us to utilize these technologies across a broad range of disease processes involving the vasculature.

Vo Dinh et al.[35] provide the simple definition of a biosensor as a device consisting of a bioreceptor, the biorecognition system, and a transducer. The bioreceptor produces a physical effect that is converted to a measurable signal by the transducer. Common bioreceptor systems utilize the interactions of antibodies and antigens, nucleic acids (DNA), enzymes, cells, or synthetic biomimetic materials. Transducer signals can be optical, electrochemical, or mass sensitive. With the advances made in micro- and nanofabrication, miniature microelectromechanical systems (MEMS) and nanoelectromechanical systems (NEMS) show great promise as potentially catheter-based and/or implantable biosensors. Three categories of nanobiosensors are discussed here: optical-based, nanoparticle-based, and nanowire-based. Each of these can utilize one or many of the receptor and transducer combinations listed.

17.3.1 Optical

Optical transducers offer the largest number of possible solutions for biosensors due to advanced development of the field of spectroscopy and the numerous types of spectroscopy possible: absorption, reflection, fluorescence, phosphorescence, refraction, and others.[35] In addition, there are a number of spectroscopic properties that can be measured: amplitude, phase, energy, polarization, decay time, etc. Out of all of these options, probably the most widely used in biosensors, especially with the creation and growing use of DNA chips, is fluorescence.

One of the most common optical techniques is to use a reporter such as green fluorescent protein (GFP) to probe specific protein expression within cells. Fluorescent molecules attached closely to active sites on cell proteins transduce protein-target interactions into fluorescent signals. In this way, fluorescence can be used to probe interactions between cell proteins and RNA, DNA, carbohydrates, lipids, and organelles.[36]

Fluorescence resonance energy transfer (FRET) is utilized to view proteolytic enzyme activity, molecular interactions, and protein folding in real time.[37] FRET is based on the concept that when two fluorophores come together, excitation of a donor fluorophore results in the transfer of energy to an acceptor fluorophore, which then fluoresces at the acceptor's characteristic wavelength.[38]

Optical technology has also been used to develop an optical nanofiber for probing single cells.[35,39] The fiberoptic probe is 40 nm in diameter at its tip and is small enough to place inside a cell and can be removed without damaging the cell. Fluorescent versions of biomolecular probes like antibodies are attached to the tip of the fiber. Once in the cell, the cell's copy of the biomolecular probe displaces the previously attached fluorescent one, thereby reducing the fluorescent signal and allowing visualization of the presence of the biomolecule in the cell.

Surface plasmon resonance (SPR) is another widely used optical technique, one that has been in use since the 1980s. SPR utilizes an evanescent wave phenomenon that occurs at the interface between a metallic surface (typically a thin gold film) and a solution. SPR devices monitor in real time the internal reflected light angle from the gold surface; this angle depends on the mass of the surface.[40] SPR biosensors are made by immobilizing receptors to the gold surface and exposing it to a solution of an analyte of interest. As the molecules in solution bind to the surface, the mass of the surface layer changes, and the reflected light angle changes proportionally. SPR accurately measures the amount of bound analyte, its affinity for the receptor, and the association and dissociation kinetics of the reaction. It is also extremely sensitive. A wide range of molecular weights can be sensed as well as interaction affinities from millimolar to picomolar strengths. SPR devices are manufactured by many companies and are widely used in drug discovery. Few applications, however, have attempted to miniaturize systems for use as implantable biosensors.

One of the most promising adaptations of optical techniques to miniaturized biosensors is the development of a multifunctional biochip (MFB) that has incorporated an electro-optical system. This chip is a self-contained integrated circuit with photodiode arrays, electronics, amplifiers, discriminators, and logic circuitry.[35,41] MFBs can function similarly to DNA chips but can include arrays of different probes for DNA, antibodies, enzymes, and cells. As examples, MFBs have been used in measurements using DNA probes specific to gene fragments of the *Mycobacterium tuberculosis* system and antibody probes targeted to the cancer-related tumor suppressor gene p53.[35]

17.3.2 Nanoparticles and Nanopores

Nanoparticles have caught scientists' attention because they easily pass through cell membranes and into remote parts of the body without adverse effect. Nanoparticles gain great potential for biosensors by coupling them with optical and other medical imaging modalities (for instance, CT or MRI). Nanoporous materials and their use as protective membranes and filters are also important for the development of implantable biosensors. Such protective membranes will help prevent the rejection of the biosensors as foreign bodies. Moreover, it enables the potential use of living cells for biosensing and as an *in vivo* cellular factory for bioactive molecules for treating disease.

"Quantum dots" are nanocrystals of cadmium selenide, silicon, and other materials that fluoresce in many different colors under white light excitation. Their colors can be controlled by their size and material, and they are much more photostable than fluorophores. These nanoparticles can be linked to biomolecules to create numerous biologic probes. Some proposed vascular applications are the screening of blood for different viruses simultaneously or for proteins associated with myocardial infarctions and aneurysms.[42,43] In addition, by tagging the appropriate molecules inside and outside cells, cellular processes can be visualized. This could be potentially used to elucidate the molecular pathogenesis of a wide range of vascular disease entities and processes.

A related technology is PEBBLES (Probes Encapsulated by Biologically Localized Embedding). PEBBLES are polymer nanoparticles with fluorescent dyes trapped in the polymer matrix.[44] The particles are ideal for sensing intracellular activity, as the polymer protects the cells from adverse effects of the dye. The particle size allows it to remain in the cytoplasm for several days. PEBBLES have been used to sense a number of physiologic parameters within cells, including levels of pH, oxygen, calcium, potassium, free zinc, glucose, and many others.

Nanoparticles in the shape of rods are being coated with metallic stripes in different patterns to create nanobarcodes. The metallic patterns can be identified using conventional light microscopy. These bar-coded rods can then be made to sense any number of biologic molecules and cells by functionalizing their surfaces. Fluorescence imaging is used to detect and quantify bound analytes while the bar codes are used to identify the analytes.[45]

Nanopores are being exploited for biosensing as well. In one example, a nanopore 1 nm in diameter is used to sequence DNA.[37] The charged DNA is drawn through the pore using a small electrical potential. By recording the current level (typically in pico-amps) and time, it may be possible to sequence DNA. This application of nanopores could potentially allow DNA sequencing of up to 1000 bases per second with very small sample volume, a considerable advancement over current sequencing technology.

Nanopore technology has also been utilized for membranes and filters. In a special case, they are implemented as immunoisolation membranes that block immunomolecules capable of cellular rejection.[46] In this application, silicon biocapsules were microfabricated with uniform pore sizes on the order of 10 nm. Pancreatic islets were placed in the biocapsules. Immune molecules with sizes >15 nm were rejected, while insulin and nutrients could pass through the pores. Islet functionality was proven by measured response of the islets to glucose stimulation. The immunoisolation properties of nanopore membranes has considerable potential for use in vascular biosensors as attention turns to developing implantable nanodevices that would otherwise be compromised by immune responses.

17.3.3 Nanowires and Nanotubes

Nanowires can serve as highly sensitive devices for detecting chemical binding to almost any chemical or biological receptor. In this type of device, a short wire tens of nanometers in diameter is coated with the desired receptor, and a binding event results in an electrical signal that can be sensed through current, voltage, or conductance measurements.[47] Such sensors can detect the binding of nucleic acids, proteins, and ions. Nanowires and nanotubes are considerably simpler than many optical sensors, requiring no special dyes or expensive detection systems. In addition, they can be adapted to work with liquids and gases and are easily placed in microfluidic channels for continuous sampling in streams of fluids. As inexpensive and portable alternatives to optical biosensors, nanowire technologies are the most promising for implantable biosensing devices.

Several nanowire detection devices are in development. For example, boron-doped silicon nanowires (SiNW) have been used as pH sensors.[48] In another example, biotin-modified SiNWs have been used to detect the ligand-receptor binding of streptavidin. These nanowires were capable of detecting streptavidin binding down to picomolar concentrations. Finally, calmodulin-coated nanowires have been shown to be capable of sensing Ca^{2+} ions.

Nanowires are also used in DNA electrochemical biosensors that may serve as alternatives to DNA microarrays.[47] The DNA electrochemical biosensor is a nanowire electrode with an oligonucleotide immobilized on its surface. The hybridization of the surface by an analyte produces an electrical signal. In a study illustrating the potential clinical application of the DNA biosensor, a DNA sensor was used to detect different genotype polymorphisms of apolipoprotein E (apoE),[49] a constituent of plasma lipoproteins such as VLDL and HDL that has been linked to the risk of developing cardiovascular disease and to familial type III hyperlipoproteinemia.

Nanowire technology is being applied to biosensors in different ways. In one approach, nanowire technology is used to create arrays small enough to probe single cells.[39] By coating each wire in the array with a different antibody or oligonucleotide, the array can potentially monitor hundreds of cellular processes. In another different but complementary approach, cell membranes are lysed, and contents are placed onto nanowires loaded with DNA.[39] While this does not allow for continuous cellular monitoring, it can probe the gene expression in the cell before it was killed. Both approaches have great potential for use in vascular implantable and catheter-based biosensors.

Finally, the quartz crystal microbalance (QCM) is another example of a biosensor using nanowire technology. It is an oscillating crystal whose resonant frequency changes with changing mass on its surface. It can be used to measure cell proliferation, differentiation, and respiration.[50] By coating the surface of a quartz nanobalance with DNA, the presence of complementary DNA in a sample can be detected by the mass change as the cDNA binds to the crystal surface.[47]

Carbon nanotubes (CNTs) are long, thin cylinders of carbon that are unique for their size, shape, and broad range of electronic, thermal, and structural properties that change depending on the diameter, length, and chirality of the tube. CNTs are used in fine-resolution atomic force microscopy (AFM) and chemical force microscopy (CFM).[37] In AFM, a carbon nanotube is grown or attached to the end of a sharp silicon tip. The deflection of this CNT is measured as it scans a surface, resulting in images with atomic resolution. CFM uses chemically modified CNTs to probe chemical and biochemical interactions. CFM can probe the local chemistry of a surface it is imaging. CFM has also been used to measure ligand binding, mechanical responses of molecules, and enzymatic activity.

Researchers at NASA Ames Research Center are using arrays of CNTs (each 30–50 nm in diameter) on a silicon chip (Figure 17.3).[51,52] DNA molecules are attached to each CNT and exposed to a solution containing DNA of interest. When the target DNA binds to a CNT, the conductance increases and is sensed. The CNT array is expected to reach the sensitivity of current fluorescence-based DNA chips and has higher potential for miniature sensors.

Overall, tremendous progress has been made in the development of useful biosensor platforms. The potential applications to vascular disease are tremendous and range from better understanding of aneurysm progression and atherosclerosis to diagnosis and control of cancer and regional angiogenesis. Focused application of biosensor technology for these applications is now beginning to be feasible in a clinically relevant way. The next decade should be marked by tremendous real-world reductions of these strategies for vascular disease, cancer, and beyond. Expect significant impact in medical understanding of disease and quality of life for patients,

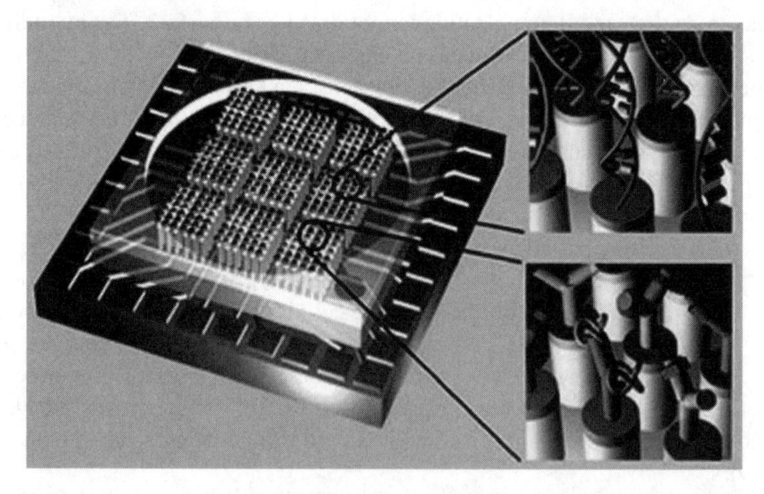

FIGURE 17.3 Carbon nanotubes (CNT) in a silicon chip. The ends of the CNTs can be functionalized to attract DNA and other analytes. Once the target molecule is bound, it is sensed through a change in the CNT's electrical conductivity. (From Smalley E. Chip senses trace DNA. *Technol. Res. News.* July 30/Aug. 6, 2003. With permission.)

as well as an evolution in the medical device industry's profit base to encompass more and more hybrid devices which include a diagnostic component.

17.4 NANOSCALE THERAPEUTIC TRANSPORT SYSTEMS FOR VASCULAR DISEASE

Vascular diseases are particularly well-suited to molecular interventions. Obviously, the very high prevalence of these diseases plays a role — in fact, the World Health Organization projects that one out of every two people on the planet will die of cardiovascular disease by the year 2020.[1] Beyond the contextual importance, vascular diseases have a marked genetic component both in susceptibility and in management. Because disease occurs frequently in diagnosable and defined locations and is accessible via advancing endovascular approaches, these diseases are particularly appealing as targets for molecular medicine.

The advent of gene therapy has afforded considerable progress in understanding functional and genetic components of vascular disease. However, clinical applications for the therapeutic potential of gene therapy remain largely undeveloped due to limitations with delivery vector systems and concerns with toxicity. The evolving paradigms of molecular medicine, particularly nanodelivery systems, offer renewed promise for understanding vascular diseases and providing clinically meaningful human interventions.

Concern over the safety of viral delivery vectors has driven considerable progress in the development of nonviral systems for gene delivery. As in a viral system, several discrete components are often necessary in the final complex. Initially, a number of investigators focused on synthetic chemical means of bringing these components together. With the advent of nanotechnology, the approach has progressively shifted to emphasize self-assembly of the discrete components into final nanoscale particles. This approach actually mirrors self-assembly of viral particles — the original template for nonviral systems — more closely. At its simplest level, charge-based interaction can allow assembly of polycation-based backbones such as activated dendrimers with nucleotide-based therapeutics which are polyanions. Considering the surface of every cell is essentially a polyanion, neutralization of the nucleotide-based therapeutics' negative charges offers functional benefits in increased efficiency of gene delivery in and of itself. As may be expected, an excess of positive charges in the complex typically affords further gains by facilitating interaction of the complex with cell surfaces.

Charge-based interactions in assembly of these particles offer a number of other advantages as well. When employed sequentially with several polyanions, charge-based interaction allows more even distribution of multiple therapeutics on chains. Essentially, the polyanions can be dispersed more evenly by repelling one another and then be assembled in a probabilistic manner into final assembled complexes (Figure 17.4). This approach has been extended to allow incorporation of polyanions tagged to targeting moieties such as Fab fragments or ligands for particular receptors and to imaging moieties such as polyanions tagged with CT, MR, or optical contrast agents. Each is dispersed, then assembled again in a probabilistic fashion into final

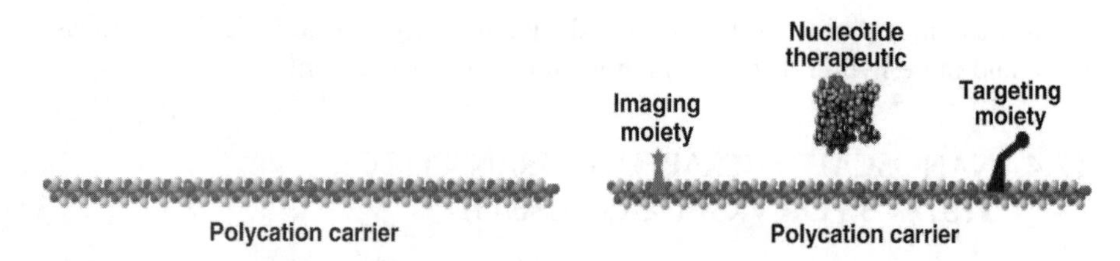

FIGURE 17.4 Nanoscale multicomponent transport system. Components for delivery include a polycation backbone carrier (left), which noncovalently self-assembles with nucleotide therapeutics and anion-tagged targeting moieties and imaging agents (right). The particle overall hosts an excess of positive charges and is nanoscale in size.

nanoscale particles, which allow delivery of the therapeutic, enhanced efficiency at a particular phenotype of cell, and imaging both for initial diagnostics and for ongoing monitoring of therapy.

Similar strategies have also been employed using nanopartitioning of composite particles containing, for example, lipophilic and hydrophilic moieties. Other nonionic interactions have also been employed to facilitate efficient loading of particular moieties at different stages of particle assembly. Perhaps simplest of these approaches is the use of avidin-biotin interactions to allow surface binding of an agent. Regardless, chemical and mechanical techniques for assembly of therapeutic nanoparticles have made tremendous progress in recent years, and continued progress will allow realization of therapeutic efficacy for molecular approaches to vascular disease.

Future development of nanotechnology for molecular medicine, at least in vascular applications, will likely be focused on scaffold assembly and controlled delivery strategies. Clinical applicability of delivery devices such as stents (discussed below) allows for a platform for assembly and delivery of nanotherapeutics for vascular disease. Recent evidence has suggested that a large portion of the delivery efficiency of nucleotide-based therapeutics in fixed durations is based on dispersion of the nanoparticles. For example, it has been known for some time that ultrasound can facilitate gene delivery, even at powers that do not allow membrane permeabilization. By performing cell culture experiments at 4°C, where endocytosis does not occur but nonspecific permeabilization does, and comparing to results from 37°C, it has become apparent that ultrasound can facilitate specific gene transfer. Other authors have demonstrated that incorporation of nanoparticles into multimers or nanoaggregates further increases efficiency of gene delivery when applied on top of cell monolayers. Mathematical modeling suggests that the efficiency in a limited period of time was thus based on the density of nanoparticles on the lamina in contact with the cells to be targeted. This is particularly important for the delivery of any nanotherapeutic, since appropriate dispersion and local delivery of the therapeutic may largely determine effectiveness. These conclusions have led many to evaluate scaffolding to incorporate and release specific nanotherapeutics over time. These topics will be considered more fully as an example of stent-based nanodelivery systems below. Due to the potential to have functional interactions with the scaffold itself, as well as improve the efficacy of any embedded nanotherapeutic, scaffolding will likely offer tremendous gains in bringing nanotechnology to a clinically relevant state. The eagerness for clinical application of drug-delivery devices has paved the way for more broadly bioactive devices to be employed. Ultimately, molecular therapeutics and devices will both improve as a result.

17.5 GUIDED REMODELING: AN *IN VIVO* VASCULAR ENGINEERING PARADIGM

Guided remodeling of vascular tissue serves as an illustrative model for many of the considerations occurring throughout nanoscale surface and scaffold engineering for *in vivo* applications.[53] Atherosclerotic disease represents the leading cause of death worldwide and is projected to exceed all other causes combined by the year 2020.[1] Angioplasty and saphenous vein bypass grafting are the two most common interventions for cardiovascular disease. However, restenosis limits the long-term efficacy of both interventions.[54,55] Clinically significant restenosis occurs in 40% of patients undergoing balloon angioplasty within 6 months[54] and in 20% of patients receiving vein graft bypasses within 1 year.[55,56] Neointima formation, whether in the balloon-injured artery or in the grafted vein, develops in a similar pathogenic sequence. In both instances, restenosis is characterized by the accumulation of vascular SMCs and extracellular matrix (ECM) in the intimal compartment.[57,58] In response to mitogenic and chemotactic cues within the vessel wall, SMCs proliferate, migrate from the media toward the intima, and deposit ECM.[57–59]

SMC migration directed from the media to the intima is essential for intimal thickening to occur in vascular disease. Many factors have been shown by *in vitro* and *in vivo* studies to be capable of providing the chemotactic impetus necessary for SMCs to undergo this migratory response.[60] However, previous attempts at disrupting this process have focused only on decreasing either the amount of stimulus or the ability of SMCs to respond to such stimulus.[61,62] Intimal thickening is a directional process, predicated on the presence of chemotactic cues within the vessel wall that can guide SMCs in the necessary media-to-lumen direction.

Elastin, an ECM component of the vessel wall, is a critical determinant of SMC activity. In various studies employing elastin knockout mice,[63] human atherosclerotic plaque specimens,[64] pig and rabbit models of cardiac transplant arteriopathy,[65,66] and rodent balloon injury models of restenosis,[67] increased endogenous elastase activity and decreased elastin content within the vessel wall have been associated with increased intimal thickening. Consistent with these roles, elastin regulates SMC migration via several mechanisms. Peptides derived from degraded elastin are directly chemotactic to SMCs.[60] In addition, SMC detachment from elastin fibers is promoted when the 67-kD elastin-binding protein on the cell surface is "shed" in the presence of galactosugar-containing ECM elements (e.g., chondroitin sulfate).[68,69] Production of fibronectin, an ECM element critical to SMC migration, also increases in the vessel wall following elastin-binding protein shedding.[70]

In prior work, perivascular elastase release achieved a pattern of elastin degradation that was preferential to the outer arterial wall (Figure 17.5A).[71,72] Since elastin-derived peptides are known to be directly chemotactic to monocytes,[73] and because

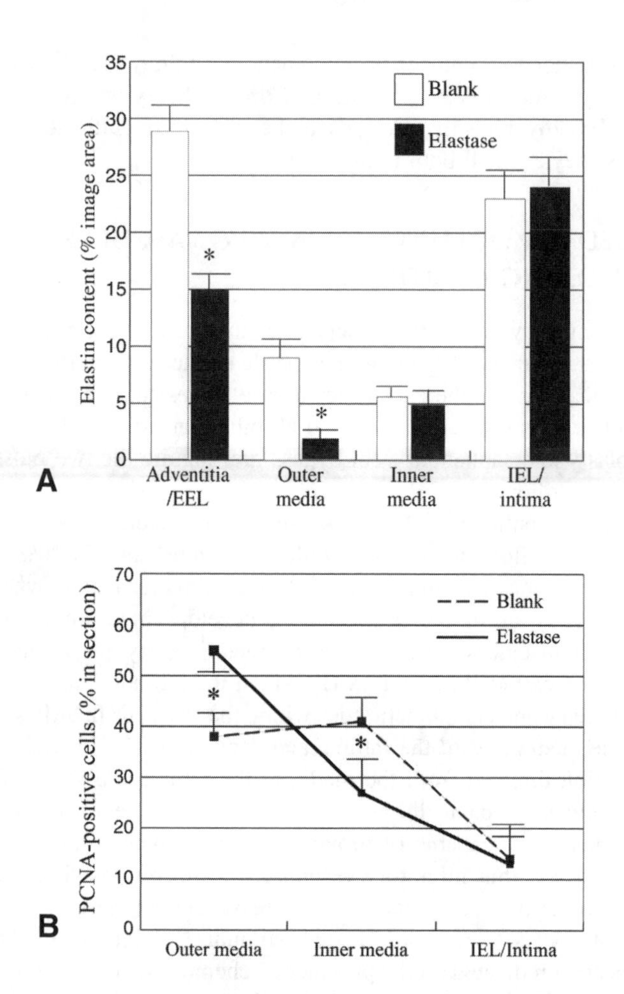

FIGURE 17.5 Elastin distribution and SMC proliferation by arterial wall zone. Bioerodable microspheres containing elastase or buffer (control) were applied perivascular to balloon-injured rabbit common femoral arteries ($n = 5$ per group). At 7 days, treated arteries were harvested, fixed in 10% formalin, and stained for elastin content and proliferating cell nuclear antigen (PCNA). (A) To confirm that elastase microspheres function *in vivo* and can establish an altered gradient of elastin degradation across the arterial wall, sections were stained with Verhoeff von Gieson (VVG)–Masson trichrome stain. Cross sections of the arterial wall were divided into four zones: adventitia including external elastic lamina (EEL), outer half of media, inner half of media, and intima including internal elastic lamina (IEL). The elastin content in each zone was determined and expressed as a percentage of total area. Results demonstrate preferential degradation of elastin in the outer zones. (B) To evaluate smooth muscle cell proliferation, sections were stained for PCNA, and the number of positively stained cells was tabulated for each section, and the percentage of positive cells for each zone was calculated. The total number of PCNA-positive cells did not differ significantly between groups (data not presented), but the distribution of proliferating cells was significantly greater in the outer media for the elastase-treated group, suggesting that perivascular elastase release led to SMC migration away from the lumen. Values represent mean ± SE; $*p < 0.05$.

neutrophils also elaborate elastases,[74] we confirmed that treatment with elastase was not associated with any increases in local inflammatory cell infiltrate. Thus, the observed elastin cleavage gradient was not due to local inflammation in the elastase group but to the elastase released. This prior work validated the notion that elastin degradation gradients could allow applied engineering of vascular responses on the substrate of native architecture *in vivo*. Additional work with regulation of elastin production further completed the demonstration.

As suggested by the work with elastin, ECM elements (e.g., collagens, elastin, proteoglycans) in the vessel wall are important regulators of SMC migration. Unperturbed and intact matrix, as present prior to vascular injury or disease, normally inhibits migration of SMCs via specific receptor signaling,[75] by presenting an architectural barrier to migration,[76] and by binding and making unavailable potentially pro-migratory factors.[77] In contrast, degraded or cleaved matrix, as present in the proteolytic milieu of injury or disease, stimulates SMC migration by breaking down physical barriers to migration, releasing previously bound pro-migratory factors, and serving as a direct chemoattractant to SMCs (Figure 17.5B).[60]

Tissue plasminogen activator (tPA) is a naturally occurring enzyme that directly cleaves nonfibronectin ECM elements[78,79] and converts plasminogen to plasmin, which in turn activates elastase and matrix metalloproteinases to cleave fibronectin, laminin, and other ECM proteins directly.[80,81] We recently demonstrated that lumen-based tPA activity, whether through viral-mediated expression or direct application of recombinant enzyme, enhances neointima formation by ECM degradation and subsequent SMC migration toward the lumen (Figure 17.6).[82] We also demonstrated that the earliest observable change after tPA exposure is ECM degradation in a graded fashion.

Since cleaved ECM is chemotactic for SMC migration, we employed ECM cleavage gradients to guide SMC migration in both post-injury arteries and vein grafts *in vivo*. In that work, perivascular delivery of tPA yielded graded ECM degradation across the vessel wall to chemotactically guide SMC migration away from the lumen and limit neointima formation. We applied this strategy first on a well-characterized *in vivo* arterial balloon angioplasty model, to provide proof of principle, and then on a more clinically relevant *in vivo* vein graft model. In both the arterial and vein graft models, we successfully modified ECM gradients, directed SMC migration outward, and limited pathologic neointima formation, demonstrating successful application of *in vivo* vascular engineering on native substrates *in situ*.

While these applications are more biochemical than nanotechnology-related, each provides important insight into considerations for applied nanotechnology, particularly of composite scaffolds. Nanoscale chemotactic and architectural cues will allow controlled population of even a complex scaffold. Translating this work to engineered scaffolds will also allow refinement of surface micropatterning to apply both mechanical and biochemical interfaces that achieve the desired effects. Device-based and tissue-based nanoscale remodeling can thus form the basis of an applied macroscale tissue engineering, whether on a substrate of native architecture as considered above, or of an exogenously provided scaffold. These concepts are particularly relevant to engineering new vascular networks as discussed below. An understanding of the key small-scale regulators of these gradients as well as the

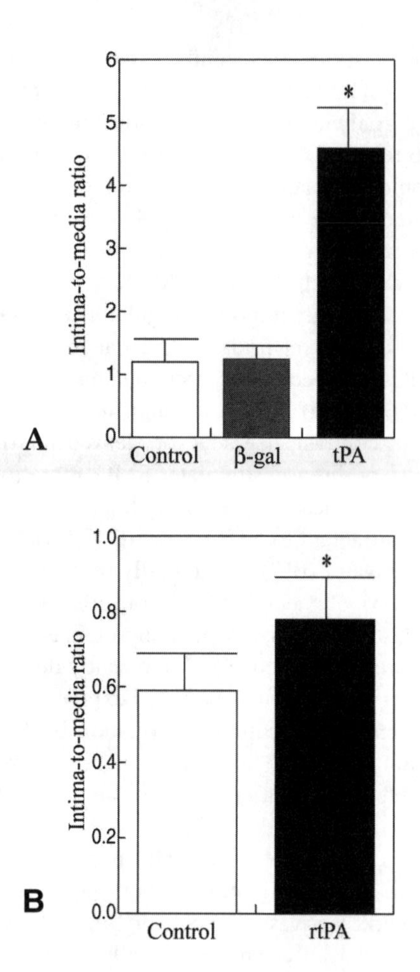

FIGURE 17.6 Intima-to-media ratio after tPA overexpression or infusion. (A) Balloon-injured rabbit common femoral arteries were transfected with adenoviral-constructs expressing either the β-galactosidase gene (β-gal) or tPA ($n = 4$ per group). At 28 days, treated vessels were harvested and examined for intima-to-media ratios. Local overexpression of tPA resulted in significant enhancement of the intima-to-media ratio. (B) In a similar experiment, balloon-injured rabbit common femoral arteries were subjected to systemic administration of buffer or recombinant tPA (r-tPA) according to clinical protocols. Systemic infusion also increased the intima-to-media ratio. Values represent mean ± SE; $*p < 0.05$.

tissue-surface interface will provide the context for problem-driven engineering solutions to vascular nanotechnology.

17.6 FROM ANGIOGENESIS TO VASCULOGENESIS

Angiogenesis is important for restoring blood flow in the heart and in the peripheral vasculature in diabetics as well as preventing or destroying blood flow to cancer cells. *In vivo*, angiogenesis relies on multiple processes and pathways, each controlled by many molecular signals and growth factors. Recent *in vivo* studies are investigating the effects of fibroblast growth factors (FGF) and vascular endothelial growth factors (VEGF) (the VIVA trial) in many different regions of the circulation, but primarily in the heart.[83–90] These studies, which rely mostly on direct injection as well as plasmid and adenovirus delivery of growth factors, have shown limited success, indicating that there is still much to be understood before angiogenesis can be a clinically viable treatment. Nanotechnology provides great potential for helping us understand and control angiogenesis.

Angiogenic nanotechnology can take the form of molecular agents like growth factors that promote specific cell activity or encourage cell migration and population of ischemic areas, cellular agents that produce desired factors in ischemic tissue,[91] or mechanical devices which provide support for cells and supply the desired time course of pro- or anti-angiogenic factors. Drug-releasing biodegradable polymer microspheres with diameters of 1–100 microns is a common example. Micromachined biogels provide another potential carrier for angiogenic stimulation. Surface patterning of gels through photolithography can provide microfeatures which are known to affect cell adhesion, arrangement, and growth.[92]

Another nanodevice is the "angiochip."[93] It is a silicon capsule with nanofilters (pores of 10- to 200-nm diameter) useful for molecular filtering as well as controlled drug release. To serve as angiogenic promoters, appropriate factors can be placed inside these capsules and the outside coated with endothelial cells. Arranging the endothelial cells into micromachined capillary-like grooves in the surface of the capsules promotes microvascular growth, which can connect to existing microvasculature. If hundreds to thousands of these capsules were implanted, they could restore blood flow to a potentially large tissue bed.

A less complicated nanoparticle, consisting of a polymerized-lipid core with 40-nm diameter, has been used as an anti-angiogenic device.[94] Hood et al. linked this cationic nanoparticle to an $\alpha_v\beta_3$-targeting ligand and to several plasmid DNA species including green fluorescent protein and ATP$^{\mu}$-Raf, an apoptotic agent. The nanoparticles were selectively delivered to tumor-related endothelial cells in mice with the positive end result of tumor regression.

17.7 BIOACTIVE ENDOVASCULAR STENTS

One of the particular strengths of vascular disease as a context for nanotechnology is the ubiquitous use of implantable devices and improved endovascular techniques for their delivery. Balloon angioplasty with stent placement is the most commonly employed primary intervention for cardiovascular occlusive disease. However, in-stent restenosis, which occurs in up to 40% of patients,[95] limits the long-term efficacy of stenting.[96–98] In-stent restenosis is the result of a multifactorial wound healing process in response to vessel injury,[58,99] characterized by neointimal hyperplasia, negative vascular remodeling, elastic recoil, and thrombosis.[100–103]

Considerable attention has recently focused on drug-eluting stents as a means of limiting in-stent restenosis. Preliminary unpublished results from commercial trials of polymer-coated rapamycin-eluting stents have reported "zero restenosis," received regulatory approval in Europe, and understandably generated tremendous enthusiasm from the cardiovascular community.[104] Like several other drugs evaluated for prevention of restenosis, rapamycin has powerful anti-proliferative and anti-inflammatory properties.[105] However, other agents with comparable anti-proliferative and anti-inflammatory actions have not achieved rapamycin's apparent early clinical success in limiting in-stent restenosis, suggesting that other rapamycin-sensitive pathways may be important as well.[106] Since rapamycin exerts a broad range of effects, the relative importance of each pathway to rapamycin's overall efficacy remains obscure.

In-stent restenosis follows a more accelerated progression than normal atherosclerotic plaque. Although the location and time-course of in-stent restenosis are well characterized, the mechanisms underlying the disease process have not been well defined. Using a stent-based controlled release platform (Figure 17.7),[107] we have delivered several endogenous factors to elucidate the underlying mechanisms of in-stent restenosis. Through controlled stent-based delivery of angiostatin, we found that limiting microvessel density leads to reductions in macrophage infiltration and SMC proliferation, which together provide a significant overall decrease in early and late plaque progression after stenting. These findings suggest that other clinically significant therapies such as stent-based rapamycin release may benefit to some degree from inhibition of plaque neovascularization. Furthermore, our findings demonstrate that angiogenesis is required for progression of atherosclerotic plaque after stenting, which raises a number of interesting issues about this disease and its potential management and may allow further refinement in the development of anti-restenotic agents. We have applied a number of stent-based controlled-release strategies to elucidate roles of nitric oxide, superoxide dismutase, fas ligand, transforming growth factor beta 1, angiotensin II inhibitors and many others in post-stenting restenosis. These factors provide insight into what bioactive stents may want to include in their scaffolding. This work must be viewed against the backdrop of work from Palmaz and others that demonstrate that surface nano- and micropatterning alter rates of preferential migration of various cell types and even preferential recruitment of marrow-derived precursors. In this perspective, this work suggests a new phase in the development of bioactive devices, where the physical structure can exert mechanical and biologic effects, and the nanotherapeutics incorporated into the scaffold can also provide both mechanical and biologic effects as well.

As one approach to the problem, our group has looked to the tremendous progress made in gene chip development. A surface can be precisely derivatized with nucleotides — affording a spatial nanopattern on the surface as well as a specific depth profile encoded by the sequence of the nucleotide itself. A nucleotide strand complementary to portions of the anchored strand can then be anchored to the surface of a nanoparticle (or smaller). The complementary strand, with appropriate attention to sterics, can then be specifically self-assembled under hybridization conditions to the prefabricated surface. Since serum contains exonucleases, but essentially low endonuclease activity, a timed surface-based delivery of the nanoparticles will be afforded. Control of surface pattern and of identity of nanoparticles can be afforded from this simple system. Obviously, similar strategies with environmentally sensitive linkers and other assembly strategies can be employed as well, with varying degrees of complexity.

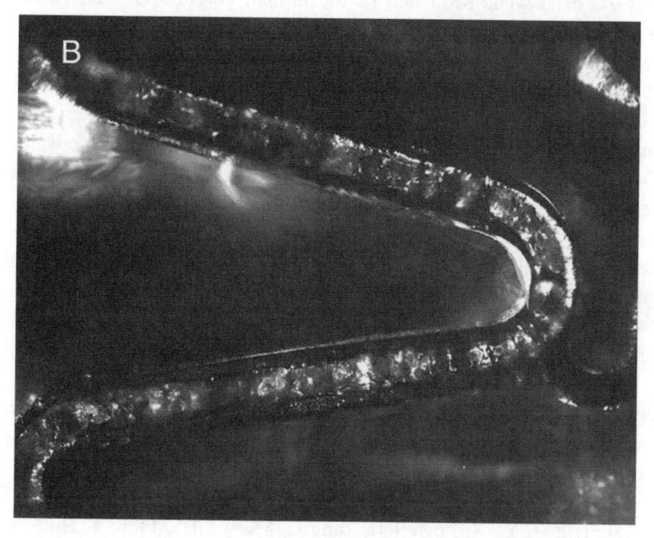

FIGURE 17.7 Stent-based controlled-release platform. Bioerodable microspheres containing drug or growth factor are loaded onto channeled stents (1.6 mm in diameter, 10 mm in length, with 10 longitudinal struts evenly spaced circumferentially, and balloon expandable up to 5 mm in diameter). Stents are shown in (A) tangential view and (B) after *in vivo* implantation and recovery.

17.8 CONCLUDING REMARKS

Micro- and nanoscale technologies promise to have a profound impact on our understanding and control of complex biological systems and the means with which to impact them in a precise and predictable manner. Genomics and proteomics are fundamental micro- and nanotechnology-oriented tools that will drive the further expansion of biomarker and drug target identification. This in turn, will provide the foundation for the pursuit of new drugs, formulations, delivery techniques, sequencing tools, and screening techniques, which will drive the growth of next generation micro- and nanotechnology tools for the diagnosis and treatment of vascular disease. Such growth will be characterized by powerful new methods of *in vivo* detection and diagnosis, monitoring, and treatment paradigms for complex vascular disease entities as we have discussed. In the coming years, expect to see translation of micro- and nanoscale tools and principles into virtually every aspect of vascular disease diagnosis and management. Successful translation of these emerging nanotechnologies promises to revolutionize the practice of vascular medicine.

REFERENCES

1. World Health Report 2000. Geneva, Switzerland: World Health Organization, 2000.
2. Lander ES et al., Initial sequencing and analysis of the human genome. *Nature*. 2001, 409(6822), 860–921.
3. Venter JC et al., The sequence of the human genome. *Science*. 2001, 291(5507), 1304–1351.
4. Schena M, Shalon D, Davis RW, Brown PO. Quantitative monitoring of gene expression patterns with a complementary DNA microarray. *Science*. 1995, 270(5235), 467–470.
5. Okamoto T, Suzuki T, Yamamoto N. Microarray fabrication with covalent attachment of DNA using bubble jet technology. *Nat. Biotechnol.* 2000, 18(4), 438–441.
6. Fodor SP, Rava RP, Huang XC, Pease AC, Holmes CP, Adams CL. Multiplexed biochemical assays with biological chips. *Nature*. 1993, 364(6437), 555–556.
7. Pease AC, Solas D, Sullivan EJ, Cronin MT, Holmes CP, Fodor SP. Light-generated oligonucleotide arrays for rapid DNA sequence analysis. *Proc. Natl. Acad. Sci. U.S.A.* 1994, 91(11), 5022–5026.
8. Hacia JG. Resequencing and mutational analysis using oligonucleotide microarrays. *Nat. Genet.* 1999, 21(Suppl. 1), 42–47.
9. Hacia JG, Brody LC, Chee MS, Fodor SP, Collins FS. Detection of heterozygous mutations in BRCA1 using high density oligonucleotide arrays and two-colour fluorescence analysis. *Nat. Genet.* 1996, 14(4), 441–447.
10. Alizadeh AA, Eisen MB, Davis RE, Ma C, Lossos IS, Rosenwald A, Boldrick JC, Sabet H, Tran T, Yu X, Powell JI, Yang L, Marti GE, Moore T, Hudson J, Jr., Lu L, Lewis DB, Tibshirani R, Sherlock G, Chan WC, Greiner TC, Weisenburger DD, Armitage JO, Warnke R, Levy R, Wilson W, Grever MR, Byrd JC, Botstein D, Brown PO, Staudt LM. Distinct types of diffuse large B-cell lymphoma identified by gene expression profiling. *Nature*. 2000, 403(6769), 503–511.

11. Golub TR, Slonim DK, Tamayo P, Huard C, Gaasenbeek M, Mesirov JP, Coller H, Loh ML, Downing JR, Caligiuri MA, Bloomfield CD, Lander ES. Molecular classification of cancer: class discovery and class prediction by gene expression monitoring. *Science*. 1999, 286(5439), 531–537.
12. Wilson M, DeRisi J, Kristensen HH, Imboden P, Rane S, Brown PO, Schoolnik GK. Exploring drug-induced alterations in gene expression in Mycobacterium tuberculosis by microarray hybridization. *Proc. Natl. Acad. Sci. U.S.A.* 1999, 96(22), 12833–12838.
13. Marton MJ, DeRisi JL, Bennett HA, Iyer VR, Meyer MR, Roberts CJ, Stoughton R, Burchard J, Slade D, Dai H, Bassett DE, Jr., Hartwell LH, Brown PO, Friend SH. Drug target validation and identification of secondary drug target effects using DNA microarrays. *Nat. Med.* 1998, 4(11), 1293–1301.
14. Staunton JE, Slonim DK, Coller HA, Tamayo P, Angelo MJ, Park J, Scherf U, Lee JK, Reinhold WO, Weinstein JN, Mesirov JP, Lander ES, Golub TR. Chemosensitivity prediction by transcriptional profiling. *Proc. Natl. Acad. Sci. U.S.A.* 2001, 98(19), 10787–10792.
15. Miki R, Kadota K, Bono H, Mizuno Y, Tomaru Y, Carninci P, Itoh M, Shibata K, Kawai J, Konno H, Watanabe S, Sato K, Tokusumi Y, Kikuchi N, Ishii Y, Hamaguchi Y, Nishizuka I, Goto H, Nitanda H, Satomi S, Yoshiki A, Kusakabe M, DeRisi JL, Eisen MB, Iyer VR, Brown PO, Muramatsu M, Shimada H, Okazaki Y, Hayashizaki Y. Delineating developmental and metabolic pathways *in vivo* by expression profiling using the RIKEN set of 18,816 full-length enriched mouse cDNA arrays. *Proc. Natl. Acad. Sci. U.S.A.* 2001, 98(5), 2199–2204.
16. Martinet W, Schrijvers DM, De Meyer GR, Thielemans J, Knaapen MW, Herman AG, Kockx MM. Gene expression profiling of apoptosis-related genes in human atherosclerosis: upregulation of death-associated protein kinase. *Arterioscler. Thromb. Vasc. Biol.* 2002, 22(12), 2023–2029.
17. Peters DG, Zhang XC, Benos PV, Heidrich-O'Hare E, Ferrell RE. Genomic analysis of immediate/early response to shear stress in human coronary artery endothelial cells. *Physiol. Genomics*. 2002, 12(1), 25–33.
18. Johnson CD, Balagurunathan Y, Lu KP, Tadesse M, Falahatpisheh MH, Carroll RJ, Dougherty ER, Afshari CA, Ramos KS. Genomic profiles and predictive biological networks in oxidant-induced atherogenesis. *Physiol. Genomics*. 2003, 13(3), 263–275.
19. Zohlnhofer D, Klein CA, Richter T, Brandl R, Murr A, Nuhrenberg T, Schomig A, Baeuerle PA, Neumann FJ. Gene expression profiling of human stent-induced neointima by cDNA array analysis of microscopic specimens retrieved by helix cutter atherectomy: Detection of FK506-binding protein 12 upregulation. *Circulation*. 2001, 103(10), 1396–1402.
20. Yajima N, Masuda M, Miyazaki M, Nakajima N, Chien S, Shyy JY. Oxidative stress is involved in the development of experimental abdominal aortic aneurysm: a study of the transcription profile with complementary DNA microarray. *J. Vasc. Surg.* 2002, 36(2), 379–385.
21. Hughes TR, Marton MJ, Jones AR, Roberts CJ, Stoughton R, Armour CD, Bennett HA, Coffey E, Dai H, He YD, Kidd MJ, King AM, Meyer MR, Slade D, Lum PY, Stepaniants SB, Shoemaker DD, Gachotte D, Chakraburtty K, Simon J, Bard M, Friend SH. Functional discovery via a compendium of expression profiles. *Cell*. 2000, 102(1), 109–126.

22. Sachidanandam R, Weissman D, Schmidt SC, Kakol JM, Stein LD, Marth G, Sherry S, Mullikin JC, Mortimore BJ, Willey DL, Hunt SE, Cole CG, Coggill PC, Rice CM, Ning Z, Rogers J, Bentley DR, Kwok PY, Mardis ER, Yeh RT, Schultz B, Cook L, Davenport R, Dante M, Fulton L, Hillier L, Waterston RH, McPherson JD, Gilman B, Schaffner S, Van Etten WJ, Reich D, Higgins J, Daly MJ, Blumenstiel B, Baldwin J, Stange-Thomann N, Zody MC, Linton L, Lander ES, Altshuler D. A map of human genome sequence variation containing 1.42 million single nucleotide polymorphisms. *Nature*. 2001, 409(6822), 928–933.
23. Johnson GC, Esposito L, Barratt BJ, Smith AN, Heward J, Di Genova G, Ueda H, Cordell HJ, Eaves IA, Dudbridge F, Twells RC, Payne F, Hughes W, Nutland S, Stevens H, Carr P, Tuomilehto-Wolf E, Tuomilehto J, Gough SC, Clayton DG, Todd JA. Haplotype tagging for the identification of common disease genes. *Nat. Genet.* 2001, 29(2), 233–237.
24. Jungblut P, Thiede B. Protein identification from 2-DE gels by MALDI mass spectrometry. *Mass Spectrom. Rev.* 1997, 16(3), 145–162.
25. Wilkins MR, Sanchez JC, Gooley AA, Appel RD, Humphery-Smith I, Hochstrasser DF, Williams KL. Progress with proteome projects: why all proteins expressed by a genome should be identified and how to do it. *Biotechnol. Genet. Eng. Rev.* 1996, 13, 19–50.
26. Yates JR, 3rd, Eng JK, McCormack AL, Schieltz D. Method to correlate tandem mass spectra of modified peptides to amino acid sequences in the protein database. *Anal. Chem.* 1995, 67(8), 1426–1436.
27. Gygi SP, Rist B, Gerber SA, Turecek F, Gelb MH, Aebersold R. Quantitative analysis of complex protein mixtures using isotope-coded affinity tags. *Nat. Biotechnol.* 1999, 17(10), 994–999.
28. Adam GC, Sorensen EJ, Cravatt BF. Proteomic profiling of mechanistically distinct enzyme classes using a common chemotype. *Nat. Biotechnol.* 2002, 20(8), 805–809.
29. Holt LJ, Enever C, de Wildt RM, Tomlinson IM. The use of recombinant antibodies in proteomics. *Curr. Opin. Biotechnol.* 2000, 11(5), 445–459.
30. Koivunen E, Wang B, Ruoslahti E. Phage libraries displaying cyclic peptides with different ring sizes: ligand specificities of the RGD-directed integrins. *Biotechnol. N.Y.* 1995, 13(3), 265–270.
31. Fields S, Song O. A novel genetic system to detect protein-protein interactions. *Nature*. 1989, 340(6230), 245–246.
32. Rain JC, Selig L, De Reuse H, Battaglia V, Reverdy C, Simon S, Lenzen G, Petel F, Wojcik J, Schachter V, Chemama Y, Labigne A, Legrain P. The protein-protein interaction map of *Helicobacter pylori*. *Nature*. 2001, 409(6817), 211–215.
33. Ho Y, Gruhler A, Heilbut A, Bader GD, Moore L, Adams SL, Millar A, Taylor P, Bennett K, Boutilier K, Yang L, Wolting C, Donaldson I, Schandorff S, Shewnarane J, Vo M, Taggart J, Goudreault M, Muskat B, Alfarano C, Dewar D, Lin Z, Michalickova K, Willems AR, Sassi H, Nielsen PA, Rasmussen KJ, Andersen JR, Johansen LE, Hansen LH, Jespersen H, Podtelejnikov A, Nielsen E, Crawford J, Poulsen V, Sorensen BD, Matthiesen J, Hendrickson RC, Gleeson F, Pawson T, Moran MF, Durocher D, Mann M, Hogue CW, Figeys D, Tyers M. Systematic identification of protein complexes in *Saccharomyces cerevisiae* by mass spectrometry. *Nature*. 2002, 415(6868), 180–183.
34. Ranish JA, Yi EC, Leslie DM, Purvine SO, Goodlett DR, Eng J, Aebersold R. The study of macromolecular complexes by quantitative proteomics. *Nat. Genet.* 2003, 33(3), 349–355.

35. Vo-Dinh T, Cullum B. Biosensors and biochips: advances in biological and medical diagnostics. *Fresenius J. Anal. Chem.* 2000, 366(6–7), 540–551.
36. Keusgen M. Biosensors: new approaches in drug discovery. *Naturwissenschaften.* 2002, 89(10), 433–444.
37. LaVan DA, Lynn DM, Langer R. Moving smaller in drug discovery and delivery. *Nat. Rev. Drug Discov.* 2002, 1(1), 77–84.
38. Gaits F, Hahn K. Shedding light on cell signaling: interpretation of FRET biosensors. *Sci. STKE.* 2003, 2003(165), PE3.
39. Zandonella C. Cell nanotechnology: the tiny toolkit. *Nature*. 2003, 423(6935), 10–2.
40. Cooper MA. Optical biosensors in drug discovery. *Nat. Rev. Drug Discov.* 2002, 1(7), 515–528.
41. Vo-Dinh T. Proceedings of the 6th Annual Biochip Technologies Conference: Chips for Hits 1999. Berkeley, CA, November 2–5, 1999.
42. Freitas RA, Jr. The future of nanofabrication and molecular scale devices in nanomedicine. *Stud. Health Technol. Inform.* 2002, 80, 45–59.
43. Quantum Dot Corporation, http://qdots.com.
44. Clark HA, Kopelman R, Tjalkens R, Philbert MA. Optical nanosensors for chemical analysis inside single living cells. 2. Sensors for pH and calcium and the intracellular application of PEBBLE sensors. *Anal. Chem.* 1999, 71(21), 4837–4843.
45. Freemantle M. Nano bar coding for bioanalysis. *CENEAR*. 2001, 79(41), 13.
46. Desai TA. Micro- and nanoscale structures for tissue engineering constructs. *Med. Eng. Phys.* 2000, 22(9), 595–606.
47. Jain KK. Current status of molecular biosensors. *Med. Device Technol.* 2003, 14(4), 10–15.
48. Cui Y, Wei Q, Park H, Lieber CM. Nanowire nanosensors for highly sensitive and selective detection of biological and chemical species. *Science*. 2001, 293(5533), 1289–1292.
49. Marrazza G, Chiti G, Mascini M, Anichini M. Detection of human apolipoprotein E genotypes by DNA electrochemical biosensor coupled with PCR. *Clin. Chem.* 2000, 46(1), 31–37.
50. Haruyama T. Micro- and nanobiotechnology for biosensing cellular responses. *Adv. Drug Deliv. Rev.* 2003, 55(3), 393–401.
51. Smith S, Nagel D. Nanotechnology-enabled sensors: possibilities, realities, and applications. *Sensors*, Nov. 2003.
52. Smalley E. Chip senses trace DNA. *Technol. Res. News.* July 30/Aug. 6, 2003.
53. Kuo MD, Waugh JM, Yuksel E, Weinfeld AB, Yuksel M, Dake MD. 1998 ARRS President's Award. The potential of *in vivo* vascular tissue engineering for the treatment of vascular thrombosis: a preliminary report. American Roentgen Ray Society. *AJR Am. J. Roentgenol.* 1998, 171(3), 553–558.
54. Tardif J-C, Cote G, Lesperance J, Bourassa M, Lambert J, Doucet S, Bilodeau L, Nattel S, de Guise P, The Multivitamins and Probucol Study Group. Probucol and multivitamins in the prevention of restenosis after coronary angioplasty. *N. Engl. J. Med.* 1997, 337(6), 365–372.
55. Motwani JG, Topol EJ. Aortocoronary saphenous vein graft disease: pathogenesis, predisposition, and prevention. *Circulation*. 1998, 97(9), 916–931.
56. Allaire E, Clowes AW. Endothelial cell injury in cardiovascular surgery: the intimal hyperplastic response. *Ann. Thorac. Surg.* 1997, 63(2), 582–591.
57. Casscells W. Migration of smooth muscle and endothelial cells: critical events in restenosis. *Circulation*. 1992, 86(3), 723–729.

58. Libby P, Tanaka H. The molecular bases of restenosis. *Prog. Cardiovasc. Dis.* 1997, 40(2), 97–106.
59. Holifield B, Helgason T, Jemelka S, Taylor A, Navran S, Allen J, Seidel C. Differentiated vascular myocytes: are they involved in neointimal formation? *J. Clin. Invest.* 1996, 97(3), 814–825.
60. Ooyama T, Fukuda K, Oda H, Nakamura H, Hikita Y. Substratum-bound elastin peptide inhibits aortic smooth muscle cell migration *in vitro*. *Arteriosclerosis*. 1987, 7(6), 593–598.
61. Forough R, Koyama N, Hasenstab D, Lea H, Clowes M, Nikkari ST, Clowes AW. Overexpression of tissue inhibitor of matrix metalloproteinase-1 inhibits vascular smooth muscle cell functions *in vitro* and *in vivo*. *Circ. Res.* 1996, 79(4), 812–820.
62. Cowan KN, Jones PL, Rabinovitch M. Elastase and matrix metalloproteinase inhibitors induce regression, and tenascin-C antisense prevents progression, of vascular disease. *J. Clin. Invest.* 2000, 105(1), 21–34.
63. Li DY, Brooke B, Davis EC, Mecham RP, Sorensen LK, Boak BB, Eichwald E, Keating MT. Elastin is an essential determinant of arterial morphogenesis. *Nature*. 1998, 393(6682), 276–280.
64. Robert L, Robert AM, Jacotot B. Elastin-elastase-atherosclerosis revisited. *Atherosclerosis*. 1998, 140(2), 281–295.
65. Cowan B, Baron O, Crack J, Coulber C, Wilson GJ, Rabinovitch M. Elafin, a serine elastase inhibitor, attenuates post-cardiac transplant coronary arteriopathy and reduces myocardial necrosis in rabbits afer heterotopic cardiac transplantation. *J. Clin. Invest.* 1996, 97(11), 2452–2468.
66. Oho S, Rabinovitch M. Post-cardiac transplant arteriopathy in piglets is associated with fragmentation of elastin and increased activity of a serine elastase. *Am. J. Pathol.* 1994, 145(1), 202–210.
67. Strauss BH, Chisholm RJ, Keeley FW, Gotlieb AI, Logan RA, Armstrong PW. Extracellular matrix remodeling after balloon angioplasty injury in a rabbit model of restenosis. *Circ. Res.* 1994, 75(4), 650–658.
68. Hinek A, Boyle J, Rabinovitch M. Vascular smooth muscle cell detachment from elastin and migration through elastic laminae is promoted by chondroitin sulfate-induced "shedding" of the 67-kDa cell surface elastin binding protein. *Exp. Cell Res.* 1992, 203(2), 344–353.
69. Hinek A, Mecham RP, Keeley F, Rabinovitch M. Impaired elastin fiber assembly related to reduced 67-kD elastin-binding protein in fetal lamb ductus arteriosus and in cultured aortic smooth muscle cells treated with chondroitin sulfate. *J. Clin. Invest.* 1991, 88(6), 2083–2094.
70. Hinek A, Molossi S, Rabinovitch M. Functional interplay between interleukin-1 receptor and elastin binding protein regulates fibronectin production in coronary artery smooth muscle cells. *Exp. Cell Res.* 1996, 225(1), 122–131.
71. Wong AH, Waugh JM, Amabile PG, Yuksel E, Dake MD. *In vivo* vascular engineering: directed migration of smooth muscle cells to limit neointima. *Tissue Eng.* 2002, 8(2), 189–199.
72. Amabile PG, Wong H, Uy M, Boroumand S, Elkins CJ, Yuksel E, Waugh JM, Dake MD. *In vivo* vascular engineering of vein grafts: directed migration of smooth muscle cells by perivascular release of elastase limits neointimal proliferation. *J. Vasc. Interv. Radiol.* 2002, 13(7), 709–715.
73. Senior RM, Griffin GL, Mecham RP, Wrenn DS, Prasad KU, Urry DW. Val-Gly-Val-Ala-Pro-Gly, a repeating peptide in elastin, is chemotactic for fibroblasts and monocytes. *J. Cell Biol.* 1984, 99(3), 870–874.

74. Thompson K, Rabinovitch M. Exogenous leukocyte and endogenous elastases can mediate mitogenic activity in pulmonary artery smooth muscle cells by release of extracellular-matrix bound basic fibroblast growth factor. *J. Cell Physiol.* 1996, 166(3), 495–505.
75. Johnson TJ. TNK-tPA: a new thrombolytic for treatment of acute myocardial infarction. *S.D. J. Med.* 2000, 53(5), 185–186.
76. Bingley JA, Campbell JH, Hayward IP, Campbell GR. Inhibition of neointimal formation by natural heparan sulfate proteoglycans of the arterial wall. *Ann. N.Y. Acad. Sci.* 1997, 811, 238–42, 242–244.
77. Batchelor WB, Robinson R, Strauss BH. The extracellular matrix in balloon arterial injury: a novel target for restenosis prevention. *Prog. Cardiovasc. Dis.* 1998, 41(1), 35–49.
78. Vassalli JD, Sappino AP, Belin D. The plasminogen activator/plasmin system. *J. Clin. Invest.* 1991, 88(4), 1067–1072.
79. Liotta LA, Goldfarb RH, Brundage R, Siegal GP, Terranova V, Garbisa S. Effect of plasminogen activator (urokinase), plasmin, and thrombin on glycoprotein and collagenous components of basement membrane. *Cancer Res.* 1981, 41(11 Pt 1), 4629–4636.
80. He CS, Wilhelm SM, Pentland AP, Marmer BL, Grant GA, Eisen AZ, Goldberg GI. Tissue cooperation in a proteolytic cascade activating human interstitial collagenase. *Proc. Natl. Acad. Sci. U.S.A.* 1989, 86(8), 2632–2636.
81. Rabbani LE, Johnstone MT, Rudd MA, Devine P, George D, Loscalzo J. PPACK attenuates plasmin-induced changes in endothelial integrity. *Thromb. Res.* 1993, 70(6), 425–436.
82. Hilfiker PR, Waugh JM, Li-Hawkins JJ, Kuo MD, Yuksel E, Geske RS, Cifra PN, Chawla M, Weinfeld AB, Thomas JW, Shenaq SM, Dake MD. Enhancement of neointima formation with tissue-type plasminogen activator. *J. Vasc. Surg.* 2001, 33(4), 821–828.
83. Henry TD, Annex BH, McKendall GR, Azrin MA, Lopez JJ, Giordano FJ, Shah PK, Willerson JT, Benza RL, Berman DS, Gibson CM, Bajamonde A, Rundle AC, Fine J, McCluskey ER. The VIVA trial: Vascular endothelial growth factor in ischemia for vascular angiogenesis. *Circulation.* 2003, 107(10), 1359–1365.
84. Simons M, Annex BH, Laham RJ, Kleiman N, Henry T, Dauerman H, Udelson JE, Gervino EV, Pike M, Whitehouse MJ, Moon T, Chronos NA. Pharmacological treatment of coronary artery disease with recombinant fibroblast growth factor-2: double-blind, randomized, controlled clinical trial. *Circulation.* 2002, 105(7), 788–793.
85. Laham RJ, Sellke FW, Edelman ER, Pearlman JD, Ware JA, Brown DL, Gold JP, Simons M. Local perivascular delivery of basic fibroblast growth factor in patients undergoing coronary bypass surgery: results of a phase I randomized, double-blind, placebo-controlled trial. *Circulation.* 1999, 100(18), 1865–1871.
86. Ruel M, Laham RJ, Parker JA, Post MJ, Ware JA, Simons M, Sellke FW. Long-term effects of surgical angiogenic therapy with fibroblast growth factor 2 protein. *J. Thorac. Cardiovasc. Surg.* 2002, 124(1), 28–34.
87. Lederman RJ, Mendelsohn FO, Anderson RD, Saucedo JF, Tenaglia AN, Hermiller JB, Hillegass WB, Rocha-Singh K, Moon TE, Whitehouse MJ, Annex BH. Therapeutic angiogenesis with recombinant fibroblast growth factor-2 for intermittent claudication (the TRAFFIC study): a randomised trial. *Lancet.* 2002, 359(9323), 2053–2058.

88. Grines CL, Watkins MW, Helmer G, Penny W, Brinker J, Marmur JD, West A, Rade JJ, Marrott P, Hammond HK, Engler RL. Angiogenic gene therapy (AGENT) trial in patients with stable angina pectoris. *Circulation*. 2002, 105(11), 1291–1297.
89. Losordo DW, Vale PR, Hendel RC, Milliken CE, Fortuin FD, Cummings N, Schatz RA, Asahara T, Isner JM, Kuntz RE. Phase 1/2 placebo-controlled, double-blind, dose-escalating trial of myocardial vascular endothelial growth factor 2 gene transfer by catheter delivery in patients with chronic myocardial ischemia. *Circulation*. 2002, 105(17), 2012–2018.
90. Hedman M, Hartikainen J, Syvanne M, Stjernvall J, Hedman A, Kivela A, Vanninen E, Mussalo H, Kauppila E, Simula S, Narvanen O, Rantala A, Peuhkurinen K, Nieminen MS, Laakso M, Yla-Herttuala S. Safety and feasibility of catheter-based local intracoronary vascular endothelial growth factor gene transfer in the prevention of postangioplasty and in-stent restenosis and in the treatment of chronic myocardial ischemia: phase II results of the Kuopio Angiogenesis Trial (KAT). *Circulation*. 2003, 107(21), 2677–2683.
91. Simons M, Ware JA. Therapeutic angiogenesis in cardiovascular disease. *Nat. Rev. Drug Discov.* 2003, 2(11), 863–871.
92. Curtis A, Wilkinson C. Topographical control of cells. *Biomaterials*. 1997, 18(24), 1573–1583.
93. Moldovan NI, Ferrari M. Prospects for microtechnology and nanotechnology in bioengineering of replacement microvessels. *Arch. Pathol. Lab. Med.* 2002, 126(3), 320–324.
94. Hood JD, Bednarski M, Frausto R, Guccione S, Reisfeld RA, Xiang R, Cheresh DA. Tumor regression by targeted gene delivery to the neovasculature. *Science*. 2002, 296(5577), 2404–2407.
95. Jacobs AK. Coronary stents — have they fulfilled their promise? *N. Engl. J. Med.* 1999, 341(26), 2005–2006.
96. Bauters C, Banos JL, Van Belle E, Mc Fadden EP, Lablanche JM, Bertrand ME. Six-month angiographic outcome after successful repeat percutaneous intervention for in-stent restenosis. *Circulation*. 1998, 97(4), 318–321.
97. Yutani C, Imakita M, Ishibashi-Ueda H, Tsukamoto Y, Nishida N, Ikeda Y. Coronary atherosclerosis and interventions: pathological sequences and restenosis. *Pathol. Int.* 1999, 49(4), 273–290.
98. Reimers B, Moussa I, Akiyama T, Tucci G, Ferraro M, Martini G, Blengino S, Di Mario C, Colombo A. Long-term clinical follow-up after successful repeat percutaneous intervention for stent restenosis. *J. Am. Coll. Cardiol.* 1997, 30(1), 186–192.
99. Ross R. Cell biology of atherosclerosis. *Annu. Rev. Physiol.* 1995, 57, 791–804.
100. Mintz GS, Popma JJ, Pichard AD, Kent KM, Salter LF, Chuang YC, Griffin J, Leon MB. Intravascular ultrasound predictors of restenosis after percutaneous transcatheter coronary revascularization. *J. Am. Coll. Cardiol.* 1996, 27(7), 1678–1687.
101. Lafont A, Guzman LA, Whitlow PL, Goormastic M, Cornhill JF, Chisolm GM. Restenosis after experimental angioplasty. Intimal, medial, and adventitial changes associated with constrictive remodeling. *Circ. Res.* 1995, 76(6), 996–1002.
102. Schwartz RS. Pathophysiology of restenosis: interaction of thrombosis, hyperplasia, and/or remodeling. *Am. J. Cardiol.* 1998, 81(7A), 14E–17E.
103. Hoffmann R, Mintz GS, Dussaillant GR, Popma JJ, Pichard AD, Satler LF, Kent KM, Griffin J, Leon MB. Patterns and mechanisms of in-stent restenosis. A serial intravascular ultrasound study. *Circulation*. 1996, 94(6), 1247–1254.

104. Sousa JE, Costa MA, Abizaid AC, Rensing BJ, Abizaid AS, Tanajura LF, Kozuma K, Van Langenhove G, Sousa AG, Falotico R, Jaeger J, Popma JJ, Serruys PW. Sustained suppression of neointimal proliferation by sirolimus-eluting stents: one-year angiographic and intravascular ultrasound follow-up. *Circulation.* 2001, 104(17), 2007–2011.
105. Marx SO, Marks AR. Bench to bedside: the development of rapamycin and its application to stent restenosis. *Circulation.* 2001, 104(8), 852–855.
106. Guba M, von Breitenbuch P, Steinbauer M, Koehl G, Flegel S, Hornung M, Bruns CJ, Zuelke C, Farkas S, Anthuber M, Jauch KW, Geissler EK. Rapamycin inhibits primary and metastatic tumor growth by antiangiogenesis: involvement of vascular endothelial growth factor. *Nat. Med.* 2002, 8(2), 128–135.
107. Elkins CJ, Waugh JM, Amabile PG, Minamiguchi H, Uy M, Sugimoto K, Do YS, Ganaha F, Razavi MK, Dake MD. Development of a platform to evaluate and limit in-stent restenosis. *Tissue Eng.* 2002, 8(3), 395–407.

18 Nanotechnology and Cancer

Jeffrey A. Norton

CONTENTS

18.1 INTRODUCTION AND DEFINITIONS

The era of molecular medicine is providing new insights into the diagnosis, treatment, and prevention of cancer. Exciting advances in technology have elicited a new discipline called *nanotechnology* that will also have a major impact on clinical cancer diagnosis and treatment. *Nanotechnology* is the ability to control, modify, and fabricate materials, structures, and devices with nanometer precision. It is the synthesis of such structures into systems of micro- and macroscopic dimensions. It encompasses the understanding of fundamental physics, chemistry, biology, and technology of nanometer-scale objects. It includes how such objects can be used in the areas of computation, sensors, structure, and biotechnology. The definition is used loosely. Small is big. A nanometer is one billionth of a meter. In general, molecules measure approximately one nanometer in diameter. A nanometer is small. It is the equivalent of ten hydrogen atoms placed side by side. It is one thousandth the size of a bacterium. It is one millionth the diameter of the head of a pin. The term nanotechnology may be used incorrectly in some instances. Investigators have used it to mean structure and fabrication at the micron scale. A micron is one millionth of a meter, which is a thousand times larger than a nanometer. Also in some cases, nanotechnology may not be technology. Rather it involves basic research on structures having at least one dimension of about one to several hundred nanometers.

Nanotechnology is a discipline in which both science and technology work together. It transcends the realm of individual atoms and molecules and descends to

0-8493-1940-4/05/$0.00+$1.50

the macroworld. Since nanotechnology works in the region of one nanometer, it bumps up against the basic building blocks of matter. It defines the smallest natural structures. Nanotechnology is the limit of technology, because it is impossible to assemble anything smaller. New tools capable of imaging and manipulating single molecules or atoms have ushered in the field of nanotechnology. The scanning probe microscope is capable of creating pictures of individual atoms and moving them from place to place. Varied approaches to fabricating nanostructures have emerged. Like sculptors, so-called top down practitioners chisel out or add bulk material to a surface. Microchips, which now boast circuit lines of little more than 100 nm, are about to become the most notable example. In contrast, bottom up manufacturers use self-assembly processes to put together larger structures, atoms, or molecules that make ordered arrangements spontaneously, given the right conditions. *Nanotubes*, which are graphite cylinders with unusual electrical properties, are a good example of self-assembled nanostructures (Figure 18.1). Nanotubes can be used to track life in real time within a living cell without affecting the function of the cell.

In this report we will consider the application of nanotechnology to cancer. Nanotechnology has tremendous potential to make important contributions in cancer prevention, detection, diagnosis, imaging, and treatment. It provides novel tools for early detection and augments existing ones. It can target a tumor, carry imaging capability to document the presence of tumor, sense pathophysiological defects in

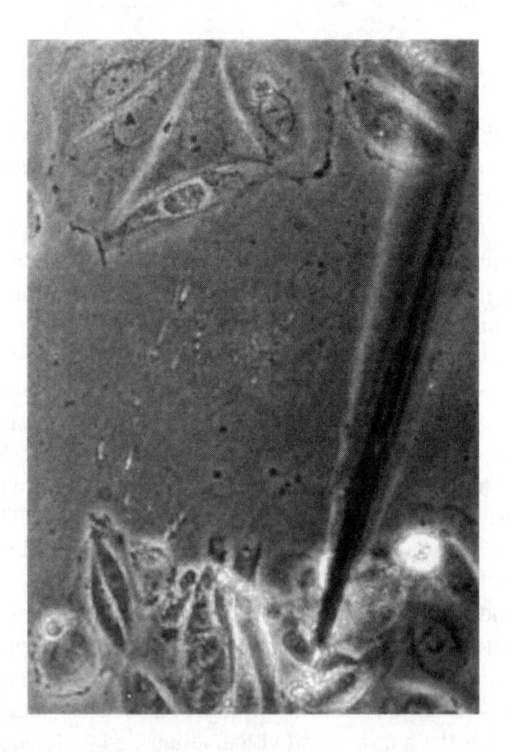

FIGURE 18.1 Nanofiber (nanotube) probes a cell without affecting cell viability. (From Zandonella, C., The tiny toolkit. *Nature* 2003, 423, 11. With permission.)

tumor cells, deliver therapeutic genes or drugs based on tumor characteristics, respond to external triggers to release the agent, document the tumor response, and identify residual tumor cells.[1] It provides tools for the real-time and direct readout of genomic and proteomic information at the single-cell or molecule level. It provides multiple simultaneous analyses of extremely small cell samples or populations of cells. It provides additional sensitivity in assays through more precise analyses. It may have profound significance on early detection, as it offers innovative tools for the understanding of differences between normal and cancer cells. It allows *in vivo* examination of normal cellular machinery. The parameters of cellular mechanics, morphology, and cytoskeleton can now be characterized. Finally, it promotes cross-cultural exchanges among different scientific and engineering disciplines that foster the study of cancer from different perspectives with new instruments.[2] Hopefully, this will allow the development of innovative new strategies for cancer management.

18.2 CELLULAR RESEARCH

As the cell is the basic unit of life, the study of cellular physiology and pathophysiology is critical to an understanding of cancer. Previously, *in vivo* study of cells has been largely impossible without disrupting cellular physiology and function. Nanotechnology is working on new tools that will allow tracking cell life in real time (Figure 18.1). These studies will be especially important to oncologists, who will discover why cancer cells divide and metastasize despite opposing signals. Several groups have reported using carbon *nanotubes* and *nanowires* to detect specific DNA oligonucleotide sequences and proteins. Approximately 1000 nanowire detectors can be arranged into a few square micrometers, roughly the size of a single cell. Potentially each of these nanowires can detect a different antibody or oligonucleotide. The oligonucleotide is a short segment of DNA that can be used to measure specific RNA sequences. These wires can then be used to determine proteins that a cancer cell secretes to allow uncontrolled growth or local invasion into adjacent matrix or blood vessels. If specific proteins can be identified, drugs and other strategies can be developed to inhibit these proteins and local invasion or distant metastases by cancer cells.

Further, *nanolabs* are being developed to determine *in vivo* cancer cell responses to treatment. For example, the gastrointestinal stromal tumor (GIST) has been shown to require the c-KIT protein kinase receptor CD-117 for growth.[3–5] Gleevec is a new drug that has been designed to specifically inhibit that receptor. Recent studies have shown that gleevec is remarkably effective at inhibiting this tumor[6] (Figure 18.2). However, eventually tumor cells become resistant to gleevec therapy and progress. The nanolab can be used to determine which genes are expressed when the GIST cell is exposed to gleevec and how the cell becomes resistant.[7] This study to determine the cellular mechanisms of resistance to gleevec may develop new drugs to inhibit the development of a resistant phenotype. Subesequently, gleevec therapy may be combined with drugs to inhibit gleevec resistance to provide longer, more durable responses to therapy.

Quantum dots are replacing traditional fluorescent probes in cellular imaging, enabling researchers to track more accurately the localization of macromolecules

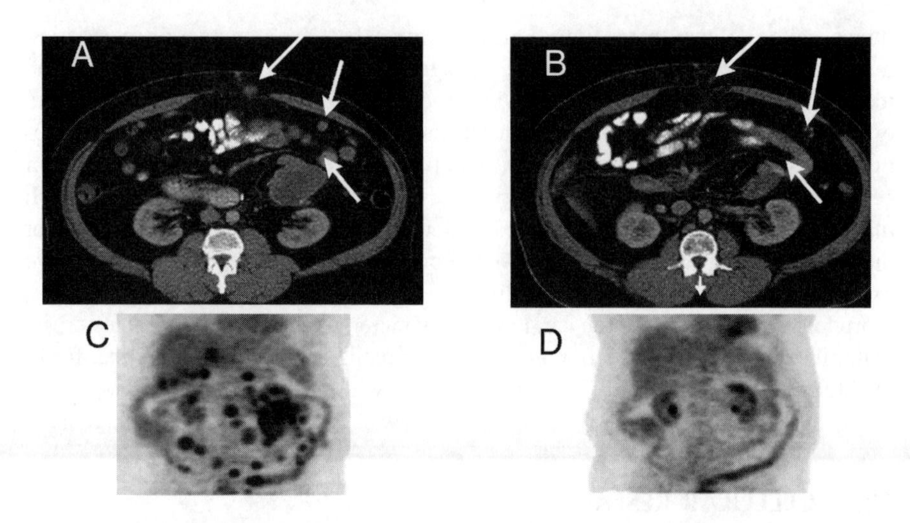

FIGURE 18.2 Computed tomography (CT) (A and B) and fluoro-deoxy glucose (FDG) positron emission tomography (PET) (C and D) of a gastrointestinal stromal tumor (GIST) before (A and C) and after (B and D) therapy with gleevec. Both CT and PET demonstrate a marked decrease in tumor following therapy.

(Figure 18.3). To visualize cells, previously scientists had to incubate them with fluorescent probes made from organic dyes conjugated either to a target antibody or a nucleotide sequence. Subsequently, unbound probe was washed away, which required fixing the cell and stopping internal machinery. Fixation eliminated the possibility of viable imaging of cells. Furthermore, using these methods, researchers could only observe a few molecules at a time. Fluorescence of conventional organic dyes and bleaches on exposure to light is neither bright nor stable enough to meet the demands of continuous viable intracellular imaging. Ideal probes must be specific enough to identify the cell of interest and adaptable enough to study many different cellular molecular processes. Further, probes must be stable to study these cell molecules over time. They must be able to pass through cell membranes and display a single molecule in a single cell, which requires a very strong fluorescence (Figure 18.4). Finally, these probes must not affect cellular viability. Quantum dots meet all of these criteria and are being used in various cancer research studies. Growing tumor cells on a layer of quantum dots can reveal whether the cells move, migrate, or clump. Because the tumor cells ingest the quantum dots near them, cells that travel leave a dot-free path of darkness. Cancer cells may fail the membrane test, in that they do not appear to penetrate a semipermeable membrane, and suggest that they are not malignant. However, cancer cells do not fail the quantum dot test. False negative results seldom occur, indicating enhanced sensitivity for the detection of migration and movement with quantum dots. Quantum dots can give information about tumor cell lineage. Each of the offspring of the original dot-containing tumor cell inherits a portion of the fluorescent grains, allowing researchers to track cell lineage and progeny over time. Researchers can conjugate the dots to a specific molecule or antibody or receptor to allow homing or migration of the particles to

FIGURE 18.3 Schematic drawing of quantum dots. (From Zandonella, C., The tiny toolkit. *Nature* 2003, 423, 11. With permission.)

tumors at various sites throughout the cell or organism (Figure 18.3).[8,9] Quantum dots will be very useful in cancer cell research.

18.3 DIAGNOSIS

At present advances in using nanotechnology for diagnostics are moving more rapidly than advances in nanoscale therapeutic agents. Semiconductor nanoparticles such as cadmium selenide (CdSe) quantum dots have the potential to replace fluorophores in many assays. Altering particle size controls the emission wavelength of quantum dots. Several bioconjugates of CdSe semiconductor nanoparticles have been developed with coatings or shells to stabilize the material in aqueous environments. Targeting the binding of these probes to specific cells or tissues can improve the discernable signal. Magnetic resonance imaging of coated or targeted superparamagnetic nanoparticles has been shown to be able to be used for imaging of gene expression. For example, HIV-1 tat peptide conjugated to superparamagnetic nanoparticles can be internalized by specific tissue types and can be used for *in vivo* analysis of gene expression.

In the genomic age of cancer, diagnosis through rapid gene sequencing will be important (Table 18.1). Nanopore sequencing is an attractive method. The technique works by applying an electric potential that causes a charged strand of DNA or other biopolymer to be drawn through a 1.5-nm pore in an α-heamolysin protein complex that is inserted into a lipid bilayer separating two conductive baths. The change in current over time is recorded, allowing the direct reporting of long DNA sequences

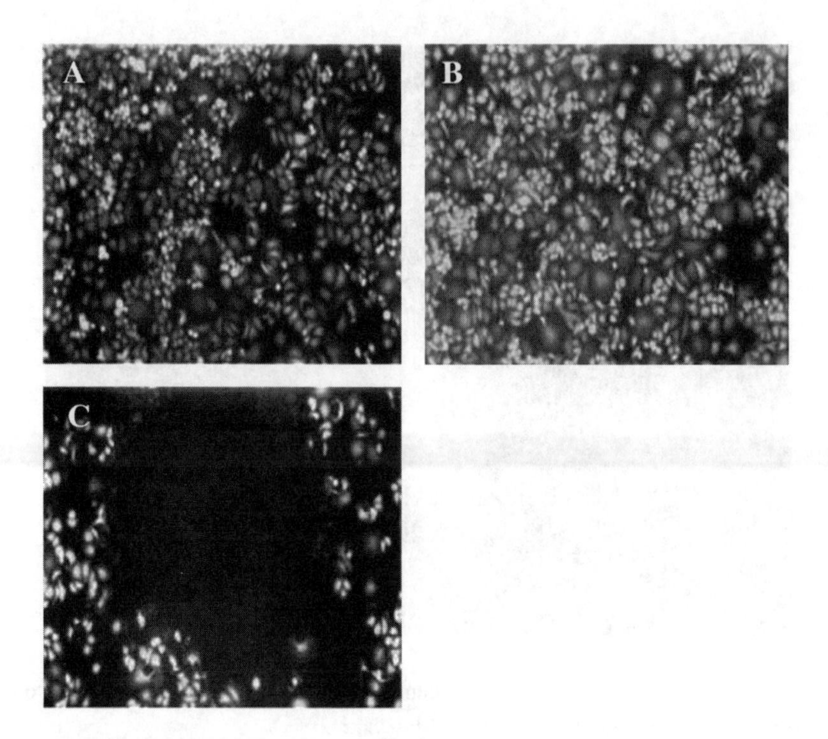

FIGURE 18.4 Breast carcinoma cells were exposed to nanoshells (A), near-infrared light (B), or both (C). As demonstrated with the fluorescent cellular viability marker, only the cells exposed to both (C) were killed. (From West JL and Halas NJ. Engineered nanomaterials for biophotonics applications: improving sensing, imaging and therapeutics. *Ann. Rev. Bioeng.* 2003, 5, 292. With permission.)

at rates of more than 1000 base pairs per second. Pores engineered to discriminate between purine and pyrimidine bases enhance resolution. Using nanotechnology, the rate of sequencing can be increased a thousand-fold, and the clinical use of DNA sequencing should be more widely available and less expensive. This method will be useful in certain diagnoses that predispose to the development of familial cancers like multiple endocrine neoplasia types 1 and 2 and Von Hippel Lindau syndrome, in which early definitive diagnosis through genetic changes leads to different treatment algorithms and screening methods for additional tumors[10] (Table 18.1).

Researchers have been experimenting with electronic diagnostics for years with minimal success. The detectors usually use *field effect transistors* (FET), devices in which a voltage applied to one electrode called a gate, changes the current flow between two other electrodes. To make a diagnostic test, researchers replace the gate with a material coated with a protein or other compound designed to capture molecules of interest. As charged target molecules latch on to the coating, there is a change in conductance between the transistor's two electrodes, creating a current that indicates the presence of the molecule of interest. Previously these devices have not been very sensitive because it takes a large number of target molecules to trigger the current. Most recently nanoscale transistors have been developed. Usually this

TABLE 18.1
Hereditary Cancer Syndromes with Associated Genetic Abnormalities

Syndrome	Mode of Inheritance	Chromosome	Gene	Gene Function	Clinical Manifestations
MEN1	Autosomal dominant	11q13	MENIN	Suppressor	HPT, pituitary tumors, NET pancreas
MEN2A	Autosomal dominant	10q11.2	RET	Oncogene	MTC, pheochromocytomas, HPT
MEN2B	Autosomal dominant	10q11.2	RET	Oncogene	MTC, Phenotype, pheochromocytoma
FMTC	Autosomal dominant	10q11.2	RET	Oncogene	MTC
VHL	Autosomal dominant	3p25-26	VHL	Suppressor	Renal cell carcinoma, tumors of brain, spine, eyes, adrenal, pancreas, inner ear, epididymis

Note: Abbreviations: MEN = multiple endocrine neoplasia; FMTC = familial medullary thyroid carcinoma; NET = neuroendocrine tumor; VHL = Von Hippel Lindau; HPT = primary hyperparathyroidism; Phenotype = marfanoid habitus, prognathism, bony abnormalities, and mucosal neuromas.

is done with silicon wires measuring as small as 1 to 2 nm. Researchers have fashioned gates from silicon nanowires coated with antibodies to prostate specific antigen (PSA) and carcinoembryonic antigen (CEA). These two antigens are markers for prostate cancer and colon cancer, respectively. After the nanowires are coated with antibodies to specific antigens, the transistors are capped with a plastic pad patterned with tiny channels that carry fluid over the devices. These nanoscale transistors can detect very low levels of antigen in tissues or body fluids. In fact, they can detect levels as low as 0.025 pg/ml. These levels are so minute that this method is one hundred times more sensitive than commercial ELISA or radioimmunoassays. Nanowire sensors can detect levels real-time in tissues so they may be able to localize tumors within the colon or prostate.[11] This method may be especially useful in patients with elevated serum levels of PSA and normal prostates on examination and ultrasound. Areas within the prostate that contain high levels of PSA should correlate with the presence of cancer.

Matrix metaloproteases (MMP) are tissue markers for invasion, as tumor cells use these enzymes during invasion into soft tissue and blood vessels for the development of metastases. *Superparamagnetic nanoparticles* have been used as magnetic relaxation switches, and protease activity is measured by the spin–spin relaxation time (T-2) of water.[12] MRI can be used with these nanoparticles to image a tumor and determine if the tumor is benign or malignant based on the presence of matrix invasion. Meade and others have developed gadolinium-based magnetic resonance imaging contrast agents that can be used to image cellular events and track MMP. If these methods can track MMP *in vivo*, they may be able to differentiate benign tumors from malignant ones and discover the margin between normal tissue and cancer where tissue invasion exists. It may be possible to observe the activity of a patient's tumor in the MRI scanner and administer a treatment to see if the antitumor drugs or agents eliminate tissue invasion.[13] Therefore, MRI with superparamagnetic nanoparticles will not only diagnose the presence of tumor, it will characterize its biology. This potential may be especially useful in cystic tumors of the pancreas that are now being commonly detected on abdominal computed tomography (CT) as part of workup for pain. Cystic pancreatic tumors may be benign or malignant, and there are CT criteria like the size of the cysts and the presence of mucin that help distinguish between the two. However, currently excision and pathology is the only unequivocal method. The problem is that pancreatic surgery is associated with potential significant morbidity and mortality. If MRI with superparamagnetic particles to assess MMP can also discriminate between benign and malignant cystic pancreatic tumors, it would be a major breakthrough and save patients much suffering and even some deaths.

18.4 IMAGING

Because molecules themselves are too small to be directly imaged, specific and sensitive site-targets are employed as beacons to image epitopes of interest. A site-targeted agent is designed to detect a selected biomarker that otherwise may not be able to be distinguished from surrounding tissue. Molecular imaging is now a clinical reality with FDG-PET and somatostatin receptor imaging are two important examples (see Figure 18.2). Further, biocompatible nanoparticles hold great promise to expand molecular imaging. The contrast medium depends on the imaging modality, the clinical problem, and accessibility. Contrast moieties are nanoparticles that are loaded with imaging agents such as paramagnetic or superparamagnetic (see Figure 18.5 below and color insert following page 204) metals, optically active (fluorescent) molecules, and radionuclides. Certain nanoparticles such as liquid perfluorocarbon nanoparticles have considerable flexibility and can be used with available imaging systems. The contrast agent needs high avidity and affinity for its target. Generally, targeting ligands are coupled directly to the carriers and comprise monoclonal antibodies or antibody fragments, peptides that bind receptors, or aptamers, each of

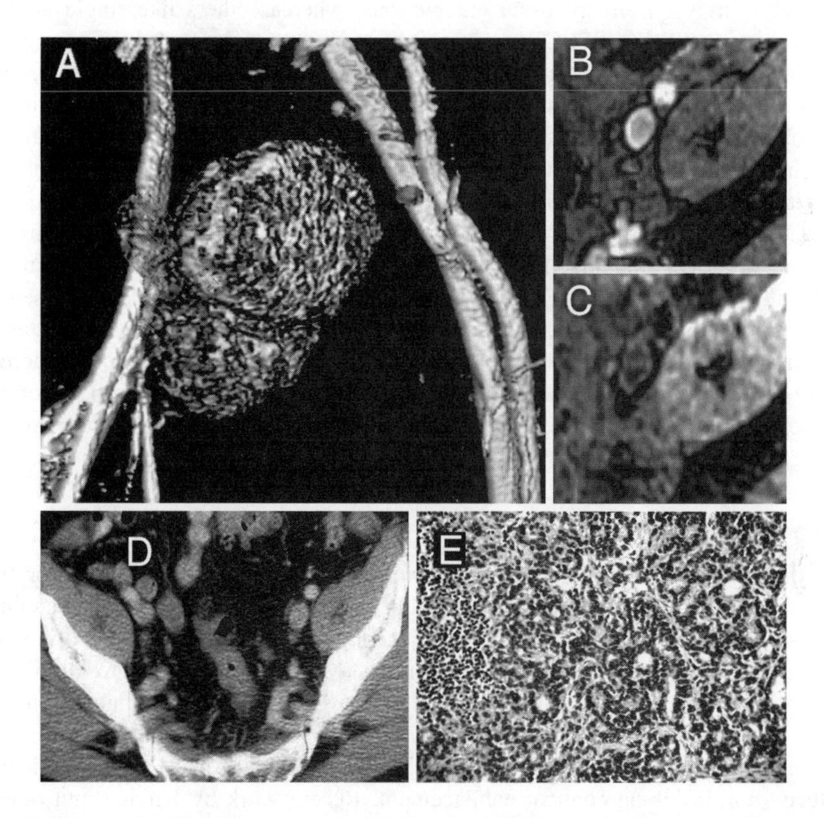

FIGURE 18.5 Three-dimensional reconstruction of pelvic lymph nodes (A), conventional MRI (B), MRI with lymphotropic superparamagnetic nanoparticles (C), abdominal CT (D), and histology (E) of lymph node metastases with prostate cancer. (See color insert following page 204.) Panel A shows 3-D reconstruction of prostate, iliac vessels, and metastatic (red) and normal (green) lymph nodes. Panel B shows that conventional MRI signal intensity is identical in positive and negative lymph nodes. Panel C shows increased activity in the metastatic nodes (arrow). Panel D shows that CT cannot discriminate between positive and normal lymph nodes. Panel E shows the histology of prostate cancer that has metastasized to a lymph node. (With permission from Harisingham MG and others. Noninvasive detection of clinically occult lymph-node metastases in prostate cancer. *N. Engl. J. Med.* 2003, 348, 2497. Copyright ©2003 Massachusetts Medical Society. All rights reserved.)

which confer specificity of binding to a target of interest. Multivalent binding is preferable to enhance avidity and reduce "off-rates," so that binding persists long enough to permit high-resolution imaging.[14] Molecular imaging through nanoparticles is a reality and will improve our ability to detect tumors (Figure 18.5).

Formerly, oncologists could only detect tumors or observe their growth through surgery or biopsy. Despite many advances in cancer research, we are limited in our ability to detect tumors at early stages, monitor tumor phenotype, identify and quantify invasion or metastases, and determine if antitumor therapy is effective or not in real-time. New approaches to tumor imaging are being developed as different modalities are being studied in mouse models. Some imaging methods (MRI and CT) rely purely on energy–tissue interactions, whereas others like single photon emission tomography (SPECT) and positron emission tomography (PET) require the administration of reporter probes or contrast agents (Figure 18.2). These probes require a targeting component like a small molecule or peptide plus a label that can be detected by a given imaging technique. MRI has considerable potential for imaging at the molecular level (Figure 18.5). It can be used to image metabolic activity in tumors, tumor phenotyping, tumor pathophysiology and tumor cell tracking. CT can be used for lung and bone tumor imaging. Ultrasound is used primarily for interventional imaging. PET scan can be used to determine metabolism of molecules such as glucose (Figure 18.2) or thymidine in tumors. SPECT is used to image through specific probes like antibodies or peptides. Fluorescence reflectance imaging can be used to screen molecular events on the surface of tumors. Fluorescence-mediated tomography is used to do targeted imaging of fluorochrome reporters in deep tumors. Bioluminescence imaging is used to determine gene expression and cell tracking.[15]

18.4.1 Brain Tumors

The determination of tumor margins is critical to complete resection of tumor and a good prognosis. This is especially critical in surgical procedures like brain or spinal cord surgery, during which removal of adjacent normal tissue means loss of neurological function. Magnetic resonance imaging (MRI) has been used to allow distinction between tumor and brain tissue. However, the widely available contrast agent gadolinium has not been satisfactory because of issues related to timing of administration, dose, and lack of specificity. This occurs because of surgically induced normal tissue contrast enhancement. Recent work by Kircher and others demonstrates that a multimodal nanoparticle contrast agent consisting of an optically detectable near-infrared fluorescence fluorochrome conjugated to a MRI-detectable iron oxide core offers a novel approach to facilitate the surgical resection of brain tumors. Brain tumor visualization by MRI is enhanced by magnetic nanoparticles. The multimodal nanoparticle probe called Cy5.5-CLIO afforded preoperative visualization of brain tumors by serving as a MRI contrast agent. It also facilitated intraoperative discrimination of tumor from adjacent normal brain tissue because of near-infrared fluorescence. The ability to use the same probe for both preoperative and intraoperative optical imaging offers a significant advantage for visualization and accurate resection of brain tumors.[16]

18.4.2 Prostate Cancer

Accurate detection of lymph node metastases in patients with prostate cancer is an essential component of the approach to treatment. Lymphotropic superparamagnetic nanoparticles have been used in conjunction with MRI to detect metastatic tumor in regional and distant lymph nodes in humans and mouse models of cancer. Recently Harisinghani and others reported the use of these particles and MRI scan in determining the presence of lymph node metastases in prostate cancer (Figure 18.5). MRI with lymphotropic superparamagnetic nanoparticles correctly identified all patients with nodal metastases, and on a node-by-node analysis had a significantly higher senstitivity than conventional MRI (90.5% compared to 35.4%, $p < 0.001$) (Table 18.2).[17] This study demonstrated that nodes smaller than 1 cm (the conventional size for positive nodes on MRI or computed tomography) are correctly detected as cancerous on MR imaging with lymphotropic nanoparticles. This study has potential for selection of patients for extended lymphadenectomy or to accurately delineate radiotherapy fields. Similar studies have been done in rectal cancer with similar results (Figure 18.6). However, additional prospective studies will be needed to determine the exact clinical modality of lymph node imaging. Further, because two sequential MRIs are needed before and after the administration of contrast material, lymphotropic nanoparticle imaging may be too expensive. Currently, PET-CT with fluorodeoxy glucose seems to be most cost effective and widely used to determine microscopic nodal metastases in colon cancer.[18]

TABLE 18.2
Sensitivity and Specificity of MRI Alone vs. MRI with Lymphotropic Superparamagnetic Nanoparticles (LSN) at Imaging Lymph Node Metastases from Prostate Cancer

Variable	N	MRI (%)	MRI + LSN (%)	P Value
Lymph nodes — all sizes	334			
sensitivity		45	100	<0.01
specificity		90	98	
Lymph nodes — 5–10 mm	45			
sensitivity		29	96	<0.01
specificity		87	99	
Patients	80			
sensitivity		45	100	<0.01
specificity		79	96	

Data are from Harisinghani MG et al., *N. Engl. J. Med.*, 2003, 348, 2491–2499, with permission.

18.4.3 Quantum Dots

Quantum dots, also called nanocrystals or nanodots, are fluorescent crystalline clumps of a few hundred atoms, less than 10 nanometers in diameter, coated with

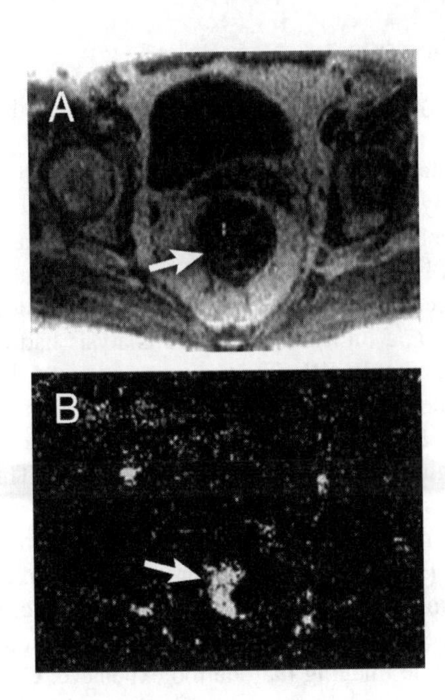

FIGURE 18.6 Tumor blood volume in rectal cancer. (A) T_2-MRI scan in patient with rectal cancer (arrow). (B) Dynamic, contrast-enhanced T_2 weighted MRI shows parametric map reflecting high blood volume (arrow). High pre-treatment blood volume is associated with good prognosis. (From Koh DM, Cook GJR and Husband JE. New horizons in oncologic imaging. *N. Engl. J. Med.* 2003, 348, 2487. With permission.)

an insulating outer shell of a different material that emits light in various hues[19] (Figure 18.3). One currently favored quantum dot has a crystal core made of a semiconductor cadmium selenide and an outer shell of zinc sulfide. Quantum dots are 1.5–2 nm in diameter. When a photon of visible light hits a quantum dot, the photon energy is confined to the crystal core and is emitted as a bright fluorescence. The emitted light is monochromatic, and the color depends on the size of the crystal. Antibodies or ligands can be conjugated to the quantum dot to confer its specificity as a molecular probe. These quantum dots have many potential applications in cancer biology for tumor imaging. For example, in cancer quantum dot imaging can be used to determine if breast tumors express Her-2 protein and are candidates for drugs designed to inhibit Her-2 like herceptin. Tumor cells are incubated with biotinylated herceptin, streptavidin-labeled quantum dots are added, and the bound quantum dots are viewed under a fluorescence microscope.[8,20]

18.5 TREATMENT

The ability to incorporate drugs or genes into detectable site-directed nanosystems represents a new paradigm in cancer therapeutics. Payloads of therapeutic agents such as genes or radionuclides or drugs can be complexed to carriers that have a site-directed molecule to ensure specificity. Drugs can be linked to or dissolved in carrier lipid coatings, deposited in subsurface oil layers, or trapped within the carriers themselves. The use of these targeted agents will be able to concentrate greater amounts of drug or therapeutic molecules within the tumor cell. *Nanoparticles* can be used to deliver pharmaceutical agents after binding to tumor cellular epitopes by a mechanism called contact-facilitated drug delivery (Figure 18.7). Binding and close apposition to the targeted tumor cell membrane permits enhanced lipid–lipid exchange with the lipid monolayer of the nanoparticle, which accelerates convective flux of lipophilic drugs (i.e., Paclitaxel) dissolved in the outer lipid membrane of the nanoparticles to the targeted tumor cells. Such nanosystems serve as drug depots, exhibiting prolonged release kinetics and persistence at the tumor site. These systems can also function as synthetic vectors for gene delivery when formulated with

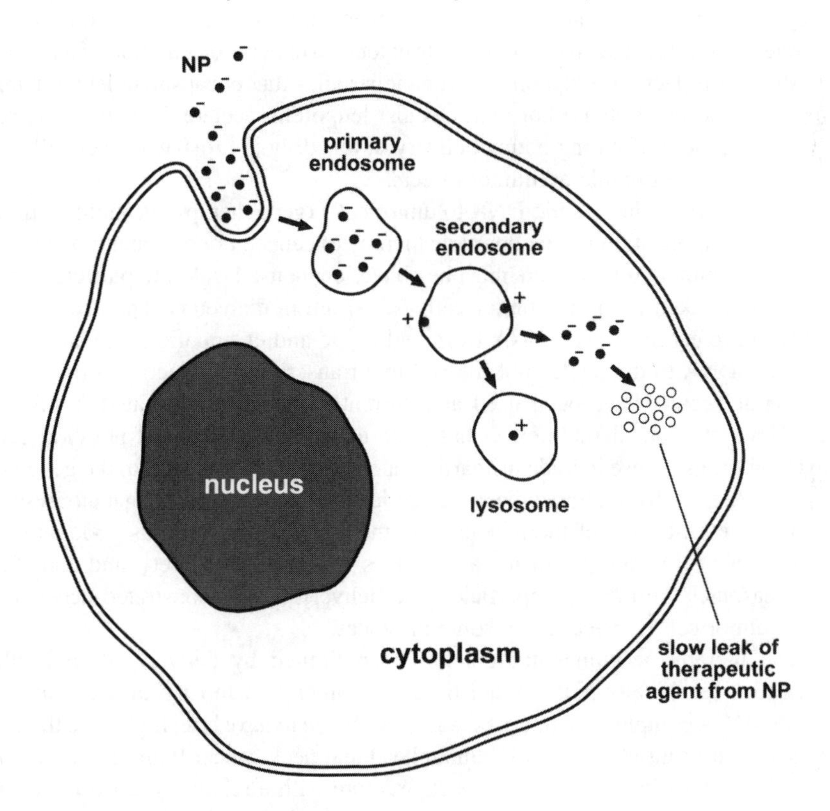

FIGURE 18.7 Schematic drawing of nanoparticle mediated drug delivery to cellular cytoplasm.

cationic lipids, much like circulating versions of conventional transfection vehicles (i.e., lipofectin). Thus, nanoparticles may be the ideal method for cancer therapy, as they can specifically deliver toxic drugs and molecules to cancer cells while sparing normal cells and minimizing toxicity.[14]

Near-infrared resonant particles called *nanoshells* can be used to enable fast whole-blood immunoassays or can be incorporated into temperature-sensitive hydrogels to synthesize a new type of composite material that collapses on laser irradiation. Plasmon excitations are quickly dampened, and the electron kinetic energy is converted into heat through electron–photon interactions. Nanoshells are small enough to navigate the human circulatory system on intravenous injection. Antibodies or receptors bound to the surface can be used to selectively bind these particles to tumor. Then using a near-infrared laser, carcinoma cells can be destroyed by local thermal heating around the nanoshells (see Figure 18.4).[21] One challenge to effective drug delivery and antitumor treatment is getting the medication to exactly the right place. To that end, researchers have been investigating myriad new methods to deliver pharmaceuticals. Tiny nanocontainers composed of polymers may be able to distribute drugs to specific sites within tumor cells. The encapsulation of drugs in a biodegradable polymer nanoparticle that are injectable can offer precise control over drug release profiles. The first nanoparticle extended-release formulation of a peptide consists of poly(lactic-co-glycolic) acid nanoparticles that encapsulate leuteinizing-hormone-releasing (LHRH) hormone agonist leuporelin acetate. This drug is used for prostate cancer. The long-acting delivery of this drug to prostate cancer cells has potential for more durable antitumor effects.

The delivery of therapeutic DNA to tumor cells (gene therapy) is another attractive potential application for nanoparticle therapy of cancer. For efficient transfection of DNA into tumor cells, vectors must be able to condense DNA into particles small enough to be taken up by the tumor cells (<200 nm in diameter). The vector must be able to protect the carrier DNA from hydrolytic and enzymatic degradation and deliver the DNA to the nucleus of the cell in a transcriptionally active form. Previously, viral vectors have been used and currently are being evaluated in clinical trials. However, considerable effort has gone to the development of nonviral gene delivery systems. These include primarily nanotechnology systems like the gene gun for gold particle delivery, liposomes, polymeric nanoparticles, and drug nanocrystals. Although the efficiency of the nanoparticle methods has not been as good as viral vectors, nanovectors are promising alternatives for economic, safety, and manufacturing reasons.[10] Further, nanoparticle gene delivery has demonstrated acceptable rates of tumor cell transfection in some instances.

Gene therapy of human tumors has been limited by delivery of molecular therapeutics specifically to the tumor tissue. A human transferring targeted cationic liposome-DNA complex, Tf-lipoplex, has been shown to have both high gene transfer rates and significant efficacy with human head and neck cancer both *in vitro* and *in vivo*.[22] Tf-lipoplex has a highly compact structure with a relatively uniform size of 50–90 nm. The nanostructure is novel in that it is a viral particle with a dense core enveloped by a membrane coated with Tf molecules spiking the surface. Compared with unliganded lipoplex, Tf-lipoplex shows enhanced stability, improved gene transfer efficiency, and long-term efficacy for systemic p53 gene therapy of human prostate cancer when used in combination with conventional radiation therapy. In nude mice experiments with human tumor xenografts, only the combination of Tf-lipoplex-p53 plus radiation resulted in cure while all other controls and treatments did not.[23]

Intracellular hyperthermia is a potential method to destroy tumor cells. If one could target nanoparticles into a tumor with a diameter of 1.1 mm and then introduce an AC magnetic field, one could achieve hyperthermic eradication of the tumor. With selective coating, nanoparticles may be introduced intravenously and targeted specifically to the tumor. In the absence of selective coating, nanoparticles can only be introduced by direct injection.[24] Nanoparticles can be used for the intracellular delivery of macromolecules like drugs and proteins. Panyam and others have used biodegradable nanoparticles formulated from the copolymers of poly(DL-lactide-co-glycolide) for the intracytoplasmic delivery of therapeutic macromolecules.[25] Further, others have demonstrated that intravenous polyethylene glycol-coated hexa-adecylcyanoacrylate nanospheres concentrate eleven times higher in brain tumors than adjacent brain tissue.[26]

Therapeutic radiation is focused biology. Radiation produces molecular events in the radiated tissue that causes cell injury and death. Radiation can kill cancer cells, by itself. It can be combined with cytotoxic or cytostatic drugs to kill tumor cells. It can induce molecular events that make targets susceptible to subsequent drugs or biological therapies.[27] Neutron capture therapy is a potential anticancer therapy that utilizes a stable, nonradioactive nuclide delivered to tumors cells that produces local cytotoxic radiation when exposed to thermal or epithermal neutrons. Nanoparticles have been used to deliver gadolinium to tumor targets for potential neutron capture radiation therapy. Tumor targeting was made possible via folate receptors on the nanoparticles. Folate-mediated endocytosis of the gadolinium nanoparticles was documented by inhibiting the endocytosis by high concentrations of folic acid.[28] Future studies with *in vivo* tumor models and neutron capture therapy are necessary to determine if this strategy will have significant antitumor effects.

Antiangiogenesis has been a strategy to inhibit tumor growth and metastases. However, efforts to influence the biology of blood vessels by gene delivery have been hampered by a lack of targeting vectors specific for endothelial cells in tumors. A recent study demonstrated that a cationic nanoparticle coupled to an integrin $\alpha_v\beta_3$-targeting ligand can selectively image blood vessels in tumors of rabbits (Figure 18.8). The therapeutic efficacy of this technique was tested in mice by generating nanoparticles conjugated to a mutant Raf-1 gene that blocks angiogenesis in response to multiple growth factors. Intravenous injection of these nanoparticles into tumor-bearing mice led to apoptosis of both primary and metastatic tumors and marked regression of established tumors in mice. The nanoparticle used had a multivalent target called integrin $\alpha_v\beta_3$ that selectively delivers genes to angiogenic blood vessels. Further, the mutant Raf-1delivered interrupts the signaling cascades of two potent angiogenic growth factors, bFGF and VEGF. Finally, since nanoparticles are less immunogenic than viral vectors, it is possible to do multiple administrations and deliver therapeutic molecules repeatedly.[29] Although the data described above are impressive, certain questions should be addressed in future studies. The fate of intravenously injected nanoparticles must be carefully documented *in vivo*. Nanoparticles are cleared by macrophages of the reticuloendothelial system, particularly Kuppfer cells in the liver. Further, once in the blood stream, the size of nanoparticles can increase significantly through aggregation. Aggregated nanoparticles may not reach the tumor target because they can be trapped in the pulmonary circulation.

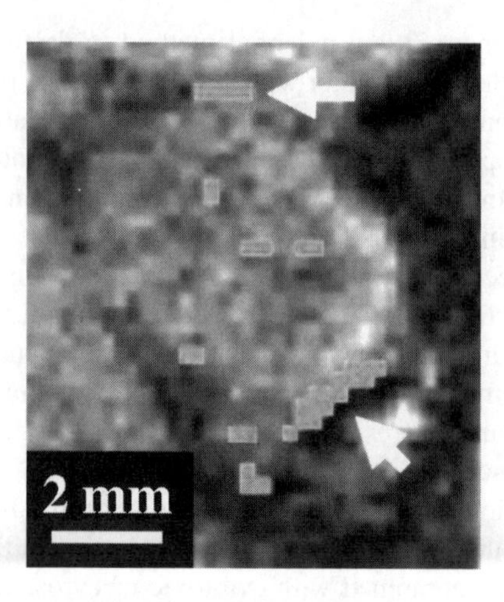

FIGURE 18.8 Enlarged section of T_1-weighted MRI showing V_x-2 tumor with signal enhancement following iv injection of $\alpha_{\varpi}B_3$-targeted nanoparticles. Enhancement indicates tumor angiogenesis. (From Winter PM and others. Molecular imaging of angiogenesis in nascent V_x-2 rabbit tumors using a novel $\alpha_{\varpi}B_3$-targeted nanoparticle and a 1.5 tesla magnetic resonance imaging. *Cancer Res.* 2003, 63, 5840. With permission.)

However, the use of the α_vB_3-specific ligand is a way to specifically target tumor endothelial tissue. The implementation of nanoparticle-mediated delivery of genes or drugs to tumor endothelium may require some refinements in particle engineering. Cancer patient studies will be necessary to be certain that the particles are safe and do, in fact, target tumor endothelium, as they appear to do in the rabbit carcinoma (Figure 18.8). Due to the possible heterogeneity of tumor vascular endothelium, it may be that different target antigens will be needed for different tumors in different patients.[30]

Researchers have developed a novel way to do radiation therapy using a "nanogenerator." They have treated nude mice bearing human cancer (lymphoma and prostate cancer) with injections of single atoms of actinium-225 (an alpha emitter) that were individually caged in specifically constructed molecules that supplied the radioactivity. These molecules were coupled to monoclonal antibodies targeted at the interior of the tumor cells. A single intravenous injection of this nanogenerator induced significant tumor regression and prolonged survival in a substantial fraction of tumor-bearing mice.[31]

The clinical use of the antitumor drug cisplatin is limited by three issues: (1) dose-limiting nephrotoxicity and neurotoxicity, (2) rapid inactivation of the drug through binding with plasma and tissue proteins, and (3) development of tumor resistance to the drug. One approach to these problems is to shield the drug from the macroenvironment by placing it in a lipid coat. However, previously this approach has failed to completely encapsulate the drug and resulted in low uptake of drug by tumor. A novel solution to this problem is the use of nanotechnology. Burger and others describe a method to allow for the efficient encapsulation of ciplatin in a lipid formulation. It is based on repeated freezing and thawing of a concentrated solution of cisplatin in the presence of negatively charged phospholipids. The method generates nanocapsules, which are small aggregates of cisplatin covered by a single lipid bilayer. The nanocapsules have an unprecedented drug-to-lipid ratio and an *in vitro* cytotoxicity more than 1000-fold higher than the free drug. This method of nanocapsule formation has applicability to other drugs showing low water solubility and lipophilicity. Since, with the use of nanocapsules, so much more drug is delivered to the tumor cells, it is much more effective than other conventional methods of cisplatin drug delivery.[32]

Irinotecan-containing nanoparticles have been prepared by coprecipitation of irinotecan with water and acetone solution of poly(DL-lactic acid), poly(ethylene glycol)-block-poly(propylene glycol)-block-poly(ethylene glycol) then subsequent evaporation of organic solvent. The nanoparticles generated demonstrated irinotecan concentrations of 4.5% by weight and small size of 80–210 nm. Irinotecan nanoparticles had significantly better antitumor effects in an experimental sarcoma than systemic irinotecan. Further, drug concentrations in tumor were higher with nanoparticles than with standard drug intravenously. These results suggest that irinotecan nanoparticles can provide higher tumor drug concentrations and better antitumor effects than standard intravenous drug.[33]

Radoslav Savic and his colleagues at McGill University tested the antitumor properties of tiny units built out of two types of polymers. The two compounds assembled into a spherical shape known as a micelle. One compound, which is hydrophobic (water fearing), aligns facing inwards, and the other, which is hydrophilic (water loving), faces outwards. Drugs can then be loaded inside the tiny molecular globs, which measure 20 to 45 nanometers in diameter. The researchers used fluorescent labeling to track the micelles. They found that the tiny containers could pass through the wall of a rat cell, but did not enter the nucleus. The micelles did, however, penetrate some cell parts, such as mitochondria and the Golgi apparatus, which are important targets for drug delivery. The scientists also determined that the micelles are very efficient at delivering their hydrophobic drug cargo once inside the cell. This property could mean that micelles may be able to deliver smaller doses of toxic medications to vital subcellular targets of tumor cells. This represents a new strategy in antitumor cell treatment with the destruction of subcellular targets like mitochondria and Golgi apparatus.

In an attempt to increase the local tumoral concentration of tamoxifen in estrogen-receptor-positive breast cancer, researchers have prepared and characterized poly(E-caprolactone) (PCL) nanoparticles. PCL nanoparticles were incubated with MCF-7 estrogen-receptor-positive experimental breast cancer to determine uptake, intracellular distribution, and localization. Results show that PCL nanoparticles were rapidly internalized in estrogen-receptor-positive breast cancer cells, and intracellular tamoxifen drug concentrations were much greater than what could be achieved with systemic administration of the drug. This approach may demonstrate improved antitumor effects by delivering the drug at high concentrations locally within the tumor for longer periods of time.[34]

Nanoparticles and other nanostructures appear to hold great promise for the future of cancer treatment. In experimental studies, primarily in animal models, nanoparticles appear to be able to selectively deliver high concentrations of antitumor drugs and genes to tumor cells. The high concentrations of toxic agents seem to persist for long periods within tumor cells and have more potent antitumor effects and less toxicity than the systemically administered counterparts.

REFERENCES

1. Omenn GS. Genetic advances will influence the practice of medicine: examples from cancer research and care of cancer patients. *Genet. Med.* 2002, 4, 15s–20s.
2. Srinivas PR, Barker P, Srivastava S. Nanotechnology in early detection of cancer. *Lab. Invest.* 2002, 82, 657–662.
3. Miettinen M, Sarlomo-Rikala M, Lasota J. Gastrointestinal stromal tumors: recent advances in understanding of their biology. *Hum. Pathol.* 1999, 30, 1213–1220.
4. Miettinen M, Sarlomo-Rikala M, Sobin LH, Lasota J. Esophageal stromal tumors: a clinicopathologic, immunohistochemical, and molecular genetic study of 17 cases and comparison with esophageal leiomyomas and leiomyosarcomas. *Am. J. Surg Pathol.* 2000, 24, 211–222.
5. Miettinen M, Sobin LH, Sarlomo-Rikala M. Immunohistochemical spectrum of GISTs at different sites and their differential diagnosis with a reference to CD117. *Mod. Pathol.* 2000, 13, 1134–1142.
6. Joensuu H, Roberts PJ, Sarlomo-Rikala M, et al. Effect of the tyrosine kinase inhibitor STI571 in a patient with a metastatic gastrointestinal stromal tumor. *N. Eng. J. Med.* 2001, 344(14), 1052–1056.
7. Zandonella C. The tiny toolkit. *Nature* 2003, 423, 10–12.
8. Wu X, Liu H, Liu J, et al. Immunofluorescent labeling of cancer marker Her2 and other cellular targets with semiconductor quantum dots. *Nat. Biotechnol.* 2003, 21, 41–46.
9. Seydel C. Quantum dots get wet. *Science* 2003, 300, 80–81.
10. LaVan DA, Lynn DM, Langer R. Moving smaller in drug discovery and delivery. *Nat. Rev.* 2002, 1, 77–84.
11. Service R. Tiny transistors scout for cancer. *Science* 2003, 300, 242–243.
12. Zhao M, Josephson L, Tang Y, Weissleder. Magnetic sensors for protease assays. *Angew. Chem. Int. Ed.* 2003, 42, 1375–1378.
13. Meade T et al. *In vivo* visualization of gene expression using magnetic resonance imaging. *Nat. Biotechnol.* 2000, 18, 321–325.
14. Wickline SA, Lanza GM. Nanotechnology for molecular imaging and targeted therapy. *Circulation* 2003, 107, 1092–1095.
15. Weissleder R. Scaling down imaging: molecular mapping of cancer in mice. *Nat. Rev.* 2002, 2, 1–8.
16. Kircher MF, Mahmood U, King RS, Weissleder R, Josephson L. A multimodal nanoparticle for preoperative magnetic resonance imaging and intraoperative optical brain delineation. *Cancer Res.* 2003, 63, 8122–8125.
17. Harisinghani MG, Barentsz J, Hahn PF et al. Noninvasive detection of clinically occult lymph-node metastases in prostate cancer. *N. Engl. J. Med.* 2003, 348, 2491–2499.
18. Koh D, Cook GJR, Husband JE. New horizons in oncologic imaging. *N. Engl. J. Med.* 2003, 348, 2487–2488.
19. Ben-Ari ET. Nanoscale quantum dots hold promise for cancer applications. *J. Natl. Cancer Inst.* 2003, 95, 502–504.

20. Mitchell P. Turning the spotlight on cellular imaging. *Nat. Biotechnol.* 2001, 19, 1013–1017.
21. Brongersma ML. Nanoshells: gifts in gold wrapper. *Nat. Mater.* 2003, 2, 296–297.
22. Xu L, Pirollo KF, Tang WH, Rait A and Chang EH. Transferrin-liposome-mediated systemic p53 gene therapy in combination with radiation results in regression of human head and neck cancer xenografts. *Hum. Gene Ther.* 1999, 10, 2941–2952.
23. Xu L, Frederik P, Pirollo KF, et al. Self-assembly of a virus-mimicking nanostructure system for efficient tumor-targeted gene delivery. *Hum. Gene Ther.* 2002, 13, 469–481.
24. Rabin Y. Is intracellular hyperthermia superior to extracellular hyperthermia in the thermal sense? *Int. J. Hyperthermia* 2002, 18, 194–202.
25. Pnayam J, Wen-Zhong Z, Prabha S, Sahoo SK, Labhasetwar V. Rapid endo-lysosomal escape of poly(DL-lactide-co-glycolide) nanoparticles: implications for drug and gene delivery. *FASEB J.* 2002, 16, 1217–1226.
26. Brigger I, Morizet J, Aubert G, et al. Poly(ethylene glycol)-Coated hexadecylcyanoacrylate nanospheres display a combined effect for brain tumor targeting. *J. Pharm. Exp. Ther.* 2002, 303, 928–936.
27. Coleman CN. Linking radiation oncology and imaging through molecular biology (or now that therapy and diagnosis have separated, it's time to get together again!). *Radiology* 2003, 228, 29–35.
28. Oyewumi MO, Mumper RJ. Engineering tumor-targeted gadolinium hexanedione nanoparticles for potential application in neutron capture therapy. *Bioconjug. Chem.* 2002, 13, 1328–1335.
29. Hood JD, et al. Tumor regression by targeted gene delivery to the neovasculature. *Science* 2002, 296, 2404–2407.
30. Reynolds AR, Moghimi SM, Hodivala-Dilke K. Nanoparticle-mediated gene delivery to tumor vasculature. *Trends Mol. Med.* 2003, 9, 2–4.
31. Randal J. Nanotechnology getting off the ground in cancer research. *J. Natl. Cancer Inst.* 2001, 93, 1836–1838.
32. Burger KNJ, Staffhorst RWHM, De Vilder HC, et al. Nanocapsules: lipid-coated aggregates of cisplatin with high cytotoxicity. *Nat. Med.* 2002, 8, 81–85.
33. Onishi H, Machida Y, Machida Y. Antitumor properties of irinotecan-containing nanoparticles prepared using poly(dl-lactic acid) and poly(ethylene glycol)-block-poly(propylene glycol)-block-poly(ethylene-glycol). *Biol. Pharm. Bull.* 2003, 26, 116–119.
34. Chawla JS and Amijii MM. Cellular uptake and concentrations of tamoxifen upon adminstration in poly(e-caprolactone) nanoparticles. *AAPS Pharm. Sci.* 2003, 5, 28–34.

19 Nanotechnology in Organ Transplantation

Stephan Busque, Hootan Roozrokh, and Minnie Sarwal

CONTENTS

19.1 INTRODUCTION

Organ transplantation offers the possibility to replace defective organs with new ones. For thousands of patients, transplantation represents the difference between life and death or significant improvement in quality of life. One of the most concerning challenges facing transplantation today is the rapidly increasing demand for organ replacement in the face of limited organ supply. More than 55,000 patients were on a wait list to receive deceased donor kidney transplantation in the United States in 2003. Only 8,500 organs were available for transplantation.[1] Only a minority of potential recipients will thus have the chance to receive a new organ. The early success rates after transplantation have improved dramatically. Potential recipients can expect graft and patient survival rates above 90% for most organs. However, the long-term results are associated with chronic dysfunction of the graft and toxicity related to the immunosuppressive agents used. A better understanding of the rejection process as well as the mechanism involved in tolerance or acceptance of the graft by the recipient could help improve the longevity of the grafts and minimize medication-related toxicity. Recent technical advances have allowed researchers to inves-

0-8493-1940-4/05/$0.00+$1.50

tigate the rejection and tolerance processes at the nanoscale of biology: genes and proteins. Microarrays technology is shedding a new light on our comprehension of the interaction between the graft and the host. The first section of this chapter will review the actual state of genomic evaluation in transplantation and envision the potential applications in the future. Organ shortage will likely continue to increase unless an unlimited source of organs is found, such as xenotransplantation. Meanwhile, the development of artificial organs may become an interesting alternative to organ transplantation. Nanotechnology would certainly be the cornerstone of such development. The second section of this chapter will describe the current progress in the development of artificial organs. The possibility of working at a nanoscale appears to be crucial to any progress in this field.

Hopefully, nanotechnology and its application in medicine will also help to reduce the need for organ transplantation by favorably altering the course of chronic diseases leading to organ failure.

19.2 MICROARRAYS IN TRANSPLANTATION

Transplantation of foreign tissues or an organ leads to global transcriptional changes, reflected in a complex cascade of activation and suppression of genes in the recipient. This process is altered by current multidrug regimens, necessary for successful engraftment of the transplanted organ. A better understanding of the rejection and tolerance processes in transplanted patients, at the genomic level, could assist in the development of new immunosuppressive medications and drug combinations that would minimize rejection and favor tolerance to the graft. Microarray technology, which reveals global patterns of gene expression[2,3,4] has been widely applied to bridge the gap between the basic and clinical sciences. This powerful tool brings with it the potential to unravel the complex immunological circuits that interplay in various immune responses, such as steroid-resistant acute rejection, vascular rejection, chronic rejection, infection, and drug toxicity. It is an example of the impact of the nanotechnology on transplantation research.

19.2.1 DNA ARRAYS

DNA arrays consist of many DNA segments arranged in a standard format on a support, and the entire array is hybridized with a labeled nucleic acid sample to measure parallel expression for each of the thousands of genes represented on the array. This parallelism of functional analysis makes this technology very powerful, specifically with the availability of structural sequence information from the Human Genome Mapping Project.[5] This technology has allowed the investigator to move rapidly from the "gene validation" approach, which has been traditionally based on analysis of genes most relevant with regard to the biological question approached, to the "gene discovery" approach using microarrays, where global expression changes of multiple genes can offer novel information on the expression profile for a clinical sample series, to develop prognostic and therapeutic information.

Expression analysis of colony filters containing cDNA clones began in the early 1980s with qualitative differential screening of duplicate membranes, hybridized in parallel with a mix of labeled radioactive cDNA mixtures isolated from two different samples; this method was instrumental in the isolation of the T-cell receptor and CTLA genes.[6,7] This is the high density filter or *macroarray*, which retains its popularity and use for analysis of experiments of moderate scope and small sample numbers. Imaging plate systems were used next for more quantitative expression analysis.[8]

DNA arrays in the 1990s evolved toward miniaturization or *microarrays*, aiming to increase the number of genes available for analysis in a single experiment, concomitantly also reducing sample usage for hybridization (with the added benefit of RNA amplification for very small sample amounts). Microarrays can also be produced on nylon membranes, but due to the intrinsic fluorescence of nylon supports, enzymatic detection[9] or radioactive labeling[10] is needed, which is relatively insensitive. Optical detection methods (fluorescence), allowing for dual labeling, and gridding DNA spots on planar supports (glass slides), made their advent, with improved results.[2,3] These microarrays can be generated in the laboratory, although the expense and logistics are not trivial (for details of microarray hardware see the Stanford Microarray Web site at http://cmgm.stanford.edu/pbrown/mguide/index.html). As more genomic sequence information and sophisticated bioinformatics tools have become available, arrays have further migrated from cDNA to genomic PCR to oligonucleotides arrays.

Oligonucleotide chips, pioneered by Affymetrix®, consist of glass chips with 20- to 25-mer oligonucleotides. Oligonucleotide arrays, in contrast to cDNA arrays, do not require the cDNA clone storage and PCR amplification; however, the melting temperature (Tm) of the oligo probes will vary sufficiently that signal intensities do not as directly reflect message abundance, and hence multiple probes must be used for each gene.

19.2.2 Application of Microarrays to Solid Organ Transplantation

Acute rejection of transplants is a complex process of allograft injury by infiltrating cells of the host immune system, which presents with complex clinical and pathological features. A preliminary analysis of acute rejection was done in a recent study,[11] and a more extensive analysis has been conducted using a microarray platform consisting of predominantly immune-related genes or "lymphochip"[12] to characterize the variations in gene expression patterns during acute rejection and related disorders in 67 pediatric renal allograft biopsies. Results from this study suggest that variations exist among the gene expression profiles between acute rejection samples.[13] Specifically, three distinct molecular signatures were identified that impact over 1,300 differentially expressed genes and cluster acute rejection (AR) cases into three distinct groups, designated AR-I, AR-II, and AR-III. AR-I has aggressive T- and B-cell, macrophage, and NK cell infiltration and activation and exhibits a strong INF- and NFκB response. In contrast, AR-II is a milder form of rejection and has a gene expression signature similar to the innate immune response to infection and to drug toxicity (DT). These patient samples cocluster with other samples with DT reactions in the absence of rejection. Finally, AR-III is an immunologically quiescent rejection, despite histologic evidence of tubulitis, and may be acute rejection captured late and on its way to spontaneous recovery. Another key discovery in this study is that AR expression overlaps with the innate immune response to infection as evidenced by cluster analysis and by differential expression of several TGF-β-modulated genes including RANTES, MIC-1, several cytokines, chemokines, and cell adhesion molecules. These acute rejections vary in the extent and cell-specific expression patterns of infiltrating lymphocytes. A surprising finding was a robust signature for immunoglobulin and B-cell specific genes; this was corroborated by an unexpected finding of B-cell dense (CD20+) clusters localizing to the interstitial compartment not causing tubulitis. A pure humoral rejection response in these samples could not be established, since these biopsies had variable C4d staining, a marker of humoral rejection. Survival analysis of graft function recovery following the rejection episode, steroid resistance (failed clinical response to first-line pulse-steroid therapy), and graft loss over the period of follow-up revealed a significant ($p < 0.001$) association with B-cell cluster density in this "learning" data set, as well as in an additional "test" data set of 31 additional acute rejection biopsies.[13] Specific gene expression differences discovered and their association with clinical events or outcomes might be considered as potential targets for individualized drug therapy. For example, specific medication such as anti-CD20 antibody (Rituximab) could be used to target B-cell infiltration of the graft, which may be efficient antigen presenting cells in aggressive acute rejection. The use of predictive analysis of microarrays (PAM)[14] has now further narrowed the analysis to only 97 unique genes,[15] represented in the 67-biopsy dataset.[13] All have >5-fold difference in expression level and classify our learning set of 26 AR samples with 96% concordance to assigned phenotype.

Novel genes of interest from this type of analysis can have their biological significance in a disease process confirmed by more traditional approaches (e.g., a novel cytotoxic T-cell molecule, granulysin, discovered during microarray analysis of patients with renal allograft rejection in our laboratory, was confirmed to be a reliable noninvasive marker of some sub-types of acute rejection and a possible predictor of steroid-resistance).[16] A recent study[17] of peripheral blood mononuclear cells from transplant recipients with concomitant anemia and acute renal allograft rejection revealed a redundancy of molecular pathways involved; many of these processes can be therapeutically manipulated to optimize treatment of anemia during the rejection process, apart from the current traditional use of recombinant human erythropoietin.

The field of organ transplantation is seeing the advent of many different immunosuppression protocols, optimizing the use of the currently available induction and maintenance immunosuppression agents. The mechanisms of action of some of these drugs is known in more detail than others. Nevertheless, little is known about the combination effect of these agents *in vivo*. A recent novel steroid-free immunosuppressive protocol in pediatric renal transplantation, consisting of daclizumab (a humanized anti-IL-2 receptor antibody), mycophenolate mofetil (an inhibitor of T- and B-lymphocyte proliferation) and tacrolimus (an inhibitor of calcineurin and cytokine production),[18] was also studied by DNA microarrays, to compare differences in expression profiling between steroid-free and steroid-based recipients. Results were validated by quantitative RT-PCR[19] and suggest a marked difference in cytokine profiles in steroid-free recipients and also suggest that a higher baseline value of T cell activation seen in stable nonrejecting steroid-free recipients may suggest a mechanism for graft accommodation rather than acute rejection. Further molecular analysis of these differences can be applied to delineate the molecular and cellular responses involved in other drug responses, thereby providing new insights into transplant immunopharmacology.

19.2.3 Limitations of Microarray Analysis

Although DNA microarrays offer several advantages, many limitations specific to the technology currently exist. Following are some of the common limitations and their solutions:

1. Data variability, especially for genes with very low expression levels/Run replicate arrays on each sample to reduce false positives.
2. Small sample amounts which limits replication/Use of amplified RNA (aRNA).
3. Unequal labeling efficiency of fluorescent dyes/Reciprocal labeling to confirm observations.
4. Prone to obtain false-positive correlations/Use high-stringency statistical cut-off and multiple validation methods to confirm associations.
5. Provides no information regarding protein expression levels and function/Confirm with other biochemical analysis for protein expression (e.g., immunohistochemistry, protein arrays).
6. Array results provide an incomplete view of the functional significance of differentially expressed genes in the experiments. mRNA expression level may not always correlate with protein expression levels.[20] Post-translational modifications that modify protein activity such as phosphorylation cannot be measured using these arrays. Techniques for protein analysis such as Western blotting, two-dimensional polyacrylamide gel electrophoresis, radioligand receptor binding, chromatographic separation and detection, as well as mass spectrometry remain indispensable for elucidating protein levels or function. Rapid advances are now being made in enabling technologies (i.e., consistent antibody library production and cost-effective slide production methods). These advances should permit development and use of protein microarrays as a key platform to complement the DNA microarrays.
7. Kidney biopsies or biopsies from other transplanted organ, contain a mixture of different cell types, and whole tissue analysis by arrays does not give an idea of cell-specific signatures. Even the infiltrating lymphocytes may not be necessarily a homogenous population. Pro-rejection and pro-tolerance cells may both be present. Thus, with the exception of cell-type-specific genes (e.g., E-selectin), the source of mRNA is unknown and limits our ability to interpret the cellular signatures relating to the gene expression patterns. To address this concern, laser-capture microdissection of cellular subtypes of interest and

microarray analysis after RNA amplification has been attempted with success.[21] An additional method is to study the gene at the protein levels by immunohistochemistry for genes of interest in specific samples of interest or by the use of tissue microarrays.[22] The latter allows for the simultaneous examination of hundreds of tissues of interest with numerous different antibodies per sample. Comprehensive systems for high-throughput analysis and storage of tissue microarray data are available at http://genome-www.stanford.edu/TMA/index.shtml.

Single-cell probing will soon be possible with the advent of nanotechnology-based laboratories (nanolabs). Single-strand DNA can be bound to a carbon nanotubule or a semiconductor nanowire and produce a signal after binding of the complementary mRNA. A new method of fabrication, the superlattice nanowire pattern transfer (Snap) has permitted the production of clusters of thousands of nanowires, each with a diameter of 8 nm on a circuit of 10×10 μm. Each nanowire could be labeled with a different DNA sequence. These nanolabs could determine the pattern of gene activation in the lymphoid infiltrating cell of an allograft in relation to the histology. This technique could also be used to analyze in real time the effect of immunosuppressive drugs on the immune system. It could thus be possible to titrate precisely how much immunosuppression is required on an individual basis to avoid rejection and overimmunosuppression.[23]

19.2.4 Concluding Remarks

The complex interplay of various genes orchestrating the immune system results in a myriad of responses in transplant recipients varying from rejection to tolerance. Currently, the lack of reliable biomarkers for individualizing therapy and directing treatment plans results in inadequate or overimmunosuppression. Unfortunately, this results in confounding clinical outcomes ranging from the rejection response to malignancy. Resolution of these issues may be possible by using either mRNA expression profiling or global genotyping assays to find genes that help predict how much immunosuppression is needed. Mutational profiles of informative genetic polymorphisms in drug-metabolizing enzymes or other genes could then be added to blood group and HLA typing panels now in the mainstay of transplant treatment. By studying differences that exist in peripheral blood or in the graft and its local environment (urine in kidney transplants, bronchial lavage in lung transplants, bile in liver transplants, etc.), informative biomarkers may be discovered to improve treatment programs. The behavior of genes orchestrated in the post-transplant setting may be informative in monitoring rejection susceptibility, ascertaining "transplant donor–recipient compatibility," assessing risk, and stratifying cases of acute rejection or chronic injury as well as for predicting long-term graft outcomes and acceptance. Protein and tissue microarrays are additional valuable tools for functional gene analysis and cell-specific expression patterns in health and disease. Ultimately, microarray technology may be most valuable in the clinical transplant setting for targeted diagnosis, treatment, monitoring, and prognosis, using a limited repertoire of genes cherry-picked for their informativeness from across the entire human genome. Single-cell probing and "nanoarray" analysis of these selected genes may be the future tools in transplantation research.

19.3 ARTIFICIAL ORGANS

19.3.1 THE KIDNEY

Since the first clinical dialyzer by Kolff in 1943, the treatment of end-stage renal failure has changed dramatically.[24] Renal failure has progressed from a fatal diagnosis to a clinically manageable entity, prolonging life many decades. Kidney transplantation is superior to dialysis, not only economically, but also from a quality-of-life standpoint.[25] The number of individuals on the transplant list outweighs the number of deceased donors available; however, the increased rate of live donation has decreased the gap, but not the void. The next major advance in the treatment of renal failure will be to mimic the results of clinical kidney transplantation.

Clinically, the only available modality to support patients with renal failure is dialysis. Analysis of peritoneal dialysis initially showed a survival benefit over traditional hemodialysis, however, reanalysis of the data revealed the benefit was from residual kidney function and not the mode of dialysis.[26,27] According to the Dialysis Outcome Quality Initiative (DOQI), the *dialysis* dose (Kt/V; k = urea dialysis clearance, t = dialysis time, and v = volume of urea distribution) of 1.25 per dialysis session has been adopted.[28] Initial work with long intermittent dialysis (8 hours at least three times a week) offers less hemodynamic variability and excellent blood pressure control with 90% of patients coming off their anti-hypertensives and improved patient survival with 10 year survival rates of 75%.[29–33] Another alternative is short daily dialysis (approximately 2 hours of dialysis 6-7 times a week) achieving a dialysis dose 50% higher than conventional dialysis (Kt/V=3). Short dialysis has had the same results of long intermittent dialysis, but has led to regression of cardiac hypertrophy,[34] improved nutritional status,[35] and improved anemia.[36] Most strikingly, there was a marked improvement in quality of life.[37,38] Daily home nocturnal hemodialysis has evolved from long intermittent and short daily dialysis (8 hours of dialysis 6–7 times a week).[39] Advantages include lower blood flow, allowing the use of only one needle, and improved hemodynamic stability, not requiring a partner to monitor hemodynamic fluxes, which in turn allows general monitoring on the Internet as well. This modality offers the highest dialysis dose (Kt/V = 5).[40] Again, there is regression of left ventricular hypertrophy and normalization of hypertension.[41] The improvement is so dramatic that patients with significant cardiac morbidity can be considered operative candidates for kidney transplantation.[42] Also, there is an improvement in anemia[43] and nutritional status,[44] but most importantly, there is an increase in quality of life.[45]

The next logical progression would be continuous daily dialysis. With the progression of nanotechnology, this may become a reality in the near future. Nanotechnology is already being used to improve the fabrication of dialysis membranes. Helixone® (Frenesius Medical Care) is a polysulfone-based membrane produced using a nanocontrol spinning procedure. The number of pores of the membrane is increased, and the spectrum of pore diameters is concentrated around the desired values. It is thus possible to increase the size of molecules to be dialyzed without increasing the leakage of molecules that should not be dialyzed. For example, this membrane allows the β_2-microglobulin (molecular weight of 13,715 Da) filtration without any leakage of serum albumin (molecular weight of 48,000 Da). This is the initial step in using nanotechnology to create membranes closer to renal physiology.[46] A step further brings us to the Human Nephron Filter-1 (HNF-1) presented by Nissenson et al. at the 2003 meeting of the American Society of Nephrology.[47] They conceptualized a wearable or implantable device that includes a pump, sensor control mechanism, display, and disposable cartridge that is the size of a native kidney. They have used nanotechnology to synthesize a series of chiral macrocyclic molecules, which function as semispecific ion channels. The membrane is <25 nm thick. The HNF-1 could provide an equivalent of 30 ml/minute of glomerular filtration rate equivalent if used 12 hours/day. This represents 50% of the filtration of the average transplanted kidney. Furthermore, this system functions more closely to the normal kidney and does not require any dialysate. Biocompatible material will also be necessary for implantation and connection to the patient's vascular system and possibly the urologic systems as well. Undoubtedly, nanotechnology will be necessary to construct fully biocompatible material. Results from clinical utilization of the HNF-1 are awaited with great impatience.

Kidney function is not limited to pure filtration. The kidney also has metabolic and endocrine functions. Another avenue being explored to improve the quality of renal replacement therapy is the addition of viable renal tubular cells to the hemodialysis circuit. Recently, a renal tubule assist device (RAD) incorporated human renal tubule cells grown in a monolayer over a porous scaffold of hollow fibers. Addition of a hemofiltration cartridge to the RAD has been shown to replace not only filtration, but also transport, metabolic, and endocrine functions of the kidney in the animal model.[48] The addition of filtration to the RAD has been coined the bioartificial kidney (BAK). The device has been evaluated in the context of acute renal failure and tubular necrosis which usually results from severe systemic diseases and reduced blood flow to the kidneys.[49] Despite renal support, acute tubular necrosis (ATN) carries a high mortality rate due to the systemic inflammatory response syndrome (SIRS) and multi-organ damage (MSOF) that develops.[50,51] The use of BAK in uremic endotoxin animal model was associated with a significantly higher cardiac output, mean arterial pressure, and a significant increase of the systemic anti-inflammatory cytokine IL-10.[52] The BAK evaluation has been brought to the clinical area recently. An NIH sponsored phase I trial is underway.

One can imagine that the combination of these two new modalities (e.g., advanced nanofabricated membranes and BAK) could be the basis of a real artificial kidney. This technology could offer a valuable alternative to organ transplantation for the majority of patients who are facing longer and longer waiting time or for the patients who have a contraindication to transplantation.

19.3.2 The Liver

The need for liver transplantation has increased dramatically over the last decade. As for kidney transplantation demand outweighs supply. In its simplest terms, the liver is responsible for detoxification and synthetic functions. The synthetic function of the liver includes carbohydrate, lipid, and protein metabolism. Many of the chemical pathways for which carbohydrates, lipids, and proteins are developed contain overlapping enzymatic pathways interlinking their function. Therefore, an inborn error in one step could interfere with the proper function of other pathways. The same can be said about bile acid, vitamin, porphyrin, and metabolism. Lastly, the detoxification (phase I or II reactions) is largely dependent on cytochrome P-450, UDP glucuronyl transferase, glutathione S-transferase, and sulfotransferase. In acute hepatic failure, hepatocyte necrosis does not allow the liver to perform its simplest functions, clinically manifesting as encephalopathy and coagulopathy. The liver is unique in that it can regenerate. Therefore, if a bridge to hepatocyte regeneration can be developed, this would allow the liver sufficient time to "heal" itself or serve as a bridge to transplantation.

One available therapeutic tool is a bioartificial liver which would take over the synthetic and detoxifying properties of the failing liver. Extracorporeal hepatocytes would serve to ultrafiltrate a patient's blood traversing through its system. One such device, HepatAssist Liver Support System developed by Circe Biomedical uses primary porcine hepatocytes. Watanabe et al. published their results of a phase I trial of the HepatAssist device.[53] Thirty-one total patients were selected. The most impressive results were with patients suffering from fulminant liver failure. Sixteen of the eighteen patients survived until liver transplantation, and three patients with primary nonfunction of their transplanted livers survived until a second transplantation was performed. Another clinically available system (extracorporeal liver-assist device [ELAD]) developed by Vitagen uses human hepatoblastoma-derived C3A hepatocytes. Although the experience has been anecdotal, four of five patients were successfully bridged to transplantation by ELAD.[54] The future of liver assist devices has found a small niche in acute fulminant liver failure; however, prospective randomized trials are in preparation to compare this therapeutic modality to the current gold standard, liver transplantation.[55] The development of the artificial liver is thus far more complex than renal replacement therapy which could be achieved in its simplest expression by dialysis. Liver replacement requires contact of the patient's systemic circulation with liver cells. The utilization of nanotechnology to create more efficient and adapted dialysis membranes to the specific physiology of liver cells is likely to also be a cornerstone of the development of an efficient artificial liver.

19.3.3 The Pancreas

The pancreas gland function is twofold: exocrine and endocrine. The exocrine pancreas secretes the enzymes responsible for food digestion. Chronic pancreatitis is a disease characterized by the inability of the pancreas to secrete enough enzymes for adequate food absorption, leading to diarrhea and malnutrition. The problem is solved by the ingestion of exogenous enzyme preparation along with meals. Clusters of endocrine cells, described as islets in 1869 by Langerhans, are associated with the endocrine function of the pancreas. These islets of cells are constituted of alpha, beta, delta, and PP cells. The beta cells secrete insulin. Insulin is the key hormone in the metabolism of glucose. Insulin deficiency is the cause of diabetes mellitus

(DM). Juvenile onset DM or type 1 DM is caused by an autoimmune destruction of the islet cells. The patients have no endogenous secretions of insulin. Adult-onset diabetes or type 2 DM is a disease caused by a relative insulin deficiency and insulin resistance. Untreated DM leads to hyperglycemia, glycosuria, dehydration, and malnutrition. The disease is fatal if not treated. The discovery of insulin by Banting and Best resulted in a cure for diabetes. Exogenous insulin treatment helps maintain the blood glucose in a safer range. However, it does not reproduce the perfect minute-by-minute glucose control of normal physiology and fails to keep the blood glucose under tight control. Chronic hyperglycemia leads to the development of secondary complications of DM; retinopathy, neuropathy, vascular disease, and chronic renal failure. Exogenous insulin treatment could also be associated with life threatening hypoglycemia if excessive insulin is given. There are close to 20 million patients in the United States who have diabetes. In 2002, the cost associated with the medical treatment of DM was $130 billion in the United States.[56] Improvement in the treatment of DM is thus highly desirable.

The first pancreatic organ transplantation was performed by Lillehei in 1966.[57] Continuous progress in surgical technique and improvement in immunosuppression protocols have contributed to establishing pancreatic transplantation as a treatment option for diabetic patients associated with a success rate higher than 85%.[58] Successful transplantation is associated with a normalization of the glycosylated hemoglobin (HbA1c), a surrogate marker of long-term blood glucose control. Pancreas transplantations are performed in the majority of cases with kidney transplantation, as these diabetic patients have reached diabetic end-stage renal failure. Pancreas transplant "alone" is performed less frequently, as it is associated with a significantly lower rate of graft survival. Full-organ pancreas transplant is currently the most effective treatment of diabetes. However, it requires a surgery of a significant magnitude and is also associated with mortality and significant morbidities.

Islet cells preparation can be obtained after digestion of the pancreatic tissue with collagenase. The islets are then purified and injected through the portal circulation in the liver. This is a more elegant way to treat diabetes than whole organ transplantation, as the exocrine portion of the organ is not required. The first successful islet cell transplantations were reported in 1980.[59] For 20 years the success rate remained dismal until Shapiro et al. from Edmonton, Canada, reported a series of seven successful cases in 2000.[60] They introduced many modifications to the current approach: reduction of cold ischemia in the preparation of islets, transplantation of islets from two donors to obtain the required mass of islets to achieve insulin independence, and the use of a less diabetogenic immunosuppression protocol, avoiding the use of corticosteroids. Despite these recent successes, islet cell transplantation remains an experimental procedure.[59] Many problems remain to be solved. One of them is the inability to detect rejection or viability of islet cells after infusion. Immunosuppression is at this point given blindly. Bulte et al. have demonstrated that incubation of stem cells with superparamagnetic iron oxide nanocomposites has allowed magnetic resonance to track the incubated cells *in vivo* for at least 6 weeks. They postulate that this noninvasive strategy could also be used to monitor a wide variety of cell transplants.[61]

Microencapsulation of islets was suggested 25 years ago by Lim et al. as a means to isolate the islet cells from the immune system, thus avoiding the need for immunosuppression.[62] Pores in the membrane would allow for the free diffusion of nutrients, sugars, and insulin. The recent progress in the field of microelectromechanical systems (MEMS) has made it possible to create nanoporous silicone-based membranes. Photolithography technology, with deposition of structural and sacrificial materials that are etched away, is used to create micropores or channels in the structure of the membrane. These pores can be made as small as 10 nm. It was initially postulated that pores in the range of 30 to 50 nm would be small enough to prevent antibodies (IgG) reaching and destroying the islets. Desai et al. demonstrated that it was required to downsize the pore to 18 nm to get a diffusion rate of IgG as low as 2% and down to 13 nm to achieve a rate of 0.001% after 4 days *in vitro*.[63] In a series of subsequent experiments, they demonstrated similar rates of insulin production from encapsulated islets with pores of 18 nm, 66 nm, or 78 nm compared to free islets. They also demonstrated that microencapsulated xeno-insulinoma cells (rat to mouse) survived after intraperitoneal injection as opposed to free nonencapsulated insulinoma cells, which were destroyed by the immune system. Encapsulated human islet cells also survived after implantation in the rat. The microencapsulation of islets cells with nanoporous silicone-based membranes seems to offer a promising solution to avoid rejection and obviate the need for immunosuppression after islet cell transplantation for diabetes.

Even if islets cell transplantation becomes successful, it does not solve the problem of supply and demand. Cell cultures, stem cell technology, or use of xeno islets (from pig) will undoubtedly be required for islet cell transplant to have a significant clinical impact.

The construct of an artificial pancreas is conceptually simpler than an artificial kidney or liver, as there is only one function to reproduce — the secretion of insulin according to blood glucose level. It is thus more appropriate to name this device an artificial beta cell. An artificial beta cell will need the integration of three fundamental components: (1) a glucose sensor, (2) an insulin delivery system, and (3) a hardware–software system to integrate the data from the sensing device and the delivery device in order to deliver the appropriate dose of insulin. Ideally, all three components should be fully implantable.[64]

The lack of availability of reliable implantable glucose sensors has been the limiting factor to the development of an artificial beta cell. Glucose oxidase enzymatic sensors are considered to be closer to clinical application in the context of artificial beta cell. The enzymatic reaction of glucose in the presence of oxigen releases two electrons at the anode. The signal is proportional to the glucose concentration. Ideally, the sensor should be implanted in the vascular system to assess accurately the blood glucose level. Potential complications could include thrombosis of the vessel, embolization of clots and deterioration of the signal secondary to fibrin or clot buildup on the catheter. Subcutaneous implantation of glucose sensors may not be as accurate as blood glucose sensing and is also associated with tissue reaction around the sensors. Updlike et al. have added an angiogenic layer over a bioprotective membrane of the glucose sensor window. The membrane protects against foreign body reactions, and neocapillary formation is stimulated by the angiogenic layer.[65]

In vivo real-time glucose sensors are thus still limited in their technology of sensing glucose rapidly and by the biocompatibility with human tissues. Progress associated with the use of the nanotechnology construction of biocompatible materiels would certainly impact the construct of glucose sensors. Utilization of platinum nanoparticules in combination with single-wall carbon nanotubles for the fabrication of glucose oxidase electrochemical sensors has been shown to increase glucose sensitivity and drastically shortened the response time.[66]

It has been established by animal and clinical experiments in the 1980s that intraperitoneal insulin infusion was the most physiologic route of insulin delivery. The insulin is absorbed by the portal circulation, mimicking the normal physiology. Implantable insulin reservoirs with intraperitoneal delivery catheters and wireless control units are currently available in Europe. The reservoir can be refilled with a needle through the skin. Catheter occlusion secondary to tissue overgrowth is frequent.[64]

Despite the lack of availability of effective glucose sensors, significant progress has been made in the investigation and validation of control strategies. To reproduce natural glucose control appears to be more complex than initially thought. The beta cell is stimulated to secrete insulin by neural signal, gut hormones, and antecedent hyperglycemia, in addition to current glycemia, and yet is inhibited by the prevailing insulin concentration in the plasma. Complex control strategies have been developed and validated using computor models. These will need to be investigated *in vivo* when a closed-loop insulin delivery (artificial beta cell) is available.[67]

Will the artificial beta cell or the islet cell transplantation prove to be a better option? It remains to be evaluated. It is clear that both treatment options could have significant impact on the care of diabetic patients.

19.3.4 Concluding Remarks

Organ transplantation is likely to benefit from the technical progress associated with nanotechnology. The possibility to work at the nanoscale may be associated with a sensible leap forward for both diagnostic and therapeutic tools. It is anticipated that the excellent results associated with organ transplantation will improve further. Longevity of the graft and quality of life for the patients are likely to be enhanced. Tailored immunosuppressive therapy according to specific patient gene profile is now foreseeable.

Nanotechnology has already opened the possibility to construct effective organ substitutes. Artificial organs may represent the only valid option to narrow the gap between organ replacement needs and organ availability. Organ transplantation will eventually become only one of the therapeutic options among the organ replacement modalities. Organ transplantation will remain the best therapeutic option until our ability to reproduce nature is improved.

REFERENCES

1. UNOS: Facts about transplantation in the United States. www.unos.org.

2. Schena, M. et al., Quantitative monitoring of gene expression patterns with a complementary DNA microarray. *Science*, 1995, 270(5235), 467–470.
3. Brown, P.O. and D. Botstein, Exploring the new world of the genome with DNA microarrays. *Nat. Genet.*, 1999, 21(Suppl. 1), 33–37.
4. Duggan, D.J. et al., Expression profiling using cDNA microarrays. *Nat. Genet.*, 1999, 21(Suppl. 1), 10–14.
5. Lander, E.S. et al., Initial sequencing and analysis of the human genome. *Nature*, 2001, 409(6822), 860–921.
6. Hedrick, S.M. et al., Isolation of cDNA clones encoding T cell-specific membrane-associated proteins. *Nature*, 1984, 308(5955), 149–153.
7. Brunet, J.F., F. Denizot, and P. Golstein, A differential molecular biology search for genes preferentially expressed in functional T lymphocytes: the CTLA genes. *Immunol. Rev.*, 1988, 103, 21–36.
8. Gress, T.M. et al., Hybridization fingerprinting of high-density cDNA-library arrays with cDNA pools derived from whole tissues. *Mamm. Genome*, 1992, 3(11), 609–619.
9. Chen, J.J. et al., Profiling expression patterns and isolating differentially expressed genes by cDNA microarray system with colorimetry detection. *Genomics*, 1998, 51(3), 313–324.
10. Bertucci, F. et al., Sensitivity issues in DNA array-based expression measurements and performance of nylon microarrays for small samples. *Hum. Mol. Genet.*, 1999, 8(9), 1715–1722.
11. Akalin, E. et al., Gene expression analysis in human renal allograft biopsy samples using high-density oligoarray technology. *Transplantation*, 2001, 72(5), 948–953.
12. Alizadeh, A. et al., The lymphochip: a specialized cDNA microarray for the genomic-scale analysis of gene expression in normal and malignant lymphocytes. *Cold Spring Harb. Symp. Quant. Biol.*, 1999, 64, 71–78.
13. Sarwal, M. et al., Molecular heterogeneity in acute renal allograft rejection identified by DNA microarray profiling. *N. Engl. J. Med.*, 2003, 349(2), 125–138.
14. Tibshirani, R. et al., Diagnosis of multiple cancer types by shrunken centroids of gene expression. *Proc. Natl. Acad. Sci. U.S.A.*, 2002, 99(10), 6567–6572.
15. Mansfield, E.S. and M.M. Sarwal, Arraying the orchestration of allograft pathology, *Am. J. Transplant.*, 2004, 4(6), 853–862.
16. Sarwal, M.M. et al., Granulysin expression is a marker for acute rejection and steroid resistance in human renal transplantation. *Hum. Immunol.*, 2001, 62(1), 21–31.
17. Chua, M.S. et al., Molecular profiling of anemia in acute renal allograft rejection using DNA microarrays. *Am. J. Transplant.*, 2003, 3(1), 17–22.
18. Sarwal, M.M. et al., Promising early outcomes with a novel, complete steroid avoidance immunosuppression protocol in pediatric renal transplantation. *Transplantation*, 2001, 72(1), 13–21.
19. Satterwhite, T. et al., Increased expression of cytotoxic effector molecules: different interpretations for steroid-based and steroid-free immunosuppression. *Pediatr. Transplant.*, 2003, 7(1), 53–58.
20. Goodstadt, L. and C.P. Ponting, Sequence variation and disease in the wake of the draft human genome. *Hum. Mol. Genet.*, 2001, 10(20), 2209–2214.
21. Kitahara, O. et al., Alterations of gene expression during colorectal carcinogenesis revealed by cDNA microarrays after laser-capture microdissection of tumor tissues and normal epithelia. *Cancer Res.*, 2001, 61(9), 3544–3549.
22. Liu, C.L. et al., Software tools for high-throughput analysis and archiving of immunohistochemistry staining data obtained with tissue microarrays. *Am. J. Pathol.*, 2002, 161(5), 1557–1565.

23. Heath, J.R., M.E. Phelps, and L. Hood, NanoSystems biology. *Mol. Imaging Biol.*, 2003, 5(5), 312–325.
24. Kolff, W.J., Lasker Clinical Medical Research Award. The artificial kidney and its effect on the development of other artificial organs. *Nat. Med.*, 2002, 8(10), 1063–1065.
25. Malchesky, P.S., Artificial organs and vanishing boundaries. *Artif. Organs*, 2001, 25(2), 75–88.
26. Adequacy of dialysis and nutrition in continuous peritoneal dialysis: association with clinical outcomes. Canada-USA (CANUSA) Peritoneal Dialysis Study Group. *J. Am. Soc. Nephrol.*, 1996, 7(2), 198–207.
27. Bargman, J.M., K.E. Thorpe, and D.N. Churchill, Relative contribution of residual renal function and peritoneal clearance to adequacy of dialysis: a reanalysis of the CANUSA study. *J. Am. Soc. Nephrol.*, 2001, 12(10), 2158–2162.
28. NKF-DOQI clinical practice guidelines for hemodialysis adequacy. National Kidney Foundation. *Am. J. Kidney Dis.*, 1997, 30(3 Suppl. 2), S15–S66.
29. Barber, S., D.R. Appleton, and D.N. Kerr, Adequate dialysis. *Nephron*, 1975, 14(2), 209–227.
30. Shaldon, S. and J.J. Oakley, Experience with regular haemodialysis in the home. *Br. J. Urol.*, 1966, 38(6), 616–620.
31. Covic, A. et al., Echocardiographic findings in long-term, long-hour hemodialysis patients. *Clin. Nephrol.*, 1996, 45(2), 104–110.
32. Laurent, G. and B. Charra, The results of an 8 h thrice weekly haemodialysis schedule. *Nephrol. Dial. Transplant.*, 1998, 13 (Suppl. 6), 125–131.
33. Scribner B, T.Z. The case for every-other-day dialysis, in *Hemodialysis Int. 2000*. 2000, San Francisco, CA.
34. Fagugli, R.M. et al., Short daily hemodialysis: blood pressure control and left ventricular mass reduction in hypertensive hemodialysis patients. *Am. J. Kidney Dis.*, 2001, 38(2), 371–376.
35. Galland, R. et al., Short daily hemodialysis rapidly improves nutritional status in hemodialysis patients. *Kidney Int.*, 2001, 60(4), 1555–1560.
36. Fagugli, R.M., U. Buoncristiani, and G. Ciao, Anemia and blood pressure correction obtained by daily hemodialysis induce a reduction of left ventricular hypertrophy in dialysed patients. *Int. J. Artif. Organs*, 1998, 21(7), 429–431.
37. Mohr, P.E. et al., The case for daily dialysis: its impact on costs and quality of life. *Am. J. Kidney Dis.*, 2001, 37(4), 777–789.
38. Kooistra, M.P. et al., Daily home haemodialysis in The Netherlands: effects on metabolic control, haemodynamics, and quality of life. *Nephrol. Dial. Transplant.*, 1998, 13(11), 2853–2860.
39. Pierratos, A., New approaches to hemodialysis. *Annu. Rev. Med.*, 2004, 55, 179–189.
40. Gotch, F.A., Is Kt/V urea a satisfactory measure for dosing the newer dialysis regimens? *Semin. Dial.*, 2001, 14(1), 15–17.
41. Chan, C.T. et al., Regression of left ventricular hypertrophy after conversion to nocturnal hemodialysis. *Kidney Int.*, 2002, 61(6), 2235–2239.
42. Chan, C. et al., Improvement in ejection fraction by nocturnal haemodialysis in end-stage renal failure patients with coexisting heart failure. *Nephrol. Dial. Transplant.*, 2002, 17(8), 1518–1521.
43. McFarlane, P.A., A. Pierratos, and D.A. Redelmeier, Cost savings of home nocturnal versus conventional in-center hemodialysis. *Kidney Int.*, 2002, 62(6), 2216–2222.

44. American Society of Nephrology 32nd Annual Meeting and the 1999 Renal Week. November 1–8, 1999, Miami Beach, FL (Abstracts). *J. Am. Soc. Nephrol.*, 1999, 10, 1A–867A.
45. American Society of Nephrology 31st Annual Meeting. Philadelphia, Pennsylvania, USA. October 25–28, 1998 (Abstracts). *J. Am. Soc. Nephrol.*, 1998, 9 Spec No, 1A–813A.
46. Ronco, C. and A.R. Nissenson, Does nanotechnology apply to dialysis? *Blood Purif.*, 2001, 19(4), 347–352.
47. Nissenson, A.R. et al., The Human Nephron Filter-1 (HNF-1): toward a continuously functioning, implantable artificial nephron system by applying nanotechnology (NT) to renal replacement therapy (RRT). *J. Am. Soc. Nephrol.*, 2003, 14, 29A.
48. Humes, H.D. et al., Replacement of renal function in uremic animals with a tissue-engineered kidney. *Nat. Biotechnol.*, 1999, 17(5), 451–455.
49. Humes, H.D., Bioartificial kidney for full renal replacement therapy. *Semin. Nephrol.*, 2000, 20(1), 71–82.
50. Humes, H.D., W.H. Fissell, and W.F. Weitzel, The bioartificial kidney in the treatment of acute renal failure. *Kidney Int. Suppl.*, 2002(80), 121–125.
51. Bone, R.C., C.J. Grodzin, and R.A. Balk, Sepsis: a new hypothesis for pathogenesis of the disease process. *Chest*, 1997, 112(1), 235–243.
52. Fissell, W.H. et al., Bioartificial kidney ameliorates gram-negative bacteria-induced septic shock in uremic animals. *J. Am. Soc. Nephrol.*, 2003, 14(2), 454–61.
53. Watanabe, F.D. et al., Clinical experience with a bioartificial liver in the treatment of severe liver failure. A phase I clinical trial. *Ann. Surg.*, 1997, 225(5), 484–491; discussion 491–494.
54. Millis, J.M. et al., Initial experience with the modified extracorporeal liver-assist device for patients with fulminant hepatic failure: system modifications and clinical impact. *Transplantation*, 2002, 74(12), 1735–1746.
55. Rosenthal, P., Is there a future for liver-assist devices? *Curr. Gastroenterol. Rep.*, 2000, 2(1), 55–60.
56. DeWitt, D.E. and I.B. Hirsch, Outpatient insulin therapy in type 1 and type 2 diabetes mellitus: scientific review. *JAMA*, 2003, 289(17), 2254–2264.
57. Kelly, W.D. et al., Allotransplantation of the pancreas and duodenum along with the kidney in diabetic nephropathy. *Surgery*, 1967, 61(6), 827–837.
58. Odorico, J.S. and H.W. Sollinger, Technical and immunosuppressive advances in transplantation for insulin-dependent diabetes mellitus. *World J. Surg.*, 2002, 26(2), 194–211.
59. Robertson, R.P., Islet transplantation as a treatment for diabetes — a work in progress. *N. Engl. J. Med.*, 2004, 350(7), 694–705.
60. Shapiro, A.M. et al., Islet transplantation in seven patients with type 1 diabetes mellitus using a glucocorticoid-free immunosuppressive regimen. *N. Engl. J. Med.*, 2000, 343(4), 230–238.
61. Bulte, J.W. et al., Magnetodendrimers allow endosomal magnetic labeling and *in vivo* tracking of stem cells. *Nat. Biotechnol.*, 2001, 19(12), 1141–1147.
62. Lim, F. and A.M. Sun, Microencapsulated islets as bioartificial endocrine pancreas. *Science*, 1980, 210(4472), 908–910.
63. Desai, T.A., MEMS-based technologies for cellular encapsulation. *Am. J. Drug Deliv.*, 2003, 1(1), 3–11.
64. Renard, E., Implantable closed-loop glucose-sensing and insulin delivery: the future for insulin pump therapy. *Curr. Opin. Pharmacol.*, 2002, 2(6), 708–716.

65. Updike, S.J. et al., A subcutaneous glucose sensor with improved longevity, dynamic range, and stability of calibration. *Diabetes Care*, 2000, 23(2), 208–214.
66. Hrapovic, S. et al., Electrochemical biosensing platforms using platinum nanoparticles and carbon nanotubes. *Anal. Chem.*, 2004, 76(4), 1083–1088.
67. Steil, G.M., A.E. Panteleon, and K. Rebrin, Closed-loop insulin delivery-the path to physiological glucose control. *Adv. Drug Deliv. Rev.*, 2004, 56(2), 125–144.

20 Crossing the Chasm: Adoption of New Medical Device Nanotechnology

Brent R. Constantz

CONTENTS

0-8493-1940-4/05/$0.00+$1.50

20.1 INTRODUCTION

New nanotechnological approaches to medical device design often produce novel new products. With new technology, whole new clinical pathways often become possible that can significantly improve clinical outcomes and destabilize the status quo. By contrast, most newly approved medical devices launched to the market simply represent iterations on a theme — a better mousetrap — and utilize common technology. For example, the number of coronary catheters that have come to market with common technology and the same clinical objective — to revascularize the coronary arteries — is enormous. Some new "gadgets" appear to offer theoretical advantages and the iterative extension of angioplasty to stents, and the cascade of solutions to in-stent restenosis have resulted in an explosion of new medical devices. A similar pattern developed in total joint replacements, which have similarly iterated on a theme a virtual explosion of marginally differentiated products intended to address the same condition. This chapter is not about adoption of this kind of conventional new medical device that only follows an established clinical approach with iterations of earlier products to produce similar outcomes to currently accepted standards of care.

Instead, in this chapter I discuss the challenges of adoption of new medical devices enabled by technological advancement in nanotechnology that facilitate a whole new clinical approach, that are not iterative extensions of earlier products, and which produce new standards of medical care. After an introduction to the stepping stones for adoption of new medical devices and the audiences addressed during new technology adoption, I will take you through an instructive case where nanotechnology was brought from concept to clinical adoption, with the potential to establish new standards of care in orthopaedic surgery. The example, Norian SRS was a fracture fixation cement allowing the fixation of fractures by cementing, which resulted from Norian Corporation, a company I founded in 1987 and led until it was acquired by the dominant market leader in bone fracture management in 1999. The Norian story will take you through most aspects of conception, development, clinical validation, commercialization, and adoption. Norian was a venture capital–backed Silicon Valley company. Many new applications of nanotechnology in medical products will become reality through the formation of venture capital–backed start-up companies, so the discussions of my experiences are particularly relevant.

20.2 STEPPING STONES TO TECHNOLOGY ADOPTION

In this section various aspects of new technology adoption are defined and discussed in the context of bringing nanotechnology to clinical practice and adoption in the medical device arena.

20.2.1 Adoption

Beginning with the end in mind, what is new technology adoption, anyway? Often the public feels new clinical advancements are marked by a media splash triggered by the acquisition of a new technology company or its initial public offering. In fact, most new approaches that have received considerable media attention and rewarded venture capital investors handsomely have never been incorporated into patient care. A new medical device is really only truly adopted when larger numbers of patients are treated with the device using the new clinical approach enabled by new technology. Adoption really means that the new approach has become the principal approach to a clinical need. This usually involves different standards of care becoming established, which destabilizes existing clinical pathways, and the reestablishment of new clinical standards of care. When medical training programs are training the new approach more than the old approach, then a new medical device is "adopted."

20.2.2 Audiences for New Technology

Unlike academic audiences such as granting agencies, journals, and tenure committees, the audiences whose blessings are necessary for getting new technology adopted represents a broad cross-section of society. It is easy to belittle the "back end" of new medical products and underestimate the challenges getting new products into hospitals and physicians' hands. To illustrate the ways that each audience plays a role in the adoption process, I discuss some aspects of each audience, based on my own experience, below. It is worth noting that most new technologies, despite their merits, never get adopted because they fail to meet the requirements of one of these "back end" audiences.

20.2.3 Intellectual Property Community

Patents on new medical devices are a critical requirement to any new medical device's eventual adoption. This aspect is important where considerable up front investment is necessary to develop the technology, which is more often the case with new technology. The more capital that is needed, the more intellectual property strength may be required to attract the capital. If the technology is truly novel, broad coverage may be available, which may figure importantly into the ability to secure the capital resources necessary to see the technology through adoption. When the intellectual property arises from a university, the simple process of out-licensing the intellectual property is perhaps the most frequent cause of failure of new technologies. For this reason alone, new technology ventures are best developed outside the university. Prior art searches of the area must be extensive and ongoing, just as would be the standard of practice in any academic pursuit. Ultimately, patents must be valid to be of value, and lack of consideration to the prior art is a common standing in patent invalidation cases. Legal costs for establishing priority dates through filings present a significant barrier to new technology births. Establishing filing dates for new inventions is critical for establishing a basis to build value that any future investment will largely be based upon.

20.2.4 Clinical and Scientific Community

About 1 billion people of the world's 6 billion people are potentially recipients of new medical technology because of the status of varying health care delivery systems. There are about 1.5 million physicians in Europe, 1.2 million physicians in the Americas, and 1.5 million physicians in the rest of the world. While physicians' influence and clout varies dramatically from hospital system to hospital system, physicians are always a key link in the healthcare delivery chain, and especially with new technologies because of potentially unanswered questions and new risks. In the United States, the number of hospitals and hospital beds has decreased over the last 12 years, while the number of hospital personnel has increased at a cumulative annual growth rate of 1.2%. The increase in the supply of physicians who have to justify their existence plays a role in their willingness to accept new technologies. Physicians' reactions to the possible introduction of new products that could significantly alter the way they treat patients is frequently mixed. Years of postgraduate training, specializing in a particular clinical method, could be supplanted by new innovations, obsolescing the value of the specialized training.

What is typically found is that research-oriented physicians and "techy" physicians are interested in new technological developments just because of the novelty. It is important for technology developers to be aware that these physicians are not primarily concerned with the eventual adoption of the technology. Often academic-based physicians are seeking research projects to support their residents and fellows to gain publications. When it comes time to incorporate new products into their routine clinical practice, these same clinicians frequently display an entirely different set of criteria for caring for their patients than for funded evaluation of new technology. Ultimately, adoption of new technologies requires that the pragmatic practicing physician mainstream accepts the new technology into their routine practice habits.

Early-stage involvement of thought-leading, but practicing physicians who are the ultimate customers for new products is necessary to ensure that the approach being taken is realistic and compelling. Many new developments lack this critical clinician-user input in their development and fail to ever become adopted because of missing clinical practicality. For this reason it is necessary to find methods to incentivize physician specialists to devote their time to provide honest and integrated input to the design and delivery of new medical products. Physician inventors provide this kind of input because they have an unambiguous interest in seeing the successful development of the new technology. It is important to remember that there will never be a shortage of skeptics and ill-wishers to the birth of new medical technology, so a few sometimes overzealous physician proponents help counterbalance the tide of naysayers.

Many successful new technological development efforts have found that relationships with key opinion-leading physicians as "product champions" is a successful approach. Concerns about conflict of interest, where the product champions utilize the new product in their practice, can be managed and are important to maintain the physicians' credibility. Often independent boards of clinical advisors or even "independent" nonprofit foundations have been formed to give new product development teams objective "arms-length" guidance, which is both critical and insightful. Many equitable and some questionable compensation structures for physicians serving in advisory capacities exist that do not present conflicts of interest.

In the United States, federal law makes it illegal for medical device and pharmaceutical companies to offer or give anything of value in exchange for purchasing any product or service that is reimbursed by the federal government (e.g., under Medicare/Medicaid). It is also unlawful for the companies' customers — primarily physicians, hospitals and healthcare institutions — to either solicit such items of value from their vendors or to receive them when they are offered, in this circumstance. Thus, the law applies directly to both the developing companies and to their healthcare customers. This "Medicare/Medicaid Anti-Kickback Law" is broadly enforced on the civil side by the Office of Inspector General, or "OIG," and on the criminal side by the Department of Justice. Significantly, penalties for violations of the anti-kickback law are severe, consisting not only of substantial criminal penalties of up to five years in prison and civil fines, but also exclusion from participation in Medicare, Medicaid, and other federally funded healthcare programs.

Several "safe harbors" are contained in the statute's implementing regulations. However, only those who structure their physician-advisor arrangements to satisfy all the criteria of a safe harbor will be protected from liability and prosecution. With new technology seeking adoption where considerable physician involvement is essential, the avoidance of the law is particularly difficult. Most physicians will not take time away from their practice to get involved in helping develop new technologies without receiving something of value in return. When a new technology is not commercially available during clinical trials, these issues are easily avoided. However, once the product can be purchased, then the law becomes a significant impairment to adoption.

Where a physician relationship does not qualify for a safe harbor, the OIG will examine the practice to determine whether it involves "remuneration" and, if so, whether the arrangement appears to involve the sort of abuse the law was designed to combat. In determining whether to institute enforcement action, the OIG will look at a variety of factors, including:

- The potential for adverse consequences to competition by freezing competing suppliers out of the marketplace. New technologies frequently obsolete "competing" technologies.
- The potential for increased charges or reported costs for items or services paid for by Medicare or Medicaid. New technologies often follow new clinical pathways that generate new charges, although often eliminating others.
- Possible encouragement of overutilization of the Medicare/Medicaid system. The "system" often is insensitive to the long-term benefit of new technologies and focus only on the near-term costs.

No one factor is dispositive, and given the interpretation of the law to date, the OIG has virtually unlimited discretion in selecting cases for enforcement. However, evidence of illegal intent of the parties is required for prosecution. Additionally, federal courts and administrative bodies considering the law in the context of actual enforcement cases have established several important interpretive principles, including the following:

- The law is violated if even one of the purposes of a payment (as opposed to its primary or sole purpose) is:
  a. To induce a decision to order, purchase, or recommend an item or service
  b. In exchange for the ordering, purchasing or recommending an item or service
  c. The referral of patients
- No actual payment by a federal health care program is necessary as long as the challenged remuneration is for an item or service that could be paid for by a federal healthcare program.
- The fact that a particular arrangement is common in the healthcare industry is not a defense to a violation of the law.
- A payment or other benefit may violate the law when the amount is sufficient to influence the customer's reason or judgment.
- The mere potential for increased costs to, or a payment to be made by, a federal healthcare program may be enough to violate the law.
- Illegal intent and violation of the law may be found even if there is no proof of an actual agreement to order, purchase, or recommend the purchase of medical items, services or referrals.
- Intent may be inferred from the circumstances of the case.

Therefore, even though it is critical that treating physicians be involved in the development of new technology, it is often difficult to negotiate arrangements that avoid the appearance of potential conflicts.

Nonphysician scientific advisors, or physicians who are not in clinical practice with the new technology, frequently play important roles as scientific advisory board members and internal consultants. External validation of their credibility in the form of awards and honors for these advisors, such as being a member of the National Academy of Sciences or having won the National Medal of Science, often is of importance when presenting the composition of a company's scientific advisory board to the various nonscientific audiences such as the FDA or the investment community. In the final analysis, new products need physicians who use the products routinely to validate the clinical utility and practical realities of the new technology.

In general, there are a large number of common arrangements in the medical device and drug industries, pursuant to which persons in a position to purchase products and services might also be hired to perform services for the manufacturer. Most obviously, physicians — who use or prescribe the company's products — are often hired to carry out research for the company. Moreover, doctors and others are paid for serving as consultants and members of scientific or medical advisory boards, for participating in focus groups, and for speaking and writing on behalf of the company. In parts of the medical device industry, it is also common for physicians to serve as trainers, demonstrating techniques for appropriate use of medical devices.

All of these practices raise issues under the anti-kickback law when the person being paid — whether a physician, pharmacist, hospital purchasing agent, or other — is in a position to influence the purchase of the manufacturer's products.

The OIG has acknowledged that it is not necessarily unlawful for manufacturers to pay physicians or healthcare institutions to perform legitimate and needed services, even if they also may be in a position to prescribe, recommend, or purchase the manufacturer's products. Because of the possibility that, when legitimate services actually are provided, the structure or amount of compensation is such as to provide an inducement to refer, and the possibility that some such service arrangements may be "shams" (paying for services that are not needed or are not actually provided), payments to individuals and healthcare institutions who are in a position to make purchasing decisions or refer patients are assured of protection from liability only when they satisfy the criteria for the personal services safe harbor. This safe harbor requires, among other things, that the aggregate amount of compensation: (a) be set in advance; (b) represents fair market value; (c) does not vary in accordance with the volume or value of business generated between the parties; and further, that (d) for part-time arrangements, the cost and intervals of service be specified in advance in a written agreement.

For successful adoption, new technologies must establish a cadre of respected physicians who are influential in the clinical community, who have experience with the product and can speak to the benefits of the new technology compared to other alternatives available to them. There is considerable development that usually needs to be accomplished between regulatory approval and eventual adoption of new technology where these physicians play a central role.

20.2.5 Health Delivery Professionals

There are 3.0 million nurses in Europe, 4.0 million nurses in the Americas, and 4.7 million nurses in the rest of the world. The nursing staff and variety of technicians and other healthcare professionals play a significant role in accepting new technology into the healthcare delivery system. There are many examples of new technologies that had the physicians' acceptance but fail due to the need for acceptance of the whole healthcare delivery team for the adoption to be a success. Separate advisory groups composed of these professionals or employing these professional as part of the product development team have proven to be effective in developing new technologies before they are adopted.

20.2.6 Healthcare Providers

There are over 30,000 hospitals worldwide, with 11,905 in Europe, 8,609 in the Americans, and 14,205 in the rest of the world. Often new technologies see their initial uses in funded clinical research efforts at teaching hospitals. Whenever a new product is deemed "experimental," hospitals have a human research committee governed by the Doctrine of Helsinki, which provides guidelines for research on humans. These committees consist of heath professional, administrators, clergy, etc. and can block the use of new technologies in their hospitals. Even where new technologies are not considered experimental, most hospitals today have a new product committee that approves the sustained use of all new products in the hospital, principally from an economic perspective. Today many hospitals and entire hospital systems enter bulk purchase agreements with single manufacturers in order to receive bulk discounts. As a result, many hospitals cannot accept new products in certain categories if they are considered alternatives to products in the bulk purchase agreements. It is often difficult for these committees to assess new technologies, which frequently cause delay or denial of new technology applications compared to conventional products. Due to the increased cost of healthcare and decreased payment for healthcare services, most hospitals are forced to prioritize cost savings as the main criteria in assessing whether to adopt any new product at their site, despite potential patient benefits.

20.2.7 Heathcare Payers

Reimbursement rates in the United States are defined under the prospective payment system, which is based on a basket of hospital costs. This index has risen at a cumulative annualized growth rate of 3% over the past decade and is expected to grow by 3.1% in 2004 again. The U.S. system of payment for the operating costs of acute care hospital in-patient stays under Medicare part A (hospital insurance) is based on predefined and annually adjusted diagnosis-related group (DRG) weights and rates. With the multiplication of weights and rates, one can determine the operating payment and the capital payment, which together represent the total DRG payment. The annual adjustments to DRG weights and tares are determined by the Center for Medicare and Medicaid Services (CMS). Depending on how CMS classifies new technologies, whether or not they fit into an existing classification has great bearing on the possibility for adoption of the new technology. Obtaining new categories for reimbursement is a long, difficult, risky process. New rates and weight are published annually in August in the Federal Register. The published rates apply from October 1 through September 30.

The operating payment is derived from the base payment, which is divided into labor- and nonlabor-related shares. The labor-related share is adjusted by the wage index applicable to the area in which the hospital is located; this based payment rate is multiplied by the DRG relative weight. Further adjustments are made for hospitals that serve a disproportionate share of low-income patients, approved teaching hospitals, and cases that involve new technologies that have been approved for special add-on payments. The capital payment represents the federal payment multiplied by the specific DRG weighting. The national adjusted operating amounts have grown in line with healthcare costs in recent years and should rise another 3.5% in 2004 based on the CMS proposals published in May 2003.

Most other payers in the United States follow Medicare's lead, so new technologies, which do not fit into the Medicare payment scheme, usually will not get reimbursed anywhere, despite peer-reviewed journal articles supporting use of the product. Some new technologies have persisted without reimbursement. These exceptions have usually been in plastic and cosmetic surgery, or in fertility procedures, where patient have always paid for the procedures themselves and insurances have never traditionally paid.

Some of the most successfully adopted new medical devices have come to market through inaccurate coding methods. For example, angioplasty was first introduced in the mid-1980s as coronary artery revascularization, a procedure which was first described as coronary artery bypass grafting, which involved a stopped-heart open-chest procedure. Despite the significant difference between the two procedures, coronary angioplasty was initially paid for at the same high level as coronary-artery bypass-grafting open-chest procedures. Such abuses of the federal reimbursement system may be, in part, responsible for today's tough Medicare abuse and fraud laws.

20.2.8 Patients

Especially in the United States, well-educated and affluent patients and direct-to-patient advertising have brought patients to the fore as a significant force in new product adoption. Patients tend now to request new technologies that they have heard about. In Europe, private-pay patients account for a considerable proportion of new technology product uses. These patients represent more affluent members of European societies, who have researched new medical options and advocate these treatments with their doctors or find physicians offering new treatments. National societies, other than professional physician societies, are gaining clout as well. For instance, the National Osteoporosis Foundation and several similar organizations have advocated for new technologies, from lobbying regulatory bodies to influencing reimbursement policies.

20.2.9 Sales Professionals

A large part of what determines which products are used, when more than one option is available, is the relationship the sales representatives and a company have with the healthcare delivery team, especially the physician. Sales professionals are well compensated and highly motivated to do whatever it takes to get new technologies and products adopted in hospitals in their sales territories. Because the relationship is as important as the merits of any new product, sales professionals are very cautious about recommending any new products to their customers, because new products, even with significant clinical data, often have varied initial success due to physician skill and trainability. Because sales representatives' livelihood depends on their relationships with their physician customers, a bad experience with a new product can nullify years of relationship development. As a result, sales representatives are extremely cautious about pushing new products whose only track records are from clinical trials or controlled testing. Often new products do not experience the same ease of acceptance in the real world that they received in funded clinical research. Physicians typically want to do "studies" with new products before adopting them into their practice. Most studies require extra effort by the medical team involved and require funding in the form of grants from the manufacturer to cover the expenses.

Recently, significant attention has been given to the OIG's application of the anti-kickback law to the marketing practices of medical device and pharmaceutical companies. In particular, the OIG on numerous occasions has expressed concern that many "research studies" funded by grants from manufacturers are actually designed to provide "remuneration" to physicians and health care institutions who purchase the manufacturer's products, and to reward them for their purchases. The OIG is especially skeptical of unrestricted research grants given in close proximity to a purchasing decision. Even if research grant money is actually used to benefit patient care, such grants can be interpreted as an inducement to the customer to purchase the manufacturer's product, in violation of the law. Also, large grants given to an institution at the time of purchase, which permit the institution to defray operating expenses, are an obvious target for close scrutiny by the OIG. Enforcement officials often find it difficult to believe that manufacturers make such grants out of disinterested generosity or for the well-being of the healthcare customer. In these cases, it is easy for the OIG to conclude that payments for "research" might in fact be intended to induce (or might have the effect of inducing) customers to purchase the manufacturer's product, thereby giving the manufacturer an unfair advantage over its competitors.

Because of the significant consequences that arise from violations of the anti-kickback law for both vendors and their healthcare-related customers, it is important for all medical device and pharmaceutical companies to establish and implement standard operating procedures that prohibit the discussion and use of research grant money in the context of sales transactions; any research grants offered to these customers must be completely separate from the sale of a product or service. Obviously, it is a fine line to walk for new technology developer who needs physicians to join in the effort to get their products adopted.

20.2.10 Distribution

Many new products are distributed through sales representatives employed by the manufacturer. In many cases, emerging companies with new products do not have the financing to fund a direct sales force and distribute instead through independent distributors or corporate distribution partners. Often the style of distribution will vary depending on geography. For instance, a company may support a direct sales force in the United States, use independent distributors in Europe, and go with a corporate partner in Japan. Typically the sales professionals working in all forms of distributors are highly trained and capable of learning new product attributes and their medical application. As the connection between the manufacturer and the sales professionals goes from very close, as the case of direct employees of the manufacturer, to fairly remote, as in the case of independent distributors, the extent of new product training varies considerably. Often new products take more extensive training than mere extensions of existing technologies. Training programs for physicians and their teams require close coordination between the sales professionals servicing the customers attending the training. This entire process is affected by the method of distribution. For example a manufacturer's ability to control the quality of sales representative and physician training through a distribution partner in Japan will differ significantly from a domestic venue where the manufacturer employs direct sales professionals. Furthermore, under the new anti-kickback laws, it is not lawful to send physicians to educational programs to learn how to properly use new products. Almost no physicians have the time or financial means to fund their own new product education, so the hurdles to training physicians on the proper use of new technologies require creative approaches.

20.2.11 Professional Societies

Almost every new product falls within the territory of one or more physician subspecialties that ultimately may strongly influence the adoption of any new technology. The professional societies for each specialty play a key role establishing "accepted" practice and can present substantial barriers to allowing for the adoption of any new technology. Often new technologies do not fall into established categories, and so the professional society leaders attempt to lump new innovations into existing categories that fit into their invested careers and status quo.

Frequently new technologies that have been successfully adopted have done so via a bypassing of the conventional medical subspecialty by taking the new technology and associated procedural billing to another physician subspecialty. There are countless examples where the accrediting powers of medical subspecialties have caused their constituency to "miss the boat" due to protectionist resistance to new technology. The most well known example is the development of interventional cardiology in treating coronary artery disease, a procedure taken from cardiac surgeons, who had traditionally controlled the problem and failed to accept interventional techniques when they began to emerge.

20.2.12 Regulatory Bodies

Generally, governmental agencies whose job it is to regulate medical products are organized by technological type and anatomy. New technologies rarely fit into any existing category in any given regulatory agency. Additionally, as new technologies emerge, there are usually several attempted variations on the new technology, and regulatory officials are given very different input from the various new technology sponsors, which confuses them in attempting to decide how to classify, categorize, and subsequently regulate new technologies. Regulatory officials seek external standards (e.g., ASTM, ANSI), which they can measure new products against in determining how to evaluate them. By definition, industry-wide standards are not established for new technology, and attempts to apply old standards to technologies frequently do not measure them properly. For instance, testing a resorbable polymer implant against a standard designed for a metal implant cannot hope to capture the critical performance factors of the new implants.

Most sponsors of new technologies do not appreciate the perspective of regulatory officials, who are receiving multiple confidential new product applications that the individual respective sponsors are unaware of, and cannot appreciate the multiple conflicting inputs the regulatory body has received. Currently, most new technologies originate in the United States, but the U.S. Food and Drug Administration is not usually the first regulatory body to review new technologies. Most American sponsors find European Community Notified Bodies are more educable and have more expeditious approval processes. This generality applies to both regulatory approvals for entering clinical trials and product commercialization approvals. European citizens get advanced access to new technologies, with their associated benefits and risks. While American citizens are barred from access to cutting-edge technology by FDA regulation, European physicians often criticize American sponsors of new technology for using their patients as test subjects. The Ministry of Health and Welfare in Japan usually follows the FDA's lead on new products, so Japanese approval typically is the most delayed.

20.2.13 Quality Concerns

Even when developed technologies are transferred from nonmedical areas to medical applications, the performance standards and tolerances necessary in medical applications are frequently much more rigorous than for nonmedical applications. As a result, there are rarely preexisting quality assurance methods for raw materials acceptance, verification, validation, release testing, stability requirements, and customer feedback. Often, routine aspects of conventional product prove to be significant hurdles for new technology. Any competent product development team addresses most of these factors.

Some technologies simply have inherent limitations that are not suitable in the medical setting and, therefore, never become adopted. An example is autologous cell culture for reimplantable tissue replacement. This concept, although extremely attractive, presents insurmountable logistical control obstacles that have so far prevented any possibility of adoption.

20.2.14 Production

Unlike conventional technologies and iterative improvements on existing technologies, where a large trained work force may be available to draw from, new technologies often involve an entirely new set of production skills. Often, production resources with new technologies consist of the research and development scientists and engineers, working in a production role until new processes and systems are established. Depending on the effort it takes to reduce new procedures to routine production processes, production itself can become a limiting feature in the adoption of new technologies. Meeting FDA and International Standards Organization (ISO) standards requires fundamental application of the regulation with a "back to basics" approach, because common practices from similar technologies are frequently nonexistent. The abdominal aortic stent graft developments are a good example of a new technology that did not have a good way of producing consistent product and went through several false starts due to manufacturing problems, and the production method is still evolving.

20.2.15 Investors

Raising money to finance any new technological development in the medical arena requires a different breed of investor than a routine medical technology investment. With new technology, the opportunity may be unfettered by competition, which is usually the biggest obstacle to any new initiative into a technological area where a technology is already adopted. Where you have a new technology that is not yet adopted, the road to adoption adds another huge hurdle, increasing the risk of failure significantly.

New technologies typically take longer and cost more to get adopted than conventional technologies. Private "angel investors" rarely have the capital resources to fund this kind of effort. Venture capitalists may have the financial resources to fund these significant efforts. It is rare to find a venture capitalist with the foresight to understand the difference between a routine investment and an investment in new technology that needs to be both developed and then be adopted. What occurs is that venture capitalists invest significant capital in new technology, expecting a typical timeline and return. When adoption take longer than average, they assume the technology is a failure and abuse their control to fold the effort and attempt to recoup their losses. Many of the pitfalls of venture investors may be avoided with corporate investors, who have a longer-term view of the world, as well as better knowledge, and the financial resources to see a technology to fruition. Unfortunately, many new technologies "die on the vine" when put in the control of larger established market players who often cannot integrate new technologies into their strategic plans or commercialization efforts.

20.2.16 Financial Institutions

Investment banks employ market analysts who evaluate new technologies and assess their market potential. Their reports have a significant influence on the ability to raise capital critical to fund adoption of new technologies. Sometimes the analysts are correct; often they are entirely wrong. Sometimes, their predictions become a "self fulfilling prophecy," and great new technologies cannot get financing to fund their adoption, and bad technologies see a fleeting moment of adoption driven by financial froth-driven financing. A good example is the area of "beating heart surgery," which was a darling of Wall Street, receiving breathtaking financings, but only received a brief period of partial adoption forced by windfall financing, but the market never adopted the technology. This is a case where a few zealous sales people forced technology into the clinic, many patients were harmed, and the surgeons were left scratching their heads.

20.2.17 Corporate Partners

Existing companies in the area being impacted by new technologies take great interest in new development in their area. It is often hard to appreciate the perspective of corporations holding significant market share in an area on new developments. One would expect them to be excited about the new possibilities for patients, when, in fact, they are usually defensive about their market share and how it may be impacted by the new development. More commonly, staff members of large-market-share companies are asked to investigate new technologies under the guise of possible investments or corporate distribution deals, even acquisition, when they are, in fact, evaluating the potential threat to their market share.

Often when corporate partner transactions are completed, the larger corporation never resolves whether they are more committed to exploiting the new technology with their vast resources or suppressing the technology to protect their market share from being eroded. A common strategy taken by corporate partners is to gain control of new technologies, but hold them in a precommercial state so they have something ready to launch in response just in case any of their principal competitors introduce something comparable to the new technology, which could otherwise erode their market share.

20.2.18 Competitors

Even new technology has competition — either similar in kind or equally as novel. The sponsors of any new technology are usually as evangelical about what they are doing as anyone else who chooses to dedicate their professional life to a risky new initiative. Often the advantages of the new technologies over the conventional clinical standards of care, as well as what the clinical need they are trying to address really is, get confused. These multiple conversations confuse all the audiences for new technology listed above.

20.3 NORIAN CORPORATION

20.3.1 Overview

In this example, the Norian story will take you through many aspects from conception, development, commercialization, and adoption. The intent here is that, by way of example, you will see the specific ways the audiences discussed above play out in the adoption of new platform technology. Obviously, this is only one example and should be recognized as such; many other scenarios have occurred and will occur in the future.

Norian SRS (Skeletal Repair System), is a nanocrystalline fracture fixation cement, which allows the fixation of bone fractures by cementing. I formed Norian Corporation in 1987 and led it until 1999, when it was taken over by the AO/AISF of Davos Switzerland and its producer, Synthes-Stratec AG, the dominating market leader in bone fracture management.

20.3.2 Seed Phase (1987–1989)

I conceived of the concept of a biological fracture cement in 1985, while working on the skeletogenetic processes of reef-building corals. Perhaps because I was coming to the problem with a fresh perspective, Norian received extremely broad intellectual property coverage, since the concept of a bone cement, which formed biomimetic bone, was completely novel. Bone is naturally a nanocrystalline material, and the attempts at synthesizing artificial bone prior to Norian involved high-temperature, thermal-process techniques, which form macrocrystalline materials with relatively inert biological properties. The Norian cement was mixed in the operating room, was implanted in the patient as a paste that hardened in place without any heat or toxicity, and formed the same mineral phase as native bone, allowing it to be remodeled and replaced by native bone over time in a mechanical stress-dependent manner.[1]

Norian Corporation was incorporated in March 1987, and between early 1987 and mid-1989, the company was able to attempt many product formulations and move from a 3000 sq. ft. laboratory to a 12,200 sq. ft. facility and get established. Although conventional acrylic bone cement had been used in orthopaedics since the 1950s, there had never been a cement for bones that formed the same materials as native bone, and it was not used to manage fractures due to toxicity. Most of the now over fifty patents on the technology had their origins in this phase. Additionally, Norian's patent attorney, Bertram Rowland, Esq., was well celebrated for having written some broad fundamental patents such as the Boyer and Cowen genetic engineering patent. A patent attorney's reputation figures strongly in the decision of other parties to challenge or infringe a patent. This strong position allowed the company to attract value-added seed investors who brought more to the effort than just money, including Dr. Rowland, Scott McNeally, the CEO of Sun Microsystems, and two business school professors.

In 1987, we formulated several cement formulations and began getting surgeon feedback about desirable properties for cement for fractures. We began to explore a variety of indications for cementing fractures with surgeon focus groups and one-on-one discussions with leading orthopaedic surgeons. We evaluated the concept of fracture-reduction balloons for reducing fractured bone fragments, using our prototype cements to fill the void created when fractures where reduced, a procedure now termed kyphoplasty. We wrote a National Science Foundation grant for developing the first kyphoplasty balloons and testing them in cadaveric spines. My founding concept of Norian in 1985 was a similar procedure performed by interventionalists, now termed, vertebroplasty, where cement is used to augment vertebral bodies collapsing due principally to osteoporosis. The company added one orthopaedic surgeon as a board member from nearby UCSF, Harold B. Skinner, M.D., Ph.D., to advise us on clinical applications of the technology. Dr. Skinner had received his Ph.D. in ceramic engineering before entering medicine.

After half a year getting organized, the company raised \$1.2 M in early 1988 in a Series A preferred investment round lead by Sutter Hill Ventures and Technology Venture Investors (TVI), two venture capital firms with proven track records for high returns on investment. TVI became the largest shareholder of the company at that time and joined the board. With this money, Norian expanded its development efforts, hiring a full technical and developmental staff, and building significant facilities. This allowed us to formulate and test many new types of cement and develop the first methods of characterization and control of this new class of biomaterial.

20.3.3 Developmental Phase (1990–1991)

While I had founded the company to develop the concept of fracture cementing for orthopaedic fractures, especially osteoporotic fractures, the venture capital investors who now effectively controlled the company saw a quicker route to commercialization, and their eventual cash out, in dental applications of the technology. An additional round of venture capital financing, a $4.5 M Series B Preferred stock offering was closed in 1989, adding another venture capitalist to the board, and an "experienced" manager was brought in to carry out the venture capitalists' wishes to lead a dental periodontal bone defect replacement pathway. This application development effort turned out to be a dead end, due to frivolous FDA regulation and low potential product margins. In essence, three years and about $5 M were wasted due to the abuse of corporate governance, as frequently occurs when short-sighted venture capital investors become involved.

The FDA had led the company to believe that approval would only entail a 510(k) application, with no clinical trials, but after spending the majority of the Series B money to develop a periodontal product, the FDA reversed their decision because a "similar technology" was being considered for a craniofacial application that worried them. The reviewers at the FDA clearly did not know how they were going to categorize this new technology, and their decision had been swayed by other "similar" new technology applications, which they were reviewing concurrently.

What had occurred was that the American Dental Association had become aware of Norian SRS and decided to copy it. They had modified a dental carries lesion remineralizing slurry they had patented earlier and then planned to use in a manner similar to Norian SRS. They licensed it to two craniofacial surgeons, who formed a company to imitate what Norian was doing. The American Dental Association had their patent reissued to sound more like Norian's technology and product applications. Not only was the American Dental Association technology patentably distinct from Norian, but also the product was far inferior and had raised significant safety concerns in the eye of the FDA. The FDA pronounced to Norian that, despite earlier written assurances, the periodontal application would require an Investigational Device Exemption (IDE) clinical trial, followed by a Premarket Approval Application (PMA), because that is what they were going to require the American Dental Association's copy of Norian to do, making it a much longer and expensive regulatory route, and making the quicker commercialization assumption no longer valid.

During this phase, the company met with all the largest orthopaedic companies, whose primary revenues where from total joint replacements. They all had difficulty seeing the application of a cement in anything except what they were already used to using the conventional acrylic cement for (i.e., fixation in total joint arthroplasty). They all considered that concept of cementing fractures way to "far out" and wanted instead to talk about ways to make a better cement for total joint replacement to give them an incremental leap up on their competitors. None of these companies showed interest in improving the standards of care in fracture management, a market area with an order of magnitude of more procedures that total joint replacement.

During this phase, we also developed composite cements which led to mineralized collagen bone growth factor delivery technology that was spun out as a company in 1994 called Orquest (now a subsidiary of Johnson & Johnson/DePuy) with a product now named "Healos." Norian also entered a partnership with Pfizer's orthopaedic subsidiary, Howmedica (now Stryker), which funded the development of a bone-like nanocrystalline coating for a $700 M line of total joint replacements, giving the implant line an incremental advantage over its competition. This deal resulted in a $2.0 M Series C Preferred financing, giving Pfizer a board seat. The deal involved another $2.0 M in development cost and significant prepaid and ongoing royalties.

20.3.4 Clinical Validation Phase (1993–1994)

The company completed a $14.7 M Series D Preferred financing in 1992 to fund the development of orthopaedic applications of the new nanocrystalline cement. At the time FDA had not considered biological cement for weight-bearing orthopaedic uses, since all their previous experience with calcium phosphate materials had been for uses as nonstructural bone graft substitutes. Considerable confusion lay in the compositionally similar synthetic bone products at various stages of review at FDA at the time, but not intended for weight-bearing applications. These products were intended to serve as substitutes for autograft bone and were measured by their healing capability measured on x-ray films compared to real transplanted autograft bone. Our cement, which hardened in situ, giving immediate strength allowing patients to load the extremity immediately, was unfamiliar to them. FDA attempted to classify our cement in as a nonweight-bearing autograft substitute, ignoring the fact that our cement changed the patient's outcome, by allowing immediate weight bearing. This would have required that the product be compared to autograft bone in a nonweight-bearing application.

FDA's only experience in weight-bearing devices had been with metallic implants and acrylic cements, which were static *in vivo* and had been subjected to fatigue tests that reflected their permanence. The "Indication for Use" we sought was "a fracture cement that allows earlier return to function" due to earlier weight bearing. FDA required that we perform an 86 canine study to show that during the replacement of the cement by new bone and the cycling of the cement by physiologic loading, it remained mechanically competent. The study took several years to conduct and cost several hundreds of thousands of dollars. At various longitudinal time points, the animal's bones were mechanically tested in the areas that had fractures created in the animal's leg and fixed with Norian SRS. This allowed us to gain approval to begin human trial in the United States.

The Company received important guidance from this phase on from its scientific advisory board on a variety of issues. The board met regularly to review scientific and medical issues, providing valuable feedback to management. The board consisted of Dennis R. Carter, Ph.D., from Stanford, Thomas Einhorn, M.D., from Boston University, Steven Goldstein, from University of Michigan — all presidents of the Orthopaedic Research Society — Thomas Bauer, M.D. Ph.D., from the Cleveland Clinic, a prominent bone pathologist, John Ross, Ph.D., from Stanford, a member of the National Academy of Sciences, and David Baylink, M.D., a prominent bone biologist.

We selected the most common fracture for clinical study, the wrist fracture, and performed a feasibility Investigational Device Exemption (IDE) in remobilizing cemented wrist fractures a few days postoperatively in 1993–1994, proceeding to a definitive prospective, multicenter randomized trial of 326 patients. Our principal investigator, Jess B. Jupiter of Massachusetts General Hospital, a prominent upper extremity surgeon, was also a trustee of the AO/AISF and considered a thought leader in wrist fracture management. We spent two years recruiting patients into the trial at 26 hospitals. The trial had a statistical design that included an interim analysis after three months, allowing the three-month data set on about half the patients to be evaluated. If the patients with the investigational device were doing significantly better, then the PMA application could be filed, even though the trial was designed with one-year follow-up. The trial was a success and demonstrated earlier return to function with wrist fractures cemented with Norian SRS compared to conventional methods with no longer-term adverse affects.

We received significant regulatory input from outside professionals, both in trial design and analysis issues, as well as regulatory strategy. One of our consultants was the former director of the Center for Devices and Radiological Health at the FDA. We also added Peter Barton Hutt, Esq., the author of the medical device regulations and their amendments, to Norian's board of directors. Our regulatory strategy included product approval as well as eventual product reimbursement studies. By gathering both clinical and economic data about the patients, as well as quality-of-life measurements, we had prepared to establish new coding with CMS (then HCFA). To broaden our eventual labeled indication for use beyond wrist fracture, we obtained IDEs for tibial plateau fractures and hip fractures and began clinical trial enrollment.

20.3.5 Commercialization (1995–date)

Norian closed a $19.8 M Series D offering in 1995 and a subsequent additional offering of $10 M in 1997; this financing was key to commercialization. Because of FDA hurdles, Norian first commercialized Norian SRS in Europe in 1994 through our Dutch subsidiary, Norian BV. Starting in Holland, the company took the cement to large medical centers staffed by opinion-leading orthopaedic surgeons and traumatologists. Because the technology was new, European physicians, who had seen many early-stage products, where particularly cautious in their adoption and usually asked to do clinical studies within their hospital for periods of up to a year before stocking the product. These physicians usually asked for educational grants to support the "evaluation" of the new technology.

Norian SRS received CE Mark approval in 1996. We opened a German subsidiary, Norian GmBh, with a direct German sales force, and Norian Ltd. in England, where I had a Vice President for Europe who established distribution throughout Europe. The Swiss AO/AISF establishment, whose Synthes-Stratec products dominate the market and especially German-speaking thought leaders, through financial relationships, took notice of Norian SRS. In fact, internal efforts had begun in 1994 by the AO to learn more about Norian and how the technology may affect their market share. I was invited to give a keynote talk at the annual meeting of the AO trustees in Davos, Switzerland in 1995 with Dr. Jupiter. We saw the AO as a potential collaborator in their dedication to fracture management training through education. A founder of the AO, Martin Allgower, M.D., a very influential figure in fracture management, approached me to say that it was the first time a new important technology had come to the field from outside of Davos. We found that our technical education efforts in Europe often utilized key members of the AO/AISF as faculty. We also entered an $18 M deal with Mochida Pharmaceuticals for distribution in Japan. All our distribution agreements required that, in order to maintain distribution rights for Norian SRS, surgeon education requirements had to be met and performed to our standards. In addition to working with the local regulatory authorities, hospitals, and reimbursement systems, our sales representative became missionaries throughout the world for a new concept in fracture management, enabled by our new technology.

The FDA received Norian's PMA application for the wrist fracture clinical trial in late 1997 and approved it in late 1998. Later, FDA down classified this category of product, because a guidance document had now been issued, making regulatory approval far quicker, not even requiring clinical trials. None of the reimbursement or quality-of-life information gathered was ever used, and the product was never presented to CMS for coding, because it simply was reimbursed like an autograft substitute, despite the compelling data showing early return to function. Norian Corporation was acquired by Stynthes-Stratec AG, the daughter organization of AO/AISF, in 1999 who had no experience with clinical trial data or economic outcome studies, so the information gathered was held until the main trial data was published in 2003.

20.3.6 Adoption (1996–date)

In 1996 Norian began to hold "Standard of Care" study sessions, where leading surgeons and therapists, reimbursement specialists, and other healthcare delivery professionals were brought together to forge new clinical pathways made possible by the new technology of biological cements. Typically these meetings were held off-site at resort locations. Applications that meetings were held around were wrist fractures, tibial plateau fractures, hip fractures, calcaneous fractures, anterior cruciate ligament replacement fixation, and spinal fixation. These meetings established new clinical treatments pathways to assist in clinical trail design and hospital adoption of the technology.

Norian established a password-protected educational Web site that could be accessed around the world in many languages, which contained all information necessary for surgeons to adopt the product into their practice, from basic science, to clinical and reimbursement information, to published studies and case histories. The content of the educational programs was essential in establishing new users and a key aspect of adoption. It was particularly helpful in resident training programs.

Norian received FDA approval for craniofacial application in 1998 and did an initial distribution deal with Synthes-Stratec AG for North American distribution.

This collaboration and the fact that Hansjorg Wyss, the owner and president of Synthes USA had invested in the Series D round and joined the Norian Board of directors, heightened discussions between the two companies. What was unknown to Norian, however, was that AO/AISF and Synthes-Stratec AG were extremely concerned about the impact that Norian SRS might have on their market, as they saw it as a destabilizing technology to a very stabile market, which they dominated. In fact, in an internal memo to the AO/ASIF trustees in 1997, the president of the AO/AISF states, "Our business plan shows no profit from this venture (Norian) but we fear to lose market if not 'controlling' this." While Norian considered Synthes-Stratec and the AO/AISF a good fit with our commitment to surgical education, we were unaware that their actual objective was to "control" our new technology, instead of bringing the new technology to physicians to improve patients' outcomes. Apparently, the profit to be made from keeping better technology off the market in this case was more attractive to this "not-for-profit" entity than bringing the new medical advances to patients.

A key aspect for adoption of Norian SRS for fracture cementing was surgical education. In orthopaedic trauma, a strong tradition of hands-on bioskills training and education had been established by the "nonprofit" organization, AO/AISF, who held approximately 100 Continuing Medical Education Accredited courses around the world every year. The AO/AISF is the parent organization for Synthes-Stratec AG, and primarily only teaches methods to use Synthes-Stratec products for which they receive royalties. Surgeons attending the AO/AISF CME courses ignorantly believe that they are attending a nonprofit course, despite the breathtaking profits Synthes-Stratec, the orthopaedic trauma market leader, which eventually took over Norian Corporation, makes from the courses' sales function. These courses have become the standard for orthopaedic trauma education.

Norian developed a similarly formatted educational program, held in conjunction with our distributors throughout the world. We established a deep worldwide pool of faculty, consisting of both basic scientists and experienced surgeons. The courses became a key aspect of having the new technology adopted. While the Norian courses were ready to incorporate into the AO/ASIF course schedule at the time Synthes-Stratec AG acquired Norian Corporation, because their real objective was to keep Norian SRS out of their large market place, the courses where never incorporated into the AO/AISF.

20.4 CONCLUSION

Through the discussion of the stepping stones to new technology adoption and the example of the Norian experience, I hope that the reader has gained an appreciation of the many facets of bringing new technology to routine clinical use. Often inventors see the formulation of the concept as the critical aspect of medical advances. In fact, new concepts are the enabling feature of new developments, but only a very small portion of the effort and resources necessary to bring a new technology to the point of adoption. Many hurdles are present, from the medical establishment, to government agencies, to large corporations attempting to protect their ongoing profits.

REFERENCE

1. Constantz, B.R., I.C. Ison, M.T. Fulmer, R.D. Poser, S.T. Smith, M. VanWagoner, J. Ross, S.A. Goldstein, J.B. Jupiter, D.I. Rosenthal. Skeletal repair by *in situ* formation of the mineral phase of bone, *Science*, 267, 1796–1799.

21 The Road to Infinitesimal . . . . . .

Ralph S. Greco

This book had its beginnings in a research collaboration headed by Friedrich Prinz and Lane Smith, which I was asked to join 18 months after I became a member of the faculty at the Stanford University School of Medicine. Prinz and Smith had conceived the idea that the micro- and nanofabrication processes used in silicon technology might be applied to biological systems. This work, which is ongoing, coincided with a request from the publisher to consider a second edition of my book, which was entitled *Implantation Biology — The Host Response and Biomedical Devices*. As I began to consider this proposal, it occurred to me that the opportunity to delve into the role that nanotechnology might play in the development of new biomedical devices would represent a more compelling topic for a textbook and one that might be achieved at a world class research university already immersed in the field.

During the last 10 years, Richard Feynman has received more notoriety for his role in the birth of nanotechnology than for his Nobel Prize in 1965 for fundamental work in quantum electrodynamics. His lecture "There is Plenty of Room at the Bottom" undoubtedly described a new field of science dealing with the manipulation and control of things on a small scale through the development of improved electron microscopy, miniature computers, and miniature machine systems.[1] However, it was the dramatic "challenges" that caught the imagination, of not only McClellan and later Newman, but also the scientific community in general. Feynman was undeniably prescient, and the development of the scanning tunneling microscope and the atomic force microscope has emboldened three generations of scientists to think that it is possible to engineer on a nanoscale. These achievements are not trivial. A micron is one millionth of a meter, one thousand times larger that a nanometer. To put it another way, a nanometer is the equivalent in size of ten hydrogen atoms and one-millionth the diameter of the head of a pin.[2]

Building on breakthroughs in microscopy and microfabrication technologies originating in the microelectronics industry, scientists have begun to expand into materials more suited for biological application, including shaping ceramics and polymers in submicron dimensions with high aspect ratios. This in turn has led to three-dimensional structures, which may now be used for both basic research in biology and for the development of new biological devices. In the last 10 years, there has been great interest in molecular machines in which molecules are assembled to apply forces in a predetermined manner with a specific objective, such as

0-8493-1940-4/05/$0.00+$1.50

the performance of work. These machines can even be based on proteins or DNA itself. Such devices can be developed by a top down approach, like that in which sculptures are chiseled out of a bulk material. More commonly of late, bottom-up manufacture uses self-assembly processes to create larger structures from atoms or molecules, which make ordered arrangements spontaneously. Nanotubes are a good example of self-assembled nanostructures made of carbon, which may have broad implications in both energy and therapeutic and diagnostic medicine.[3]

The development of biomaterials for human application was rooted in industrial materials developed without any biological application in mind. The history of the second half of the twentieth century in this regard is discussed in the first chapter of this book by Russell Woo, Denny Jenkins, and myself, together with the advent of new biomaterials with two distinctly new features. The first is the development of biomaterials specifically intended for human application, and the second is the development of materials at a micro- and nanoscale. What is particularly interesting is what effect these new materials will have on the host response. The second chapter in the book reviews the host response at a cellular and humoral level. In it, we speculate on the impact of nanoscale architectures on the various defense mechanisms that are involved in the healing process when a foreign body is implanted into a human whose immune system was developed to destroy things seen as foreign.

In Chapters 3, 4, and 5, we begin to explore the new field of bionanotechnology. Peter Wagner, a former academic at Stanford, who is now an industrial officer, outlines the realm of bionanotechnology, as well as initial applications to the human condition. My coeditor, Friedrich Prinz, and his associates focus on micronanofabrication in Chapter 4, and in Chapter 5 on the physical methods of cell isolation and manipulation, which will be critical to the burgeoning field of tissue engineering. In Chapter 6, Lidan You and Christopher Jacobs discuss the broad subject of cellular mechanotransduction. Mechanical factors play a critical role in cell metabolism, but how cell signals are translated into cell behaviors is unclear. Advances in nanoscale fabrication will be an important factor in understanding this micromechanical environment. Martin Morf discusses nanoarchitectures and creates a comparison between biological and computer systems in Chapter 7. In Chapter 8, Jim Spudich, one of the seminal figures in the bottom up approach to molecular technology, provides the reader with an analysis of molecular motors, which should be of great interest to those in biology and engineering alike. In Chapters 9 and 10, two cutting-edge scientists review fundamental research in biomineralization (Brent Constantz) and DNA polyelectrolyte behavior (Mary Tang).

Beginning with Chapter 11, we attempt to move into the clinical realm in a more direct manner. Admittedly, application of nanotechnology to the clinical arena is in its infancy, and much of what is reviewed in these chapters is preliminary, or in some cases speculative. In Chapter 11, Thomas Krummel and his associate review the use of micro- and nanoelectromechanical systems in medicine and surgery, after which Chris Contag reviews new concepts in imaging *in vivo*. In Chapters 13 and 14, Lane Smith, one of my coeditors, reviews tissue engineering, as well as artificial organs and stem cell biology, two fascinating chapters with widespread importance in this emerging field. In Chapter 15, Alfred Spormann reviews molecular genetics of biofilms and matrix macromolecules, which will undoubtedly impact on the future application of nanotechnology to biomaterials. Finally, we attempt in the next four chapters to look at potential applications of nanobiology in cardiology and cardiac surgery, vascular disease and vascular surgery, cancer, and transplantation. Chapter 20 by Brent Constantz discusses crossing the divide between a concept and the clinic, thereby framing the future for readers immersed in the field. We can only speculate on what the future of nanotechnology will be like. Most experts believe that nanoscale technology will have its greatest initial impact on energy production. But, there is no question that the clinical applications have sparked the imaginations of biomedical engineers and clinicians as well as those in engineering, physics, biocomputation, and imaging.

NEMS technology may be applicable to diabetes through a variety of methods of delivering insulin, including closed-loop diabetes management systems in which glucose is detected and insulin is delivered in the same system — a so-called artificial pancreas.[4] In cardiovascular disease, a smart stent, which not only maintains diameter, but also delivers drugs to prevent restenosis can be easily conceived.[5] The capsule colonoscope is the only true MEMS device in clinical use today. This ingenious diagnostic imaging system is in a one-inch package containing batteries and a metal oxide detector array, which transmits 50,000 images over the seven hours in which the device passes through the gastrointestinal tract.[6] Recently, a new NEMS trial was reported in which iron oxide nanoparticles may be used to detect metastatic lymph nodes in patients with prostate cancer.[7] Micro- and nanoscale technologies will likely have a profound effect on our understanding and control of complex biological systems in a precise and predictable manner.

Work in our laboratory has focused on protyping biomaterials with features and aspect ratios just above the nanoscale. These surfaces, when bonded together, produce architectures conducive to cell attachment, protein absorption, microfluidics, and drug delivery. We hope to utilize surfaces like this to create a variety of "smart" biomaterials for use in a broad array of clinical circumstances.

Nanoscale fabrication will build the foundation for the development of a contractile bioartificial muscle with performance viability and biosensing properties and may offer an alternative to surgical or interventional therapy in congenital heart defects.[8] Molecularly manipulated nanostructured biomedical materials will play an important role in drug delivery systems, medical interventions to induce and maintain and replace the injured heart with living cells and accelerate tissue regeneration. Quantum dots may play a role in specific genetic sequences of proteins in high throughput biological assays. They may then be tagged with short genetic sequences to make them bind to specific DNA sequences or to messenger RNA produced by active genes.[9] Recent success with drug-eluting stents confirms that local therapy with appropriate antiproliferative agents can prevent restenosis of blood vessels.[10] Nanoparticle drug systems releasing similar agents directly from within the injured vessel wall may be equally effective, particularly if the vessel is less amenable to stent treatment. Roles of the respirocyte and clottocyte theoretically will be part of a major alteration in the way we replace blood loss and heal large wounds.[11,12]

Nanotubes have been reported by several groups to be able to detect specific DNA nucleotide sequences in proteins. Nanowire detectors can be arranged in an area roughly the size of single cell.[13] Quantum dots may be able to reveal whether cancer cells move, migrate, or clump. Tumors ingest quantum dots, and cellular traffic can then be determined. Imaging techniques will become based on such fluoroprobes. Molecular imaging will become part of the diagnostic and therapeutic approaches to clinical oncology.[14,15] Field effect transistors (FETs) may be useful in diagnostic testing, and silicon nanowires coated with antibodies of specific proteins may be utilized as markers much more sensitive than traditional techniques. Nanolabs may be able to determine which genes are expressed when tumors are exposed to chemotherapeutic agents and how cells thereafter become resistant to those agents. This, in turn, will allow for more specific therapy, with drugs targeting specific cells that are known to be sensitive, and others utilized when resistance is predicted or diagnosed at its inception. Cancer treatments based on nanoparticles and nanostructures will impact the future of cancer treatment, and nanoparticles may be able to selectively deliver high concentrations of antitumor drugs and genes to tumor cells.

Many of the leaders of this emerging field have just been born or are in grade school or high school. Much of the progress that will be made is speculative, but highly likely to evolve in a positive way and improve the lives of patients on a regular basis. What is certain at this early stage in the clinical applications of nanotechnology is that diagnostics and interventions will become more targeted, less invasive, and increasingly based on technologies at a molecular and even an atomic level. The challenges are endless, the opportunities compelling, and the ultimate outcome utterly unpredictable. This book only scratches the surface. Hopefully, it gives a perspective that readers will find valuable in sorting through the complexities and nuances of nanotechnology and nanobiology.

REFERENCES

1. Feynman, R., Plenty of room at the bottom, *Engineering and Science*. February 1960.
2. Norton, J.A., Nanotechnology and cancer, in *Nanobiology: Nanoscale Fabrication of a New Generation of Biomedical Devices,* 2004.
3. Kircher, M.F., Mahmood, U., King, R.S., Weissleder, R., and Josephson, L., A multimodal nanoparticle for preoperative magnetic resonance imaging and intraoperative optical brain delineation, *Cancer Res.*, 2003, 63, 8122–8125.
4. Burrin, J. and Price, C., Measurement of blood glucose, *Ann. Clin. Biochem.*, 1985, 22, 327–342.
5. Nugent, H. and Edelman, E., Tissue engineering therapy for cardiovascular disease, *Circ. Res.*, 2003, 92, 1068–1078.
6. Mylonaki, M., Fritscher-Ravens, A., and Swain, P., Wireless capsule endoscopy: a comparison with push enteroscopy in patients with gastroscopy and colonoscopy negative gastrointestinal bleeding, *Gut*, 2003, 52, 1122–1126.
7. Harisinghani, M. et al., Noninvasive detection of clinically occult lymph-node metastases in prostate cancer, *N. Engl. J. Med.*, 2003, 2003, 2491–2499.

8. Ito, E., Suzuki, K., Yamato, M., Yokoyama, M., Sakurai, Y., and Okano, T., Active platelet movements on hydrophobic/hydrophilic microdomain-structured surfaces, *J. Biomed. Mater. Res.*, 1998, 42, 148–155.
9. Schwartz, D.A., Norberg, N.S., Nguyen, Q.P., Parker, J.M., Gamelin, D.R., Magnetic quantum dots: synthesis, spectroscopy, and magnetism of Co(2+)- and Ni(2+)-doped ZnO nanocrystals, *J. Am. Chem. Soc.*, 2003, 125(43), 13205–13218.
10. Wieneke, H., Dirsch, O., Sawitowski, T., Gu, Y.L., Brauer, H., Dahmen, U., Fischer, A., Wnendt, S., and Erbel, R., Synergistic effects of a novel nanoporous stent coating and tacrolimus on intima proliferation in rabbits, *Catheter. Cardiovasc. Interv.*, 2003, 60(3), 399–407.
11. Freitas, R.A., Jr., Exploratory design in medical nanotechnology: a mechanical artificial red cell. *Artif. Cells Blood Substit. Immobil. Biotechnol.*, 1998, 26(4), 411–430.
12. Awadhiya, R.P., Vegad, J.L., and Kolte, G.N., Demonstration of the phagocytic activity of chicken thrombocytes using colloidal carbon, *Res. Vet. Sci.* 1980, 29, 120–122.
13. Zandonella, C., The tiny toolkit, *Nature*, 2003, 423, 10–12.
14. Seydel, C., Quantum dots get wet, *Science*, 2003, 300, 80–81.
15. Service, R., Tiny transistors scout for cancer, *Science*, 2003, 300, 242–243.

Index

C

D

E

F

G

H

I

J

K

L

M

N

O

P

Q

R

S

T

U

V

W

X

Y